M-GOV

United Nations Conference on Trade and Development

INFORMATION ECONOMY REPORT 2007-2008

Science and technology for development:

the new paradigm of ICT

Prepared by the UNCTAD secretariat

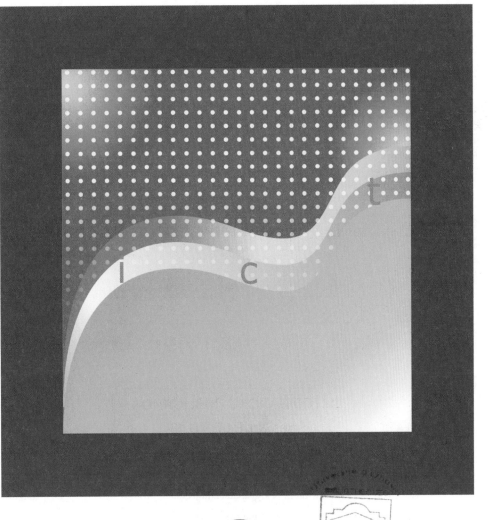

UNITED NATIONS
New York and Geneva, 2007

Note

Symbols of United Nations documents are composed of capital letters with figures. Mention of such a symbol indicates a reference to a United Nations document.

The designations employed and the presentation of the material in this publication do not imply the expression of any opinion whatsoever on the part of the Secretariat of the United Nations concerning the legal status of any country, territory, city or area, or of its authorities, or concerning the delimitation of its frontiers or boundaries.

Material in this publication may be freely quoted or reprinted, but full acknowledgement is requested, together with a reference to the document number. A copy of the publication containing the quotation or reprint should be sent to the UNCTAD secretariat at: Palais des Nations, CH-1211, Geneva 10, Switzerland.

The English version of the full report and the English, French and Spanish versions of its Overview section are currently available on the Internet at the address indicated below. Versions in other languages will be posted as they become available.

www.unctad.org/ecommerce

UNCTAD/SDTE/ECB/2007/1

UNITED NATIONS PUBLICATION
Sales No. E.07.II.D.13
ISBN 978-92-1-112724-9

Foreword

The world economy is increasingly driven by technological innovations. If developing countries are to seize the opportunities this implies, and also to address emerging global challenges, they will have to harness those innovations and the knowledge that comes with them.

UNCTAD's *Information Economy Report 2007-2008: Science and technology for development: the new paradigm of ICT* analyses the current – and potential – contribution of ICT to knowledge creation and diffusion. It looks at how developing countries use technology to generate innovations that improve the livelihoods of the poor and support enterprise competitiveness. And it examines the impact of ICTs on productivity and growth, international trade and employment in developing countries.

But mastering technology is not enough; it must, as the Report also stresses, be complemented by government policies to develop the ICT sector – by providing incentives, for example, and by building national institutional capacity for knowledge creation and diffusion. At the international level, more effective technology transfer and knowledge-sharing are needed, which can be achieved through more flexible intellectual property rights regimes, open access to knowledge and international partnerships. Development partners, in turn, can also help close the digital divide, including through technical assistance and funding of ICT infrastructure.

We are now at the midpoint in the timeline set by the international community for achieving the Millennium Development Goals. The eighth goal – developing a global partnership for development – included making available the benefits of new technologies, especially information and communication technologies, in cooperation with the private sector. This Report makes an important contribution to our understanding of how to achieve that.

Ban Ki-moon
Secretary-General of the United Nations

Acknowledgements

The *Information Economy Report 2007-2008* was prepared under the overall direction of Anh-Nga Tran-Nguyen, Director of UNCTAD's Division for Services Infrastructure for Development and Trade Efficiency (SITE). It was written by a team supervised by Geneviève Féraud, Chief of the ICT and E-Business Branch, until her change of functions on 1 October 2007, and consisting of the following UNCTAD staff members: Cécile Barayre, Dimo Calovski, Scarlett Fondeur Gil, Angel González Sanz, Muriel Guigue, Rouben Indjikian, Diana Korka, Marta Pérez Cusó and Susan Teltscher. The Introduction was drafted by Anh-Nga Tran-Nguyen.

The UNCTAD team acknowledges the contribution to chapter 8 by Chris Connolly of the consulting company Galexia, and by the ASEAN secretariat, whose material was based on a project funded by the ASEAN Australia Development Cooperation Program (AADCP) through the Australian Government (AusAID).

The team also wishes to thank the following individuals for their contributions to, and comments on, various parts of the Report: Carolin Averbeck, Claire Buré, Carlo Cattani, Florencio Ceballos, José Luis Cervera Ferri, Diane Coyle, Charles Geiger, Benedict Hugenschmidt, Martina Metzger and Ugo Panizza, as well as staff from telecentre. org. Participants in the expert meeting "In support of the implementation and follow-up of WSIS: using ICTs to achieve growth and development", which was held jointly by UNCTAD, OECD and ILO in Geneva on 4 and 5 December 2006, provided useful inputs that were considered in the preparation of this Report.

UNCTAD appreciates the sharing of statistical data by the National Statistical Office of Thailand, the Egyptian Information Technology Industry Development Agency (ITIDA), ITU, OECD and Eurostat, as well as the National Statistical Offices that responded to the UNCTAD 2007 questionnaire on ICT use by enterprises and on the ICT sector. The UNCTAD Central Statistics and Information Retrieval Branch provided valuable advice and support in the analysis of trade data.

Research assistance was provided by Bjorn Griesbach, Laura Harbidge, Jackie Lee, Giulia Quaia, Iulia Teodoru and Lidia Villalba during their internships with UNCTAD.

Administrative and secretarial support was provided at different times by Karima Aoukili, Elvira Chudzinski, Marie Kamara, Monica Morrica and Janaki Venkatchalam.

The Report was edited by Graham Grayston. The overall layout and desktop publishing were done by Christophe Manceau. Diego Oyarzùn Reyes designed the cover and formatted the charts.

The financial support of the Governments of Finland and France is gratefully acknowledged.

Contents

List of boxes

List of charts

Chart   Page

List of Tables

Table

Page

List of abbreviations

ACH	automated clearing house
ALADI	Asociación Latinoamericana de Integración
ASEAN	Association of Southeast Asian Nations
ATM	automated teller machine
BDT	Bangladesh taka
BIS	Bank for International Settlements
BOP	balance of payments
BPO	business process outsourcing
B2B	business to business
B2C	business to consumer
CAGR	compound annual growth rate
CAP	community access programme
CB	Central Bank
CD	compact disc
CDI	Comité para la Democratización de la Informática
CIS	Commonwealth of Independent States
CISP	community Internet service provider
CLIC	community learning and information centres
CLMV	Cambodia, Lao People's Democratic Republic, Myanmar and Viet Nam
CONACYT	National Council for Science and Technology (Mexico)
CPC	Central Product Classification
CTIC	Centro Tecnológico de la Información y Comunicación
DFID	Department for International Development (United Kingdom)
DISK	Dairy Information Service Kiosk
DNS	deferred net settlements
DVD	digital versatile disc
EAC	East African Community
EBA	Euro Banking Association
ECA	Economic Commission for Africa
ECB	European Central Bank
ECDR	*E-commerce and Development Report*
ECLAC	Economic Commission for Latin America and the Caribbean
EDI	electronic data interchange
EISI	Egyptian information society initiative
ENRD	e-network research and development
EU	European Union
Eurostat	Statistical Office of the European Communities

EU15	The 15 countries members of the European Union until May 2004
FDI	foreign direct investment
FTTH	fibre (-optic cable) to the home
FX	foreign exchange
GDP	gross domestic product
GPCIC	Grameen Bank Community Information Centre
GPTK	gabungan pemuela terampil dan kreatif
G10	Group of 10
HFC	hybrid fibre-coax (cable)
HS	harmonized commodity description and coding system
IAI	initiative for ASEAN integration
ICTs	information and communication technologies
IER	*Information Economy Report*
ILO	International Labour Organization
IMF	International Monetary Fund
IPRs	intellectual property rights
IPO	initial public offering
ISIC	International Standard Industrial Classification
ISIC Rev.3	International Standard Industrial Classification of all economic activities, third revision
ISP	Internet service provider
IT	information technology
ITA	information technology agreement
ITCR	Instituto Tecnológico de Costa Rica
ITES	IT-enabled services
ITIDA	Information Technology Industry Development Agency
ITU	International Telecommunication Union
Kbps	kilo bytes per second
LAN	local area network
LDC	least developed country
MCIT	Ministry of Communication and Information Technology
MFN	most favoured nation
MFI	microfinance institution
MIS	management information system
MMS	multimedia message service
MNEs	multinational enterprises
MNO	mobile network operators

MOSTI	Ministry of Science, Technologies and Innovation (Malaysia)
M&As	mergers and acquisitions
M1	narrow money supply concept
NASDAQ	National Association of Securities Dealers Automated Quotations system
NASSCOM	National Association of Software and Service Companies
NGO	non-governmental organization
NSO	National Statistical Office
ODA	official development assistance
ODR	online dispute resolution
OECD	Organisation for Economic Co-operation and Development
OEM	original equipment manufacturer
PAA	Pan-Asian E-commerce Alliance
Pe-PP	partnership for e-prosperity for the poor
PIF	Pacific Islands Forum
PKI	public key infrastructure
PMC	Philippine multiDoor courthouse
POS	point of sale
P2P	peer to peer
R&D	research and development
ROE	return on equity
ROA	return on assets
RTGS	real-time gross settlement systems
SAARC	South Asian Association for Regional Cooperation
SADC	Southern African Development Community
SEPA	Single Euro Payments Area
SMEs	small and medium-sized enterprises
SPV	special purpose vehicle
SWIFT	Society for Worldwide Interbank Financial Telecommunication
STPI	software technology parks of India
S&D	special and differential treatment
TARGET	Trans-European Automated Real-time Gross-settlement Express Transfer system
TELMIN	telecommunications and ICT ministers (ASEAN)
TELSOM	telecommunications and ICT senior officials meeting (ASEAN)
TFP	total factor productivity
TNC	transnational corporation
UCR	University of Costa Rica
UK	United Kingdom

UN	United Nations
UNCITRAL	United Nation Commission on International Trade Law
UNCTAD	United Nations Conference on Trade and Development
UNDP	United Nations Development Programme
UNESCAP	United Nations Economic and Social Commission for Asia and the Pacific
UNSD	United Nations Statistics Division
US	United States
VoIP	Voice over Internet Protocol
WACB	West African Central Bank
WiBro	wireless broadband
WITSA	World Information Technology and Services Alliance
WTO	World Trade Organization
xDSL	any of several types of digital subscriber lines

EXPLANATORY NOTES

The term "dollars" ($) refers to United States dollars unless otherwise stated. The term "billion" means 1,000 million.

Two dots (..) indicate that the data are not available or are not separately reported.

A hyphen (-) indicates that the amount is nil or negligible.

Because of rounding, details and percentages do not necessarily add up to totals.

OVERVIEW

Harnessing knowledge for development

It is now well established that technological progress and innovation are the long-term drivers of economic growth. In the context of a global knowledge economy fuelled by the fast pace of technological innovation, it is important for developing countries to lay good foundations for building their capacity to acquire and create knowledge and technology in order to take advantage of the opportunities offered by globalization and, at the same time, to address emerging global challenges. The challenge is therefore to harness knowledge for development – by providing an enabling environment for the production of ideas and innovations, as well as for their dissemination and use by different actors, directly or indirectly involved in the production process.

In this broader context of science and technology for development, the *Information Economy Report 2007/2008* is analysing the contribution of information and communication technology (ICT) to growth and development. As in the case of other technologies, the ICT contribution is determined by factors such as the role of human capital, externalities and spillovers (notably through learning and complementary innovation), and appropriate policies and institutions supporting innovations. The analysis highlights the importance of open access to knowledge and, hence, the importance of diffusion and sharing of knowledge and technology, especially in the case of developing countries.

The institutional framework should ensure a good flow of knowledge between scientific research and technological applications, as well as a good flow of information among researchers and users, at the national and the international level. Governments play a crucial role, because knowledge creation cannot rely on market mechanisms alone. Policies to support knowledge creation (such as government funding, government procurement, tax subsidies and intellectual property protection) as well as knowledge diffusion (establishment of libraries, communication networks, access cost subsidies, etc.) are examples of government measures in this area. A clear legal and regulatory framework in many areas touching upon the interactions and transactions among different actors is also necessary.

The market for knowledge is often characterized by imperfections - that is to say, social and private returns derived from knowledge can widely differ. In the area of knowledge creation, this "market failure" may lead to private underinvestment in knowledge: that is why Governments have taken measures to provide incentives for individual agents to create knowledge, namely through intellectual property rights (IPRs), tax rebates and subsidies, and full or partial funding of research. In particular, intellectual property protection through patents, trademarks, copyrights or trade secrets confers the right to appropriate the income derived from the application of proprietary research in order to recover the high fixed cost of research. However, the approach to IPRs needs to strike a balance between incentives for creativity and society's interest in maximizing the dissemination of knowledge and information.

Knowledge is increasingly being privatized and commercialized, and the use of IPRs to protect knowledge has restricted access to information and technologies. Furthermore, the international governance of knowledge has moved towards tighter and more harmonized IPRs at the international level, with a view to minimizing the free-rider problem. This has been achieved through the World Trade Organization's Agreement on Trade-Related Aspects of Intellectual Property Rights (TRIPS), the "TRIPS-plus" provisions of regional and bilateral trade agreements negotiated by the European Union (EU) and the United States, and new treaties negotiated under the auspices of the World Intellectual Property Organization.

In the area of knowledge diffusion and technology transfer, externalities and spillovers can bring enormous benefits for the economy as a whole, and for the rest of world, derived from technology flows among countries. In some instances knowledge can be considered to be a public good (non-rivalrous and non-excludable). Many inventions were built on earlier inventions ("standing on shoulders of giants") and the benefits of technological progress are not just limited to one firm or one sector, because of complementarities in the application and generation of knowledge, and are thus extended to the economy as a whole. The benefits of externalities and spillovers may not be fully reaped because of high access costs or other obstacles. In the case of cross-border flows of knowledge, high access costs and barriers resulting from a harmonization and tightening of IPR standards at the international level may be harmful for poorer countries with limited human and capital capacity. While market failures in the area of technology diffusion may be important, measures to address them have not been clearly identified, especially with regard to the transfer of technology from developed to less developed countries.

Many approaches to encourage more effective transfer of technology to developing countries have been suggested:

- Improving flexibilities in IPRs, in terms of calibration of standards and norms for countries with varying levels of development.

- Open access systems. The key feature of open access systems is that the knowledge is free for use by the general public. In some areas involving extensive cumulative innovation, such as computer software, biotechnology or other domains of common knowledge, these systems may be the most efficient to advance knowledge.

- International partnerships for generating and sharing information. Many global initiatives have been launched, with the financial support of public and private sectors, to enhance global research and information capabilities, so as to overcome crucial problems in the areas of rural development, environment and health in the poorest developing countries. International partnerships could be reinforced in those areas, as well as in other areas, in order to allow more effective participation by poor countries in sharing the benefits of common knowledge.

- Global support for building capacity in developing countries, especially the least developed countries, to enhance human capital, infrastructure and institutions in order to develop those countries' capacity to absorb and create scientific and technical knowledge. There is a strong case for donors to increase "knowledge aid" and aid for science and technology.

The new paradigm of ICT: implications for innovations and development policies

---

ICT is a general-purpose technology and as such has a pervasive impact on the economy. It introduces a new paradigm for the configuration of economic activities, radically changing the approach to technology for development. The main aspects of this new paradigm can be summarized as follows:

- First, the economic impact of ICT could be more important, in terms of externalities and spillovers through its use and applications in different sectors of the economy, than its direct contribution to gross domestic product as a production sector.

- Second, one of the most important externalities is a new mode of organization of production and consumption, which results in cost-saving transactions and faster and better communication between economic agents. With regard to the developing countries, those innovations have provided new opportunities for their insertion to the global value chains and for diversifying production activities and exports. At the same time, ICTs facilitate the creation of networks and increase the exchange of information locally and globally.

- Third, the rapid pace of innovation in the ICT sector itself has considerably reduced the costs of access to ICTs. This has allowed a democratization of ICT use, including use by poorer people to support their livelihoods, and it has also facilitated the adoption of ICT in poverty reduction programmes.

- Fourth, ICT has generated new services in the form of e-commerce, e-finance, e-government, and so forth. These new services can contribute to greater economic efficiency. However, other challenges may arise concerning questions of trust and security in the transactions that these new e-services generate.

- Fifth, ICT requires skills, and education and training are ever more important in building a knowledge economy in which ICT represents an indispensable tool.

- Lastly, ICT has given rise to new models of sharing knowledge and collective production of ideas and innovations, which often bypass the proprietary system provided by IPRs. These "open access" models, whether in activities such as open source software, open innovations or common knowledge associations, have become very widespread and are promising in terms of the rapid diffusion of knowledge to less advanced countries.

The ICT revolution is spreading to the developing world and brings with it the promise of major technology leapfrogging, which will contribute to rapid modernization of the economies of developing countries. In order to reap the opportunities offered by ICT, countries may find it necessary to identify a set of policies to encourage the creation, diffusion and use of knowledge, which should form the basis of a sustained growth strategy. The ICT contribution to building knowledge capital should be taken fully into account in designing innovation policies. Innovation patterns are country-and industry-specific, and countries with different level of development will have different approaches according to their capacities and priorities. Taking for granted a general enabling policy framework for investment and enterprise development, specific innovation policies should aim at promoting national knowledge systems to support the competitiveness of national economies.

Within this broad policy framework to encourage innovations, the particular role of ICT as an enabler of innovations should be recognized and encouraged. Given the strong links between ICT use by enterprises, competitiveness and innovation, there is a need for better integration of policies to promote ICT use by enterprises within general innovation policies. One way to achieve that integration is to systematically coordinate the policies of different ministries and to do so at different levels. Many of the developed countries have entrusted overall policymaking for innovation and e-business to the same organizations, which formulate ICT policy as an integral part of science, technology and innovation policies.

It should be stressed that ICTs enable faster dissemination and better coordination of knowledge, thus encouraging open access to sources of innovations. An innovation policy framework that takes fully into consideration the changes generated by ICT may give prominence to open approaches to innovation, which could present significant advantages for developing countries.

ICT policy should also address the digital divide between rich and poor countries, and the national digital divide between different income groups in the population. Technological progress in ICTs is moving fast, and at the same time, costs are declining and many kinds of software have become available through the free and open source software networks. Although some new ICT applications (Wi-Fi, Semantic Web, to name but a few) and the continuing fall in access costs will allow developing countries to leapfrog on the technology trail, a number of challenges remain to be tackled in order to close the digital divide. The first is to invest in the development of human capital capable of rapidly absorbing and effectively using those new technologies. The second is to regulate e-commerce and provide protection and security to users under cyber laws. The third is the financing of infrastructure, taking into account the costs of adjustment of displaced technologies. In all three areas, the international community of development partners can make a significant contribution.

The chapters of this Report illustrate the applications of the ICT new paradigm for the economic development of developing countries, in an international context characterized by a still substantial digital divide between developed and developing nations, and at the same time by the dynamism of some developing countries that are becoming competitive in a few ICT sectors.

1. Trends in ICT access and use

The diffusion of ICT in developing countries is growing steadily, but except for East Asian countries that straddle the line between developed and developing status (notably the Republic of Korea and Singapore), developing countries remain far behind developed ones in the adoption of ICTs and their use by enterprises.

Among ICTs, mobile phones are most widely spread in the developing world. The number of mobile phone subscribers in developing countries has almost tripled in the last five years, and they now account for 58 per cent of mobile phone subscribers worldwide. This marked increase suggests that mobile telephony serves as a "digital bridge", which will help many developing countries reduce the connectivity divide. In Africa, where the increase in terms of the number of mobile phone subscribers and penetration has been greatest, this technology can improve the economic life of the population as a whole.

Mobile phones, being the main communication tool for many entrepreneurs (particularly small entrepreneurs) in developing countries, have great potential for small and medium-sized enterprises in those countries. For example, in Africa mobile phones were the most commonly used ICT for communicating with clients and for ordering supplies. Small and medium-sized enterprises (SMEs) that export agricultural products receive daily price quotes and are alerted to business opportunities through their mobile phones. M-commerce (the buying and selling of goods and services using wireless hand-held devices) of digital products such as mobile content is starting in most developing countries and is expected to grow. Payments and banking by means of mobile phones are likely to promote the growth of m-commerce, provided that there is an enabling regulatory environment.

The Internet has continued growing worldwide in terms of users and penetration. Although developed economies still account for the majority of Internet users and have the highest Internet penetration, developing economies are slowly catching up. While in 2002 Internet penetration in developed economies was ten times higher than in developing economies, in 2006 it was six times higher. Countries with economies in transition had the highest annual Internet penetration growth rates between 2002 and 2006. Several developing country Governments are taking steps to improve Internet penetration, through a combination of ICT for development policies to improve ICT access and skills, regulatory reforms to increase the offer of services and competition, and fostering investments in infrastructure and in the ICT sector. Internet access by enterprises in developing countries continues to grow, as does the number of employees using the Internet in their daily work. The number of enterprises with websites is also slowly increasing. The adoption of ICT by enterprises goes hand in hand with the investments they make in ICT, and it is the larger enterprises that invest more often.

While available data show that the number of broadband subscribers has grown rapidly worldwide, developed countries still dominate subscriptions, and the gap between those countries and developing countries in terms of penetration has widened since 2002. Broadband access to the Internet can enable or enhance the adoption of certain applications that have an impact on enterprise productivity, and the technology is changing fast (ever-increasing access speeds). A more competitive environment brought about broadband growth in developed countries, while continuous improvements and diversification in infrastructure contributed to increasing the bandwidth. In developing countries, however, different infrastructure and market conditions resulted in price policies that still hindered the wider adoption of broadband. Although data on broadband adoption by enterprises in developing countries are still scarce, there is some indication that it is growing.

The use of ICT for business processes can also contribute to income generation and increased labour productivity. ICT can reduce the cost of transactions and increase market access. However, wider adoption of ICTs by developing country enterprises is still limited by a lack of awareness of the potential benefits of ICT use, and by investment and implementation costs. Few enterprises in developing countries have an intranet or extranet, which are often the first steps towards the automated integration of business processes − that is, automatic linking between computer systems to manage orders that have been placed or received and other internal systems (reordering of supplies, invoicing and payment, and management of production logistics or service operations). On average, 34 per cent of European enterprises have automated integration of internal business processes. This figure cannot be determined at present for developing countries.

There are, however, encouraging signs that some enterprises in developing countries are realizing the benefits of ICT adoption. In some cases, Governments can help by encouraging enterprises' use

of e-government services to improve the efficiency of their operations. The Government of the Republic of Korea, for example, provides firms with information on export-import logistics and Customs, and offers an electronic documentation service for private companies that have a high level of document exchanges with the Government. In India, the government of West Bengal is launching an electronic trading programme for agricultural products.

2. The ICT-producing sector and the emerging South

The ICT sector is a dynamic and fast-changing market, with an important growth potential in developing countries. As a key technology producer, it contributes to total factor productivity and GDP growth and can play an important role in the development of a competitive information economy in developing countries. The strong growth in ICT production, trade and investment observed since the mid-1990s has continued over the past few years, in particular in developing countries and countries with economies in transition, where ICT sector supply and markets have grown much faster than in member countries of the Organisation for Economic Co-operation and Development (OECD). These shifts from developed to developing countries are likely to continue, and the ICT sector will, therefore, play an increasing role in the emerging South-South trade.

ICT sector value added as a share of total business sector value added continues to increase globally. ICT services account for more than two thirds of ICT sector value added in the OECD countries, with growth sectors being communications services and software services. Between 2003 and 2005, in the EU countries, high ICT-sector value-added shares were seen in Finland and the United Kingdom, whereas shares were falling in Ireland and Austria. On the other hand, new EU member countries such as Hungary, Slovakia, Romania and Estonia, have increasing shares of ICT sector value added. Among the developing countries, ICT sector value-added share is still small, with the exception of some Asian countries. Growing shares can be observed in some smaller economies, such as Mauritius and Cuba.

Employment in the ICT sector is also increasing. The share of the ICT sector workforce in total business

sector workforce is highest in the Republic of Korea, accounting for more than 10 per cent in 2003. As in the case of ICT sector value added, ICT workforce shares are increasing in EU countries such as Finland and the United Kingdom, but are decreasing in others, for example Ireland, Sweden and the Netherlands. Available figures from developing countries point to small but increasing shares in countries such as Egypt, India, the Philippines and Sri Lanka.

The past decade has witnessed strong growth in ICT-related trade flows and full recovery from the crisis in 2000, with growth rates for trade in ICT goods equal to those in overall manufacturing trade and above the average growth of trade in ICT services. In 2004, exports from developing to developing countries (i.e. South – South) exceeded those from developing to developed countries. The ($ 410 billion) value of South – South trade in ICT goods almost equalled the ($ 450 billion) value of North – North trade, and is likely to have exceeded it in 2006, given the strong growth of South – South ICT trade and the relatively weaker growth of North – North trade. This confirms the increasing importance of trade among developing countries, and the overall shift of ICT production and trade from developed to developing countries. It also demonstrates the growth of the ICT market in developing countries, where the potential for ICT uptake is considerable and hence the demand for ICT goods is high. Although the developing world ICT market is concentrated in a few Asian economies, a number of small economies (including some least developed countries) have succeeded in building some competitive advantage and increasing their shares of exports of ICT goods and services.

Exports of ICT-enabled services grew faster than total services exports during the period 2000–2005. In 2005, the $ 1.1 trillion value of ICT-enabled services represented about 50 per cent of total services exports, compared with only 37 per cent in 1995. This has created new export opportunities for developing countries. Up until 2004 the top 10 exporters of ICT-enabled services were all from developed countries, but in 2005 India joined their ranks as the first developing economy. Computer and information services exports grew six times faster than total services exports between 1995 and 2004, and the share of developing countries in this export sector increased from 4 per cent in 1995 to 28 per cent in 2005.

Foreign direct investment (FDI) in the ICT sector is growing strongly, especially in ICT manufacturing and services, with developing countries increasingly becoming a destination for FDI flows. While most of those flows are directed towards Asian emerging economies, they represent larger shares of GDP in smaller developing countries. South Asia, East Asia and South-East Asia are the main magnet for FDI inflows into developing countries, which reached $165 billion in 2005, representing 18 per cent of world inflows. Manufacturing FDI has been increasingly attracted to South, East and South-East Asia, although specific locations have changed as countries have moved up the value chain. It has included large inflows into the electronics industry. In particular, South – South investment inflows into the telecommunications sector are increasing, driven by large transnational corporations from such countries as South Africa, Malaysia and Mexico.

China and India are the world's largest players in the export of ICT goods and services, respectively. The strong growth of the ICT sector has played a critical role in the expansion of the two economies. China has overtaken the United States as the world's number one producer and exporter of ICT goods in 2004. India is the world's largest exporter of ICT services and ICT-enabled services and the main supplier for business process outsourcing (BPO). Foreign investment and international sourcing play an important role in the economic growth of China and India. In the next few years, not only will China and India continue to be major recipients of FDI and international sourcing,

but also international sourcing by those two countries to other locations in developing countries will increase. Both countries are in the process of shifting from labour-intensive to knowledge-intensive goods and services. It is to be expected that they will develop huge domestic markets, and as a result foreign trade is likely to become relatively less important than in smaller economies. The two countries will generate a large pool of knowledge, as well as developing new technologies, and will therefore further contribute to global shifts in ICT production, trade and employment.

The trend observed in the international spreading of production of ICT goods and services will most likely continue, with a huge potential for developing countries to host this production, while the impact on employment in ICT sector in developed countries is insignificant overall, although more pronounced in some lower-skilled sectors. At the same time, competition will increase and countries wishing to attract FDI and BPO contracts will need to invest more in their domestic labour skills and telecommunications infrastructure, and improve their investment climate.

Government policy can be instrumental in the development of the ICT sector. In particular, in the area of telecommunications infrastructure and services, government policy can contribute to creating a more competitive market with a view to lowering prices and improving the quality of services. Governments can also strengthen technical education and training in order to create a high-skilled workforce for the information technology (IT) industry, and provide a stable regulatory and enabling environment to attract BPO contracts and promote call centres. Creating an investment-friendly environment is also critical in this process.

At the international level, the WTO international technology agreement has contributed to facilitating trade in ICT goods, 93 per cent of which are now imported duty-free. A revision of the agreement to harmonize the product coverage on the basis of international classifications and to take into consideration the fast-changing nature of the ICT market should fully assess the implications for developing countries, and especially the least developed countries.

3. Measuring the impact of ICT on production efficiency

The positive macroeconomic impact of ICT on GDP growth in the case of developed countries has been well demonstrated and researched. There are, however, only a few studies for developing countries, but they confirm that in recent years those countries have benefited from a positive contribution by ICT investment to GDP growth. This positive contribution comes mostly from the ICT production sector, although the role of ICT use by enterprises in increasing labour productivity is also recognized. Gains for labour productivity from ICT derive from two principal sources: capital-deepening through investment in ICT, and technological progress in the ICT-producing industry. In countries with a low level of ICT use, the effect of ICT investment on GDP growth remained similarly low.

At the firm level, a number of studies using developed country statistical data measure the effect of specific ICTs on business productivity. The magnitude of this effect depends very much on the business environment. For example, a 10 per cent higher share of employees using computers was found to generate 1.8 and 2.8 per cent more labour productivity in Finnish manufacturing firms and services firms respectively, while in Sweden the estimated effect was 1.3 per cent for a mixed sample of businesses. Additionally, total factor productivity gains from computer capital become significant only after a series of organizational changes and restructuring of the business process. Estimates show that factors such as firm age, foreign ownership or industry affiliation have an influence on the relationship between ICT and labour productivity.

The Thai National Statistical Office and UNCTAD have conducted a joint research project to assess the link between ICT use and labour productivity in Thai manufacturing firms. The study is part of a broader global initiative to improve ICT measurement and the quality of data on ICT uptake, promoted by UNCTAD through the Partnership on Measuring ICT for Development,[1] and is the first developing country analysis undertaken in this context on the impact of ICT on labour productivity at the firm level.

The study confirms that the use of ICT by Thai enterprises is associated with significantly higher sales per employee. Unlike similar studies on developed country firms, this study reveals that in Thailand the use of basic ICTs such as computers still accounts for large differences in terms of productivity between enterprises. While in developed countries computer penetration rates are close to saturation levels, in some developing countries the share of firms using at least one computer remained considerably lower (60 per cent in manufacturing in Thailand in 2002).

Also, variations in the intensity of computer use in Thailand were reflected in larger productivity differentials between firms. A 10 per cent increase in the share of employees using computers was associated with 3.5 per cent higher labour productivity, compared with 1.8 per cent in the case of Finnish firms. On the other hand, Internet access and website presence were found to be correlated with higher sales per employee in Thailand, with a coefficient similar to that estimated by other studies in developed countries (between 4 and 6 per cent in Thailand from the Internet and 5 per cent in the United States from computer networks).

The Thai example confirms the hypothesis that developing countries can benefit as much as developed countries from the use of ICT. Moreover, results show that even the use of ICTs such as computers can make a substantial and positive difference to the economic performance of developing countries' firms.

Given the positive impact of ICT on productivity, developing countries should encourage the use of ICT on a wider scale. For that purpose, it is necessary to gather information and constantly monitor the evolution of ICT use in order to assess the impact of ICT on economic growth. Further analysis is needed to identify the complementary factors that lead to superior productivity gains from ICT in developing country firms. In addition, ICT policymakers should ensure that domestic businesses have access to information about the best practices whereby ICT use can enhance production efficiency.

---

[1] For more information see *http://measuring-ict.unctad.org.*

4. ICT, e-business and innovation policies

Technological progress generates productivity gains through product or business process innovation. Hence, it represents the main source of long-term improvement in per capita income. ICT is the technology that powered the strong wave of innovation that transformed the global economy during the last quarter of the 20th century. In particular, the application of ICT to financial, manufacturing and marketing and distribution activities has helped enterprises to become more efficient through process innovations, and it has resulted in the emergence of entirely new products or services.

Economic globalization has significantly increased the competitive pressure on enterprises in many sectors. This comes as a result of, among other factors, the emergence of new, lower-cost producers, fast-changing demand patterns, increased market fragmentation and shortened product life cycles. In such an environment, product and/or process innovation becomes crucial for the long-term competitiveness and survival of enterprises. Innovation also enables them to climb the value ladder, a particularly important consideration for the enterprises of many developing countries. At the same time, the enterprises of developing countries, particularly SMEs, face serious difficulties in benefiting from ICT-led innovation. For example, since research and development (R&D) involves high fixed costs, it is a high-risk activity and is subject to economies of scope that favour larger firms. Other general features of SMEs, such as greater vulnerability to the essentially unpredictable market responses to innovative activity, or the greater difficulties they face in accessing financial and human capital, put them at a disadvantage with regard to engaging in innovative activities. When it comes to ICT-based innovation, policymakers need to take into account the general difficulties encountered by enterprises in developing countries, particularly in connection with access to ICT and its use by enterprises.

ICT has profoundly changed the techno-economic paradigm within which innovation takes place today. Whereas in the past innovation revolved around concepts of mass production, economies of scale and corporate-dominated R&D, in the last three decades of the 20th century this was replaced to a large extent by an emphasis on economies of scope, exploiting the benefits of interconnected, flexible production facilities, and greater flexibility and decentralization of R&D. Flexibility, interconnectedness and collaboration rely on ICT, which also plays a fundamental role in facilitating research diversification and collaborative, interdisciplinary approaches.

ICT also enables faster cross-border knowledge dissemination, particularly within transnational corporations, but also through networking and partnering among smaller players. By investing in ICTs enterprises improve their capacity to combine disparate technologies in new applications. This is important not only from the point of view of ensuring that firms achieve an adequate spread of internal technological undertakings but also from the point of view of the need to engage in R&D partnerships. In that regard, the major benefit of adopting ICTs may not necessarily derive from the technology per se but from its potential to facilitate technological recombination and change.

Government policies aimed at supporting long-term growth need to recognize and exploit the dynamic relationship between the use of ICT and innovation. A growing number of initiatives at all governmental levels now aim at supporting ICT-driven innovation. The links between innovation policies and policies to promote the use of ICT and e-business by enterprises are becoming stronger and in many countries they are now being placed within the same institutional framework and under the same overall political responsibility. However, even when innovation and policies related to the use of ICT by enterprises share the same institutional framework, they are not necessarily envisioned as a single set of policy objectives with a coherent arsenal of policy instruments to achieve them. The border lines are uncertain, and ministries and agencies dealing with matters such as industry, SMEs, education and scientific research may be involved at various levels. As developing countries adapt their national innovation systems to benefit from the dynamic interplay between ICT use in economic activities and innovation-led competitiveness they need to be aware of available experience in that regard and adapt the lessons from it to their specific needs and concerns. There are several institutional issues to be addressed, such as the establishment of a development-friendly intellectual property regime and competition policies, the reinforcement of education and research systems, the creation of public knowledge structures, the development of IT infrastructures, the creation of a trust environment for ICT use, and well-functioning capital markets.

Beyond those general issues, developing countries need to reinforce the complementarity between ICT and innovation policies. An important consideration for them is to put in place instruments to support ICT-enabled innovation among SMEs. This involves identifying the specific contribution that e-business can make to their innovation and competitiveness strategies, which is not something that can be done quickly or at low cost. But it is essential, for ICT-enabled innovation policies to succeed among the SMEs, that the latter understand the long-term competitive implications of ICT adoption and what skills they need to equip themselves with if they wish to be able to engage in process innovation and – probably in a second stage – product innovation.

Policies should aim at helping SMEs integrate ICT and e-business considerations as a fundamental element of their enterprise development plans. To that end, it is necessary to speak to enterprises in a language they understand, namely the financial performance benchmarks that they are used to. When SMEs can make a clear connection between their performance benchmarks against those of their competitors and their relative position in terms of ICT, e-business use and innovation, the vital importance of ICT

integration in their business will become evident. This calls for adequate outreach strategies. It also means that policies should be implemented with sufficient continuity and be solidly integrated into sectoral approaches. Policies should facilitate the emergence of ICT-enabled alliances and networks for R&D. Equally important from the developing country perspective is the need to support open, user-centred innovation approaches: development-friendly intellectual property regimes are particularly useful in that regard, as it is often observed in markets where intellectual property regimes are weak, that open access models are more likely to expand.

An important lesson for countries that are considering putting in place support programmes in this field is that for initiatives to succeed they need to remain in place for a reasonable period of time. The value of any set of ICT innovation support measures can be judged on a rational basis only once an impact measurement has been undertaken, and this takes time. However, it is not uncommon to see programmes in that field terminated before their effects on enterprises can be assessed. This makes it difficult to replicate and scale up successful initiatives and to accumulate and disseminate best practice. At the same time, it is important that policies adapt and change in response to practical experience. Striking the right balance between the need for policy stability and the need for flexibility and evolution requires mechanisms that allow feedback from end-users to policymakers and frequent and meaningful interaction between all stakeholders.

5. E-banking and e-payments: implications for developing and transition economies

Being one of the most information-intensive services, finance is at the forefront in the use of modern ICTs as a means to achieve efficiency gains in every step of the financial supply chain. In the case of banking, Internet banking or e-banking and e-payments are becoming one of the main delivery channels as they make it possible to dramatically decrease the unit costs of financial operations, and to make the latter much faster and in many cases safer. Consequently, commercial banks and other financial service providers are increasing the share of Internet-based operations and services and

are establishing a constant and sustainable relationship with their clients through online communications.

ICTs have thus accommodated the explosion of large international financial transactions mainly between banks, thanks to the introduction of new online payments protocols and real-time gross settlement systems. In retail payments traffic, actively used electronic means of payments include payment cards, automated teller machines, telephone banking and mobile banking (m-banking) or m-payments. The latter can use the Internet Protocol and other communication protocols and are relatively more important in the context of developing countries.

Banks and payment card providers remain at the core of e-banking and e-payments. At the same time, relatively new players such as non-bank money transfer operators, mobile phone operators and e-payment technology vendors are trying to develop niches or special value-added operations via the main players, or to make various cooperative arrangements with them.

Innovative e-banking and e-payments, both corporate and retail, proved to be less costly and more convenient for commercial banks and their clients, including businesses, Governments and households. They encouraged further use of bank money and reduced the role of cash (notes and coins). However, Internet banking created another set of security challenges, such as the need for protection from emerging cybercrime. Consequently, further innovations were introduced to allow more secure methods of e-finance. The intensive use of ICTs also facilitated the transformation of traditional bank-related loans into securities that are floated on capital markets. As a result, banks' activities in securities trading have increased, while their role as deposit-taking institutions has become less important in comparison.

Making e-banking and e-payments more affordable for banks and their clients in developing countries is still a major challenge. Furthermore, providing SMEs, microenterprises, and individuals (a proportion of whom are "unbanked", i.e. have no bank accounts) with better access to simple forms of e-banking and e-payments or m-payments is also an issue that is being addressed in many developing countries.

The financial flows between developed and developing countries take place mainly in the framework of major online inter-bank transfer systems. While those systems facilitate the transmission of the main private and public financial flows such as bank credits, FDI, portfolio investments and official development assistance, ICTs are no less important for retail or small-volume financial transfers destined for households and small businesses. The most important of small-scale private financial transfers are migrant remittances; these are increasingly relying on online money transfer systems, which are cost-saving for both originators and end-users of those funds.

SMEs and microenterprises in developing countries are still largely excluded from formal financial intermediation. This long-standing issue could be addressed by introducing e-finance techniques. For example, one of the obstacles to lending to those small-scale enterprises is the lack of information about their credit risks and the high cost involved in keeping credit risks on record. Building up, at much lower unit costs, online databases and scoring systems concerning small clients' credit risks can provide solutions for overcoming traditional information asymmetry barriers to their access to finance. In that respect, the introduction of a business-friendly and streamlined regulatory and institutional environment will help those enterprises move out of the informal sector and start creating their credit histories. To provide credit to SMEs, banks will increasingly need either to rate borrowers' credit risks internally or to rely on external acceptable credit-rating institutions.

E-banking and e-payments have achieved quite a high level of penetration in developed economies and in a number of emerging ones. However, they are still at an early stage in the overwhelming majority of developing countries and countries with economies in transition. To exploit that potential, the financial sector of developing and transition economies will need the capacity to move rapidly towards modern ICT-based systems.

6. Mobile telephony for business connections for the poor

Mobile telephony has become the most important mode of telecommunications in developing countries. While Internet access has become a reality for many businesses and public institutions, and for individuals with higher levels of education and income, for the vast majority of low-income populations mobile telephony is likely to be the sole tool connecting them to the information society in the short to medium term.

Mobile telephony has grown remarkably fast in developing countries and continues to be the only ICT-use sector where developing countries are catching up quickly or have in some ways even overtaken developed countries. Mobile connectivity has a clear advantage over fixed-line infrastructure, which faces high costs and difficulties in connecting remote areas. As a result, mobile phone subscriptions since 2001 have doubled globally, and quadrupled in Africa. However, for many developing countries and regions, improving both rural penetration and access is a more daunting task because of the lack of commercial distribution channels, the low level of education and widespread poverty.

Mobile telephony can lead to economic growth in several ways. Investment in network infrastructure and related services creates direct and indirect employment opportunities. The use of mobile telephony in the conduct of business reduces the costs and increases the speed of transactions. Those effects will be more pronounced in economic activities that have a greater need for information or where added information enables increasing returns to scale. The growing availability of mobile services and their constantly decreasing cost further increase the use of mobile phones in the business communities in developing countries, even in the informal sector.

Mobile telephony services are often provided on a prepayment basis, which helps to avoid problems of non-payment. This is important in regions where large groups of the population are poor and are thus "unbankable". Once the network is in place, there is no waiting time for new mobile subscriptions. In many countries prepaid services are used to provide mobile public payphones, and this improves connectivity and accessibility in rural areas.

Examples of innovative and productive use of mobile telephony in small businesses among the poor in developing countries abound. Mobile telephony provides market information for, and improves the earnings of, various communities, such as the fishermen of Kerala, the farmers of Rajasthan, the rural communities in Uganda, and the small vendors in South Africa, Senegal and Kenya.

Mobile telephony is a dynamic technology growing in sophistication. Short message services have introduced simple wireless text and data transfers. Mobile handsets have new functionalities, such as digital photography, multimedia messaging and other programmes and utilities many of which were previously available only on personal computers with Internet connections. Mobile telephony provides a gateway to digital literacy. For many individuals and communities, once the initial hurdle of ICT acceptance has been overcome, the adoption of subsequent higher-level technologies may be less intimidating. In that sense, mobile telephony is the most useful ICT tool for low-income populations.

In addition to policies to encourage competition with a view to reducing costs and improving the servicing of mobile telephony, Governments may explore locally relevant policies to extend mobile services and networks to remote rural areas and to poor communities.

7. Promoting livelihoods through telecentres

Telecentres – that is, public facilities where people can access ICTs, communicate with others and develop digital skills – have become a key programme and policy instrument to extend the benefits of ICTs to poorer communities. They can support the livelihoods, or means of living, of people living in poverty by providing access to key information, supporting the development of technical and business skills, facilitating access to government services and financial resources, and supporting micro-entrepreneurs. For example, telecentres such as the rural information centres in Bangladesh and the Partnership for E-prosperity for the Poor in Indonesia are providing farmers with access to valuable farming knowledge to combat crop insects and improve breeding techniques.

To understand how telecentres are currently supporting livelihoods, UNCTAD has surveyed a number of telecentre networks. The survey assesses what services telecentres are providing, who benefits from them, and what are the key environmental and institutional factors influencing the ability of telecentres to support livelihoods. The results show that most telecentres are concentrating their efforts on providing access to ICTs and developing basic ICT skills. In line with the type of services offered, telecentres are primarily used for informational and educational purposes. However, wider access to ICTs and general training in ICT skills are not sufficient to support the livelihoods of people living in poverty. For instance, few telecentres provide specific training in how to use ICTs for the development of economic opportunities, such as e-business training, or training to support the development of business and/or occupational skills.

There are some good examples of how telecentres provide access to business-related services, most notably access to government services, employment-related information (in more developed economies), sector-specific information and business communications. However, there is limited support for crucial business-related services such as banking or access to finance.

Access to relevant information and knowledge is crucial for livelihoods, and telecentre network leaders believe that improvements would be most valuable in increasing the availability of relevant content. To facilitate access to content, some telecentres are repackaging information in formats accessible to their illiterate customers. Other telecentres are facilitating access to content through the development of local content and user-generated content or by supporting accessibility through help desks and infomediaries. The quality of the general infrastructure and the broader economic and business conditions are two other important external factors influencing telecentres capacity to support livelihoods.

Questionnaire responses show that, institutionally, the majority of telecentre networks support economic activities where possible, but this is not their main objective. Telecentre networks work with social and educational institutions and to a lesser extent with organizations that promote economic activities, such as professional associations or business supporting organizations. Therefore, there is scope for working with business-supporting organizations in order to, for instance, share/provide training programmes and business-related services.

One efficient approach to support economic opportunities is to embed the activities of telecentres in existing economic activities. For example, e-Choupal is a commodity services programme supporting farmers in India through information kiosks that provide real-time information on commodity prices, customized agricultural knowledge, a supply of farm inputs and a direct marketing channel for farm produce. Because the network is strongly embedded in a specific economic activity, it enables its participants to derive economic

opportunities. However, the downside is that those not part of such activities will be excluded.

Another approach that some telecentres are successfully using to support livelihoods is to develop niches of economic opportunities. For example, a telecentre in an impoverished community in Nunavut (Canada), as a result of increasing demand, is supporting two specific sectors, namely film production and scientific research. It acquired filming equipment through additional funding and is providing training in film production. As a result, film companies are interested in filming in this community because of the availability of trained personnel. The network has also developed a programme to support research work by offering services for visiting scientific researchers (i.e renting of equipment) and by training community members in basic research methods.

Telecentres can better support economic activities when providing value-added services, and not only connectivity. For instance, the availability of training to develop skills important for undertaking economic activities (such as e-business skills) is still limited. Telecentres should provide a continuum of training, from basic ICT skills training to more specialized skills training, and support customers in putting those skills into practice. There is also scope for providing a wider range of services, such as access to finance or to expertise in specific sectors.

Special efforts are needed to support those in weaker positions. Such support may be provided by an intermediary who can offer adequate information, specific programmes targeted to disadvantaged groups and special services to support their economic activities. For example, in Indonesia, each telecentre of the Partnership for e-Prosperity for the Poor has an infomobilizer - that is, a person who supports the development of the community by using and promoting the use of relevant information. The infomobilizer helps the community/village identify its needs and opportunities for improving livelihoods (such as acquiring new agricultural skills or expanding the marketing of village products).

Policy-makers and telecentre managers may consider some useful measures to ensure that telecentres support the livelihoods of people living in poverty. In particular, recommendations are made to Governments for promoting relevant e-government content and services, supporting the development of e-business skills and providing strategic financial support for telecentre networks.

Similarly, recommendations are made to managers of telecentre networks for providing value-added services, offering e-business skills training programmes, supporting the economic activities of those in weaker positions by employing community infomediaries, and collaborating with other organizations that support economic activities, such as business associations or microcredit institutions.

8. Harmonizing cyber legislation at the regional level: the case of ASEAN

The Association of Southeast Asian Nations (ASEAN) is the first regional organization in the developing world to have set in motion the harmonization of its members' e-commerce legal framework. By the end of 2008, all ASEAN member countries will have enacted consistent national e-commerce legislation. In that connection, ASEAN has commissioned a project, the ASEAN E-Commerce Project, to help its ten member countries develop and implement a harmonized e-commerce legal infrastructure.

An increasing number of developing countries are adapting their legislation to e-commerce in order to remove barriers to online services and provide legal certainty for business and citizens. The impact of the introduction of legislation on the expansion of e-commerce activities is reported by countries to be positive, leading to increased ICT-related business opportunities and FDI, according to an UNCTAD survey on e-commerce legislation in developing countries carried out in 2007.

Developing countries within their region and subregion are also considering the development of a harmonized legal framework for e-commerce to make their region competitive and help boost e-business and economic growth. The harmonization of e-commerce legal frameworks is expected to lead to larger internal or external consumer and business markets by facilitating cross-border e-commerce and the cross-border recognition of digital signatures.

Drawing on the ASEAN experience, the chapter presents the benefits of the implementation of law reform, as well as possible options and potential challenges awaiting countries in the development of a common regional and national e-commerce legal framework. Such challenges include different e-readiness levels and the development stage of e-commerce legislation, which can vary from one country to another.

The experience of the ASEAN member countries may be useful for other regional associations in the developing world that are currently considering the harmonization of e-commerce legal infrastructure. Harmonization projects are designed to align individual member country laws in order to remove unwanted gaps, overlaps and duplication, the aim being to increase legal certainty for parties engaged with more than one member country – for example, multinational businesses that are attempting to expand their business in a new region.

Harmonization projects usually fall into one of two categories – "soft harmonization" (based on training and capacity-building) or "hard harmonization" (based on model or uniform laws). Most e-commerce legal harmonization projects are "soft harmonization" projects, in that there is no intention or requirement that countries adopt the same (or even model) laws and regulatory systems. All that is undertaken is training and capacity development activities, so as to ensure a common (or harmonized) understanding of e-commerce legal requirements. However, the ASEAN E-commerce Project is an example of a "hard harmonization" project, based on implementation guidelines that build on the common objectives and principles for e-commerce legal infrastructure in ASEAN, rather than simple capacity-building. Although the ASEAN E-Commerce Project has enjoyed some success in generating rapid progress in the development of a harmonized e-commerce legal infrastructure in ASEAN, it also reinforces some of the significant challenges faced by organizations in implementing harmonization projects of this type, or in implementing domestic e-commerce legal infrastructure.

ASEAN member countries have identified a number of implementation challenges, and these are likely to be common to many other countries, especially developing countries. Key challenges include securing government policy support, identifying sufficient funding, and obtaining relevant training and assistance. Furthermore, many developing nations may not be able to develop effective e-commerce legal infrastructure without some form of external assistance. Several ASEAN member countries have benefited from external assistance, including training programmes and advisory services concerning the legal aspects of e-commerce provided by United Nations organizations such as the United Nations Commission on International Trade Law and UNCTAD.

Part of the success of the ASEAN E-Commerce Project is due to the fact that it focused on global harmonization and international interoperability, rather than simply focusing on regional harmonization. This focus on international interoperability included the selection of international models and templates, especially the United Nations Convention on Electronic Contracting, for the implementation of domestic e-commerce law in ASEAN member countries. This ensured that ASEAN's e-commerce legal infrastructure would also be compatible with international developments, providing greater certainty for consumers and greater

consistency for businesses. Another important factor for success was the strong focus on trade facilitation in this project. This resulted in constant testing of the project outputs against trade facilitation objectives.

There is a need for detailed implementation tools to help developing countries implement e-commerce legal infrastructure, rather than just high-level recommendations or generic discussion papers. The implementation tools used in the ASEAN project included regional implementation guides, country-specific implementation guides, implementation progress checklists and implementation timelines.

The ASEAN project shows the importance of developing comprehensive legal infrastructure – not just written laws – and of aligning domestic and international e-commerce laws to avoid overlaps and inconsistencies. It is important for countries to minimize inconsistencies and duplications in order to create a smooth, consistent legal platform for businesses engaging in e-commerce in their region.

Introduction

SCIENCE AND TECHNOLOGY FOR DEVELOPMENT: THE NEW PARADIGM OF ICT

It is now well established that the capacity to generate, assimilate, disseminate and effectively use knowledge to enhance economic development is crucial for sustainable growth and development, since knowledge forms the basis of technology innovations. In the context of a global economy driven by technological innovations, it is important for developing countries to lay proper foundations for building their capacity to acquire and create knowledge and technology in order to take advantage of the opportunities offered by globalization and, at the same time, to address emerging global challenges.

In the last part of the 20th century, the world economy witnessed an enormous increase in the generation of knowledge, resulting from the growth of research budgets and the availability of powerful research tools created by the rapid development of ICTs. This process was supported by global opportunities for accessing and disseminating knowledge, following the opening of borders to international trade and movements of persons and significant progress in transportation and communication technologies. Consequently, knowledge has become more important economically, in terms of investment in and production of knowledge-based goods and services. The adoption of new technologies and the improvement in human capital through knowledge have enhanced economic performance and increased factor productivity in many countries. At the same time, the fast pace at which new technologies are being developed, but also become obsolete, has profoundly changed the process of knowledge creation and acquisition, with sustained efforts being required for continuous upgrading of knowledge and virtually lifelong learning.

The challenge is therefore to harness knowledge for development by providing an enabling environment for the production of ideas and innovations, as well as for their dissemination and use by different actors, directly or indirectly involved in the production process. The creation and the use of knowledge to generate innovations – to improve or upgrade existing technologies or to introduce new technologies and business processes – depend on a number of prerequisites. The existence of supporting policies and institutions, such as government regulations and measures to encourage the creation and application of knowledge, financial institutions including venture capital, and institutions for standard- and norm-setting, is of the utmost importance in this regard. And so is the availability of qualified human resources and local training and research institutes – schools that train technicians and research institutes that are sources of technological innovations, as well as specialized institutes that prepare qualified businesspersons and policymakers. Also, efforts are needed at the international level to share knowledge and transfer technology for the benefit of less advanced countries.

Recognizing the power of knowledge, the IER 2007/2008 analyses the contribution of ICT to growth and development in the broader context of science and technology by considering ICT as one particular type of technology and applying the same economic analysis as in the case of technology for development. This analysis highlights the role of human capital, externalities and spillovers (notably through learning) and appropriate policies and institutions in the creation of innovations, which support the competitiveness of enterprises and of the economy in general. It also highlights the importance of open access to knowledge and hence the importance of diffusion of knowledge and technology, especially in the case of developing countries.

ICT is a general-purpose technology and as such has a pervasive impact on the economy. It introduces a new paradigm to the configuration of economic activities, changing in a radical way the approach to technology for development, because of the importance of spillover effects of ICT economic applications.

- First, the use and economic applications of ICT, more than the production of ICT, can make a powerful contribution to economic development. The economic impact of ICT could be more important, in terms of externalities and spillovers through its use and applications in different sectors of the economy, than its direct contribution to GDP as a production sector.

- Second, one of the most important externalities is a new mode of organization of production and consumption, which results in cost-saving transactions and faster and better communication between economic agents.

- Third, the rapid pace of innovation in the ICT sector itself has considerably reduced the costs of access to ICTs. This has allowed a democratization of ICT use, to be spread to poorer people in support of their livelihoods, and it has also facilitated the adoption of ICT in poverty reduction programmes.

- Fourth, ICT has generated new services in the forms of e-commerce, e-finance, e-government, and so forth. These new services can contribute to greater economic efficiency. However, other challenges may arise concerning questions of trust and security in the transactions that these new e-services generate.

- Fifth, ICT is skills-demanding, and the importance of education and training to build a knowledge economy in which ICT represents an indispensable tool continues to grow.

- Finally, ICT has given rise to new models for sharing knowledge and collective production of ideas and innovations, which often bypass the incentive system provided by intellectual property rights. These "open access" models, whether in activities such as open source software, open innovations or common knowledge associations, can be an efficient channel for rapid diffusion of knowledge to less advanced countries and deserve greater attention.

This chapter will analyse the factors underpinning the contribution to economic development of science and technology and ICTs. It will then introduce the subsequent chapters of the Report which, in the light of this analysis, will examine in detail the different aspects of a strategy to harness ICTs for development.

A. Science and technology in development

1. Role of technology in growth and development

Science and technology are at the heart of economic growth and development. Different schools of growth theory, whether of the neoclassical or endogenous types, have all recognized the essential role of technological progress and its corollary, knowledge, in sustaining the growth process and increasing the level of per capita income. What are the factors which promote technological progress? What should be the policy and institutional arrangements which are most propitious for the creation of knowledge and innovations, leading to constant technological progress? Since knowledge is non-rivalrous,[1] its diffusion should be rapid. However, the process of technology diffusion and transfer, which is one of the most important issues for the economic development of developing countries, is not straightforward. Identifying the factors which encourage or hinder technology diffusion is the first step in helping developing countries secure access to, absorb or transform, and use technologies developed by technology leaders.

Understanding the mechanics of economic growth has always been the single most important objective of economists, from Adam Smith to Thomas Malthus, Karl Marx, and the more modern neoclassical economists and "endogenous growth" economists. The evolution of economic thought has led to a universal acceptance of the role of technology and knowledge as an engine of growth.[2] While capital and labour are necessary inputs, the production of goods and services is driven by technology, which is the technical process, method or knowledge associated with an appropriate combination of inputs to produce and enhance the quality of goods or services. Technology has a direct impact on the productivity of firms, but has also an impact at the macro level, insofar as some platform technologies provide the means for better organization of work and better communications among people. Economic growth cannot be sustained by capital accumulation alone, as the contribution of capital, without technological progress, will be subject to diminishing returns.

The two schools, neoclassical and endogenous, differ in the analysis of the duration of the economic impact of technology. The neoclassical approach takes the view that technological progress has only a transitory effect on the rate of growth, but a lasting effect on the level of per capita income, which will move to a higher new steady state level. The endogenous growth theory (or at least some versions of it) implies a permanent effect on the long-term rate of growth. Under the neoclassical approach, the economic development of countries at different levels of development will converge towards the same steady state level (the catch-up phenomenon), given conditions of perfect competition and free flow of technology between countries, while under the endogenous growth approach, structural characteristics implying different endogenous technological capabilities result in a persistent divergence of growth paths, which require government intervention to address structural problems, going beyond the simple recipe of increasing savings and investment rates.

Behind technology is the accumulation of science and knowledge which leads to innovations and their applications to new technologies.[3] The endogenous growth school has highlighted the importance of human capital, research and development for technology production. It has also studied the policy implications of knowledge externalities and spillovers (Romer, 1990, 1994). In order to harness knowledge for technological progress, growth and development, there should be good institutions to coordinate the activities of different actors (from researchers to entrepreneurs and including intermediaries and consumers), and to legislate for, and support (through provision of finance and infrastructure), the creation of knowledge and technology, their diffusion and access to them by the public. Many authors (e.g. North, 1990; Mokyr, 2002) maintain that the interaction between institutions and knowledge is central to economic progress.[4]

For poorer countries, the most important challenge is the mobilization of resources to build up institutions and industries and build capacity to absorb and use imported technology, as well as to create local innovations.[5] The question of technology diffusion and transfer from lead technology producers to less advanced countries is of crucial importance for developing countries in general.

Measuring the economic impact of technology

What is technology? Technology is broadly defined as the way in which inputs to the production process are transformed into outputs (goods and services), and is basically knowledge. Technological progress can take the form of *product innovation,* which increases the quality

of output or *process innovation,* which in turn improves efficiency in the production of output. Technology can be embodied in capital goods (and is thus directly measurable) or in human capital (and captured as tacit knowledge or know-how which can or cannot be codified). It is thus, in practical terms, very difficult to measure technology.[6] This is why technology is not entered as an independent input in empirical studies on sources of economic growth: ideas and knowledge are not quantifiable magnitudes.

Instead, it is assumed by economists that technology increases the productivity of the production function or its factor inputs, and that the economic impact of technology can be therefore measured by changes in productivity. The productivity effect of technology can be "Hicks-neutral" (i.e. it has an equal impact on factors of production and is exogenous to the production function), in which case it is captured by the concept of total factor productivity[7] (TFP) (also known as multifactor productivity), or technology increases the productivity of capital (capital-augmenting technology) or labour (labour-augmenting technology).

At the macro level, the concept of TFP has been used to measure the contribution of technology to economic growth. It should be noted, however, that TFP might capture other factors, including policy and institutional changes, variations in capacity utilization or other inefficiencies or efficiencies not due to technology. Nevertheless, a number of empirical studies have used TFP to analyse income gaps or long-term growth. For example, Klenow and Rodriguez-Clare (1997) and Hall and Jones (1999) found that a large part of the differences in per capita income between rich and poor countries is explained by differences in TFP. Other studies estimated the contribution of TFP growth, as compared with the growth of capital and labour (or sometimes human capital), to the growth of GDP. In the case of East Asian countries, a number of studies, including Young (1994, 1995), Kim and Lau (1994) and Collins and Bosworth (1996), using the TFP approach, found that TFP explained a small part of the growth of East Asian countries (covering the period 1960–1990) and that the role of capital accumulation was preponderant in the strong growth experience of those countries. Those findings led to the conclusion that growth in East Asian countries was not led by technological progress, but by high rates of capital accumulation. This was a surprising conclusion, given the well-known fact that economies such as the Republic of Korea, Taiwan Province of China, Singapore, Malaysia and Thailand have sustained their economic growth by rapidly moving up on the technological ladder. Government policies in those economies have always attached importance to the development of knowledge and human capital, to the absorption of foreign technology and to the production of indigenous technology. There are thus obvious shortcomings in the TFP approach based on the neoclassical aggregate production function, and many authors have questioned the applicability of the assumptions of that model to the analysis of real economic life (see Felipe, 1997; Nelson and Pack, 1997). It has been pointed out that the spectacular growth achievements of the East Asian economies are closely associated with the entrepreneurship, innovation and learning that occurred in those economies before they were able to master the new technologies imported from industrialized countries. Thus, technology matters and high investment rates are not the only explanation, the more so since the contributions from capital accumulation and technological progress to growth can be closely interdependent: technological progress increases capital and labour productivity, thus encouraging capital-deepening in the production process, which in turn allows the introduction of more innovations (for example, in the case of investment in ICT). Thus, the impact of technology on economic growth should not be assessed by using only the concept of TFP, since technology impact, through

capital-deepening, on the productivity of capital and labour is equally important.

The concept of TFP is also used at the industry and firm levels to determine the factors which increase the TFP of the economy as a whole or of industrial firms in different sectors. For reasons of data availability, most studies have been done in the context of industrialized countries. The role of ICT in particular in increasing TFP has been extensively studied (see chapter 3 of this Report). Other studies found that research and development (both public and private), as well as technology spillover effects, contribute to increasing TFP in the manufacturing and services sector.[8]

2. Technology and institutions

In analysing the history of industrial revolutions in Europe and the United States, as well as the "Asian Miracle", scholars have identified the crucial role of institutions and policies in supporting the production and diffusion of ideas and knowledge, which are the foundations of technological progress. What should be the appropriate policy and institutional environment to encourage the production, diffusion and use of scientific and technological knowledge for the benefit of the economy as a whole? It is often said that

Chart i.1

Science and engineering in first university degrees, by selected region, 2002 (or more recent year)

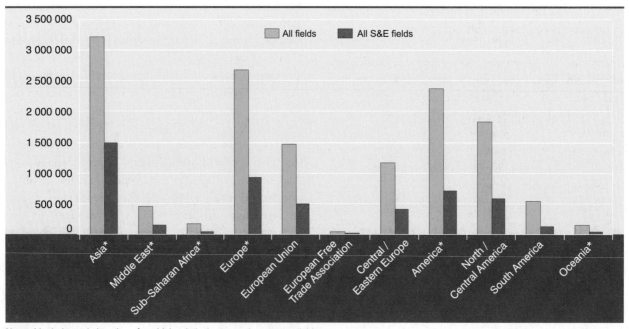

Note: * Includes only locations for which relatively recent data are available.

Source: Data complied by the National Science Foundation from the OECD Education database, and UNESCO, Institute for Statistics.

without entrepreneurs, venture capital and schools that train the technicians who can build and maintain new technologies, innovations will not yield economically significant results.

Human capital and research and development are crucial to the development of ideas and their applications to technological innovations. In that respect, countries should attach the utmost importance to the quality of the national education system and research institutions, aiming at building up an "army" of good researchers. Recent statistics show that there is still a wide gap between developed and developing countries with regard to research and development, although the newly industrializing countries in East Asia are narrowing it.

As can be seen from chart i.1, the proportion of university graduates in the field of science and engineering in the total number of university graduates is the highest in Asia, followed by Europe and North America. In other regions, this proportion is negligible.

Table i.1 shows statistics concerning the number of researchers and the amount of gross expenditure on R&D (GERD). It appears from that table that the developed countries have a very high concentration of researchers (3,272.7 per million inhabitants and 70.8 per cent of the total number of researchers in the world), while developing countries have much lower magnitudes (314.3 researchers per million inhabitants and 29.2 per cent of the total number of researchers). Likewise, the percentage in GDP of gross expenditure on R&D (GERD/GDP) is 2.3 per cent for developed countries, which is more than double the level of 1.0 per cent for developing countries. Among the developing countries, however, the newly industrializing countries of Asia are above the developing country average, showing a concentration of 777.2 researchers per million inhabitants, and GERD/GDP equivalent to 2.3 per cent of GDP.

The institutional framework should also ensure a good flow of knowledge between scientific research and technological applications (in both directions), as well as a good flow of information among researchers and users, at the national and international levels. This framework has become known as the "national innovation system", which can be defined as a complex network of agents, policies and innovations supporting technical progress in an economy.[9] Within that framework, and in the context of developed countries where more information is available, the increasing

frequency of collaborative research and public–private partnerships is to be noted. Industry clusters and technology parks are also interesting models of organizations that allow a good flow of information, complementarities and spillover between different firms along the value chain and in different sectors (production, business services, financial services, transportation etc.). Likewise, domestic and foreign networks reinforce each other through international research cooperation or strategic alliances for R&D.

Governments play a crucial role, because knowledge creation cannot rely on market mechanisms alone. Policies to support knowledge creation (such as tax subsidies, intellectual property protection, government funding and government procurement) as well as knowledge diffusion (establishment of libraries, communication networks, access cost subsidies, etc.) are examples of government measures in that area. A clear legal and regulatory framework in many areas, touching upon the interactions and transactions among different actors, is also necessary. The approach to intellectual property rights needs to strike a balance between incentives for creativity and society's interest in maximizing the dissemination of knowledge and information.

Lastly, the question of financing of innovation and technology is of equal importance. Developing countries may explore the applicability of the venture capital market model of developed countries. Development assistance from multilateral and bilateral agencies needs to attach more importance to the science and technology development. Taxes and subsidies can also play a role.

3. Knowledge diffusion and technology transfer

The market for knowledge[10] is often characterized by imperfections, whereby social and private returns derived from knowledge can differ widely. In the area of knowledge creation, this "market failure" may lead to private underinvestment in knowledge: this is why Governments have taken measures to provide incentives for individual agents to create knowledge through intellectual property rights, tax rebates and subsidies,[11] and full or partial funding of research. In particular, intellectual property protection through the establishment of patents, trademarks, copyrights or trade secrets confers the right to appropriate the income derived from the application of proprietary research in order to recover the high fixed cost of research or to enjoy monopoly rights for a certain period.

Table i.1

Researchers worldwide, 2002

	Researchers (thousands)	Researchers worldwide (%)	Researchers per million inhabitant	GERD / GDP (%)
World	**5 521.4**	**100.0**	**894.0**	**1.7**
Developed countries	3 911.1	70.8	3 272.7	2.3
Developing countries	1 607.2	29.1	314.3	1.0
LDCs	3.1	0.1	4.5	0.1
Americas	**1 506.9**	**27.3**	**1 773.4**	**2.2**
North America	1 368.5	24.8	4 279.5	2.7
Latin America and the Caribbean	138.4	2.5	261.2	0.6
Europe	**1 843.4**	**33.4**	**2 318.8**	**1.7**
European Union	1 106.5	20.0	2 438.9	1.8
Comm. of Ind. States in Europe	616.6	11.2	2 979.1	1.2
Central, Eastern and other Europe	120.4	2.2	895.9	1.1
Africa	**60.9**	**1.1**	**73.2**	**0.3**
Sub-Saharan countries	30.9	0.6	48.0	0.3
Arab States (Africa)	30.0	0.5	159.4	0.2
Asia	**2 034.0**	**36.8**	**554.6**	**1.5**
Comm. of ind. States in Asia	83.9	1.5	1 155.0	0.4
Newly indus. Asia	291.1	5.3	777.2	2.3
China	810.5	14.7	633.0	1.2
India	117.5	2.1	112.1	0.7
Japan	646.5	11.7	5 084.9	3.1
Israel	9.2	0.2	1 395.2	4.9
Arab States (Asia)	9.7	0.2	93.5	0.1
Other Asia	65.5	1.2	100.2	0.1
Oceania	**76.2**	**1.4**	**2 396.5**	**1.4**

Source: Adapted from UNESCO.[12]

It has been noted that the past 20 years have witnessed four major trends in knowledge.[13] First, there has been an enormous increase in the creation of knowledge, resulting from the growth of research budgets and the availability of powerful scientific research tools. Second, knowledge has become more important economically. It represents an increasingly important product or service, as in marketed information; an increasing share of competitive investment, in the information economy; and an increasing share of physical products. Third, the increasing openness of borders to products and people and the development of transportation and communication (particularly digital information technologies) have created new global opportunities for accessing and disseminating knowledge. Fourth, knowledge is increasingly privatized and commercialized, and the use of intellectual property rights to protect knowledge has restricted access to information and technologies. Furthermore, the international governance of knowledge has moved towards tighter and more harmonized IPRs at the international level, with a view to minimizing the free-rider problem. That has been achieved through the WTO's Agreement on Trade-Related Aspects of Intellectual Property Rights (TRIPS), the "TRIPS-plus" provisions of regional and bilateral trade agreements negotiated by the European Union and the United States, and a number of new treaties negotiated under the auspices of the World Intellectual Property Organization (WIPO).

In the area of knowledge diffusion and technology transfer, externalities and spillovers can yield enormous benefits for the economy as a whole, and for the rest of world in the presence of technology flows among countries. In some instances knowledge can be considered a public good (non-rivalrous and non-excludable). Many inventions were built on earlier inventions ("standing on the shoulders of giants"), and the benefits of technological progress are not limited to one firm or one sector, because of complementarities in the application and generation of knowledge, and are thus extended to the economy as a whole. The benefits of externalities and spillovers may not be fully reaped because of high access costs or other obstacles. In the case of cross-border flows of knowledge, high access costs and barriers resulting from a harmonization and tightening of IPR standards at the international level may be harmful for poorer countries with limited human and capital capacity.[14] While market failures may be important, measures to address them have not been clearly identified, especially with regard to the transfer of technology from developed to less developed countries.

Market-based mechanisms for technology transfer through trade, foreign direct investment or licensing have always been used by developing countries to acquire new technologies. Many of the newly industrializing countries in East Asia have relied on those mechanisms to develop and move their productive technological capacity from OEM (original equipment manufacture) to ODM (own-design manufacture), and finally to OBM (own-brand manufacture).[15] However, with the restrictions in the international IPR regimes (especially after the adoption of the TRIPS Agreement in the WTO), costs for access to foreign technology are increasing, and many learning by-doing-methods, for example reverse engineering, may not be possible any more. Foreign enterprises in high-technology sectors may prefer to establish their own subsidiaries

or to acquire local firms in order to restrict the transfer of technology. Other mechanisms for technology transfer involve arm's length arrangements in the form of inter-firm strategic alliances for R&D, public–private partnership projects (for example, between public research institutes in developing countries and foreign firms, mostly subsidiaries of transnational corporations), labour migration (skilled inputs of expatriates), and so forth. In the case of low-income countries, those mechanisms are used less because of weak local capacities.

An international debate has been engaged with regard to finding the right balance between production and dissemination of knowledge. Finding an optimal international IPR regime is a complex problem: in addition to a careful evaluation of externalities and spillovers from both the producer and user perspectives, account should be taken of the differential development and social needs of developing countries. One of the most visible results of that debate was the decision taken by the WTO in Doha to interpret TRIPS in a way that recognizes the urgent public health needs of poor countries. However, worldwide, IPR regimes in developed countries have become more restrictive.[16]

Many approaches have been suggested in order to encourage more effective transfer of technology to developing countries.[17] These can be grouped under a few main headings, and deserve more attention and research at the policymaking level:

- Improving flexibilities in intellectual property rights, in terms of calibration of standards and norms for countries at varying levels of development. For example, it has been suggested that patent holders from developed countries could decide to apply a multi-price strategy in order to reduce access costs for low-income countries. The lower price could be maintained through price control or compulsory licensing. Another flexibility is the distinction between basic research and commercially applied research. The former, including related databases, may be accessible free of charge. It should be noted that open-access basic scientific research and commercial research have traditionally operated in a complementary fashion. Increasing transaction and access costs in scientific research can be harmful to dynamic industrial competition, as scientific spillovers to potential commercial technology innovations risk being minimized. Flexibility could also be applied in the form of exemptions or exceptions for acute public health, environmental and social needs of poor countries. The TRIPS Agreement also provides scope for flexibilities (in terms of limitations, exceptions or extensions) that can be exploited by the least developed countries.

- Open access regimes. These are arrangements whereby researchers deliberately forgo the pecuniary benefits attached to the proprietary protection of their inventions and collectively contribute to improvement of the research product (see box i.1). In some areas involving extensive cumulative innovation, such as computer software,[18] biotechnology or other public domains of common knowledge, those arrangements may be the most efficient forms for advancing knowledge. The key feature of open access models is that either the knowledge is put in the public domain or its use is free of restrictions, as specified in the terms and conditions of licences, and open access projects may arise where traditional intellectual property incentives are weak.[19]

- International partnerships for generating and sharing information. Many global initiatives have been launched, with the financial support of the public and private sectors, to enhance global research and information capabilities, so as to overcome crucial problems in the areas of rural development, environment and health in the poorest developing countries. International partnerships could be reinforced in those areas, as well as in other areas, in order to allow more effective participation by poor countries in sharing the benefits of common knowledge.

- Global support for building capacity in developing countries, especially the least developed countries, to enhance human capital, infrastructure and institutions in order to develop their scientific and technical knowledge. There is a strong case for donors to increase "knowledge aid" and aid for science and technology.[20]

B. ICT as a general-purpose technology

The ICT revolution has been compared to earlier industrial revolutions in modern economic history, and ICT has been classified as a general-purpose technology (GPT) to the same extent as power delivery systems (electricity, steam) or transport innovations (railways and motor vehicles). The widespread use of ICT in economic and social spheres has shaped what is usually called as the "New Economy". What is the "new paradigm" that ICTs have brought to the economic landscape, and what are the broad policy implications that can be drawn from this new trend? An analysis based on the proliferating literature on ICT as a GPT and its economic impact provides some answers.

1. New paradigm

The role of technology in growth and development as reviewed in the earlier section is crucial. When it comes to the GPT, this role is pervasive, as GPT interacts with any single sector of the economy and opens up and facilitates the emergence of new opportunities, entailing innovational complementarities in downstream sectors. According to Lipsey, Bekar and Carlaw (in Helpman, 1988), GPTs are technologies that have four characteristics:

1. Wide scope for improvement and elaboration;

2. Applicability across a broad range of uses;

3. Potential for use in a wide variety of products and processes;

4. Strong complementarities with existing and potential new technologies.

All those characteristics contribute to a substantial increase in the productivity of capital and labour for the economy as a whole. As reported by David and Wright (1999), GPTs are variously characterized in terms of inter-industry linkages, R&D investments, economies of scale, better coordination, spillover and so forth. But GPT phenomena can also generate alternating phases of slow and rapid productivity: the slowdown phase may be attributed to the diversion of resources into knowledge investment during the gestation of new GPT. Reference is made to the famous "Solow paradox" concerning the lack of impact of ICT investment on productivity in the United States in the 1970s and early 1980s. Another general conclusion is that GPT diffusion may be strongly conditioned by the supply of complementary productive inputs, especially high by skilled labour.

As a GPT with the characteristics mentioned above, ICT, as much as electrical power, has a profound and

Box i.1

Copyright and open access to knowledge

A copyright is a statutory (i.e. designated and imposed by law) time-bound monopoly right that is assigned to the creator of a copyrightable work at the time of the creation of that work. A copyright is tradable; it can be sold, leased or rented. Copyrighted works do not have to be registered with an authority, but if there is an intent to sell, lease or rent, registration secures the confidence of contracting parties in such a transaction. After a copyright expires, the work falls into the public domain. The chart below describes the interaction of several types of copyright and access, while the table organizes these concepts into a grid.

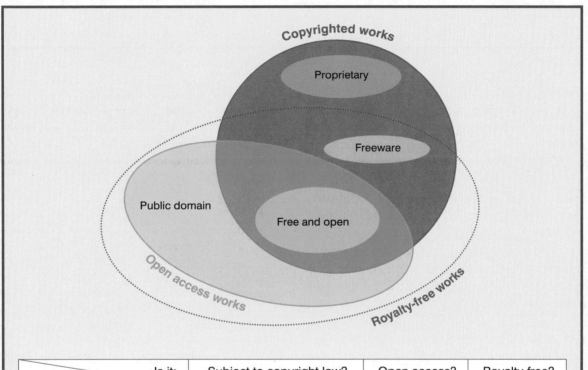

Is it: The content is:	Subject to copyright law?	Open access?	Royalty-free?
Proprietary	Yes	No	No
Freeware	Yes	No	Yes
Free and open	Yes	Yes	Yes
Public domain	No	Yes	Yes

If a work does not carry an explicitly stated copyright licence, or carries only a "*copyright-date-owner*" designation, the default copyright is *proprietary copyright*. A proprietary licence may also be explicitly stated. The owners of such works have exclusive rights to use them and to authorize use by others. "Use" is defined as any reproduction (copying), public performance (e.g. theatrical or musical works), recording of the work (e.g. a composer's song or a writer's screenplay is recorded and distributed as a song or film on a CD or DVD), broadcasting (radio, cable or satellite), or translation of the work into other languages or its adaptation (e.g. a novel into a screenplay).

Works under a *freeware copyright* have the same restrictions as proprietary works except that the owner will explicitly grant a royalty-free licence for non-commercial use of the work as it is without allowing subsequent copying, broadcasting or adaptation.

The terms and conditions of a *free and open copyright* (FOC) are expressed explicitly in an accompanying FOC licence that promotes content and knowledge sharing without letting them fall into the public domain. An FOC licence will prevent or discourage the copying and redistribution of the work under a proprietary copyright licence, but will oblige or encourage users to copy, modify and redistribute the work or derivative works under a FOC licence. FOC licenses for software are developed or accredited by the Free Software Foundation (FSF) and the Open Source Initiative; the most common licence is the FSF "GNU General Public License". Free and Open licences for text works, music and audiovisual expressions are developed and maintained by the Creative Commons project. The FSF "GNU Free Document License" is also in use.

Box i.1 (continued)

The **public domain** consists of works whose copyright owners have explicitly and publicly relinquished any copyrights they may have had on those works. Works whose copyright duration has expired, as a result of which the terms and conditions of the copyright no longer apply, also fall into the public domain. Works in the **public domain** can be freely copied, altered and used in derivative works without any obligation for any of its prior copyright owners. However, derivative works based on **public domain** content can be distributed under restrictive proprietary copyright – something that FOC generally avoids.

Open access designates all works that are free of restrictions on use, sharing, modification and incorporation, entirely or in part, into derivative works, and include works under FOC and the public domain. **Royalty-free** works are those that either generally or under certain conditions (e.g. non-commercial use) do not require users to pay a royalty for the right to access the content or knowledge. These include open access works and works under a freeware proprietary copyright.

Source: UNCTAD.

---

widespread impact on the structural transformation of all sectors of the economy as well as on society as a whole. It also has a widespread impact on the consumption side, since it is also a consumption product. The ICT revolution is often associated with the advent of the "New Economy". By the 1990s, the United States had grown at a surprisingly rapid pace, with record employment, low inflation, and acceleration of productivity growth. This was accompanied by significant investment in, and diffusion and widespread application of, ICTs. The term "New Economy" was first adopted in the context of the United States, but was then applied to the rest of the world when economies were characterized by dynamic growth led by adoption of new ICTs, and accompanied by significant changes in the organization of work which are different from those of the "old economy".

What is the new paradigm that ICTs have introduced? ICT is first of all a powerful technology for information processing, from both the quantitative point of view (astronomical size of data that can be stored and processed) and the qualitative point of view (adapted to a great variety of uses, rapid and wireless connections, distanceless, constantly upgraded to respond to changing needs). Moreover, ICTs are diverse instruments (computers, phones, other audio-visual instruments) that can be used by customers with different degrees of skills.

ICT applications allow not only a significant increase in productivity (total factor productivity, labour productivity, capital productivity), but also a reconfiguration of work organization, within the firm, among firms, among all market participants (consumers and producers), and between the Government and the rest of the economy. The innovations that accompany

ICT applications are numerous and will most likely multiply in the future because of the "wide scope for improvement and elaboration" of ICTs. The following is an enumeration (which is not exhaustive) of the implications related to the new paradigm introduced by the ICT revolution.

- First, ICTs have significant *spillover* effects on the productivity of the economy as a whole. They contribute to economic growth[21] in three ways:

- As ICT production: contribution of the ICT product sector to GDP (chapter 2 will review latest developments in the ICT sector);

- As capital input: ICT equipment is used for the production of other industries and services as a capital good;

- As a special capital input: ICTs produce benefits that go beyond those accruing to the ICT sector or the investment in ICT as capital input in other sectors. Those benefits are derived from externalities or spillovers which benefit indirectly all market participants, thus, contributing, to overall productivity and aggregate income growth.

In sum, ICTs contribute directly to GDP growth as a production sector in themselves: the rapid technological progress in the production of ICT goods and services contributes to more rapid growth of TFP in the ICT-producing sector. Second, investment in ICT as a capital input contributes to overall capital- deepening in other sectors, thus helping to increase labour productivity. Third, greater use of ICTs creates "intangible assets" (in the form, for example, of organizational or managerial improvements), which contribute to increasing the overall efficiency of all sectors of production, thus increasing the TFP. ICT can be instrumental in generating complementary innovations, which boost the productivity in industries/services using ICT. Several econometric studies[22] in the context of industrialized countries have shown the importance of ICT's spillover effect on GDP. If the externalities or spillover effects are substantial, there is a large discrepancy between social and private returns, which may warrant the introduction of measures to encourage wider diffusion of ICT use by all market participants.

- Second, wireless and distanceless communications allow more *flexibility* and *networking* in the organization of work. It was reported that ICT use in developed countries is associated with changes in organizational practices, including the transition from mass production to flexible manufacturing technologies, faster and more direct interactions with suppliers and customers, decentralized decision- making, greater ease of coordination, and enhanced communication within the firm.[23] This has given rise to cost-saving management techniques such as just-in-time inventory control, fragmentation and internationalization of the production chain, and outsourcing of services and certain production tasks, to name but a few. For developing countries, those innovations have provided new opportunities for insertion in the global value chains and for diversifying production activities and exports. At the same time, ICTs facilitate the creation of networks, which enhance teamwork and increase the exchange of information among people belonging to the same professional associations and the same communities. It can be said that ICT networking facilities were at the origin of open source systems (UNCTAD, 2003). The tremendous increase in information flows and the availability of ICT tools to analyse that information have encouraged, and accelerated the pace of, business innovations, which in turn enhance the productivity and competitiveness of firms (see chapter 3 of this Report).

- Third, the rapid pace of innovation in the ICT sector has contributed to the decline in the costs of access to ICTs. This opens up opportunities for poorer people to use ICTs for income-earning activities as well as for upgrading

their own knowledge. Chapter 6 will review the role of mobile telephony in the digital empowerment and inclusion of poor people in Africa. Recognizing the useful role of ICT in poverty reduction strategies, many Governments of developed and developing countries have developed policies for "e-inclusion". The approach to "e-inclusion" should not be limited to the problem of access, but should include the concept of supporting livelihoods through ICT. Chapter 7 will analyse that concept in the light of the experiences of telecentres for the poor.

- Fourth, the services sector has been particularly affected by information technology. The last decade has seen the creation of many forms of business (B2B or B2C), which rely on electronic communications for business transactions and organization of work, with names such as e-commerce, e-finance, e-banking, and e-insurance. Governments, through e-government, have also increasingly relied on ICT to organize their services and to deal with the public. Chapter 5 will review the implications of e-banking and e-payments for developing countries and economies in transition.

- Fifth, ICTs affect employment and wages for low- and high-skilled workers. They influence the evolution of the composition of the labour force: the unskilled workers are the main losers in sectors where investment in technology and greater productivity are high, and this phenomenon seems to be manifesting itself in both developed countries and developing countries (UNCTAD, 2006). Computer-based networks are changing the way people work and the way they are paid, in the sense that the rewards for multi-tasking are increasing and employers seem to prefer employees with broad-based education and conceptual and problem-solving skills. Skills can also be acquired on the job through on site-training. The returns to higher education are rising and the benefits for lifelong learning are increasingly being recognized.

2. Some policy implications of ICT for development

The ICT revolution is spreading to the developing world and brings with it the promise of a major technology leapfrog, which will contribute to rapid modernization of the economies of developing countries. In order to reap the opportunities offered by ICT, countries may find it necessary to identify a set of policies to encourage the creation, diffusion and use of knowledge, which should form the basis of a sustained growth strategy. The ICT contribution to building knowledge capital should be taken fully into account in designing innovation policies (see chapter 4). Innovation patterns are country- and industry-specific, and countries at different levels of development will have different approaches according to their capacities and priorities. Taking for granted a general enabling policy framework for investment and enterprise development, specific innovation policies should aim at promoting national knowledge systems to support the competitiveness of national economies. The building blocks of such an innovation policy framework can include the following:

- Enhancing human capital, by upgrading the education system in line with the needs of the economy, and by encouraging well-targeted R&D programmes;

- Providing adequate infrastructure for supporting the creation, diffusion and exchange of knowledge: finance (banking, venture capital), ICTs and business services (including institutions dealing with standards and norms);

- Encouraging partnerships (between public and private sectors, among firms, between domestic and foreign partners, regional initiatives etc.);

- Facilitating networking through the creation of clusters and technology parks;

- Special programmes to support start-ups (for example, business incubators);

- Regulations which provide clear and transparent rules for conducting business (intellectual property protection, Internetgovernance, ICT policies, labour policies, etc.);

- Promoting an "inclusive" policy of technology diffusion in order to encourage low-skill innovations in support of the livelihoods of the poor;

- Other government contributions, such as the launching of large R&D projects, tax rebates or subsidies for R&D activities by the private sector, a special government-sponsored technology fund, and so forth.

Against this broad policy framework to encourage innovations, the particular role of ICT as an enabler of innovations should be recognized and encouraged. Given the strong links between ICT use by enterprises, competitiveness and innovation, there is a need for better integration of policies to promote ICT use by enterprises within general innovation policies. One way to achieve that integration is to systematically coordinate policies from different ministries and at different levels. Many of the developed countries have entrusted overall policymaking for innovation and e-business to the same organizations, which formulate ICT policy as an integral part of science, technology and innovation policies.

It should be stressed that ICTs enable more rapid dissemination and better coordination of knowledge, thus encouraging open access to sources of innovation. An innovation policy framework that fully takes into consideration the changes generated by ICT must give prominence to open approaches to innovation, which present significant advantages for developing countries. Chapter 4 will set out in some detail the policies needed in that respect.

ICT policy should also address the question of *digital divide* (in comparison with the rest of the world, but also the internal digital divide faced by the underprivileged strata of the population). The basic dimensions of the digital divide include issues of access (connectivity, costs), skills (digital literacy) and content (localization of content). ICT policies have been extensively reviewed in earlier IERs, especially the IER 2006.

The digital divide between rich and poor countries remains significant (chapter 1). ICTs are evolving rapidly, and at the same time, costs are declining and many kinds of software have become available through the free and open source software networks. Although some new ICT applications (Wi-Fi, Semantic Web, to name just two) and continuing falls in access costs will allow developing countries to leapfrog on the technology trail, a number of challenges remain to be tackled in order to close the digital divide. The first is to invest in the development of human capital capable of rapidly absorbing and effectively using the new technologies. The second is to regulate e-commerce and provide protection and security to users under cyberlaws (chapter 8 will address the question of harmonising cyber legislation in ASEAN). The third is the financing of infrastructure, taking into account the costs of adjustment of displaced technologies. In all three areas, development partners can make a significant contribution.

C. Harnessing ICT for development

Against this backdrop of introductory analysis of the role of science and technology, and ICTs in development, the IER 2007/2008 examines in detail different aspects mentioned earlier concerning the role of ICT.

Chapter 1 reviews the recent trends in the diffusion of ICT in developing countries. Except for countries in East Asia, developing countries in general remain far behind developed countries in the adoption of ICTs and their use by enterprises. Statistical information is presented to illustrate the recent trends in ICT access and use by individuals, households and enterprises worldwide.

Chapter 2 reviews recent developments in the ICT sector worldwide. It is noted that production, trade and investment in the ICT sector will continue to increase, and the production sites will tend to shift from developed to developing countries. South–South trade in ICT goods exceeded South–North trade in 2004, and is expected to exceed North–North trade in 2007. This corresponds to the rapid catching-up by several developing countries in terms of ICT uptake and the development of their information economies. China and India are the world's largest exporters of ICT goods and services respectively. The expanding ICT industry and international service outsourcing offer a huge potential for developing countries. At the same time, competition will increase and countries wishing to attract FDI and BPO contracts will need to invest in education and telecommunication infrastructure, and to improve their investment climate.

Chapter 3 assesses the impact of ICTs on production efficiency. While in the case of developed countries the relationship between ICT and economic growth is well documented, there are very few studies in the case of developing countries. In general, the favourable impact of ICT on labour productivity is observed in both developed and developing countries. UNCTAD, in cooperation with the Government of Thailand, has undertaken an econometric study of the impact of ICT use on the labour productivity of Thai firms, and SMEs in particular. The results are very encouraging and demonstrate the favourable impact of computer use on the productivity of those firms.

Chapter 4 looks at ICT-driven innovation processes, and the policy framework encouraging innovations. Government policies aimed at supporting and facilitating long-term growth need to recognize and exploit the dynamic relationship between the use of ICT and innovation. A growing number of initiatives at all governmental levels now aim at supporting ICT-driven innovation. The links between innovation policies and policies to promote the use of ICT and e-business by enterprises are becoming stronger and in many countries they are now being placed within the same institutional framework and under the same overall political responsibility. There are, as mentioned earlier in the present chapter, a number of "institutional" matters to be addressed, such as the establishment of a development-friendly intellectual property regime and competition policies, the reinforcement of education and research systems, the creation of public knowledge structures, the development of IT infrastructures, the creation of a trust environment for ICT use, and well-functioning capital markets. Beyond those general issues, countries need to reinforce the complementarity between ICT and innovation policies. An important consideration for them is to put in place instruments to support ICT-enabled innovation among SMEs. This means that adequate outreach strategies need to be developed and that policies should be implemented with enough continuity and integration with sectoral approaches. These should facilitate the emergence of ICT-enabled alliances and networks for research and development. Equally important from the developing country perspective is the need to support open, user-centred innovation approaches: development-friendly intellectual property regimes are particularly useful in that regard.

Chapter 5 deals with the application of ICT in one service sector, namely financial services. The use of ICT in banking and payments is one of the main channels for innovation in the financial services industry, making it possible to drastically decrease the unit transaction costs of repetitive and high-volume operations and to automate the process of credit risk management. Given the relatively higher share of banking as a source of finance for development, the chapter stresses the importance of introducing modern e-banking and e-payment methods in developing and transition economies. In addition to traditional e-banking and e-payments operations, it reviews highly important issues such as access of SMEs and micro-enterprises to e-banking and e-payments facilities. Special attention in that respect is paid to the phenomenon of remittances as a major source of finance for enterprises and households, and the use ICT as a means of reducing the costs of delivering those small-scale private financial transfers.

Chapter 6 examines the experience of using mobile telephone to increase economic opportunities in Africa. While Internet access has become a reality for many businesses and public institutions, and for individuals with higher levels of education and income, for the vast majority of the low-income population, mobile telephony is likely to be their sole tool connecting them to the information society in the short to medium term. Mobile telephony can lead to economic growth in several ways. Investment in network infrastructure and related services creates direct and indirect employment opportunities. The use of mobile telephony in the conduct of business reduces the costs and increases the speed of transactions. Those effects will be more pronounced in economic activities that have a greater need for information or where added information enables increasing returns to scale. While methodologies and estimates will vary, current research indicates that mobile telephony has important economic effects for developing countries. The growing availability and the constant decrease in the price of mobile services are, in themselves a welfare gain.

In Africa, mobile phones have proved so successful that in many cases they have replaced fixed lines. As a result, mobile phone subscriptions since 2001 have quadrupled. However, improving rural penetration and achieving universal access remain a daunting task because of inherent difficulties caused by a lack of commercial distribution channels, insufficient education and poverty.

Chapter 7 analyses the experiences of telecentres in promoting the livelihoods of people living in poverty. Telecentres – namely, public facilities where people can access ICTs, communicate with others and develop digital skills – have become a key programme and policy instrument for supporting wider ICT access, thus broadening the social and economic benefits of ICT. Telecentres can support the livelihoods of people living in poverty by providing access to key information, supporting the development of technical and business skills, facilitating access to government services and financial resources, and supporting micro-entrepreneurs.

A number of policy recommendations are addressed to Governments in order to enhance the favourable impact of telecentres:

- Develop and promote relevant e-government content and services that support economic activities and livelihoods;

- Support the development of e-business skills;

- Provide strategic financial support for telecentre networks to provide value- added services and develop expertise in economic activities, as in the example of the telecentre in Nunavut (Canada).

Likewise, some recommendations are addressed to telecentre networks:

- Provide value-added services that have a direct impact on the livelihoods of the local community;

- Mainstream e-business skills training programmes;

- Enhance the understanding of the local context and strategies for supporting livelihoods;

- Support the economic activities of those in weaker positions by employing community infomediaries, as well as by providing targeted training and services; and

- Further engage with organizations that support economic activities, such as business associations or micro-credit institutions, in order to provide business- supporting services through the telecentres.

Chapter 8 examines the experience of the Association of Southeast Asian Nations (ASEAN) in the adoption of a regional harmonized e-commerce legal framework consistent across jurisdictions. The ASEAN region is the first region in the developing world to harmonize its e-commerce legal framework. By the end of 2008, all ASEAN member countries will have enacted consistent national e-commerce legislation. An increasing number of developing countries are adapting their legislation to e-commerce in order to remove barriers to online services and provide legal certainty to business and citizens. The impact of the introduction of legislation on the expansion of e-commerce activities is reported by countries to be positive, leading to increased ICT-related business opportunities and foreign direct investment, according to an UNCTAD survey on e-commerce legislation in developing countries carried out in 2007. The ASEAN project shows the importance of aligning domestic and international e-commerce laws to avoid overlaps and inconsistencies and create a smooth, consistent legal platform for businesses engaging in e-commerce at the regional level. Part of the success of the ASEAN E-Commerce Project is due to its focus on trade facilitation and global harmonization and international interoperability, rather than merely on regional harmonisation.

References and bibliography

Barton JH (2006). Knowledge. In: Expert Paper Series Six: Knowledge, International Task Force on Global Public Goods, Stockholm, Sweden, http://www.gpgtaskforce.org.

Basu S and Fernald J (2006). Information and communications technology as a general-purpose technology: evidence from US industry data. Federal Reserve Bank of San Francisco, Working Paper Series, Working Paper 2006-29, http://wwwfrbsf.org/publications/economics/paper/2006/up 06-29k.pdf.

Bresnahan TF, Brynjolfsoon E and Hitt LM (2001). Information technology, workplace organization, and the demand for skilled labor: firm level evidence. *Quarterly Journal of Economics*, vol. 116, pp. 339–376.

Collins S and Bosworth BP (1996). Economic growth in East Asia: accumulation versus assimilation. *Brookings Papers in Economic Activity*, vol. 1996, no. 2, pp. 135–203.

Comin D, Hobijn B and Rovito E (2006). Five facts you need to know about technology diffusion. National Bureau of Economic Research Working Paper Series, Working Paper 11928, http://www.nber.org/papers/w 11928.

David PA and Wright G (1999). *General purpose technologies and surges in productivity: historical reflections on the future of the ICT revolution.* Paper presented to the International Symposium on Economic Challenges of the 21st Century in Historical Perspective, Oxford, England, 2–4 July 1999.

Felipe J (1997). Total factor productivity growth in East Asia: a critical survey. Economics and Development Resource Center Report Series No. 65, Asian Development Bank.

Guerrieri P and Padoan PC (2007). *Modeling ICT as a General Purpose Technology.* Collegium, Special Edition, No. 35, European Commission and College of Europe. www.coleurope.eu/research/modelling ICT.

Hall RE and Jones C (1999). Why do some countries produce so much more output per worker than others? *Quarterly Journal of Economics*, vol. 114, February, pp. 83–166.

Helpman E (ed.) (1998). *General Purpose Technologies and Economic Growth,* Cambridge, Mass. MIT Press.

Hobday M (1995). *Innovation in East Asia.* Edward Elgar, Aldershot, England, and Brookfield, Vermont.

International Task Force on Global Public Goods (2006). *Meeting Global challenges.* Stockholm, Sweden, http://www.gpgtaskforce.org.

Kim JI and Lau L (1994). The sources of economic growth of the East Asian newly industrialized countries. *Journal of the Japanese and International Economies*, 8, pp. 235–271.

Klenow P and Rodriguez-Clare A (1997). The neoclassical revival in growth economics: has it gone too far? *NBER Macroeconomic Annual 1997*, Cambridge, Mass.: MIT Press.

Lipsey RG, Bekar C and Carlaw K (1998). What requires explanation. In: Helpman   E. (ed.), *General Purpose Technologies and Economic Growth*, pp 14–54.

Maskus KE (2006). Information as a global public good. In: Expert Paper Series Six: Knowledge. International Task Force on Global Public Goods, Stockholm, Sweden.

Maurer SM and Scotchmer S (2006). Open source software: the new intellectual property paradigm. National Bureau of Economic Research, Working Paper Series, No. 12148, March 2006, http://www.nber.org/papers/ w12148.

Mokyr J (2002). *The Gifts of Athena: Historical Origins of the Knowledge Economy.* Princeton: Princeton University Press.

Mokyr J (2003). The knowledge society: theoretical and historical underpinnings. Paper presented to the Ad Hoc Expert Group on Knowledge Systems, United Nations, New York, 4–5 September 2003.

Nelson RR and Pack H (1997). The Asian miracle and modern growth theory. World Bank Policy Research Working Paper No. 1881, World Bank.

North DC (1990). Institutions, Institutional change, and economic performance. Cambridge University Press, Cambridge.

Romer PM (1990). Endogenous technological change. *The Journal of Political Economy*, vol. 98, no. 5, Part 2, pp. S71–S102.

Romer PM (1994). The origins of endogenous growth. *Journal of Economic Perspectives*, vol. 8, no. 1, Winter, pp.3–22.

Schreyer P (2000). *The Contribution of Information and Communication Technology Output Growth: A Study of the G7 Countries.* OECD, Directorate for Science, Technology and Industry, STI Working Paper 2000/2.

Solow RM (1957). Technical change and the aggregate production function. *Review of Economics and Statistics*, 39, August, pp. 312–320.

UNCTAD (2003). *E-Commerce and Development Report 2003.* United Nations Publication, Sales No. E.03.11.D.30.

UNCTAD (2006). *Information Economy Report 2006.* United Nations Publication, Sales No. E.06.11.D.8.

UNCTAD (2007). *The Least Developed Countries Report 2007.* United Nations Publication, Sales No. E.07.11.D.8.

Young A (1994). Lessons from the NICs: a contrarian view. *European Economic Review*, 38, pp. 964–973.

Young A (1995). The tyranny of numbers: confronting the statistical realities of the East Asian growth experience. *Quarterly Journal of Economics*, August, pp. 641–680.

Notes

1. Knowledge is non-rivalrous because its use by an individual will not prevent anyone else from using it. The typical example is a mathematical formula, which can be used by an unlimited number of people.

2. Many pioneering works, including by Kaldor and Schumpeter, highlighted the importance of technological progress and innovations for economic growth. More recently, two schools of analysis became prominent. The approach of the neoclassical school on economic growth was exemplified by the seminal work of Robert Solow and the endogenous growth approach was introduced by Paul Romer. Both schools recognized technological progress as the engine of growth, but the neoclassical school takes technology as an exogenous factor, while the endogenous school makes technology endogenously determined by economic activities, including research and development.

3. The chain from science and knowledge to innovations, and finally to technology, is not necessarily linear and runs in both directions: many technologies, especially the general-purpose technologies such as ICTs, contribute to the creation of knowledge and innovations, which in turn can lead to improvements in those technologies and others.

4. These authors point out that even though technology may be the engine of economic growth, institutions are the drivers of that engine. Institutions, formal or informal, matter more: "the trustworthiness of government, the functionality of the family as the basic unit, security and the rule of law, a reliable system of contract enforcement, and the attitudes of the elite in power toward individual initiative and innovation" (Mokyr, 2003).

5. The special case of the least developed countries is extensively analysed in UNCTAD (2007).

6. Comin, Hobijn and Rovito (2006) made a major effort to build a comprehensive database of technologies, which are classified as capital goods and new production techniques. They measure the diffusion of technology by measuring the use of capital goods or the share of output produced with the technique. Their data cover 115 technologies in over 150 countries during the last 200 years. Although very comprehensive, this database may not capture all the technological product or process innovations in every single sector.

7. Using an aggregate production function (most of the time, the Cobb–Douglas function), total factor productivity (TFP) is defined as the portion of growth in output which cannot be explained by the contribution of growth in inputs (capital and labour) and is thus attributed to the effect of technology. This is often measured as the residual after subtracting output growth from growth in inputs (multiplied by their respective shares in income), using a growth accounting method and assuming perfect competition in factor markets. It is often referred to as the "Solow residual".

8. A review of the role of scientific and technological knowledge in productivity at the firm level, in industrialized countries was undertaken for UNCTAD by Donald Siegel (Science and technology for development, mimeograph, 2007).

9. For more detailed analysis of the "national innovation system" and its applicability to developing countries, and especially to the LDCs, see UNCTAD (2007 chapter 3).

10. Reference is made here to commercial innovations leading to industrial technologies, while basic research is often funded by Governments in the context of university or other government-sponsored research programmes.

11. The WTO Agreement on Subsidies and Countervailing Measures prohibits subsidies contingent upon export performance or use of domestic goods, and classifies as actionable subsidies those that cause injury to the domestic industry of another Member or cause serious prejudice to the interests of another Member.

Subsidies for research activities conducted by firms or by higher education or research establishments on a contract basis with firms are not actionable (up to a limit of 75 per cent of the cost of industrial research or 50 per cent of the costs of pre-competitive development activity). The agreement does not apply to fundamental research activities conducted independently by higher education or research establishments.

12. http://www.unesco.org/science/psd/publications/wdsciencereptable2.pdf.

13. Barton (2006).

14. It has been argued that, in addition to the loss of benefits in terms of forgone externalities and spillovers, high access costs and barriers reduce policy space, insofar as poor countries cannot encourage uncompensated knowledge spillovers through weak IPR regimes and other technology-related policies, such as the performance requirements for technology transfer.

15. See Hobday (1995) for a detailed analysis of technological upgrading in East Asian countries.

16. The European Union has introduced more protection of databases, while the United States has expanded the scope of patentable research findings, including what otherwise would be considered discoveries of nature rather than inventions. Other developed countries have also increased protection of plant and animal varieties, genetic sequences and biotechnological research tools. See International Task Force on Global Public Goods (2006).

17. There have been many studies and proposals on this issue. The approaches suggested in this chapter are adapted from two main recent publications – UNCTAD (2007, chapter 3) and Maskus (2006).

18. On free and open source software, see UNCTAD (2003, chapter 4).

19. There are many instances where open access to new ideas and inventions/research is preferred: when open source projects are complementary to proprietary projects; when personal motivations are prevalent, such as intellectual challenge, education, blocking market dominance by others, or achieving common standards; when intellectual property incentives are weak, because of the small size of markets; or when intellectual property benefits are slight. Open access projects are often protected by a general public licence that obligates contributors to maintain open access and prevents improvers from taking ideas private through proprietary copyright licences. A good review of market conditions pertaining to open source can be found in Maurer and Scotchmer (2006).

20. UNCTAD (2007, chapter 5).

21. Paul Schreyer (2000).

22. See, for example, Basu and Fernald (2006) and Guerrieri and Padoan (2007).

23. Bresnahan, Brynjolfsoon and Hitt (2001).

Chapter 1

TRENDS IN ICT ACCESS AND USE

A. Introduction

The diffusion of ICTs in developing countries is growing steadily, but except for some East Asian countries that are straddling the line between developed and developing status (notably the Republic of Korea and Singapore), developing countries remain far behind developed ones in the adoption of ICTs and their use by enterprises.[1] Among ICTs, mobile phones are often threshold technologies in many developing countries, and broadband Internet will be essential for developing an information economy (and integrating it at the regional and global levels).

This chapter presents trends since 2002 in ICT access and use by individuals and enterprises worldwide, paying special attention to the situation in developing countries and presenting illustrative examples of those trends. Section B will examine the trends in ICT access and use by individuals, focusing on the technologies of mobile telephony and the Internet, which have been recognized as having the greatest impact on developing countries. Mobile phones have lowered the threshold of ICT access for developing countries, while the Internet and in particular broadband Internet exponentially increase the availability of access to information and the ability to exchange it.

Section C will examine trends related to the use of basic ICT in enterprises, including for electronic transactions (e-commerce) and for conducting business (e-business). The use of the Internet for e-commerce and e-business helps enterprises improve efficiency and can have a positive impact on productivity. Section C also presents the highlights of the results of UNCTAD's annual collection of data on ICT in business in developing countries, which is based on the core list of ICT indicators developed by the Partnership on Measuring ICT for Development, of which UNCTAD is a leading member.[2] The policy implications of trends in ICT access and use will be summarized in the final part of the chapter, section D. The statistical annex at the end of the chapter provides more detailed country tables on the evolution of basic ICT access and use worldwide.

Using statistics on ICT access and use, or ICT diffusion, developing countries can analyse trends in their information economies, and formulate and assess their ICT for development policies. Furthermore, in the context of a growing global information society, and of development initiatives such as the Millennium Development Goals, data on ICT allow countries to better assess their position in terms of the digital divide. For those reasons, there should be continued efforts to improve the availability of data on ICT in order to assess the information economy. The World Summit on the Information Society (WSIS) has called for the periodic evaluation of information society developments on the basis of appropriate indicators and benchmarking.

B. ICT access and use by individuals

Among the ICTs to be examined in this section, the differences in access to and use of ICT by individuals (mobile and broadband subscribers) can help countries compare themselves with other countries, or categories of users within national borders, in order to assess the progress in their ICT diffusion policies.[3] The statistical annex to this chapter shows the data available by country on mobile phone subscribers and penetration, Internet users and penetration, and broadband subscribers and penetration.

1. Mobile phones as the breakthrough ICT in developing countries

In the past couple of years, mobile telephony has emerged as the most important ICT for low-income countries, and as the principal gateway to increased ICT access and use. The continued and substantial growth in mobile phone coverage in developing countries has been confirmed again in 2007 (see tables 1.1 and 1.2).

Table 1.1. Mobile phone subscribers by level of development and region

	2002	% change 2002–2003	2003	% change 2003–2004	2004	% change 2004–2005	2005	% change 2005–2006	2006
World	1 166 620 215	21.0	1 412 020 934	24.5	1 757 737 968	23.0	2 161 999 103	23.0	2 658 551 657
Developed economies	606 945 165	9.5	664 725 049	11.3	740 018 120	8.9	805 873 152	9.5	882 647 414
Asia	87 452 320	6.5	93 154 960	5.9	98 661 436	3.9	102 545 000	7.4	110 101 800
Europe	349 980 073	9.3	382 606 705	10.6	423 027 952	9.5	463 043 252	7.1	495 694 514
North America	154 488 772	11.3	172 017 384	15.6	198 852 732	9.8	218 334 900	16.1	253 561 100
Oceania	15 024 000	12.8	16 946 000	14.9	19 476 000	12.7	21 950 000	6.1	23 290 000
Developing economies	520 151 801	30.6	679 319 888	31.6	893 760 760	31.0	1 170 638 544	32.1	1 546 324 643
Africa	36 918 573	39.4	51 456 107	50.8	77 608 792	69.9	131 863 273	43.7	189 497 105
Asia	382 884 203	31.2	502 288 259	27.7	641 318 745	24.6	798 880 468	32.0	1 054 509 700
Latin America and the Caribbean	100 079 725	25.1	125 232 228	39.2	174 347 694	37.2	239 249 946	26.1	301 640 938
Oceania	269 300	27.5	343 294	41.4	485 529	32.8	644 857	5.0	676 900
Transition economies	39 523 249	72.0	67 975 997	82.4	123 959 088	49.6	185 487 407	23.8	229 579 600

Source: UNCTAD calculations based on the ITU World Telecommunication/ICT Indicators database, 2007.

Table 1.2. Mobile phone penetration by level of development and region

Mobile phone subscribers per 100 inhabitants

	2002	% change 2002–2003	2003	% change 2003–2004	2004	% change 2004–2005	2005	% change 2005–2006	2006
World	18.8	19.5	22.5	22.7	27.6	21.2	33.4	21.4	40.6
Developed economies	64.1	8.6	69.6	10.8	77.1	8.1	83.3	9.0	90.8
Asia	65.2	6.3	69.3	5.8	73.3	3.8	76.1	7.2	81.6
Europe	75.2	8.3	81.4	10.3	89.7	9.0	97.8	6.8	104.4
North America	47.7	10.3	52.6	14.5	60.3	8.2	65.2	15.1	75.0
Oceania	63.7	11.6	71.1	14.6	81.5	11.4	90.7	5.1	95.3
Developing economies	10.6	28.7	13.6	29.4	17.6	28.9	22.7	30.1	29.5
Africa	4.5	36.3	6.1	46.9	9.0	64.0	14.7	39.2	20.5
Asia	10.7	29.5	13.9	25.9	17.5	23.1	21.5	30.5	28.1
Latin America and the Caribbean	18.9	23.4	23.3	36.7	31.8	35.3	43.1	24.1	53.5
Oceania	3.4	24.0	4.2	37.4	5.7	31.4	7.5	2.8	7.7
Transition economies	12.0	72.5	20.6	82.4	37.6	48.6	55.9	23.9	69.3

Source: UNCTAD calculations based on the ITU World Telecommunication/ICT Indicators database, 2007.

The marked increase in mobile phone penetration rates in developing countries points to the role of mobile telephony as a "digital bridge", which will help many developing countries reduce the connectivity divide that separates them from others with more developed fixed-line infrastructure.

The number of mobile phone subscribers in developing countries has almost tripled in the last five years, and they now make up 58 per cent of mobile phone subscribers worldwide (see chart 1.1). Although mobile phone subscribers in developing countries are almost twice as many as in developed economies, there is a wide and inverse gap in terms of mobile phone penetration. At one extreme there are several developed countries where mobile phone penetration has even exceeded 100 per cent, while at the other extreme there are some 40 developing countries with a penetration rate of under 10 per cent. The gap has been persistent for years (see chart 1.2), although the divide could be expected to narrow in the next few years, as mobile markets reach saturation point in developed economies and continue their spectacular growth in developing countries.

Chart 1.1

Mobile phone subscribers

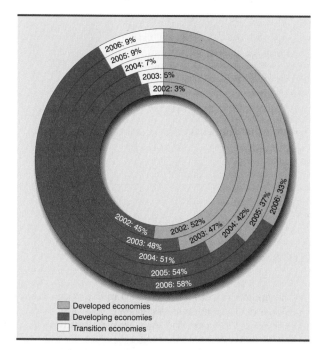

Source: UNCTAD calculations based on the ITU World Telecommunication/ICT Indicators database, 2007.

In Africa, where the increases in terms of mobile phone subscribers and penetration have been highest, this technology's potential also holds the greatest promise.

Chart 1.2

Mobile phone penetration

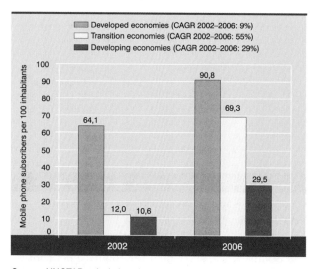

Source: UNCTAD calculations based on the ITU World Telecommunication/ICT Indicators database, 2007.

South Africa and Nigeria are the leading countries in terms of number of subscribers in the continent (mobile phones in Africa, and in these two countries in particular, are also the subject of a more in-depth analysis in chapter 6 of this report), followed by Algeria, Egypt and Morocco. In some countries, while growth in the number of mobile phone subscribers might seem impressive, the high rate is mostly due to the low starting levels. Other countries might have a higher penetration but relatively low growth. However, those countries that reflect recent high growth (from 2005 to 2006) and high penetration levels (in 2006) better reflect the popularity (and widespread adoption) of mobile phones in Africa. For example, Algeria, Mauritania, Morocco, Seychelles and Tunisia all have mobile phone penetration of more than 50 per cent, with the most recent penetration growth rate being more than 20 per cent.

Developing Asia is the next region after Africa with the highest growth in the number of mobile phone subscribers and mobile phone penetration. It also accounts for nearly 40 per cent of mobile phone subscribers worldwide. The number of subscribers is not surprising since China and India are the two most populous countries in Asia and in the world. Although in terms of penetration China and India are still well below many other Asian countries, there has been strong growth (see chart 1.3), which is expected to continue. Growth expectations are based on current trends of prepaid mobile telephony, ICT spending in those countries – the so-called emerging BRICS economies (Brazil, Russian Federation, India, China

Chart 1.3

Mobile phone penetration in China and India

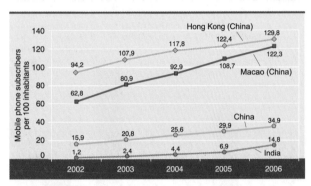

Source: UNCTAD calculations based on the ITU World Telecommunication/ICT Indicators database, 2007.

general, those countries that have implemented reforms to increase competition in the telecommunications market (for example, Mongolia) and that have brought about or enabled infrastructure development (such as China) have also experienced strong recent growth of the mobile phone market. The political (policies, regulatory environment) and economic (infrastructure, investment) elements that determine the uptake of mobile telephony in Asian countries also hold true for countries in other regions.

and South Africa) accounted for about 10 per cent of worldwide spending on communications equipment in 2006 – and the constant improvement of infrastructure and the competitive and regulatory environment (OECD, 2007a).

The widely differing mobile phone penetration levels (and other ICT diffusion) in the other Asian countries could be a reflection of their political and economic diversity. Asia encompasses conflict and post-conflict countries (such as Afghanistan and Iraq) and least developed countries (such as Myanmar and Nepal) with very low levels of mobile phone penetration. But it also includes two OECD member countries (Japan and the Republic of Korea), oil-rich countries (such as Kuwait and Qatar), and other Asian economic "tigers" (such as Singapore and Taiwan Province of China) with very high levels of mobile phone penetration. Despite countries' different circumstances, it can be said that, in

European countries account for more than half of mobile phone subscribers in developed economies, and together have a penetration rate of more than 100 per cent. As the market reaches saturation point, overall growth in the number of subscribers and in penetration has slowed down in recent years. The highest recent growth was in Eastern European countries (Croatia, Estonia, Latvia and Poland), which have now reached levels of mobile penetration above the developed country average. Despite saturation, however, revenue from mobile phones has increased, in part thanks to the increased offer of non-voice services such as Internet access and Short Message Service (SMS) (OECD, 2007a). In Germany and the United Kingdom, for example, non-voice services accounted for about 20 per cent of telecommunications revenues in 2006. In the EU, 12 per cent of households that accessed the Internet in 2006 used a mobile phone to do so, reflecting the growing trend of fixed-mobile convergence (see box 1.1).[4] Noteworthy in that connection is Latvia, where 67 per cent of households reported that the device for Internet access at home is a mobile phone.

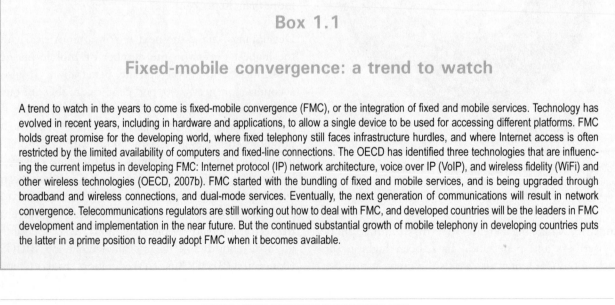

Box 1.1

Fixed-mobile convergence: a trend to watch

A trend to watch in the years to come is fixed-mobile convergence (FMC), or the integration of fixed and mobile services. Technology has evolved in recent years, including in hardware and applications, to allow a single device to be used for accessing different platforms. FMC holds great promise for the developing world, where fixed telephony still faces infrastructure hurdles, and where Internet access is often restricted by the limited availability of computers and fixed-line connections. The OECD has identified three technologies that are influencing the current impetus in developing FMC: Internet protocol (IP) network architecture, voice over IP (VoIP), and wireless fidelity (WiFi) and other wireless technologies (OECD, 2007b). FMC started with the bundling of fixed and mobile services, and is being upgraded through broadband and wireless connections, and dual-mode services. Eventually, the next generation of communications will result in network convergence. Telecommunications regulators are still working out how to deal with FMC, and developed countries will be the leaders in FMC development and implementation in the near future. But the continued substantial growth of mobile telephony in developing countries puts the latter in a prime position to readily adopt FMC when it becomes available.

In North America, the United States (the dominant market in this region) has had recent strong growth in the number of mobile phone subscribers and in increased penetration – from 49 per cent in 2002 to almost 78 per cent in 2006 – while Canada's penetration rate has remained roughly stable at 52 per cent. Latin America and the Caribbean had the third largest growth of all regions in the number of mobile phone subscribers, with Brazil, Mexico, Argentina, Colombia and Venezuela together accounting for 75 per cent of new subscribers in the region. The growth in the number of subscribers in those countries from 2005 to 2006 brought their level of mobile phone penetration to over the 50 per cent mark (except for Argentina, which already had a much higher penetration rate). While the average for the whole region is about 53 per cent, the Caribbean has the countries with both the highest and the lowest penetration (several of the small Caribbean islands have penetration rates of above 100 per cent, while Cuba and Haiti have rates of 1.4 per cent and 5.8 per cent respectively).

Oceania has the fewest mobile phone subscribers, with Australia, the largest country in the region, naturally accounting for most new subscribers. The low numbers of mobile phone subscribers compared with other regions is understandable, since Oceania represents less than 1 per cent of the total world population. On the other hand, mobile phone penetration in Australia and New Zealand (both OECD member countries) is very high, and several of the smaller Pacific islands are catching up.

Among transition economies, the Commonwealth of Independent States (CIS) dominates mobile phone subscriber numbers (the Russian Federation alone accounts for about 60 per cent) and growth, and has caught up with South-East Europe in terms of average penetration. There are, however, wide differences between individual countries: on the one hand, Ukraine has experienced spectacular recent growth and reached over 100 per cent penetration, while on the other hand, Tajikistan, Turkmenistan and Uzbekistan have penetration rates of below 5 per cent.

2. The Internet gap is closing slowly

The Internet has continued growing worldwide in terms of users and penetration (see tables 1.3 and 1.4). Although developed economies still account for the majority of Internet users and are still very much ahead in terms of Internet penetration, developing economies are slowly catching up. While in 2002

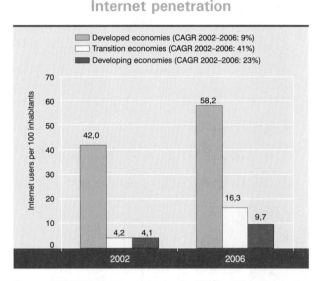

Chart 1.4

Internet penetration

Source: UNCTAD calculations based on the ITU World Telecommunication/ICT Indicators database, 2007.

Internet penetration in developed economies was 10 times higher than in developing economies, in 2006 it was only 6 times higher. Transition economies had the highest compound annual growth rate in Internet penetration between 2002 and 2006 (see chart 1.4).

Although the number of Internet users in Africa continues to grow strongly, penetration continues to be extremely low. Four out of 53 countries (Nigeria, Morocco, Egypt and South Africa) account for almost 60 per cent of Internet users in the region. While the small island State of Seychelles has the highest Internet penetration (33.3 per cent), Morocco has the highest penetration on the mainland at only 19.9 per cent. African countries still have much work to do in order to improve Internet penetration, but several are strenuously pursuing ICT diffusion and are slowly but surely improving the situation through a combination of ICT for development policies to improve ICT access and skills, regulatory reforms to increase the offer of services and competition, and the fostering of investments in infrastructure and in the ICT sector.

As in the case of mobile phones, Asia is the region contributing the largest share of Internet users, mostly because of China and India, which together account for nearly 200 million users (more than half of users in the region). The Republic of Korea has the third largest number of Internet users (16 million) and the highest Internet penetration (70.4 per cent) in Asia. Japan follows it closely in terms of penetration. It is interesting to note that a number of West Asian countries (Brunei Darussalam, United Arab Emirates,

Table 1.3 Internet users by level of development and region

	2002	% change 2002–2003	2003	% change 2003–2004	2004	% change 2004–2005	2005	% change 2005–2006	2006
World	**618 514 417**	**16.0**	**717 707 944**	**20.7**	**866 364 366**	**16.2**	**1 006 429 544**	**12.4**	**1 131 078 697**
Developed economies	**397 605 044**	**7.3**	**426 435 696**	**15.6**	**492 755 839**	**7.5**	**529 869 769**	**6.8**	**566 077 247**
Asia	60 345 200	4.2	62 904 500	28.7	80 976 600	7.4	86 975 900	2.8	89 439 100
Europe	149 899 844	13.4	169 994 796	12.0	190 421 239	7.7	205 089 269	10.7	227 077 547
North America	174 952 000	3.0	180 126 400	14.4	206 008 000	7.2	220 860 600	4.6	231 060 600
Oceania	12 408 000	8.1	13 410 000	14.5	15 350 000	10.4	16 944 000	9.2	18 500 000
Developing economies	**207 465 892**	**28.8**	**267 302 205**	**26.1**	**337 136 119**	**28.6**	**433 560 239**	**17.9**	**511 035 250**
Africa	10 290 156	45.3	14 953 500	48.5	22 206 421	48.8	33 032 605	31.4	43 397 500
Asia	153 538 659	29.9	199 488 152	25.8	250 869 483	24.0	311 164 987	16.1	361 391 800
Latin America and the Caribbean	43 411 477	21.2	52 597 353	21.2	63 756 215	39.6	89 022 947	18.9	105 864 150
Oceania	225 600	16.7	263 200	15.5	304 000	11.7	339 700	12.4	381 800
Transition economies	**13 443 481**	**78.3**	**23 970 043**	**52.2**	**36 472 408**	**17.9**	**42 999 536**	**25.5**	**53 966 200**

Source: UNCTAD calculations based on the ITU World Telecommunication/ICT Indicators database, 2007.

Table 1.4 Internet penetration by level of development and region

	2002	% change 2002–2003	2003	% change 2003–2004	2004	% change 2004–2005	2005	% change 2005–2006	2006
World	**10.0**	**14.6**	**11.4**	**19.0**	**13.6**	**14.5**	**15.6**	**10.9**	**17.3**
Developed economies	**42.0**	**6.3**	**44.7**	**15.0**	**51.3**	**6.7**	**54.8**	**6.3**	**58.2**
Asia	45.0	4.0	46.8	28.6	60.2	7.3	64.6	2.6	66.3
Europe	32.2	12.3	36.2	11.7	40.4	7.2	43.3	10.4	47.8
North America	54.1	2.0	55.1	13.3	62.4	5.6	66.0	3.7	68.4
Oceania	52.6	6.9	56.3	14.1	64.2	9.1	70.0	8.1	75.7
Developing economies	**4.2**	**27.0**	**5.3**	**24.0**	**6.6**	**26.5**	**8.4**	**16.1**	**9.7**
Africa	1.3	42.1	1.8	44.6	2.6	43.5	3.7	27.3	4.7
Asia	4.3	28.3	5.5	24.0	6.8	22.6	8.4	14.8	9.6
Latin America and the Caribbean	8.2	19.5	9.8	19.0	11.6	37.7	16.0	17.1	18.8
Oceania	2.8	13.5	3.2	12.2	3.6	10.5	4.0	10.1	4.4
Transition economies	**4.1**	**78.8**	**7.3**	**52.2**	**11.1**	**17.1**	**13.0**	**25.6**	**16.3**

Source: UNCTAD calculations based on the ITU World Telecommunication/ICT Indicators database, 2007.

Qatar, Kuwait, Lebanon, Islamic Republic of Iran and Saudi Arabia) experienced recent high growth in Internet penetration and reached a 2006 penetration rate above the world average.

Among developed regions, European countries had the highest growth in the number of Internet users overall, but the lowest average Internet penetration. For the first time, the Netherlands became the European country with the highest Internet penetration rate in 2006, overtaking Sweden. Several Eastern European countries (Hungary, Latvia, Lithuania and Slovakia) have shown remarkable growth in the past five years. The only European country that has not significantly increased its Internet diffusion has been Greece, possibly because of the combined effect of a lack of competition (the Government's national telecommunications company has a monopoly over end-user networks) and individuals' lack of awareness about the benefits of Internet access and use (Yannopoulos, 2006).

Both in developed North America and in developing Latin America and the Caribbean, the numbers of Internet users and Internet penetration have continued to increase, although their growth rates have slowed down. Many of the small island States have the highest Internet penetration rates, while Brazil and Mexico, the two largest Internet markets in Latin America and the Caribbean (60 per cent of the region's Internet users), have shown healthy growth and above-average Internet penetration.

Developed and developing Oceania had moderate recent growth in the number of Internet subscribers and in Internet penetration. While Australia and New Zealand account for the highest average Internet penetration among in Oceania's developed regions, most of the small island States still have very low penetration. Only French Polynesia, the Federated States of Micronesia and New Caledonia have had meaningful increases in Internet penetration rates since 2002.

Transition economies have had remarkable growth in the number of Internet users and in Internet penetration in the past five years. In general, the Internet user population has grown significantly in CIS countries, with the Russian Federation as the single largest contributor to Internet user growth. However, Internet penetration in South-East Europe remains higher than in the CIS, and has grown remarkably since 2002.

Chart 1.5

Broadband subscribers

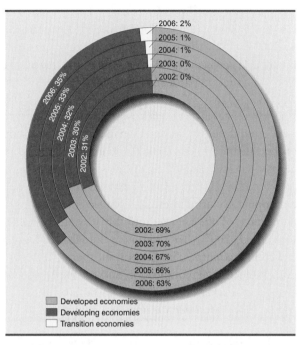

- Developed economies
- Developing economies
- Transition economies

Source: UNCTAD calculations based on the ITU World Telecommunication/ICT Indicators database, 2007.

Chart 1.6

Broadband penetration

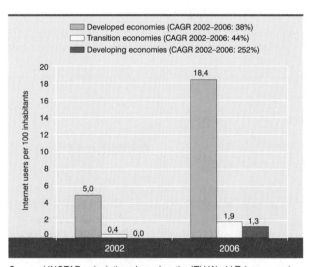

Source: UNCTAD calculations based on the ITU World Telecommunication/ICT Indicators database, 2007.

The promise of broadband

Although available data show that the number of broadband Internet subscribers has grown rapidly worldwide, developed economies still dominate subscriptions (see chart 1.5), and the gap in terms of penetration has widened since 2002 (see chart 1.6). As mentioned in the IER 2006, increased competition has determined broadband growth in almost all developed countries, while continuous improvements and diversification in infrastructure have resulted in increased bandwidth. The European Commission has noted that "the most significant factor enabling broadband growth is the existence of alternative infrastructures, in particular cable".[5]

The differences in infrastructure and markets result into price conditions that are less than ideal for the adoption of broadband by developing country consumers. In developed, high-income economies, the average cost of a broadband connection is significantly less than in developing countries, both in nominal terms and as a percentage of the average monthly income (ITU and UNCTAD, 2007). To promote affordable access in developing countries, it is important that government policies encourage competition and the expansion of consumer markets, as well as promote investment in, and develop, infrastructure and connectivity.

The ITU reports that from 2002 to 2005 the number of countries where fixed-line broadband services were available grew from 81 to 166 (and about 60 already offered mobile broadband services). While overall trends indicate that broadband Internet is gaining ground and is likely to become ubiquitous in a few years' time, it is important to keep in mind that, in the meantime, Internet users without broadband will be disadvantaged with respect to broadband users, as they will be unable to use certain applications. A study recently found that broadband users report higher levels of activity and spend more time on the Internet than dial-up users (although other demographic factors are also taken into account, such as age, education, race and sex) (Davison and Cotten, 2003). In particular, a broadband connection will usually mean that users will engage in activities that require high bandwidth, such as downloading music or playing games online, and be more inclined than dial-up users to engage in e-payments and e-banking.

In Africa, Morocco has the highest number of broadband Internet subscribers (390,000), but a penetration of barely 1.3 per cent. The highest penetration in the continent is in Mauritius, with only 1.7 per cent. It is evident that Africa still has a long way to go in order to reap the potential benefits of broadband connectivity. Plans to install undersea fibre-optic cable systems (see box 1.2) should help Africa achieve better connectivity, although the meaningful adoption of broadband is likely to take still several years.

In Asia, the average levels of broadband subscribers and penetration are better than in Africa mainly because of a handful of countries. China has of course the majority of broadband subscribers in the region (62 per cent), but still a low penetration rate of 3.9 per cent. The best performers in terms of broadband penetration are above the developed country average

---

Box 1.2

Ambitious broadband plans for Africa

In order to benefit fully from the Internet, Africa needs to improve its connectivity infrastructure in order to achieve affordable broadband. Initiated in January 2003, the Eastern Africa Submarine Cable System (EASSy) is aimed at providing broadband connection to 19 African countries (including 11 landlocked countries) via an undersea fibre-optic cable running from South Africa to Sudan, and could be offering commercial service by the end of 2008. The EASSy project coordinator estimates that it could cut telecommunication costs in Africa at least by half.

However, during EASSy's slow-starting initial phase, there were disagreements about whether it should be controlled by the private sector or by an open access system (where investors and non-investors are given international bandwidth access at the same price). Kenya withdrew from EASSy and announced the launch of an East African Marine System, and the South African-led New Partnership for Africa's Development (NEPAD) has launched the NEPAD Broadband Infrastructure Network. At the same time a private company called SEACOM (with African, European and United States capital) is also planning an undersea cable.

Whether African countries support one initiative or the other, it is important that these initiatives result in affordable connectivity soon. This will be particularly significant for landlocked countries, which currently depend on expensive satellite communications. The high prices that Africa continues to pay for broadband owing to the current situation can only be negative for the development of its information society.[6]

(18.4 per cent): the Republic of Korea, Hong Kong (China), Israel and Taiwan Province of China. Singapore and Macao (China) are not far behind. On the other hand, there are still a considerable number of countries (at least 19) with broadband penetration rates of under 1 per cent.

Europe is the broadband leader among regions in terms of subscribers, with the bulk of subscribers coming from the same countries as the bulk of Internet users (Germany, United Kingdom, France and Italy). The significant growth of overall penetration in the past five years, from 3 to 17 per cent, is a reflection of the active pro-broadband policies in the EU (and Norway and Iceland). As in the case of the Internet, the Netherlands took the lead for the first time in terms of Internet penetration, with broadband subscribers accounting for more than one third of Internet users in that country.

In North America, the United States and Canada are lagging behind Europe in terms of the number of broadband subscribers and the growth rate of the broadband subscriber base. And although broadband penetration in the United States and Canada in 2006 was higher than the EU-25 average (14.8 per cent as at July 2006), it is still lower than that of the more industrialized EU members. Furthermore, the compound annual growth rate of broadband penetration in Europe was of 54 per cent, compared with 28 per cent in North America. The IER 2005 had pointed out that the broadband market in the United States in particular was insufficiently competitive.

In Latin America and the Caribbean the growth of broadband Internet subscriber numbers has been very rapid, but penetration remains very low at 2.6 per cent. Broadband penetration in Brazil and Mexico, the dominant countries in terms of subscribers, is above the regional average but still low. The highest levels of penetration are found in the small island States, while Chile maintains the highest rate of penetration among the larger countries. Chile's position is partly due to its active ICT promotion policy, which includes the introduction of ICT in schools, its e-government strategy and its national programme for the support of telecentres (see chapter 7).

Both developed and developing Oceania have experienced strong growth in the number of broadband subscribers in recent years, but while developed Oceania (Australia and New Zealand) is doing very well in terms of penetration (18.3 per cent), the small island economies of Oceania have a very low penetration rate (0.4 per cent). French Polynesia and New Caledonia

are the exceptions, with penetration rates of 7 and 4 per cent respectively, but those figures do not indicate an improvement for other economies.

Despite continued strong growth in broadband subscriptions, the average penetration in transition economies (1.3 per cent) is even lower than the average penetration in developing economies (1.9 per cent). South-East Europe is slightly ahead of CIS countries, mostly because of a recent increase in broadband subscriptions in Bulgaria.

C. ICT access and use by enterprises

Increased access to ICTs by enterprises in developing countries means that they have better access to information, and the use of ICTs for business processes can also help increase productivity (see chapter 3). A study of small and medium-sized enterprises in 14 African countries concluded that the use of ICTs makes a significant contribution to income generation and is linked to increased labour productivity (RIA!, 2006). They help reduce the cost of transactions and increase market access, thereby increasing commercial efficiency.

Despite the increasingly evident advantages of ICTs for business, their adoption by developing country enterprises is still limited. The lack of awareness of the potential benefits of ICT use, and investment and implementation costs, are important constraints on wider ICT adoption. This chapter looks specifically at the adoption of mobile phones, the Internet (and broadband Internet) and e-business (and e-commerce) by enterprises.

With regards to computer use by businesses, chapter 3 makes an in-depth analysis of the firm-level impact of ICTs, including the use of computers by enterprises in Thailand, and shows that computer use, Internet access and web presence are linked to higher sales per employee. The use of computers by enterprises in developed countries is nearly ubiquitous, but for enterprises in poorer countries they can make a substantial difference in the efficiency of managerial processes. When computers provide access to networks such as the Internet, their impact on labour productivity is even higher. A review of other studies supports the finding that an increase in the share of employees using computers has a positive effect on labour productivity, although the magnitude of the impact is determined

Table 1.5.

Argentina: use of ICT in the manufacturing sector (%)

Indicators	2001	2004	2005	% change 2004–2005
Enterprises that invested in ICT	..	43.3	45.1	4.2
Enterprises with access to the Internet	86.9	92.4	95.2	3.0
Enterprises with a website (of those with the Internet)	56.5	60.2	73.5	22.1
Enterprises with an intranet	35.6	42.1	44.2	5.0
Enterprises receiving orders over the Internet (of those with the Internet)	14.6	40.0	45.5	13.8
Enterprises using the Internet for:				
Sending and receiving e-mail	..	89.1	92.4	3.7
Getting information about goods or services	..	70.2	83.4	18.8
Internet banking or accessing other financial services	..	67.1	78.6	17.1
Transacting with government organizations/public authorities	..	45.1	53.4	18.4
Providing customer services	..	34.7	40.7	17.3
Getting information for research and development activities	..	29.0	37.9	30.7
Delivering products online	..	4.4	5.7	29.5

Note: See after table 1.20. in the Statistical Annex for notes to data.
Source: Instituto Nacional de Estadística y Censos de Argentina (INDEC).

by the business environment and variables such as enterprise age, origin of capital, industry affiliation and enterprise size.

The adoption of ICTs by enterprises goes hand in hand with the investments they make in ICTs, and it is the larger enterprises that invest more often. Despite the apparent benefits of ICTs, many SMEs are unsure of what are the appropriate ICT solutions in which to invest (Kotelnikov, 2007). ICT investment can help improve productivity, although it should be noted that in countries with very low ICT diffusion the effects of ICT investment at the macroeconomic level are limited (see chapter 3 of this Report and OECD, 2006). At firm level, an Argentine survey of the manufacturing sector showed that the growth in ICT investments (including system and software development), often combined with training and organizational changes, was accompanied by growth in the use of ICTs and e-business, as seen in table 1.5.

1. Mobile phones are valuable business tools

Mobile phones are being increasingly used by individuals not only for personal purposes, but also for business. The use of mobiles for business purposes has been classified by an Australian study (O'Donnell et al., 2007) into mCommerce, mEnterprise (enterprise solutions enabled through mobiles such as logistics)

and mServices (mobile applications that do not imply a transfer of goods such as mobile banking).[7]

Enterprises will provide their salespeople with mobile phones to ensure they are always connected and able to seize the latest business opportunities. Entrepreneurs without a physical office are permanently in touch. A Spanish study finds that mobile phones are valuable interactive marketing tools, since they complement other direct sales systems such as the Internet and television (Bigné, Ruiz and Sanz, 2007). Internet-enabled mobile phones (including a new generation that will allow high-speed Internet access) will also have a role to play in e-commerce (see also section C.3 below).[8] As consumers (in particular the younger generations of consumers) are increasingly exposed to ubiquitous and more sophisticated mobile phones, they will probably become more comfortable with, and make more use of, mobile commerce; "the closer an individual's relationship with a medium, the greater the probability of purchase" (Bigné, Ruiz and Sanz, 2007).

Mobile phones are the main communication tool for many entrepreneurs (particularly small entrepreneurs) in developing countries, especially in Africa (see chapter 6). A study of 14 African countries found that more than 80 per cent of SMEs use a mobile phone for business purposes, and only 51 per cent use fixed lines (RIA!, 2006). Mobiles were the most commonly used ICT (more than faxes, fixed-line phones, the Internet or mail) for communicating with clients and

---

Box 1.3

Trade at Hand: mobile telephony to increase exports

The Trade at Hand project pilot was launched in sub-Saharan Africa in 2006 by the International Trade Centre (ITC) of UNCTAD/WTO (funded by various donors) to provide real-time market information through mobile phones to exporting SMEs. In Africa, where there is a great need for better business and market information, mobile phones can help meet this need as the most common ICT in the continent.

Trade at Hand sends, via SMS and in real time, product prices on international markets to agricultural exporters from Burkina Faso and Mali. A second module of the project sends to mobile phones information about business opportunities, contacts and markets, in collaboration with the ITC's existing network of local trade support institutions. With better information on daily international prices for mangoes and green beans, for example, exporters are able to negotiate better deals for their produce and improve revenues. Better information on business opportunities also contributes to increasing exports. With time, Trade at Hand should become self-sustainable and will be managed by local organizations. The ITC plans to further expand Trade at Hand in Mali and Mozambique, and also in Benin, Liberia and Senegal.

Source: http://www.tradeathand.info/mozambique/.

---

for ordering supplies. Furthermore, more than 95 per cent of SMEs in that study stated that mobile phones where either important or very important for their business operations. A project in sub-Saharan Africa uses mobile phones to overcome the market information asymmetry suffered by SMEs that want to export (see box 1.3).

The level of use of mobile phones for business purposes, apart from contacting customers and suppliers, and keeping employees in touch, is likely to depend on the development of applications that are relevant for enterprises and provide value added. This means that mobile business applications should not be mere substitutes for applications that are already available and work well through fixed lines, computers or the Internet. Examples of applications are inventory tracking, the provision of cash flow and income statements, and m-payments to replace cash transactions and «build up transaction histories» (which in turn facilitate the access of unbanked enterprises to formal financial services) (RIA!, 2006). However, the Australian study found that while many enterprises want to adopt m-commerce, they find it difficult to make a strong business case that will outweigh the financial and legal risks taken due to the lack of relevant laws and regulations (O'Donnell et al., 2007).

Despite increasing anecdotal evidence about their potential benefits, there are no official statistics on the use of mobile phones by enterprises or by individual entrepreneurs. Countries should collect such statistics in order to take into account the impact of that technology when formulating their ICT for development policies and strategies. There is a need for more empirical studies with firm-level data to determine how the use of mobile phones impacts business productivity, especially in poor countries.

2. Internet use by businesses is slowly growing

While growth in the number of Internet users in developing countries has been linked to growth in exports (World Bank, 2006), Internet access by enterprises in developing countries continues to grow, as does the number of employees using the Internet in their daily work. The number of enterprises with websites is also slowly increasing.

Firm-level surveys by the World Bank[9] revealed that service-sector enterprises use websites and computers more than the manufacturing sector, and have a higher proportion of employees that use computers regularly. Within the service sector, it is not surprising that firms in the telecommunications and information technology industries are the main users of ICT; they are followed by the real estate and the hotel and restaurant sectors. Exporters also use the Internet (websites and electronic mail) much more than non-exporters.

In the European Union, Internet access by enterprises with 10 or more employees has continued to grow slowly but surely, reaching an average of 92 per cent in 2007 (including in the two new member countries, Bulgaria and Romania), compared with 86 per cent in 2003. Internet connectivity in individual countries is shown in table 1.6. Approximately one third of employees in enterprises located in EU countries use (computers connected to) the Internet in their

Table 1.6

Enterprises with Internet and website

	Reference year	Proportion of:				Reference year	Proportion of:		
		Enterprises using Internet	Employees using Internet	Enterprises with a website (of those using Internet)			Enterprises using Internet	Employees using Internet	Enterprises with a website (of those using Internet)
		B3	B4	B5			B3	B4	B5
Argentina	2005	95.8	24.8	74.4	Italy	2006	93.0	28.0	61.0
Australia	2005	87.3	..	51.8	Japan	2005	97.7	..	85.6
Austria	2006	98.0	..	80.0	Latvia	2005	77.1	20.9	43.5
Azerbaijan	2006	8.7	1.7	32.6	Lithuania	2006	88.0	23.0	47.0
Belarus	2005	37.6	..	27.2	Luxembourg	2006	93.0	32.0	65.0
Belgium	2006	95.0	41.0	72.0	Macao (China)	2003	53.3	..	26.3
Bermuda	2005	71.0	37.0	56.3	Malta	2005	90.0	..	61.0
Brazil	2006	94.3	36.5	49.6	Mauritius	2006	87.4	..	46.1
Bulgaria	2006	74.8	14.5	43.8	Netherlands	2006	97.0	45.0	81.0
Canada	2006	94.9	..	71.2	New Zealand	2006	94.5	97.4	62.7
Chile	2005	48.8	..	38.8	Norway	2006	94.0	50.0	76.0
China	2005	47.4	..	23.7	Panama	2006	80.1	20.3	..
Cuba	2006	70.9	29.6	23.7	Poland	2006	89.0	28.0	60.0
Cyprus	2006	86.0	31.0	50.0	Portugal	2006	83.0	25.0	42.0
Czech Republic	2006	95.0	29.0	74.0	Qatar	2005	68.4	90.0	99.0
Denmark	2006	98.0	61.0	85.0	Republic of Korea	2005	95.9	..	58.9
Egypt	2006	53.2	9.9	71.0	Romania	2005	58.4	15.9	41.2
Estonia	2006	92.0	33.0	63.0	Russian Federation	2005	53.3	12.4	27.8
Finland	2006	99.0	59.0	81.0	Singapore	2006	87.2	..	69.6
France	2006	94.0	34.0	65.0	Slovakia	2006	93.0	29.0	65.0
Germany	2006	95.0	39.0	77.0	Slovenia	2006	96.0	35.0	65.0
Greece	2006	94.0	26.0	64.0	Spain	2006	93.0	35.0	50.0
Hong Kong (China)	2006	82.8	45.9	51.5	Sweden	2006	96.0	53.0	90.0
Hungary	2006	80.0	21.0	53.0	Switzerland	2005	98.2	47.6	91.6
Iceland	2006	99.3	51.6	77.5	Thailand	2006	69.6	..	50.5
Ireland	2006	94.0	37.0	67.0	Turkey	2005	80.4	34.1	59.9
					United Kingdom	2006	93.0	42.0	81.0

Note: See after table 1.20. in the Statistical Annex for notes to data.
Source: UNCTAD information economy database, 2007.

normal work routine (at least once a week). However, only about 60 per cent of enterprises have interactive websites; as usual, the level of adoption is higher in larger enterprises.

Broadband Internet is essential to the information economy

Last year's report argued that broadband access to the Internet can enable or enhance the adoption of certain applications that have an impact on enterprise productivity. In OECD countries, affordable broadband

connectivity has been linked to the increased use of ICTs by SMEs (OECD, 2004). Although there is still not enough information about broadband adoption in developing countries, it is clear that it is growing and the technology is changing fast. Access speeds are continuously higher in the more advanced countries, and pioneers such as the Republic of Korea are already looking forward to wireless broadband (WiBro). [10]

Unlike Internet access, whose growth has slowed down as the market becomes saturated, broadband access in the EU by enterprises with 10 or more employees has increased rapidly in the past few years and is expected to continue growing. On average, 73 per cent of enterprises had broadband access in 2007 (including the two new member countries, Bulgaria and Romania), up from 40 per cent in 2003. In OECD countries, broadband is quickly becoming the basic medium for service delivery on both fixed and wireless networks (OECD, 2007a).

Broadband increases the capacity of enterprises to benefit from, and deliver through, the Internet, and can enable the adoption of certain applications that have an impact on enterprise productivity. Although broadband particularly enhances multimedia applications and can have evident benefits for the ICT-enabled services and media sectors, it is increasingly being used in non-ICT-intensive economic sectors. In the European manufacturing industry, it is a key enabler of online procurement, which helps enterprises manage their supply chain. Voice over Internet Protocol applications have cost-saving potentials for firms in all economic sectors, while marketing and sales applications can also be applied to all economic sectors. In general, industries can enhance e-business solutions through broadband. Despite growing recognition of the positive impact that broadband can have on productivity, the awareness of its potential among developing country enterprises needs to increase.

3. E-Business is facilitated by government

E-business often requires networked business functions through LANs, intranets and/or extranets. The adoption of LANs in developing countries is more widespread than the use of intranets. In those countries where data for more than one year are available, both indicators have increased (see table 1.8).

In Europe, enterprises are increasingly and rapidly adopting LANs (60 per cent in small enterprises, 85 per cent in medium-sized enterprises and 95 per cent in large enterprises, in 2005). As in the case of the introduction of an intranet, enterprises in the services sector are the more frequent users of LANs, mostly in media and then in business services.

The levels of intranet and extranet use in EU enterprises have remained stable since last year. On average, 35 per cent of EU enterprises have an intranet, while 16 per cent have an extranet, but these figures conceal a much higher level of use in large enterprises (80 per cent of enterprises with 250 or more employees have an intranet and 47 per cent have an extranet).

A further step to the introduction of intranets and extranets is the automated integration of business processes – that is, automatic linking between computer systems to manage orders that have been placed or received and other internal systems (reordering of supplies, invoicing and payment, and management of production logistics or service operations). On average, 34 per cent of European enterprises have automated integration of internal business processes, with the distributive trade sector having the highest level of integration (45 per cent of companies). This sector also had the highest level of integration between different companies, particularly in the sale, maintenance and repair of vehicles (the European average was 15 per cent of enterprises in 2005).[11]

Some developing countries are not far behind Europe in terms of integrating business processes. In Brazil, for example, 36 per cent of enterprises use information technology to manage orders and purchases, with automated linking to several related activities (inventory control, invoicing and payment, production and logistics). The sector with the highest level of integration is the automobile trade and repair sector. Integration with supplier or customer commercial systems (outside the enterprise) is less common.

In terms of the use of the Internet for different business purposes, data from developing countries is scarce and not always comparable. Table 1.9 sets out the available information for the core indicators on Internet use by activity.

For enterprises in developing countries that are starting to use the Internet, the mere fact of having increased access to information can have immediate positive effects. In Ghana, SMEs that export non-traditional products use the Internet to find information about international best practices and exporting opportunities. Although online payments are almost non-existent at present, as SMEs increase exports they will also become more likely to use the Internet for receiving orders (Hinson, Sorensen and Buatsi, 2007).

The use of e-government services by enterprises is of particular importance for encouraging businesses to make further use of ICT to improve the efficiency of their operations. The Government of the Republic of Korea, for example, provides firms with information on export-import logistics and Customs, and has started to implement an electronic documentation service for private companies that have a high level of document exchanges with it. In Mexico, the taxation agency recently launched online tax returns and e-

Table 1.7
Enterprises with Internet by mode of access

	Reference year	Proportion of:	Proportion of enterprises accesing the Internet by:				
		Enterprises using Internet	Analogue modem	ISDN	Fixed line connection under 2 Mbps	Fixed line connection of 2 Mbps or more	Other modes of access
		B3	B9.a	B9.b	B9.c	B9.d	B9.e
Argentina	2005	95.8	15.4	..	..	..	9.6
Australia	2005	87.3	24.2	7.1	..	68.7	..
Azerbaijan	2006	8.7	67.3	4.8	12.1	5.1	17.8
Belgium	2006	95.0	17.0	28.0	26.0	89.0	12.0
Brazil	2006	94.3	13.9	..	68.7	4.7	4.8
Bulgaria	2006	74.8	26.4	10.9	26.5	6.0	19.1
Canada	2006	94.9	..	..	..	92.2	..
Chile	2005	48.8	16.4	..	19.4	80.6	..
China	2005	47.4	13.6	6.0	..	..	..
Cuba	2006	70.9	51.1	0.0	32.9	0.2	0.1
Cyprus	2006	86.0	33.0	26.0	11.0	63.0	5.0
Czech Republic	2006	95.0	12.0	27.0	41.0	73.0	31.0
Denmark	2006	98.0	4.0	17.0	17.0	84.0	10.0
Egypt	2006	53.2	6.3	1.8	71.9	22.8	0.9
Estonia	2006	92.0	10.0	17.0	20.0	82.0	9.0
Finland	2006	99.0	19.0	22.0	21.0	90.0	22.0
France	2006	94.0	..	22.0	22.0	92.0	..
Germany	2006	95.0	4.0	38.0	13.0	77.0	7.0
Greece	2006	94.0	28.0	43.0	10.0	62.0	5.0
Hong Kong (China)	2006	82.8	3.6	..	93.4	10.2	10.6
Hungary	2006	80.0	14.0	26.0	26.0	77.0	13.0
Iceland	2006	99.3	4.5	5.7	33.0	44.1	5.7
Ireland	2006	94.0	27.0	33.0	30.0	64.0	9.0
Italy	2006	93.0	23.0	30.0	8.0	75.0	8.0
Japan	2005	97.7	9.6	16.1	16.0	63.5	68.1
Latvia	2005	77.1	10.3	20.6	..	73.7	12.6
Lithuania	2006	88.0	33.0	14.0	21.0	65.0	17.0
Luxembourg	2006	93.0	18.0	42.0	16.0	81.0	14.0
Macao (China)	2003	53.3	9.9	..	8.1	..	78.8
Malta	2005	90.0	20.0	7.0	21.0	87.0	7.0
Netherlands	2006	97.0	6.0	23.0	15.0	84.0	5.0
New Zealand	2006	94.5	35.0	..	81.6	..	21.1
Norway	2006	94.0	6.0	16.0	33.0	91.0	15.0
Panama	2006	80.1	8.2	4.0	61.7	..	36.8
Poland	2006	89.0	39.0	34.0	16.0	52.0	14.0
Portugal	2006	83.0	25.0	18.0	24.0	79.0	5.0
Republic of Korea	2005	95.9	0.7	0.8	..	98.2	0.2
Romania	2005	58.4	33.8	11.8	32.1	7.8	42.4
Singapore	2006	87.2	19.3	12.8	66.7	18.3	33.3
Slovakia	2006	93.0	19.0	32.0	18.0	65.0	29.0
Slovenia	2006	96.0	10.0	23.0	14.0	78.0	9.0
Spain	2006	93.0	16.0	19.0	8.0	94.0	9.0
Sweden	2006	96.0	18.0	17.0	30.0	92.0	28.0
Switzerland	2005	98.2	..	..	54.8	42.2	..
Thailand	2006	69.6	52.2	5.3	..	39.4	19.4
Turkey	2005	80.4	35.3	6.8	52.3	27.4	14.1
United Kingdom	2006	93.0	37.0	33.0	16.0	83.0	2.0

Note: See after table 1.20. in the Statistical Annex for notes to data.
Source: UNCTAD information economy database, 2007.

Box 1.4

E-business in Mexico

Mexico's overall ICT policy aims to increase the country's competitiveness and keep foreign investments from moving to other countries with lower labour costs. A main objective of that policy is to increase the diffusion of ICTs to business, creating a regulatory environment conducive to e-commerce, supporting the development of local ICT human resources and industry, and developing a domestic ICT market.

While Internet penetration among Mexico's larger businesses (those with more than 50 employees) is high (90 per cent in 2003), they represent only 8 per cent of all businesses. Only 2.2 per cent of enterprises are engaged in (mostly B2B) e-commerce (although the percentage is much higher among retail companies: 16 per cent). B2C e-commerce is limited owing to lack of demand, although Internet penetration in Mexico grew from 10.7 per cent in 2001 to 20.3 per cent in 2006, and ISP development has continued. Mexico City, Guadalajara and Monterrey are much more advanced than other cities: Guadalajara and Monterrey host important technology parks, and Monterrey also has Mexico's leading technological institute. To increase overall demand, the e-Mexico initiative that was launched in 2002 is establishing a network of digital community centres (also used by micro businesses) and has a portal focused on the subjects of learning, health, economy and government.

Regarding SMEs, Mexico is challenged by the lack of information on ICT diffusion among those businesses, which hinders the formulation of policies and analysis of ICTs' potential impact on them. The large informal economy in Mexico also "flies under the radar" of ICT diffusion assessments, however, this does not preclude the Government from offering most of its research and development tax incentives to SMEs. A preliminary evaluation of that programme showed that those tax incentives had contributed to increased production, new product sales as well as total sales, exports and profits among the beneficiaries. A public–private project has focused on integrating value chains through ICTs and developing the domestic market in four industries (food retailing, processed foods, maquiladoras[12] and hotels).

In order to increase the diffusion of ICTs to business and its positive impact on Mexico's economic performance and competitiveness, an OECD peer review recommended steps in several ICT-related policy areas. The highest priority is given to the coordination of ICT policy actions, which are currently implemented by various ministries and should be centrally monitored. The peer review also suggests that Mexico increase (i) research and development, (ii) its levels of innovation, and (iii) activities to develop locally relevant ICT goods, services and solutions. Equally important is to focus ICT initiatives for SMEs on increasing their productivity and competitiveness. The next priority should be the building of a competitive (and affordable) ICT infrastructure, the continuation of public–private partnerships to develop ICT skills, and better standards and security initiatives that will facilitate e-commerce and increase trust among enterprises. Finally, venture capital, local content creation, and evaluation (including impact studies) should increase.

Source: OECD (2006); Curry and Kenney (2006).

---

invoices to encourage online transactions (see box 1.4). E-government for businesses is also proving successful in Brazil; more than 60 per cent of enterprises with the Internet use online income tax services and State registries, and varying but significant proportions gather information about other government services, imports and exports, obtain licences and make online payments. About a quarter of Brazilian enterprises use the Internet to submit e-procurement (Government to business, or G2B) tenders, as e-procurement continues to expand in developing countries because of the increased transparency and efficiency it affords both Governments and suppliers.

4. E-Commerce remains limited in developing countries

Despite the relatively high levels of Internet (and broadband Internet) access by enterprises in developed countries, not all enterprises use the Internet for e-commerce, and there are wide gaps between countries and industries.

In the EU, Denmark has the leading position with 34 per cent of enterprises receiving online orders in 2007, compared with 2 per cent in Bulgaria (the lowest percentage). The average for the EU is 15 per cent in 2007, up from 9 per cent in 2003. During the same period, online payments and the average share of e-commerce in total turnover in EU enterprises also continued to grow. Of all enterprises that received online orders in 2006, 28 per cent received online payments, and e-commerce represented 12 per cent of total turnover. The EU average for online purchases is higher than for online sales, growing from 13 per cent in 2003 to 27 per cent in 2007. In 2007, the highest percentage of enterprises that purchased online was in Ireland (53 per cent) and the lowest in Bulgaria and Latvia (3 per cent in each case).[13]

Table 1.8

Enterprises with intranet, extranet and LAN

	Reference year	Proportion of enterprises with:		
		An intranet	An extranet	LAN
		B6	B11	B10
Argentina	2005	44.2	19.3	76.8
Austria	2006	41.0	19.0	..
Azerbaijan	2006	..	..	11.5
Belarus	2005	..	..	41.1
Belgium	2006	45.0	28.0	..
Bermuda	2005	34.0	..	..
Brazil	2006	39.0	20.3	89.0
Bulgaria	2006	35.0	4.1	52.8
Canada	2006	38.9	16.7	..
Chile	2005	..	3.4	12.6
China	2005	..	..	16.3
Cuba	2006	34.0	..	57.4
Cyprus	2006	21.0	7.0	..
Czech Republic	2006	23.0	7.0	..
Denmark	2006	35.0	22.0	..
Egypt	2006	34.0	2.4	78.9
Estonia	2006	35.0	12.0	..
Finland	2006	39.0	25.0	..
France	2006	40.0	22.0	..
Germany	2006	41.0	24.0	..
Greece	2006	39.0	11.0	..
Hong Kong (China)	2006	28.9	10.1	60.7
Hungary	2006	17.0	4.0	..
Iceland	2003	36.1	30.1	50.3
Ireland	2006	46.0	18.0	..

In terms of economic sectors, EU enterprises engaged in computer and related activities lead the way in terms of making online purchases (56 per cent of enterprises), with the lowest level being in enterprises manufacturing food, beverages, tobacco, textiles, leather, wood, and publishing and printing products (14 per cent). On the other hand, the hotel and accommodation sector leads in terms of online sales (31 per cent of enterprises), and the construction sector has the lowest level (3 per cent).

Few developing countries have information on online sales and purchases (see table 1.10) or are currently keeping track of their volumes of e-commerce, although it is clear that B2B trade continues to largely exceed B2C. There are exceptions in some countries, however, where B2C has been fostered (see box 1.5). For example, B2B trade volume is 10 times that of B2C in the Republic of Korea, where e-commerce continues to show healthy growth. That country's B2B trade volume grew by 14.7 per cent from 2005 to 2006 (to about $400.3 billion), and B2C grew by 17.5 per cent in the same period (to about $3.9 billion),[14] while

Table 1.8 *(continued)*

	Reference year	Proportion of enterprises with:		
		An intranet	An extranet	LAN
		B6	B11	B10
Italy	2006	33.0	13.0	..
Japan	2005	89.5	60.1	39.6
Latvia	2005	21.6	7.6	65.9
Lithuania	2006	57.0	8.0	..
Luxembourg	2006	44.0	25.0	..
Malta	2005	43.0	23.0	..
Mauritius	2006	37.3	..	..
Netherlands	2006	36.0	15.0	..
New Zealand	2006	22.4	7.5	61.6
Norway	2006	34.0	16.0	..
Panama	2006	28.0	13.7	53.3
Poland	2006	30.0	7.0	..
Portugal	2006	33.0	20.0	..
Qatar	2005	38.2	..	..
Republic of Korea	2005	37.3	..	66.5
Romania	2005	23.2	19.1	45.1
Russian Federation	2005	..	..	52.4
Singapore	2006	28.5	12.4	60.1
Slovakia	2006	31.0	12.0	..
Slovenia	2006	27.0	13.0	..
Spain	2006	28.0	13.0	..
Sweden	2006	43.0	20.0	..
Switzerland	2005	61.4	33.1	79.6
Turkey	2005	38.9	7.6	64.6
United Kingdom	2006	34.0	10.0	..

Note: See after table 1.20. in the Statistical Annex for notes to data.
Source: UNCTAD information economy database, 2007.

a decrease in the number of e-commerce websites and e-marketplaces indicated a tougher competitive environment. Companies based in the Republic of Korea are increasingly exporting their ICT-related services, namely search companies (Internet services), the electronic game industry, information security companies (anti-spam, anti-virus products, digital rights protection and PC security), and the software and content industry. In China, the Alibaba.com B2B Marketplace has more than 6 million SME users for domestic trade, and more than 24 million members for international trade.[15]

E-commerce is also restricted by issues of trust on the demand side. A 2007 survey in Thailand showed that despite recent growth in the number of Internet users and improvements in Internet infrastructure (and thus services), there was no concomitant increase in e-commerce. Seventy-one per cent of Thai Internet users had never made an online purchase, mainly because they distrust online merchants (and merchandise) that

Table 1.9 Use of the Internet for e-business activities

	Reference year	Proportion of enterprises using the Internet for:								
		Sending and receiving e-mail	Information about goods or services	Information from public authorities	Other information searches or research	Internet banking or financial services	Transacting with public authorities	Providing customer services	Delivering products online	Other types of activity
		B12.a	B12.b.i	B12.b.ii	B12.b.iii	B12.c	B12.d	B12.e	B12.f	B12.g
Argentina	2005	97.2	88.1	74.9	40.2	83.7	56.6	43.0	6.1	54.9
Australia	2005	..	..	..	..	..	50.4	..	..	..
Austria	2006	..	..	..	47.0	88.0	81.0	..	..	..
Azerbaijan	2006	..	..	25.8	..	25.5	26.4	..	..	..
Belgium	2006	..	..	..	62.0	88.0	59.0	..	..	..
Brazil	2006	97.7	78.4	59.4	82.4	80.1	84.1	30.9	13.6	..
Bulgaria	2006	..	..	57.3	46.9	53.4	61.4	..	..	..
Canada	2006	92.6	..	..	..	..	..	..	..	..
Chile	2005	99.2	..	..	..	..	..	..	..	..
China	2005	80.4	65.0	46.1	38.9	..	37.4	35.2	11.0	..
Cyprus	2006	..	..	..	67.0	57.0	44.0	..	..	..
Czech Republic	2006	..	..	..	71.0	91.0	76.0	..	..	..
Denmark	2006	..	..	..	53.0	94.0	87.0	..	..	..
Egypt	2006	93.3	58.9	..	58.9	26.8	5.8	36.2	..	0.4
Estonia	2006	..	..	..	49.0	98.0	69.0	..	..	..
Finland	2006	..	..	..	71.0	93.0	93.0	..	..	..
France	2006	..	..	..	..	77.0	66.0	..	..	..
Germany	2006	..	..	..	67.0	77.0	49.0	..	..	..
Greece	2006	..	..	..	62.0	74.0	84.0	..	..	..
Hong Kong (China)	2006	96.9	96.0	72.6	..	42.2	..	23.4	43.2	53.0
Hungary	2006	..	..	..	52.0	68.0	45.0	..	..	..
Iceland	2006	..	..	85.3	..	91.0	100.0	..	..	..
Ireland	2006	..	..	..	46.0	86.0	84.0	..	..	..
Italy	2006	..	..	..	66.0	81.0	87.0	..	..	..
Latvia	2005	..	..	46.3	..	89.2	50.1	..	..	..
Lithuania	2006	..	..	..	61.0	94.0	76.0	..	..	..
Luxembourg	2006	..	..	..	53.0	76.0	83.0	..	..	..
Macao (China)	2003	88.8	..	20.1	68.5	..	..	14.8	..	3.8
Malta	2005	..	..	..	56.0	67.0	68.0	..	..	..
Netherlands	2006	..	..	..	52.0	76.0	70.0	..	..	..

Table 1.9 *(continued)*

	Reference year	Proportion of enterprises using the Internet for:								
		Sending and receiving e-mail	Information about goods or services	Information from public authorities	Other information searches or research	Internet banking or financial services	Transacting with public authorities	Providing customer services	Delivering products online	Other types of activity
		B12.a	B12.b.i	B12.b.ii	B12.b.iii	B12.c	B12.d	B12.e	B12.f	B12.g
New Zealand	2006	..	..	68.0	..	87.2	76.8	29.9	..	..
Norway	2006	..	..	..	55.0	92.0	74.0	..	..	..
Panama	2006	97.3	80.6	67.9	60.7	70.1	35.8	39.1	..	69.6
Poland	2006	..	..	..	56.0	75.0	61.0	..	..	..
Portugal	2006	..	..	..	40.0	75.0	60.0	..	..	..
Republic of Korea	2005	88.6	60.9	53.5	77.5	67.4	43.4	34.5	13.2	2.4
Romania	2005	93.9	..	64.9	65.3	51.7	10.2	8.7	3.6	..
Russian Federation	2005	91.6	54.7	42.5	..	14.9	..	4.5	5.3	..
Singapore	2006	95.1	90.6	82.3	90.6	55.5	85.0	..	35.0	..
Slovakia	2006	..	..	..	64.0	84.0	77.0	..	..	..
Slovenia	2006	..	..	..	77.0	93.0	75.0	..	..	..
Spain	2006	..	..	..	33.0	85.0	58.0	..	..	..
Sweden	2006	..	..	..	72.0	92.0	80.0	..	..	..
Switzerland	2005	..	97.7	..	59.5	85.0	56.5	21.3	22.4	..
Thailand	2006	80.7	..	..	65.0	9.5	..	24.0	20.5	14.2
Turkey	2005	..	..	56.3	..	75.4	62.5	15.5	38.0	..
United Kingdom	2006	..	..	..	53.0	73.0	52.0	..	..	..

Note: See after table 1.20. in the Statistical Annex for notes to data.

Source: UNCTAD information economy database, 2007.

Table 1.10

Enterprises receiving and placing orders over the Internet

	Reference year	Proportion of enterprises (of those using Internet)	
		Receiving orders over the Internet B7	Placing orders over the Internet B8
Argentina	2005	45.6	44.6
Australia	2004	19.9	45.7
Austria	2006	18.0	51.0
Belgium	2006	15.0	44.0
Bermuda	2005	14.1	40.8
Brazil	2006	50.3	52.2
Bulgaria	2006	4.7	8.4
Canada	2005	10.7	60.8
Canada	2006	12.5	61.6
Chile	2005	4.2	6.7
China	2005	12.4	9.6
Cuba	2006	1.0	3.7
Cyprus	2006	6.0	21.0
Czech Republic	2006	9.0	27.0
Denmark	2006	35.0	59.0
Egypt	2006	34.8	21.0
Estonia	2006	14.0	25.0
Finland	2006	12.0	56.0
France	2006	16.0	26.0
Germany	2006	19.0	54.0
Greece	2006	8.0	14.0
Hong Kong (China)	2006	2.9	21.6
Hungary	2006	11.0	12.0
Iceland	2006	31.5	58.2
Ireland	2006	23.0	56.0

	Reference year	Proportion of enterprises (of those using Internet)	
		Receiving orders over the Internet B7	Placing orders over the Internet B8
Italy	2006	3.0	27.0
Japan	2005	15.6	20.6
Latvia	2005	3.7	15.3
Lithuania	2006	15.0	22.0
Luxembourg	2006	..	40.0
Macao (China)	2003	14.8	21.0
Malta	2005	14.0	47.0
Mauritius	2006	32.9	34.8
Netherlands	2006	28.0	45.0
New Zealand	2006	36.8	60.3
Norway	2006	25.0	66.0
Panama	2006	39.0	43.5
Poland	2006	7.0	23.0
Portugal	2006	5.0	20.0
Qatar	2005	50.9	41.3
Republic of Korea	2005	7.9	33.9
Romania	2005	4.1	11.0
Russian Federation	2005	23.6	30.7
Singapore	2006	37.6	42.8
Slovakia	2006	7.0	22.0
Slovenia	2006	12.0	22.0
Spain	2006	8.0	16.0
Sweden	2006	23.0	70.0
Switzerland	2005	23.2	58.0
Thailand	2006	11.1	13.9
United Kingdom	2006	19.0	62.0

Note: See after table 1.20. in the Statistical Annex for notes to data.
Source: UNCTAD information economy database, 2007.

they cannot see, electronic payment systems and credit card security. A Brazilian survey has also identified a lack of consumer trust as an obstacle to e-commerce (see box 1.5).

Governments can encourage e-commerce in their economies, particularly among SMEs. In India, for example, the government of West Bengal is planning to launch an electronic trading programme for fruits: through an electronic auction, farmers can set prices for their crop and sell it even before its physical transportation.[16] In Tunisia, it is an objective of the Ministry for Trade and Handicrafts that e-commerce

generate 2 per cent of all export revenues by 2011 (it was estimated at 0.05 per cent in 2006); in order to achieve that goal, the ministry supports private initiatives (raising awareness and training SMEs) and invests in sectoral e-marketplaces.[17]

B2C digital content e-commerce is also growing. For example, total worldwide revenues in 2006 from online music sales (legal downloads and other online music services) were estimated at approximately \$1.7 billion, and are expected to more than triple by 2011; currently, 14 per cent of B2C online music services are in non-OECD countries.[18] E-commerce of digital products

Box 1.5

E-commerce in Brazil

In Brazil in 2006, almost 95 per cent of enterprises (in a survey of the formal sector with 10 or more employees) had access to the Internet. Of these, almost 89 per cent had broadband. The high level of enterprise Internet access contrasts with the fact that only half of Brazilian enterprises have a website, which usually serves as a portal for e-commerce. In fact, only 52 per cent of enterprises have placed orders over the Web (through e-mail and Web forms, and many use online payment methods), and about 50 per cent of enterprises have received orders (67 per cent of enterprises that have websites use them to sell products). Among enterprises engaging in e-commerce, online purchases represent 23 per cent of total purchases, and online sales represent about 30 per cent of their revenue, which is quite significant. About 60 per cent of online sales are B2C, and the vast majority are local (items sold are destined for Brazil) and consist of goods and services ordered online and delivered offline.

The Internet is used for other activities related to sales and purchases in addition to e-commerce (see charts 1.7 and 1.8).

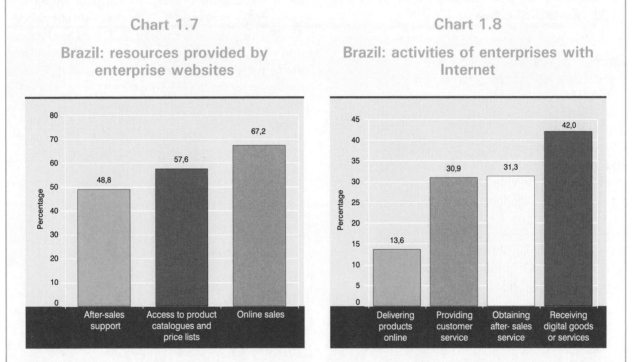

Chart 1.7

Brazil: resources provided by enterprise websites

Chart 1.8

Brazil: activities of enterprises with Internet

Those companies that are already engaged in e-commerce recognize that the time and cost of transactions are reduced, and more than half have increased their sales volume or clientele thanks to e-commerce. Brazil sees a great potential for growth in e-commerce on both the supply and the demand sides (only 6 per cent of Internet users made Internet purchases in 2006). To increase e-commerce, trust in the safety of electronic transactions must be increased among consumers and small and medium-sized enterprises, including through awareness-raising and also through a better regulatory environment. The Brazilian Internet Steering Committee asks the private sector, represented by industry and trade associations, to kick-start the adoption of e-commerce among their members, including through the establishment of e-marketplaces.

Source: Brazilian Internet Steering Committee (2006).

includes content for mobile phones (and other wireless handheld devices such as personal digital assistants), such as games, multimedia messaging services, music, ringtones, and wallpapers. In the future, real products (event tickets, retail products) and mobile payments and banking (see chapter 5) are likely to generate the largest growth in this market.

D. Conclusions

Figures show that developing countries continue to increase their access to and use of ICTs, although there is still a gap between them and developed countries. The gap is being reduced more significantly with respect to mobile telephony, which has come to replace fixed lines in several developing country areas. In the future, wireless technologies are likely to have a more prominent role in ICT diffusion in the developing world, including in business.

Mobile telephony is currently the most significant ICT for developing countries

Among ICTs, mobile phones are worthy of special attention as threshold technologies in many developing countries. For enterprises, the increased availability of computers has been proved to have a significant impact in productivity. However, connections to the Internet remain expensive and thus limited, although broadband Internet is essential for developing an information economy (and integrating the information economy at the regional and global levels). However, ICTs cannot really become widespread and effect a change unless they are accessible and affordable.

Government policy can encourage ICT diffusion

The diffusion of ICT in developing countries depends on adequate policymaking and may positively benefit from the involvement of government in the implementation of policies, particularly in building up the telecommunication infrastructure and ensuring a regulatory environment that will allow competition in the telecommunications and ICT sector, as well as facilitate trade and e-commerce. Government can also set an example and pioneer ICTs for e-government. Finally, government can raise awareness among enterprises, particularly SMEs, on the potential of ICTs for business use.

A competitive telecommunications market encourages ICT diffusion

Increased competition can lower prices, and this in turn facilitates the adoption of ICT by lower-income individuals and enterprises. For example, SMEs in Africa find that costs are a major obstacle to increased Internet access.

More data on ICT use by businesses in developing countries are needed in order to better assess the impact of ICT policies

National, regional and international efforts to improve the availability of data on ICT in order to assess the information economy should continue. Official statistical data on ICT in developing countries, particularly on the use of ICT by business and on the ICT sector, are essential so that policymakers can assess progress in their ICT for development policies. The Partnership on Measuring ICT for Development (http://measuring-ict.unctad.org) provides guidance in that respect. More research is needed on ICT use by businesses in developing countries to better assess the impact of ICT on firm-level productivity.

STATISTICAL ANNEX

Table 1.11

Mobile phone subscribers: economies by level of development and region

	2002	% change 2002–2003	2003	% change 2003–2004	2004	% change 2004–2005	2005	% change 2005–2006	2006
DEVELOPED ECONOMIES									
ASIA									
Israel	6 334 000	2.6	6 500 000	10.6	7 187 500	8.5	7 800 000	7.7	8 403 800
Japan	81 118 320	6.8	86 654 960	5.6	91 473 936	3.6	94 745 000	7.3	101 698 000
EUROPE									
Andorra	32 790	58.3	51 893	24.4	64 560	0.0	64 560	0.1	64 600
Austria	6 736 000	5.3	7 094 502	12.6	7 989 955	2.1	8 160 000	13.4	9 255 000
Belgium	8 101 777	6.2	8 605 834	6.1	9 131 705	3.6	9 460 000	2.1	9 659 700
Cyprus	417 933	32.0	551 752	16.1	640 515	12.2	718 842	8.2	777 500
Czech Republic	8 610 177	12.8	9 708 683	10.9	10 771 270	9.3	11 775 878	3.2	12 149 900
Denmark (incl. Faroe Islands)	4 508 461	6.6	4 805 917	7.5	5 165 546	6.7	5 511 878	6.9	5 890 900
Estonia	881 000	19.2	1 050 241	19.6	1 255 731	15.1	1 445 300	14.8	1 658 700
Finland	4 516 772	5.1	4 747 126	5.1	4 988 000	4.9	5 231 000	8.4	5 670 000
France (incl. Guadeloupe. French Guiana. Martinique and Réunion)	39 805 800	6.1	42 248 100	7.3	45 348 800	6.0	48 058 400	7.5	51 662 000
Germany	59 128 000	9.6	64 800 000	10.1	71 316 000	11.1	79 200 000	6.4	84 300 000
Greece	9 314 260	11.0	10 337 000	6.8	11 044 232	-7.1	10 260 400	8.2	11 097 500
Guernsey	36 580	13.5	41 530	5.5	43 824	..	..	..	..
Hungary	6 886 111	15.4	7 944 586	9.9	8 727 188	6.8	9 320 000	6.9	9 965 000
Iceland	260 438	7.4	279 670	4.2	291 372	4.3	304 001	8.1	328 500
Ireland	3 000 000	16.7	3 500 000	8.0	3 780 000	11.4	4 210 000	11.4	4 690 000
Italy	54 200 000	4.7	56 770 000	10.5	62 750 000	14.0	71 535 000	0.0	71 500 000
Jersey	61 400	32.2	81 200	3.3	83 900	..	..	..	..
Latvia	917 196	33.0	1 219 550	26.0	1 536 712	21.8	1 871 602	16.7	2 183 700
Liechtenstein	11 402	119.3	25 000	2.0	25 500	7.8	27 500	..	..
Lithuania	1 645 568	31.9	2 169 866	57.7	3 421 538	27.2	4 353 447	8.4	4 718 200
Luxembourg	473 000	14.0	539 000	19.9	646 000	11.5	720 000	-0.9	713 800
Malta	276 859	4.7	289 992	5.6	306 100	5.8	323 980	7.0	346 800
Monaco	7 200	109.7	15 100	4.6	15 800	8.9	17 200	..	..
Netherlands	12 300 000	9.7	13 491 000	9.9	14 821 000	6.8	15 834 000	..	..
Norway	3 911 136	6.4	4 163 381	8.7	4 524 800	5.1	4 754 453	6.0	5 040 600
Poland	13 898 471	25.2	17 401 222	32.7	23 096 064	26.7	29 260 000	25.6	36 745 500
Portugal	8 528 900	17.6	10 030 000	2.7	10 300 000	11.1	11 447 670	6.8	12 226 400
San Marino	16 759	0.8	16 900	1.2	17 100	0.6	17 200	1.1	17 390
Slovakia	2 923 383	25.8	3 678 774	16.2	4 275 164	6.2	4 540 374	7.8	4 893 200
Slovenia	1 667 234	4.3	1 739 146	6.3	1 848 600	-4.8	1 759 232	3.4	1 819 600
Spain	33 530 996	11.0	37 219 840	3.8	38 622 584	7.0	41 327 911	11.7	46 152 000
Sweden	7 949 000	10.7	8 801 000	-0.2	8 785 000	3.4	9 087 000	..	..
Switzerland	5 736 303	7.9	6 189 000	1.4	6 275 000	9.1	6 847 000	8.3	7 418 000
United Kingdom (incl. Gibraltar)	49 689 167	6.7	52 999 900	15.3	61 118 392	7.1	65 471 700	6.4	69 656 600

Table 1.11 *(continued)*

	2002	% change 2002–2003	2003	% change 2003–2004	2004	% change 2004–2005	2005	% change 2005–2006	2006
DEVELOPED ECONOMIES									
NORTH AMERICA									
Bermuda	30 000	33.3	40 000	22.5	49 000	7.6	52 700	14.0	60 100
Canada	11 872 000	11.4	13 228 000	13.3	14 984 396	10.8	16 600 000	2.5	17 017 000
Greenland	19 924	37.5	27 400	17.5	32 200	..	..	..	..
United States (incl. Puerto Rico and Guam)	142 566 848	11.3	158 721 984	15.8	183 787 136	9.7	201 650 000	17.3	236 451 800
OCEANIA									
Australia	12 575 000	14.1	14 347 000	14.7	16 449 000	12.0	18 420 000	7.3	19 760 000
New Zealand	2 449 000	6.1	2 599 000	16.5	3 027 000	16.6	3 530 000	..	..
DEVELOPING ECONOMIES									
AFRICA									
Algeria	400 000	260.4	1 441 400	224.9	4 682 690	191.7	13 661 000	53.7	20 998 000
Angola	130 000	156.0	332 800	182.5	940 000	16.4	1 094 115	106.9	2 264 200
Benin	218 770	8.0	236 175	63.7	386 700	93.9	750 000	..	..
Botswana	435 000	20.2	522 840	7.8	563 782	46.0	823 070	19.0	979 800
Burkina Faso	113 000	100.9	227 000	75.3	398 000	43.8	572 200	77.7	1 016 600
Burundi	52 000	23.1	64 000	57.2	100 600	52.1	153 000	..	..
Cameroon	701 507	53.5	1 077 000	42.7	1 536 594	46.6	2 252 500	..	..
Cape Verde	42 949	24.2	53 342	23.3	65 780	24.2	81 721	33.3	108 900
Central African Republic	12 600	217.5	40 000	50.0	60 000	66.7	100 000	..	..
Chad	34 200	90.1	65 000	89.2	123 000	70.7	210 000	122.0	466 100
Comoros	..	..	2 000	370.0	9 400	70.9	16 065	0.2	16 100
Congo	221 800	48.8	330 000	16.3	383 653	27.7	490 000	..	..
Côte d'Ivoire	1 027 058	24.7	1 280 696	19.6	1 531 846	43.0	2 190 000	85.6	4 065 400
Democratic Republic of the Congo	560 000	78.6	1 000 000	99.1	1 990 700	37.9	2 746 000	..	..
Djibouti	15 000	53.3	23 000	50.0	34 500	27.8	44 100	..	..
Egypt	4 494 700	29.0	5 797 530	31.8	7 643 060	78.3	13 629 602	32.1	0.4
Equatorial Guinea	32 000	29.7	41 500	33.7	55 500	74.6	96 900	..	..
Eritrea	..	..	..	..	20 000	102.2	40 438	53.3	62 000
Ethiopia	50 369	94.2	97 827	82.0	178 000	130.7	410 600	111.1	866 700
Gabon	279 289	7.4	300 000	63.1	489 367	32.8	649 807	17.7	764 700
Gambia	100 000	49.3	149 300	17.2	175 000	41.4	247 478	63.4	404 300
Ghana	386 775	105.7	795 529	113.1	1 695 000	4.1	1 765 000	195.0	5 207 200
Guinea	90 772	22.8	111 500	38.9	154 900	22.0	189 000	..	..
Guinea-Bissau	..	..	1 275	3170.6	41 700	60.7	67 000	41.8	95 000
Kenya	1 187 122	34.0	1 590 785	60.1	2 546 157	81.1	4 611 970	40.6	6 484 800
Lesotho	96 843	4.8	101 474	56.7	159 000	54.1	245 052	1.9	249 800
Liberia	2 000	2262.5	47 250	99.8	94 400	69.5	160 000	..	..
Libyan Arab Jamahiriya	..	..	..	..	..	..	..	..	3 927 600
Madagascar	163 010	74.0	283 666	17.7	333 888	51.1	504 660	107.2	1 045 900
Malawi	86 047	57.0	135 114	64.4	222 135	93.3	429 305	..	..
Mali	52 639	365.3	244 930	63.3	400 000	117.4	869 576	..	..
Mauritania	247 238	41.9	350 954	48.9	522 400	42.7	745 615	42.2	1 060 100
Mauritius	348 137	-6.3	326 033	56.4	510 000	39.9	713 300	8.3	772 400

Table 1.11 *(continued)*

	2002	% change 2002–2003	2003	% change 2003–2004	2004	% change 2004–2005	2005	% change 2005–2006	2006
DEVELOPING ECONOMIES									
Mayotte	21 700	65.9	36 000	5.6	38 000	26.6	48 100	..	..
Morocco	6 198 670	18.7	7 359 870	26.9	9 336 878	32.7	12 392 805	29.1	16 004 700
Mozambique	254 759	71.0	435 757	62.5	708 000	72.3	1 220 000	91.7	2 339 300
Namibia	150 000	49.1	223 671	27.9	286 095	73.0	495 000	..	..
Niger	16 648	360.0	76 580	93.6	148 276	102.3	299 899	8.0	323 900
Nigeria	1 607 931	95.9	3 149 473	190.4	9 147 209	103.3	18 600 000	73.8	32 322 200
Rwanda	82 391	58.7	130 720	6.1	138 728	109.0	290 000	..	..
Sao Tome and Principe	1 980	143.4	4 819	59.8	7 700	55.8	12 000	..	..
Senegal	455 645	26.4	575 917	78.5	1 028 061	68.3	1 730 106	72.4	2 982 600
Seychelles	44 731	10.1	49 229	0.0	49 230	15.8	57 003	23.3	70 300
Sierra Leone	67 000	69.0	113 214	0.0	113 200	..	..	..	..
Somalia	100 000	100.0	200 000	150.0	500 000	..	..	..	..
South Africa	13 702 000	23.0	16 860 000	15.7	19 500 000	59.0	31 000 000	9.5	33 960 000
Sudan	190 778	176.4	527 233	98.9	1 048 558	89.4	1 986 000	135.8	4 683 100
Swaziland	68 000	25.0	85 000	32.9	113 000	77.0	200 000	25.0	250 000
Togo	170 000	29.4	220 000	51.2	332 600	33.4	443 635	59.6	708 000
Tunisia	574 334	233.9	1 917 530	85.8	3 562 970	59.4	5 680 726	29.2	7 339 000
Uganda	393 310	97.3	776 169	50.1	1 165 035	30.9	1 525 125	31.7	2 008 800
United Republic of Tanzania	760 000	36.9	1 040 640	57.6	1 640 000	106.7	3 389 800	84.1	6 240 800
Zambia	139 092	73.3	241 000	24.5	300 000	145.0	735 000	29.2	949 600
Zimbabwe	338 779	7.3	363 365	9.4	397 500	75.8	699 000	19.1	832 500
ASIA									
Afghanistan	25 000	700.0	200 000	200.0	600 000	100.0	1 200 000	110.0	2 520 400
Bahrain	388 990	13.9	443 109	46.6	649 764	15.2	748 703	20.1	898 900
Bangladesh	1 075 000	27.0	1 365 000	217.0	4 327 516	108.0	9 000 000	112.6	19 131 000
Bhutan	..	..	7 998	122.6	17 800	112.6	37 842	117.0	82 100
Brunei Darussalam	153 600	15.5	177 400	14.1	202 500	15.0	232 900	9.1	254 000
Cambodia	380 000	31.2	498 388	72.9	861 500	23.3	1 062 000	7.3	1 140 000
China	206 004 992	31.0	269 952 992	24.0	334 824 000	17.5	393 428 000	17.2	461 058 000
Dem. People's Republic of Korea	..	..	..	..	..	..	..	..	..
Hong Kong (China)	6 395 725	14.9	7 349 202	10.9	8 148 685	6.0	8 635 532	8.3	9 356 400
India	12 687 637	106.1	26 154 404	80.8	47 300 000	60.7	76 000 000	118.5	166 050 000
Indonesia	11 700 000	60.7	18 800 000	59.6	30 000 000	56.4	46 909 972	36.0	63 803 000
Iran (Islamic Republic of)	2 186 958	54.4	3 376 526	27.3	4 300 000	68.0	7 222 538	89.1	13 659 100
Iraq	20 000	300.0	80 000	617.5	574 000	..	..	..	..
Jordan	1 219 597	8.7	1 325 313	20.3	1 594 513	96.8	3 137 700	38.4	4 343 100
Kuwait	1 227 000	15.7	1 420 000	40.8	2 000 000	19.0	2 379 811	..	..
Lao PDR	55 160	103.5	112 275	81.9	204 191	212.6	638 202	..	..
Lebanon	775 104	5.8	820 000	8.3	888 000	11.5	990 000	11.5	1 103 400
Macao (China)	276 138	31.8	364 031	18.8	432 450	23.2	532 758	19.4	636 300
Malaysia	9 253 387	20.2	11 124 112	31.4	14 611 902	33.8	19 545 000	-0.4	19 463 700
Maldives	41 899	58.6	66 466	70.4	113 246	35.5	153 393	71.2	262 600
Mongolia	216 000	47.7	319 000	34.4	428 700	30.0	557 207	..	..
Myanmar	47 982	38.6	66 517	38.3	92 007	99.4	183 434	..	..

Table 1.11 *(continued)*

	2002	% change 2002–2003	2003	% change 2003–2004	2004	% change 2004–2005	2005	% change 2005–2006	2006
DEVELOPING ECONOMIES									
Nepal	21 881	130.2	50 367	255.6	179 126	26.9	227 300	358.3	1 041 800
Oman	464 896	27.7	593 450	35.6	805 000	65.6	1 333 225	36.4	1 818 000
Pakistan	1 698 536	41.6	2 404 400	108.9	5 022 908	154.3	12 771 203	170.2	34 506 600
Palestine	320 000	50.0	480 000	103.0	974 345	12.3	1 094 640	..	..
Philippines	15 383 001	46.3	22 509 560	46.3	32 935 876	5.6	34 779 000	19.6	41 600 000
Qatar	266 703	41.2	376 535	30.2	490 333	46.2	716 763	28.3	919 800
Republic of Korea	32 342 492	3.9	33 591 760	8.9	36 586 052	4.8	38 342 323	4.8	40 197 100
Saudi Arabia	5 007 965	44.5	7 238 224	26.8	9 175 764	44.9	13 300 000	47.8	19 662 600
Singapore	3 344 800	4.0	3 477 100	11.0	3 860 600	13.6	4 384 600	9.2	4 788 600
Sri Lanka	931 580	49.6	1 393 403	58.7	2 211 158	52.0	3 361 775	61.0	5 412 500
Syrian Arab Republic	400 000	196.3	1 185 000	97.9	2 345 000	25.8	2 950 000	58.5	4 675 000
Taiwan Province of China	24 390 520	5.8	25 799 840	-11.8	22 760 144	-2.6	22 170 702	4.9	23 249 300
Thailand	16 117 000	54.3	24 864 020	10.1	27 379 000	13.7	31 136 500	31.1	40 815 500
Turkey	23 323 118	19.6	27 887 536	24.5	34 707 548	25.6	43 608 965	..	..
United Arab Emirates	2 428 071	22.4	2 972 331	23.9	3 683 117	23.1	4 534 480	21.7	5 519 300
Viet Nam	1 902 388	44.1	2 742 000	80.9	4 960 000	81.5	9 000 000	72.3	15 505 400
Yemen	411 083	70.3	700 000	53.1	1 072 000	86.6	2 000 000	..	..
LATIN AMERICA AND THE CARIBBEAN									
Antigua and Barbuda	38 205	20.7	46 100	17.1	54 000	59.3	86 000	18.6	102 000
Argentina	6 566 740	19.4	7 842 233	72.3	13 512 383	63.6	22 100 000	42.6	31 510 400
Aruba	61 800	13.3	70 000	40.6	98 400	10.0	108 200	..	..
Bahamas	121 759	-4.5	116 267	60.0	186 007	22.5	227 800	..	..
Barbados	97 193	44.0	140 000	43.0	200 138	3.0	206 190	..	..
Belize	51 729	16.8	60 403	61.8	97 755	22.3	119 600	-1.1	118 300
Bolivia	1 023 333	25.0	1 278 844	40.8	1 800 789	34.5	2 421 402	..	..
Brazil	34 880 964	32.9	46 373 264	41.5	65 605 000	31.4	86 210 000	15.9	99 918 600
Cayman Islands	..	..	..	..	..	..	..	..	..
Chile	6 445 698	16.7	7 520 280	27.2	9 566 581	10.5	10 569 572	17.8	12 450 800
Colombia	4 596 594	34.6	6 186 206	68.1	10 400 578	109.6	21 800 000	36.5	29 762 700
Costa Rica	502 478	54.9	778 299	18.6	923 084	19.3	1 101 035	31.1	1 443 700
Cuba	17 851	98.1	35 356	114.4	75 797	77.4	134 480	13.5	152 700
Dominica	12 173	73.3	21 099	98.3	41 838	..	..	..	..
Dominican Republic	1 700 609	24.8	2 122 543	19.4	2 534 063	43.0	3 623 289	27.1	4 605 700
Ecuador	1 560 861	53.6	2 398 161	89.5	4 544 174	37.5	6 246 332	35.8	8 485 000
El Salvador	888 818	29.4	1 149 790	59.4	1 832 579	31.6	2 411 753	59.7	3 851 600
Grenada	7 553	459.9	42 293	2.4	43 313	8.3	46 900	..	..
Guatemala	1 577 085	29.0	2 034 776	55.7	3 168 256	42.4	4 510 100	..	..
Guyana	87 300	35.9	118 658	21.3	143 945	73.7	250 000	12.6	281 400
Haiti	140 000	128.6	320 000	25.0	400 000	25.1	500 200	..	..
Honduras	326 508	16.2	379 362	86.4	707 201	81.2	1 281 462	74.9	2 240 800
Jamaica	1 187 295	34.8	1 600 000	37.5	2 200 000	22.7	2 700 000	3.9	2 804 400
Mexico	25 928 266	16.1	30 097 700	27.8	38 451 136	23.4	47 462 108	20.1	57 016 400
Netherlands Antilles	..	..	200 000	0.0	200 000	0.0	200 000	..	..
Nicaragua	237 248	96.7	466 706	58.3	738 624	51.5	1 119 379	63.5	1 830 200

Table 1.11 (continued)

	2002	% change 2002–2003	2003	% change 2003–2004	2004	% change 2004–2005	2005	% change 2005–2006	2006
DEVELOPING ECONOMIES									
Panama	525 845	58.6	834 031	2.6	855 852	58.0	1 351 924	25.3	1 693 500
Paraguay	1 667 018	6.2	1 770 345	-0.1	1 767 824	6.7	1 887 000	71.3	3 232 800
Peru	2 306 944	27.0	2 930 343	39.7	4 092 558	36.4	5 583 356	52.2	8 500 000
Saint Kitts and Nevis	5 000	0.0	5 000	100.0	10 000	..	..	..	..
Saint Lucia	14 313	-0.1	14 300	550.3	93 000	13.7	105 700	..	..
Saint Vincent and the Grenadines	9 982	530.2	62 911	-9.5	56 950	24.0	70 620	24.0	87 600
Suriname	108 363	55.5	168 522	26.3	212 819	9.4	232 785	37.5	320 000
Trinidad and Tobago	361 911	34.3	485 871	33.3	647 870	23.5	800 000	106.9	1 654 900
Uruguay	513 528	-3.1	497 530	20.6	600 000	92.5	1 154 900	101.7	2 330 000
Venezuela	6 463 561	8.5	7 015 735	20.0	8 420 980	48.4	12 495 721	50.4	18 789 500
Virgin Islands (United States)	45 200	9.1	49 300	30.2	64 200	25.1	80 300	..	..
OCEANIA									
American Samoa	2 000	5.0	2 100	4.8	2 200	..	..	..	..
Fiji	89 900	22.2	109 900	29.4	142 200	44.2	205 000	..	..
French Polynesia	52 200	15.1	60 100	59.7	96 000	25.0	120 000	26.7	152 000
Kiribati	495	6.3	526	14.1	600	..	..	..	..
Marshall Islands	552	8.3	598	0.3	600	..	..	..	..
Micronesia (Fed. States of)	100	5 769.0	5 869	117.8	12 782	10.3	14 100	..	..
Nauru	..	..	..	..	..	..	..	..	..
New Caledonia	80 000	21.4	97 113	19.9	116 443	15.3	134 265	0.0	134 300
Northern Mariana Islands	17 100	8.8	18 600	10.2	20 500	..	..	..	..
Palau	..	..	..	..	..	..	..	..	..
Papua New Guinea	15 000	16.7	17 500	176.0	48 300	55.3	75 000	..	..
Samoa	2 700	288.9	10 500	52.4	16 000	50.0	24 000	..	..
Solomon Islands	999	48.9	1 488	101.6	3 000	100.0	6 000	..	..
Tonga	3 354	233.9	11 200	46.4	16 400	82.3	29 900	..	..
Tuvalu	..	..	..	..	..	..	..	..	..
Vanuatu	4 900	59.2	7 800	34.7	10 504	20.8	12 692	0.1	12 700
TRANSITION ECONOMIES									
Albania	851 000	29.3	1 100 000	14.5	1 259 600	21.5	1 530 200	..	..
Armenia	71 300	60.4	114 400	77.7	203 300	56.4	318 000	..	..
Azerbaijan	794 000	33.1	1 057 000	68.7	1 782 900	25.8	2 242 000	48.2	3 323 500
Belarus	462 630	141.7	1 118 000	100.3	2 239 300	83.0	4 097 997	45.4	5 960 000
Bosnia and Herzegovina	748 780	40.2	1 050 000	34.0	1 407 400	13.3	1 594 367	18.4	1 887 800
Bulgaria	2 597 548	34.8	3 500 869	35.1	4 729 731	32.0	6 244 693	32.2	8 253 400
Croatia	2 340 000	9.1	2 553 000	11.1	2 835 500	5.2	2 983 900	49.8	4 469 700
Georgia	503 619	41.2	711 224	18.2	840 600	73.6	1 459 180	16.8	1 703 900
Kazakhstan	1 027 000	29.6	1 330 730	107.3	2 758 940	79.6	4 955 200	58.0	7 830 400
Kyrgyzstan	53 084	160.5	138 279	117.0	300 000	80.6	541 652	..	..
Moldova	338 225	40.7	475 942	65.4	787 000	38.5	1 089 800	24.6	1 358 200
Montenegro	..	..	..	..	..	..	..	..	821 800
Romania	5 110 591	37.8	7 039 898	45.1	10 215 388	30.7	13 354 138	..	..
Russian Federation	8.5	107.3	36 500 000	103.9	74 420 000	61.2	120 000 000	..	..
Serbia	2 750 397	32.1	3 634 613	30.1	4 729 629	16.5	5 510 700	20.6	6 643 700

Table 1.11 *(continued)*

	2002	% change 2002–2003	2003	% change 2003–2004	2004	% change 2004–2005	2005	% change 2005–2006	2006
TRANSITION ECONOMIES									
Tajikistan	13 200	260.7	47 617	183.5	135 000	96.3	265 000	..	..
TFYR Macedonia	365 346	112.4	776 000	27.0	985 600	28.0	1 261 300	12.3	1 417 000
Turkmenistan	8 173	12.4	9 187	445.3	50 100	109.6	105 000	..	..
Ukraine	3 692 700	76.0	6 498 423	111.4	13 735 000	25.3	17 214 280	185.1	49 076 200
Uzbekistan	186 900	71.7	320 815	69.6	544 100	32.3	720 000	..	..

Source: UNCTAD calculations based on the ITU World Telecommunication/ICT Indicators database, 2007.

Table 1.12

Mobile phone penetration: economies by level of development and region

Mobile phone subscribers per 100 inhabitants

	2002	% change 2002–2003	2003	% change 2003–2004	2004	% change 2004–2005	2005	% change 2005–2006	2006
DEVELOPED ECONOMIES									
ASIA									
Israel	95.4	0.7	96.0	9.1	104.8	7.9	113.0	4.0	117.5
Japan	63.7	6.7	67.9	5.5	71.6	3.5	74.2	7.3	79.6
EUROPE									
Andorra	48.8	47.0	71.8	17.0	84.0	-2.1	82.2	-3.2	79.5
Austria	83.2	4.8	87.2	11.7	97.3	1.1	98.4	13.3	111.5
Belgium	78.2	5.8	82.7	5.6	87.4	3.5	90.5	1.9	92.2
Cyprus	52.9	30.4	69.0	13.3	78.1	10.1	86.0	5.8	90.9
Czech Republic	84.4	12.8	95.2	10.8	105.5	9.2	115.2	2.5	118.1
Denmark (incl. Faroe Islands)	83.2	6.0	88.2	7.3	94.6	6.3	100.6	6.5	107.1
Estonia	64.7	19.7	77.4	20.0	92.9	15.4	107.3	15.0	123.4
Finland	86.7	4.9	90.9	5.1	95.6	4.3	99.6	8.2	107.8
France (incl. Guadeloupe. French Guiana. Martinique and Réunion)	66.7	5.7	70.5	6.4	75.0	5.9	79.4	7.1	85.1
Germany	71.6	9.6	78.5	10.1	86.4	10.8	95.8	6.4	101.9
Greece	84.5	6.7	90.2	10.3	99.5	-7.3	92.3	8.0	99.6
Guernsey	61.0	13.5	69.2	5.5	73.0	..	..	..	..
Hungary	67.9	15.6	78.5	10.1	86.4	6.8	92.3	7.2	99.0
Iceland	89.8	7.4	96.4	4.2	100.5	4.3	104.8	4.5	109.5
Ireland	76.3	15.2	87.9	6.4	93.6	8.4	101.4	9.8	111.4
Italy	96.0	2.2	98.1	10.3	108.2	13.8	123.1	-0.1	123.0
Jersey	70.1	32.2	92.7	3.2	95.7	..	..	..	..
Latvia	39.4	33.5	52.6	27.7	67.1	20.7	81.0	17.2	94.9
Liechtenstein	35.6	112.6	75.8	-1.0	75.0	4.8	78.6	..	..
Lithuania	47.3	33.0	62.9	57.7	99.2	28.0	126.9	8.7	138.0
Luxembourg	105.1	14.0	119.8	17.2	140.4	11.5	156.5	-3.0	151.9
Malta	71.6	4.2	74.6	5.1	78.4	5.5	82.7	3.1	85.2
Monaco	22.2	108.4	46.3	4.0	48.2	8.2	52.1	..	..
Netherlands	75.9	9.1	82.8	10.3	91.3	6.4	97.1	..	..
Norway	86.0	5.8	90.9	8.2	98.4	4.6	102.9	5.6	108.6
Poland	36.4	25.3	45.6	32.8	60.5	26.7	76.7	25.7	96.4
Portugal	81.9	16.9	95.8	2.1	97.8	11.6	109.1	6.3	116.0
San Marino	59.9	-0.9	59.3	-.6	59.0	-2.8	57.3	-2.2	56.1
Slovakia	54.3	25.8	68.4	16.1	79.4	6.2	84.3	7.7	90.8
Slovenia	83.5	4.3	87.1	6.3	92.5	-5.1	87.8	2.1	89.6
Spain	81.6	6.9	87.2	2.5	89.4	3.4	92.4	10.7	102.3
Sweden	88.9	10.2	98.0	-0.5	97.5	3.1	100.5	..	..
Switzerland	78.8	6.7	84.1	.6	84.6	8.5	91.8	8.0	99.1
United Kingdom (incl. Gibraltar)	84.1	5.9	89.1	14.7	102.2	7.4	109.7	6.1	116.4
NORTH AMERICA									
Bermuda	49.2	31.2	64.5	20.6	77.8	5.9	82.3	12.3	92.5

Table 1.12 *(continued)*

	2002	% change 2002–2003	2003	% change 2003–2004	2004	% change 2004–2005	2005	% change 2005–2006	2006
DEVELOPED ECONOMIES									
Canada	37.7	10.4	41.6	12.1	46.7	9.7	51.2	2.0	52.2
Greenland	35.3	37.0	48.4	17.1	56.7	..	..	..	..
United States (incl. Puerto Rico and Guam)	48.8	10.3	53.8	14.7	61.7	8.0	66.7	16.2	77.5
OCEANIA									
Australia	64.0	12.7	72.2	14.5	82.6	10.6	91.4	6.2	97.0
New Zealand	62.2	5.9	65.8	15.3	75.9	15.5	87.6	..	..
DEVELOPING ECONOMIES									
AFRICA									
Algeria	1.3	254.8	4.5	219.8	14.5	186.9	41.5	51.7	63.0
Angola	0.9	148.5	2.3	188.1	6.7	5.3	7.0	103.9	14.3
Benin	3.2	4.4	3.4	58.5	5.3	87.7	10.0	..	..
Botswana	24.7	18.2	29.2	7.2	31.3	43.1	44.8	16.1	52.1
Burkina Faso	0.9	96.0	1.9	60.5	3.0	45.5	4.3	72.5	7.5
Burundi	0.7	20.0	0.9	60.3	1.4	42.4	2.0	..	..
Cameroon	4.4	49.5	6.6	42.3	9.4	46.4	13.8	..	..
Cape Verde	9.5	21.5	11.6	20.7	14.0	21.6	17.0	23.0	20.9
Central African Republic	0.3	203.7	1.0	58.8	1.5	61.3	2.5	..	..
Chad	0.4	85.1	0.8	72.8	1.4	55.0	2.2	115.8	4.6
Comoros	..	..	0.3	364.1	1.2	68.8	2.0	-2.2	2.0
Congo	6.7	40.3	9.4	6.5	10.0	22.0	12.3	..	..
Côte d'Ivoire	6.2	23.6	7.7	17.7	9.1	33.1	12.1	82.6	22.0
Democratic Republic of the Congo	1.1	73.6	1.8	93.3	3.6	33.9	4.8	..	..
Djibouti	2.1	49.2	3.1	45.2	4.5	23.2	5.5	..	..
Egypt	6.7	26.5	8.4	29.3	10.9	75.0	19.1	24.9	23.9
Equatorial Guinea	7.0	26.9	8.9	30.9	11.7	71.0	20.0	..	..
Eritrea	..	..	..	..	0.5	93.9	0.9	47.9	1.4
Ethiopia	0.1	88.6	0.1	74.3	0.2	115.7	0.5	106.1	1.1
Gabon	21.5	4.2	22.4	61.9	36.2	29.9	47.1	15.2	54.2
Gambia	7.5	46.0	11.0	9.2	12.0	35.8	16.3	59.2	25.9
Ghana	1.9	100.3	3.7	112.1	7.9	0.7	8.0	189.1	23.1
Guinea	1.2	21.4	1.4	38.0	2.0	19.0	2.4	..	..
Guinea-Bissau	..	..	0.1	3095.7	3.2	57.1	5.0	16.6	5.8
Kenya	3.8	33.1	5.0	54.7	7.8	73.5	13.5	37.2	18.5
Lesotho	5.1	4.2	5.3	53.5	8.2	50.9	12.4	0.6	12.4
Liberia	0.1	2171.4	1.4	98.0	2.8	67.4	4.6	..	..
Libyan Arab Jamahiriya	..	..	..	..	..	..	..	..	65.8
Madagascar	1.0	69.2	1.6	14.6	1.8	47.1	2.7	101.9	5.5
Malawi	0.8	56.3	1.3	39.8	1.8	85.2	3.3	..	..
Mali	0.5	355.4	2.3	59.8	3.6	112.8	7.7	..	..
Mauritania	9.2	38.3	12.8	37.4	17.5	38.5	24.3	38.1	33.5
Mauritius	28.8	-7.1	26.7	55.2	41.5	37.6	57.1	7.4	61.3
Mayotte	13.6	56.1	21.2	5.6	22.4	26.6	28.3	..	..
Morocco	21.3	17.1	25.0	25.2	31.2	30.9	40.9	27.4	52.1

Table 1.12 *(continued)*

	2002	% change 2002–2003	2003	% change 2003–2004	2004	% change 2004–2005	2005	% change 2005–2006	2006
DEVELOPING ECONOMIES									
Mozambique	1.4	67.1	2.4	58.6	3.7	65.1	6.2	88.2	11.6
Namibia	8.0	46.0	11.6	22.2	14.2	71.3	24.4	..	..
Niger	0.1	339.8	0.6	91.6	1.2	79.9	2.1	4.5	2.2
Nigeria	1.3	90.7	2.6	181.7	7.2	96.5	14.1	70.1	24.1
Rwanda	1.0	54.3	1.6	5.1	1.6	96.1	3.2	..	..
Sao Tome and Principe	1.4	138.4	3.3	56.6	5.1	52.8	7.8	..	..
Senegal	4.5	23.0	5.6	78.9	9.9	49.2	14.8	68.4	25.0
Seychelles	53.9	8.7	58.6	-1.2	57.9	14.4	66.3	21.9	80.8
Sierra Leone	1.4	61.4	2.2	-4.1	2.1	..	..	..	..
Somalia	1.3	93.5	2.6	142.1	6.3	..	..	..	..
South Africa	29.4	22.1	35.9	14.9	41.3	58.2	65.4	9.2	71.4
Sudan	0.6	172.8	1.6	91.8	3.0	86.3	5.7	123.6	12.7
Swaziland	6.6	23.8	8.2	28.0	10.5	69.9	17.8	23.2	21.9
Togo	3.1	24.9	3.9	43.5	5.6	26.1	7.1	51.2	10.8
Tunisia	5.9	230.5	19.4	83.9	35.7	57.7	56.3	27.7	71.9
Uganda	1.5	90.7	2.9	45.0	4.2	26.4	5.3	27.1	6.7
United Republic of Tanzania	2.1	34.3	2.8	54.6	4.4	102.9	8.8	80.8	16.0
Zambia	1.3	70.4	2.1	22.4	2.6	145.0	6.4	25.0	8.0
Zimbabwe	2.9	6.1	3.1	8.2	3.3	59.4	5.3	17.1	6.2
ASIA									
Afghanistan	0.1	661.2	0.7	185.9	2.1	91.4	4.0	101.8	8.1
Bahrain	55.6	12.3	62.4	44.6	90.2	13.6	102.6	18.4	121.5
Bangladesh	0.8	24.6	1.0	211.1	3.1	104.1	6.3	108.7	13.2
Bhutan	..	..	1.3	108.9	2.8	101.4	5.6	113.3	12.0
Brunei Darussalam	45.2	12.2	50.7	11.0	56.3	11.9	62.9	6.2	66.8
Cambodia	2.9	28.2	3.7	69.0	6.3	20.1	7.6	3.7	7.9
China	15.9	30.2	20.8	23.3	25.6	16.8	29.9	16.8	34.9
Dem. People's Republic of Korea	..	..	..	..	..	..	..	..	..
Hong Kong (China)	94.2	14.6	107.9	9.1	117.8	3.9	122.4	6.1	129.8
India	1.2	103.0	2.4	78.1	4.4	58.3	6.9	115.3	14.8
Indonesia	5.5	58.5	8.7	54.2	13.5	56.2	21.1	34.4	28.3
Iran (Islamic Republic of)	3.3	52.2	5.1	24.2	6.3	64.5	10.4	84.4	19.2
Iraq	0.1	289.1	0.3	591.4	2.2	..	..	..	..
Jordan	22.9	5.5	24.2	17.5	28.4	93.7	55.0	35.1	74.4
Kuwait	52.0	10.1	57.3	37.0	78.4	12.8	88.5	..	..
Lao PDR	1.0	98.2	2.0	78.4	3.5	205.7	10.8	..	..
Lebanon	22.7	3.4	23.4	6.8	25.0	10.6	27.7	10.5	30.6
Macao (China)	62.8	28.9	80.9	14.9	92.9	17.0	108.7	12.5	122.3
Malaysia	37.7	17.7	44.4	28.6	57.1	33.4	76.2	-1.0	75.4
Maldives	15.0	53.2	22.9	64.7	37.7	34.1	50.6	69.5	85.8
Mongolia	8.9	45.9	13.0	25.7	16.3	29.0	21.0	..	..
Myanmar	0.1	38.3	0.1	35.4	0.2	95.4	0.4	..	..
Nepal	0.1	125.5	0.2	240.3	0.7	15.8	0.8	349.2	3.8
Oman	18.7	26.6	23.6	34.6	31.8	63.0	51.9	34.3	69.7

Table 1.12 (continued)

	2002	% change 2002–2003	2003	% change 2003–2004	2004	% change 2004–2005	2005	% change 2005–2006	2006
DEVELOPING ECONOMIES									
Pakistan	1.2	38.8	1.6	105.0	3.3	151.9	8.3	165.0	22.0
Palestine	9.2	43.4	13.3	99.7	26.5	11.7	29.6	..	..
Philippines	19.4	43.5	27.8	43.5	39.8	3.6	41.3	19.2	49.2
Qatar	39.8	33.2	53.0	24.9	66.3	38.7	91.9	19.2	109.5
Republic of Korea	67.9	3.4	70.2	8.4	76.1	4.3	79.4	4.4	82.9
Saudi Arabia	22.1	40.6	31.0	23.5	38.3	41.3	54.1	44.2	78.1
Singapore	80.4	3.7	83.4	10.8	92.4	9.1	100.8	8.5	109.3
Sri Lanka	4.9	47.2	7.2	57.0	11.4	42.7	16.2	59.7	25.9
Syrian Arab Republic	2.3	189.2	6.8	90.6	12.9	20.4	15.5	54.7	24.0
Taiwan Province of China	108.3	5.4	114.2	-12.1	100.3	-2.9	97.4	4.7	102.0
Thailand	25.8	52.9	39.4	9.2	43.0	12.8	48.5	30.0	63.0
Turkey	33.2	17.9	39.1	22.8	48.1	24.0	59.6	..	..
United Arab Emirates	64.7	13.6	73.6	17.0	86.1	17.1	100.8	17.5	118.4
Viet Nam	2.4	42.2	3.3	78.5	6.0	79.0	10.7	70.1	18.2
Yemen	2.1	64.8	3.5	48.9	5.2	84.3	9.5	..	..
LATIN AMERICA AND THE CARIBBEAN									
Antigua and Barbuda	47.8	17.7	56.2	15.7	65.1	57.4	102.4	17.2	120.0
Argentina	17.5	18.3	20.7	70.7	35.3	62.0	57.3	40.6	80.5
Aruba	61.8	12.1	69.3	39.2	96.5	8.9	105.0	..	..
Bahamas	39.3	-4.5	37.5	55.0	58.1	22.5	71.2	..	..
Barbados	34.0	43.0	48.6	42.0	69.0	2.3	70.6	..	..
Belize	19.9	16.8	23.2	61.8	37.6	17.8	44.3	-4.6	42.3
Bolivia	11.8	22.4	14.5	38.2	20.0	32.0	26.4	..	..
Brazil	19.5	31.1	25.6	39.5	35.7	29.7	46.3	14.4	52.9
Cayman Islands	..	..	..	..	..	..	..	..	..
Chile	42.8	15.3	49.4	25.7	62.1	9.2	67.8	11.5	75.6
Colombia	10.6	33.1	14.1	62.4	22.9	108.4	47.8	34.5	64.3
Costa Rica	12.5	49.3	18.7	16.4	21.7	17.1	25.4	29.0	32.8
Cuba	0.2	97.5	0.3	114.2	0.7	76.9	1.2	13.3	1.4
Dominica	17.9	73.3	31.0	98.3	61.5	..	..	..	..
Dominican Republic	19.9	23.1	24.5	17.9	28.8	41.1	40.7	25.6	51.1
Ecuador	12.0	52.5	18.3	88.2	34.5	37.0	47.2	33.9	63.2
El Salvador	13.8	25.9	17.3	60.1	27.7	26.4	35.1	57.0	55.0
Grenada	7.4	454.5	41.1	1.4	41.6	7.3	44.7	..	..
Guatemala	13.1	25.7	16.5	51.5	25.0	43.0	35.8	..	..
Guyana	11.9	35.7	16.1	21.1	19.5	73.4	33.8	12.7	38.1
Haiti	1.7	139.2	4.0	20.7	4.9	20.4	5.9	..	..
Honduras	4.9	13.1	5.6	81.1	10.1	76.2	17.8	71.1	30.4
Jamaica	45.3	33.7	60.6	35.4	82.1	24.1	101.9	3.5	105.4
Mexico	25.8	14.4	29.5	24.3	36.6	21.0	44.3	18.7	52.6
Netherlands Antilles	..	..	108.7	-0.5	108.1	-0.5	107.5	..	..
Nicaragua	4.6	93.4	8.8	55.3	13.7	49.8	20.5	59.4	32.7
Panama	17.5	53.0	26.7	1.0	27.0	55.0	41.9	23.0	51.5
Paraguay	28.8	3.5	29.9	-1.6	29.4	4.3	30.6	67.5	51.3

Table 1.12 (continued)

	2002	% change 2002–2003	2003	% change 2003–2004	2004	% change 2004–2005	2005	% change 2005–2006	2006
DEVELOPING ECONOMIES									
Peru	8.6	23.9	10.7	38.0	14.7	35.4	20.0	50.0	30.0
Saint Kitts and Nevis	10.9	-2.1	10.6	95.8	20.8	..	..	..	..
Saint Lucia	9.2	-1.4	9.1	542.2	58.5	12.2	65.7	..	..
Saint Vincent and the Grenadines	8.6	524.9	53.8	-10.2	48.3	23.0	59.3	23.0	73.0
Suriname	24.6	54.1	38.0	25.2	47.5	8.4	51.5	35.7	69.9
Trinidad and Tobago	27.7	33.7	37.0	32.8	49.2	22.8	60.4	105.5	124.1
Uruguay	15.9	-3.4	15.4	20.2	18.5	91.9	35.5	87.9	66.8
Venezuela	25.6	6.5	27.3	17.8	32.2	45.2	46.7	47.8	69.0
Virgin Islands (United States)	42.6	8.1	46.1	28.3	59.1	23.5	73.0	..	..
OCEANIA									
American Samoa	3.4	1.5	3.5	1.4	3.5	..	..	..	..
Fiji	11.2	21.0	13.5	28.1	17.3	42.8	24.8	..	..
French Polynesia	21.8	12.8	24.5	56.5	38.4	22.1	46.9	24.8	58.5
Kiribati	0.6	3.8	0.6	11.5	0.7	..	..	..	..
Marshall Islands	1.1	6.3	1.1	-3.3	1.1	..	..	..	..
Micronesia (Fed. States of)	0.1	5 714.7	5.4	115.8	11.7	9.3	12.8	..	..
Nauru	..	..	..	..	..	..	..	..	..
New Caledonia	35.8	19.1	42.6	17.6	50.1	13.2	56.8	-1.6	55.9
Northern Mariana Islands	23.1	5.9	24.5	7.4	26.3	-2.5	25.6	..	..
Palau	..	..	..	..	..	..	..	..	..
Papua New Guinea	0.3	12.9	0.3	166.5	0.8	54.0	1.3	..	..
Samoa	1.5	284.6	5.8	50.7	8.8	48.4	13.0	..	..
Solomon Islands	0.2	45.7	0.3	97.2	0.6	95.8	1.3	..	..
Tonga	3.4	233.3	11.4	46.1	16.6	81.6	30.2	..	..
Tuvalu	..	..	..	..	..	..	..	..	..
Vanuatu	2.5	55.3	3.8	31.5	5.0	18.0	5.9	-4.8	5.6
TRANSITION ECONOMIES									
Albania	27.6	29.7	35.8	10.2	39.5	23.8	48.9	..	..
Armenia	2.3	61.0	3.8	78.3	6.7	56.9	10.5	..	..
Azerbaijan	9.6	32.3	12.7	67.7	21.4	24.9	26.7	47.2	39.2
Belarus	4.7	142.9	11.3	101.5	22.8	83.9	42.0	46.3	61.4
Bosnia and Herzegovina	19.6	39.9	27.4	32.7	36.4	12.1	40.8	18.4	48.3
Bulgaria	33.1	35.6	44.9	35.8	61.0	32.5	80.8	33.2	107.6
Croatia	53.5	9.1	58.4	9.8	64.2	2.2	65.6	49.5	98.0
Georgia	10.9	42.8	15.6	19.2	18.6	75.5	32.6	17.8	38.5
Kazakhstan	6.9	29.8	9.0	107.6	18.6	79.8	33.4	58.1	52.9
Kyrgyzstan	1.1	158.9	2.7	114.8	5.9	74.7	10.3	..	..
Moldova	8.0	41.4	11.3	64.2	18.5	40.5	25.9	24.6	32.3
Montenegro	..	..	..	..	..	..	..	..	132.5
Romania	23.5	38.5	32.5	45.2	47.1	31.1	61.8	..	..
Russian Federation	12.1	108.3	25.2	104.9	51.7	61.2	83.4	..	..
Serbia	36.7	32.1	48.5	30.1	63.1	-11.4	55.9	17.1	65.5
Tajikistan	0.2	251.9	0.7	194.3	2.1	90.0	4.1	..	..

Table ·1.12 *(continued)*

	2002	% change 2002–2003	2003	% change 2003–2004	2004	% change 2004–2005	2005	% change 2005–2006	2006
TRANSITION ECONOMIES									
TFYR Macedonia	18.1	111.4	38.2	27.0	48.6	28.0	62.1	11.8	69.5
Turkmenistan	0.2	11.9	0.2	437.6	1.0	114.4	2.2	..	..
Ukraine	7.7	77.9	13.7	112.2	29.0	26.4	36.7	186.8	105.2
Uzbekistan	0.7	69.5	1.3	64.0	2.1	31.8	2.7	..	..

Source: UNCTAD calculations based on the ITU World Telecommunication/ICT Indicators database, 2007.

Table 1.13

Economies ranked by 2006 mobile phone penetration

Mobile phone subscribers per 100 inhabitants

		2002	% change 2002–2003	2003	% change 2003–2004	2004	% change 2004–2005	2005	% change 2005–2006	2006
1	Luxembourg	105.1	14.0	119.8	17.2	140.4	11.5	156.5	-3.0	151.9
2	Lithuania	47.3	33.0	62.9	57.7	99.2	28.0	126.9	8.7	138.0
3	Montenegro	..	..	..	..	..	..	..	..	132.5
4	Hong Kong (China)	94.2	14.6	107.9	9.1	117.8	3.9	122.4	6.1	129.8
5	Trinidad and Tobago	27.7	33.7	37.0	32.8	49.2	22.8	60.4	105.5	124.1
6	Estonia	64.7	19.7	77.4	20.0	92.9	15.4	107.3	15.0	123.4
7	Italy	96.0	2.2	98.1	10.3	108.2	13.8	123.1	-0.1	123.0
8	Macao (China)	62.8	28.9	80.9	14.9	92.9	17.0	108.7	12.5	122.3
9	Bahrain	55.6	12.3	62.4	44.6	90.2	13.6	102.6	18.4	121.5
10	Antigua and Barbuda	47.8	17.7	56.2	15.7	65.1	57.4	102.4	17.2	120.0
11	United Arab Emirates	64.7	13.6	73.6	17.0	86.1	17.1	100.8	17.5	118.4
12	Czech Republic	84.4	12.8	95.2	10.8	105.5	9.2	115.2	2.5	118.1
13	Israel	95.4	0.7	96.0	9.1	104.8	7.9	113.0	4.0	117.5
14	United Kingdom (incl. Gibraltar)	84.1	5.9	89.1	14.7	102.2	7.4	109.7	6.1	116.4
15	Portugal	81.9	16.9	95.8	2.1	97.8	11.6	109.1	6.3	116.0
16	Austria	83.2	4.8	87.2	11.7	97.3	1.1	98.4	13.3	111.5
17	Ireland	76.3	15.2	87.9	6.4	93.6	8.4	101.4	9.8	111.4
18	Iceland	89.8	7.4	96.4	4.2	100.5	4.3	104.8	4.5	109.5
19	Qatar	39.8	33.2	53.0	24.9	66.3	38.7	91.9	19.2	109.5
20	Singapore	80.4	3.7	83.4	10.8	92.4	9.1	100.8	8.5	109.3
21	Norway	86.0	5.8	90.9	8.2	98.4	4.6	102.9	5.6	108.6
22	Finland	86.7	4.9	90.9	5.1	95.6	4.3	99.6	8.2	107.8
23	Bulgaria	33.1	35.6	44.9	35.8	61.0	32.5	80.8	33.2	107.6
24	Denmark (incl. Faroe Islands)	83.2	6.0	88.2	7.3	94.6	6.3	100.6	6.5	107.1
25	Jamaica	45.3	33.7	60.6	35.4	82.1	24.1	101.9	3.5	105.4
26	Ukraine	7.7	77.9	13.7	112.2	29.0	26.4	36.7	186.8	105.2
27	Spain	81.6	6.9	87.2	2.5	89.4	3.4	92.4	10.7	102.3
28	Taiwan Province of China	108.3	5.4	114.2	-12.1	100.3	-2.9	97.4	4.7	102.0
29	Germany	71.6	9.6	78.5	10.1	86.4	10.8	95.8	6.4	101.9
30	Greece	84.5	6.7	90.2	10.3	99.5	-7.3	92.3	8.0	99.6
31	Switzerland	78.8	6.7	84.1	.6	84.6	8.5	91.8	8.0	99.1
32	Hungary	67.9	15.6	78.5	10.1	86.4	6.8	92.3	7.2	99.0
33	Croatia	53.5	9.1	58.4	9.8	64.2	2.2	65.6	49.5	98.0
34	Australia	64.0	12.7	72.2	14.5	82.6	10.6	91.4	6.2	97.0
35	Poland	36.4	25.3	45.6	32.8	60.5	26.7	76.7	25.7	96.4
36	Jersey	70.1	32.2	92.7	3.2	95.7	-0.6	95.1	0.0	95.1
37	Latvia	39.4	33.5	52.6	27.7	67.1	20.7	81.0	17.2	94.9
38	Bermuda	49.2	31.2	64.5	20.6	77.8	5.9	82.3	12.3	92.5
39	Belgium	78.2	5.8	82.7	5.6	87.4	3.5	90.5	1.9	92.2
40	Cyprus	52.9	30.4	69.0	13.3	78.1	10.1	86.0	5.8	90.9
41	Slovakia	54.3	25.8	68.4	16.1	79.4	6.2	84.3	7.7	90.8
42	Slovenia	83.5	4.3	87.1	6.3	92.5	-5.1	87.8	2.1	89.6
43	New Zealand	62.2	5.9	65.8	15.3	75.9	15.5	87.6	-0.7	86.9

Table 1.13 *(continued)*

		2002	% change 2002–2003	2003	% change 2003–2004	2004	% change 2004–2005	2005	% change 2005–2006	2006
44	Maldives	15.0	53.2	22.9	64.7	37.7	34.1	50.6	69.5	85.8
45	Malta	71.6	4.2	74.6	5.1	78.4	5.5	82.7	3.1	85.2
46	France (incl. Guadeloupe. French Guiana. Martinique and Réunion)	66.7	5.7	70.5	6.4	75.0	5.9	79.4	7.1	85.1
47	Republic of Korea	67.9	3.4	70.2	8.4	76.1	4.3	79.4	4.4	82.9
48	Seychelles	53.9	8.7	58.6	-1.2	57.9	14.4	66.3	21.9	80.8
49	Argentina	17.5	18.3	20.7	70.7	35.3	62.0	57.3	40.6	80.5
50	Japan	63.7	6.7	67.9	5.5	71.6	3.5	74.2	7.3	79.6
51	Andorra	48.8	47.0	71.8	17.0	84.0	-2.1	82.2	-3.2	79.5
52	Saudi Arabia	22.1	40.6	31.0	23.5	38.3	41.3	54.1	44.2	78.1
53	United States (incl. Puerto Rico and Guam)	48.8	10.3	53.8	14.7	61.7	8.0	66.7	16.2	77.5
54	Chile	42.8	15.3	49.4	25.7	62.1	9.2	67.8	11.5	75.6
55	Malaysia	37.7	17.7	44.4	28.6	57.1	33.4	76.2	-1.0	75.4
56	Jordan	22.9	5.5	24.2	17.5	28.4	93.7	55.0	35.1	74.4
57	Guernsey	61.0	13.5	69.2	5.5	73.0	0.0	73.0	0.0	73.0
58	Saint Vincent and the Grenadines	8.6	524.9	53.8	-10.2	48.3	23.0	59.3	23.0	73.0
59	Tunisia	5.9	230.5	19.4	83.9	35.7	57.7	56.3	27.7	71.9
60	South Africa	29.4	22.1	35.9	14.9	41.3	58.2	65.4	9.2	71.4
61	Suriname	24.6	54.1	38.0	25.2	47.5	8.4	51.5	35.7	69.9
62	Oman	18.7	26.6	23.6	34.6	31.8	63.0	51.9	34.3	69.7
63	TFYR Macedonia	18.1	111.4	38.2	27.0	48.6	28.0	62.1	11.8	69.5
64	Venezuela	25.6	6.5	27.3	17.8	32.2	45.2	46.7	47.8	69.0
65	Brunei Darussalam	45.2	12.2	50.7	11.0	56.3	11.9	62.9	6.2	66.8
66	Uruguay	15.9	-3.4	15.4	20.2	18.5	91.9	35.5	87.9	66.8
67	Libyan Arab Jamahiriya	..	..	..	..	..	..	..	..	65.8
68	Serbia	36.7	32.1	48.5	30.1	63.1	-11.4	55.9	17.1	65.5
69	Colombia	10.6	33.1	14.1	62.4	22.9	108.4	47.8	34.5	64.3
70	Ecuador	12.0	52.5	18.3	88.2	34.5	37.0	47.2	33.9	63.2
71	Thailand	25.8	52.9	39.4	9.2	43.0	12.8	48.5	30.0	63.0
72	Algeria	1.3	254.8	4.5	219.8	14.5	186.9	41.5	51.7	63.0
73	Dominica	17.9	73.3	31.0	98.3	61.5	0.0	61.5	1.5	62.4
74	Belarus	4.7	142.9	11.3	101.5	22.8	83.9	42.0	46.3	61.4
75	Mauritius	28.8	-7.1	26.7	55.2	41.5	37.6	57.1	7.4	61.3
76	French Polynesia	21.8	12.8	24.5	56.5	38.4	22.1	46.9	24.8	58.5
77	San Marino	59.9	-0.9	59.3	-0.6	59.0	-2.8	57.3	-2.2	56.1
78	New Caledonia	35.8	19.1	42.6	17.6	50.1	13.2	56.8	-1.6	55.9
79	El Salvador	13.8	25.9	17.3	60.1	27.7	26.4	35.1	57.0	55.0
80	Gabon	21.5	4.2	22.4	61.9	36.2	29.9	47.1	15.2	54.2
81	Brazil	19.5	31.1	25.6	39.5	35.7	29.7	46.3	14.4	52.9
82	Kazakhstan	6.9	29.8	9.0	107.6	18.6	79.8	33.4	58.1	52.9
83	Mexico	25.8	14.4	29.5	24.3	36.6	21.0	44.3	18.7	52.6
84	Canada	37.7	10.4	41.6	12.1	46.7	9.7	51.2	2.0	52.2
85	Morocco	21.3	17.1	25.0	25.2	31.2	30.9	40.9	27.4	52.1
86	Botswana	24.7	18.2	29.2	7.2	31.3	43.1	44.8	16.1	52.1
87	Panama	17.5	53.0	26.7	1.0	27.0	55.0	41.9	23.0	51.5

Table 1.13 *(continued)*

		2002	% change 2002–2003	2003	% change 2003–2004	2004	% change 2004–2005	2005	% change 2005–2006	2006
88	Paraguay	28.8	3.5	29.9	-1.6	29.4	4.3	30.6	67.5	51.3
89	Dominican Republic	19.9	23.1	24.5	17.9	28.8	41.1	40.7	25.6	51.1
90	Philippines	19.4	43.5	27.8	43.5	39.8	3.6	41.3	19.2	49.2
91	Bosnia and Herzegovina	19.6	39.9	27.4	32.7	36.4	12.1	40.8	18.4	48.3
92	Belize	19.9	16.8	23.2	61.8	37.6	17.8	44.3	-4.6	42.3
93	Azerbaijan	9.6	32.3	12.7	67.7	21.4	24.9	26.7	47.2	39.2
94	Georgia	10.9	42.8	15.6	19.2	18.6	75.5	32.6	17.8	38.5
95	Guyana	11.9	35.7	16.1	21.1	19.5	73.4	33.8	12.7	38.1
96	China	15.9	30.2	20.8	23.3	25.6	16.8	29.9	16.8	34.9
97	Mauritania	9.2	38.3	12.8	37.4	17.5	38.5	24.3	38.1	33.5
98	Costa Rica	12.5	49.3	18.7	16.4	21.7	17.1	25.4	29.0	32.8
99	Nicaragua	4.6	93.4	8.8	55.3	13.7	49.8	20.5	59.4	32.7
100	Moldova	8.0	41.4	11.3	64.2	18.5	40.5	25.9	24.6	32.3
101	Lebanon	22.7	3.4	23.4	6.8	25.0	10.6	27.7	10.5	30.6
102	Honduras	4.9	13.1	5.6	81.1	10.1	76.2	17.8	71.1	30.4
103	Peru	8.6	23.9	10.7	38.0	14.7	35.4	20.0	50.0	30.0
104	Palestine	9.2	43.4	13.3	99.7	26.5	11.7	29.6	0.0	29.6
105	Indonesia	5.5	58.5	8.7	54.2	13.5	56.2	21.1	34.4	28.3
106	Gambia	7.5	46.0	11.0	9.2	12.0	35.8	16.3	59.2	25.9
107	Sri Lanka	4.9	47.2	7.2	57.0	11.4	42.7	16.2	59.7	25.9
108	Senegal	4.5	23.0	5.6	78.9	9.9	49.2	14.8	68.4	25.0
109	Nigeria	1.3	90.7	2.6	181.7	7.2	96.5	14.1	70.1	24.1
110	Syrian Arab Republic	2.3	189.2	6.8	90.6	12.9	20.4	15.5	54.7	24.0
111	Egypt	6.7	26.5	8.4	29.3	10.9	75.0	19.1	24.9	23.9
112	Ghana	1.9	100.3	3.7	112.1	7.9	0.7	8.0	189.1	23.1
113	Côte d'Ivoire	6.2	23.6	7.7	17.7	9.1	33.1	12.1	82.6	22.0
114	Pakistan	1.2	38.8	1.6	105.0	3.3	151.9	8.3	165.0	22.0
115	Swaziland	6.6	23.8	8.2	28.0	10.5	69.9	17.8	23.2	21.9
116	Cape Verde	9.5	21.5	11.6	20.7	14.0	21.6	17.0	23.0	20.9
117	Iran (Islamic Republic of)	3.3	52.2	5.1	24.2	6.3	64.5	10.4	84.4	19.2
118	Kenya	3.8	33.1	5.0	54.7	7.8	73.5	13.5	37.2	18.5
119	Viet Nam	2.4	42.2	3.3	78.5	6.0	79.0	10.7	70.1	18.2
120	United Republic of Tanzania	2.1	34.3	2.8	54.6	4.4	102.9	8.8	80.8	16.0
121	India	1.2	103.0	2.4	78.1	4.4	58.3	6.9	115.3	14.8
122	Angola	0.9	148.5	2.3	188.1	6.7	5.3	7.0	103.9	14.3
123	Bangladesh	0.8	24.6	1.0	211.1	3.1	104.1	6.3	108.7	13.2
124	Sudan	0.6	172.8	1.6	91.8	3.0	86.3	5.7	123.6	12.7
125	Lesotho	5.1	4.2	5.3	53.5	8.2	50.9	12.4	0.6	12.4
126	Bhutan	..	..	1.3	108.9	2.8	101.4	5.6	113.3	12.0
127	Mozambique	1.4	67.1	2.4	58.6	3.7	65.1	6.2	88.2	11.6
128	Togo	3.1	24.9	3.9	43.5	5.6	26.1	7.1	51.2	10.8
129	Armenia	2.3	61.0	3.8	78.3	6.7	56.9	10.5	0.3	10.6
130	Afghanistan	0.1	661.2	0.7	185.9	2.1	91.4	4.0	101.8	8.1
131	Zambia	1.3	70.4	2.1	22.4	2.6	145.0	6.4	25.0	8.0
132	Cambodia	2.9	28.2	3.7	69.0	6.3	20.1	7.6	3.7	7.9

Table 1.13 *(continued)*

		2002	% change 2002–2003	2003	% change 2003–2004	2004	% change 2004–2005	2005	% change 2005–2006	2006
133	Burkina Faso	0.9	96.0	1.9	60.5	3.0	45.5	4.3	72.5	7.5
134	Uganda	1.5	90.7	2.9	45.0	4.2	26.4	5.3	27.1	6.7
135	Zimbabwe	2.9	6.1	3.1	8.2	3.3	59.4	5.3	17.1	6.2
136	Guinea-Bissau	..	..	0.1	3095.7	3.2	57.1	5.0	16.6	5.8
137	Vanuatu	2.5	55.3	3.8	31.5	5.0	18.0	5.9	-4.8	5.6
138	Madagascar	1.0	69.2	1.6	14.6	1.8	47.1	2.7	101.9	5.5
139	Chad	0.4	85.1	0.8	72.8	1.4	55.0	2.2	115.8	4.6
140	Nepal	0.1	125.5	0.2	240.3	0.7	15.8	0.8	349.2	3.8
141	Niger	0.1	339.8	0.6	91.6	1.2	79.9	2.1	4.5	2.2
142	Comoros	..	..	0.3	364.1	1.2	68.8	2.0	-2.2	2.0
143	Eritrea	..	..	..	..	0.5	93.9	0.9	47.9	1.4
144	Cuba	0.2	97.5	0.3	114.2	0.7	76.9	1.2	13.3	1.4
145	Ethiopia	0.1	88.6	0.1	74.3	0.2	115.7	0.5	106.1	1.1
146	Netherlands Antilles	..	..	108.7	-0.5	108.1	-0.5	107.5	..	..
147	Aruba	61.8	12.1	69.3	39.2	96.5	8.9	105.0	..	..
148	Sweden	88.9	10.2	98.0	-0.5	97.5	3.1	100.5	..	..
149	Netherlands	75.9	9.1	82.8	10.3	91.3	6.4	97.1	..	..
150	Kuwait	52.0	10.1	57.3	37.0	78.4	12.8	88.5	..	..
151	Russian Federation	12.1	108.3	25.2	104.9	51.7	61.2	83.4	..	..
152	Liechtenstein	35.6	112.6	75.8	-1.0	75.0	4.8	78.6	..	..
153	Virgin Islands (United States)	42.6	8.1	46.1	28.3	59.1	23.5	73.0	..	..
154	Bahamas	39.3	-4.5	37.5	55.0	58.1	22.5	71.2	..	..
155	Barbados	34.0	43.0	48.6	42.0	69.0	2.3	70.6	..	..
156	Saint Lucia	9.2	-1.4	9.1	542.2	58.5	12.2	65.7	..	..
157	Romania	23.5	38.5	32.5	45.2	47.1	31.1	61.8	..	..
158	Turkey	33.2	17.9	39.1	22.8	48.1	24.0	59.6	..	..
159	Monaco	22.2	108.4	46.3	4.0	48.2	8.2	52.1	..	..
160	Albania	27.6	29.7	35.8	10.2	39.5	23.8	48.9	..	..
161	Grenada	7.4	454.5	41.1	1.4	41.6	7.3	44.7	..	..
162	Guatemala	13.1	25.7	16.5	51.5	25.0	43.0	35.8	..	..
163	Tonga	3.4	233.3	11.4	46.1	16.6	81.6	30.2	..	..
164	Mayotte	13.6	56.1	21.2	5.6	22.4	26.6	28.3	..	..
165	Bolivia	11.8	22.4	14.5	38.2	20.0	32.0	26.4	..	..
166	Fiji	11.2	21.0	13.5	28.1	17.3	42.8	24.8	..	..
167	Namibia	8.0	46.0	11.6	22.2	14.2	71.3	24.4	..	..
168	Mongolia	8.9	45.9	13.0	25.7	16.3	29.0	21.0	..	..
169	Equatorial Guinea	7.0	26.9	8.9	30.9	11.7	71.0	20.0	..	..
170	Cameroon	4.4	49.5	6.6	42.3	9.4	46.4	13.8	..	..
171	Samoa	1.5	284.6	5.8	50.7	8.8	48.4	13.0	..	..
172	Micronesia (Fed. States of)	0.1	5 714.7	5.4	115.8	11.7	9.3	12.8	..	..
173	Congo	6.7	40.3	9.4	6.5	10.0	22.0	12.3	..	..
174	Lao PDR	1.0	98.2	2.0	78.4	3.5	205.7	10.8	..	..
175	Kyrgyzstan	1.1	158.9	2.7	114.8	5.9	74.7	10.3	..	..
176	Benin	3.2	4.4	3.4	58.5	5.3	87.7	10.0	..	..
177	Yemen	2.1	64.8	3.5	48.9	5.2	84.3	9.5	..	..

Table 1.13 *(continued)*

		2002	% change 2002–2003	2003	% change 2003–2004	2004	% change 2004–2005	2005	% change 2005–2006	2006
178	Sao Tome and Principe	1.4	138.4	3.3	56.6	5.1	52.8	7.8	..	..
179	Mali	0.5	355.4	2.3	59.8	3.6	112.8	7.7	..	..
180	Haiti	1.7	139.2	4.0	20.7	4.9	20.4	5.9	..	..
181	Djibouti	2.1	49.2	3.1	45.2	4.5	23.2	5.5	..	..
182	Democratic Republic of the Congo	1.1	73.6	1.8	93.3	3.6	33.9	4.8	..	..
183	Liberia	0.1	2171.4	1.4	98.0	2.8	67.4	4.6	..	..
184	Tajikistan	0.2	251.9	0.7	194.3	2.1	90.0	4.1	..	..
185	Malawi	0.8	56.3	1.3	39.8	1.8	85.2	3.3	..	..
186	Rwanda	1.0	54.3	1.6	5.1	1.6	96.1	3.2	..	..
187	Uzbekistan	0.7	69.5	1.3	64.0	2.1	31.8	2.7	..	..
188	Central African Republic	0.3	203.7	1.0	58.8	1.5	61.3	2.5	..	..
189	Guinea	1.2	21.4	1.4	38.0	2.0	19.0	2.4	..	..
190	Turkmenistan	0.2	11.9	0.2	437.6	1.0	114.4	2.2	..	..
191	Burundi	0.7	20.0	0.9	60.3	1.4	42.4	2.0	..	..
192	Papua New Guinea	0.3	12.9	0.3	166.5	0.8	54.0	1.3	..	..
193	Solomon Islands	0.2	45.7	0.3	97.2	0.6	95.8	1.3	..	..
194	Marshall Islands	1.1	6.3	1.1	-3.3	1.1	-3.5	1.1	..	..
195	Myanmar	0.1	38.3	0.1	35.4	0.2	95.4	0.4	..	..

Developed economy.

Developing economy.

Transition economy.

Source: UNCTAD calculations based on the ITU World Telecommunication/ICT Indicators database, 2007.

Table 1.14

Internet users: economies by level of development and by region

	2002	% change 2002–2003	2003	% change 2003–2004	2004	% change 2004–2005	2005	% change 2005–2006	2006
DEVELOPED ECONOMIES									
ASIA									
Israel	1 125 200	12.4	1 264 500	18.4	1 496 600	12.6	1 685 900	12.6	1 899 100
Japan	59 220 000	4.1	61 640 000	28.9	79 480 000	7.3	85 290 000	2.6	87 540 000
EUROPE									
Andorra	7 000	43.6	10 049	9.5	11 000	99.3	21 922	5.8	23 200
Austria	3 340 000	11.7	3 730 000	4.6	3 900 000	2.6	4 000 000	5.0	4 200 000
Belgium	3 400 000	17.6	4 000 000	5.0	4 200 000	14.3	4 800 000	..	..
Cyprus	210 000	19.0	250 000	19.2	298 000	9.4	326 000	9.4	356 600
Czech Republic	2 600 180	-7.9	2 395 000	7.6	2 576 000	7.1	2 758 000	28.4	3 541 300
Denmark (incl. Faroe Islands)	2 415 500	3.9	2 509 000	9.9	2 757 000	4.7	2 887 000	11.0	3 205 200
Estonia	444 000	35.1	600 000	11.7	670 000	3.0	690 000	10.1	760 000
Finland	2 529 000	1.2	2 560 000	4.7	2 680 000	4.5	2 800 000	4.5	2 925 400
France (incl. Guadeloupe. French Guiana. Martinique and Réunion)	18 342 000	20.6	22 119 000	9.2	24 159 000	10.2	26 631 000	13.0	30 100 000
Germany	28 000 000	17.9	33 000 000	6.7	35 200 000	6.5	37 500 000	2.9	38 600 000
Greece	1 485 281	15.7	1 718 435	13.8	1 955 000	2.4	2 001 000	2.4	2 048 100
Guernsey	30 000	10.0	33 000	9.1	36 000	..	..	..	..
Hungary	1 600 000	50.0	2 400 000	12.5	2 700 000	11.1	3 000 000	16.7	3 500 000
Iceland	150 000	10.7	166 000	1.1	167 800	9.1	183 000	6.0	194 000
Ireland	1 102 000	14.3	1 260 000	-4.9	1 198 000	16.9	1 400 000	2.6	1 437 000
Italy	19 800 000	15.6	22 880 000	18.8	27 170 000	3.1	28 000 000	3.1	28 855 400
Jersey	..	..	20 000	35.0	27 000	..	..	..	..
Latvia	310 000	80.6	560 000	44.6	810 000	27.2	1 030 000	4.0	1 070 800
Liechtenstein	20 000	0.0	20 000	10.0	22 000	..	..	..	..
Lithuania	500 000	39.1	695 700	10.2	767 000	15.1	882 900	22.7	1 083 000
Luxembourg	165 000	3.0	170 000	59.3	270 810	16.3	315 000	7.6	339 000
Malta	80 410	19.4	96 022	16.3	111 634	14.0	127 247	..	..
Monaco	15 500	3.2	16 000	6.3	17 000	5.9	18 000	11.1	20 000
Netherlands	8 200 000	3.7	8 500 000	17.6	10 000 000	20.6	12 060 000	20.6	14 544 400
Norway	1 398 600	13.2	1 583 300	13.2	1 792 000	89.7	3 400 000	19.8	4 074 100
Poland	8 880 000	1.0	8 970 000	0.3	9 000 000	11.1	10 000 000	10.0	11 000 000
Portugal	2 267 200	17.9	2 674 000	-3.7	2 575 700	14.1	2 939 000	9.3	3 213 000
San Marino	14 340	1.0	14 481	3.6	15 000	0.0	15 000	2.7	15 400
Slovakia	862 833	59.5	1 375 809	20.1	1 652 200	15.3	1 905 200	18.4	2 255 600
Slovenia	750 000	6.7	800 000	18.8	950 000	14.7	1 090 000	14.7	1 250 600
Spain	7 856 000	24.6	9 789 000	46.4	14 332 800	5.5	15 119 000	22.9	18 578 000
Sweden	5 125 000	10.3	5 655 000	20.2	6 800 000	1.3	6 890 000	1.3	6 981 200
Switzerland	3 000 000	13.3	3 400 000	2.9	3 500 000	5.7	3 700 000	17.8	4 360 000
United Kingdom (incl. Gibraltar)	25 000 000	4.1	26 025 000	8.0	28 100 295	1.5	28 515 000	17.6	33 534 000
NORTH AMERICA									
Bermuda	..	..	36 000	8.3	39 000	7.7	42 000	..	..
Canada	15 200 000	15.8	17 600 000	13.6	20 000 000	10.0	22 000 000	..	..
Greenland	25 000	24.0	31 000	22.6	38 000	..	..	..	..

Table 1.14 *(continued)*

	2002	% change 2002–2003	2003	% change 2003–2004	2004	% change 2004–2005	2005	% change 2005–2006	2006
United States (incl. Puerto Rico and Guam)	159 727 000	1.7	162 459 400	14.4	185 931 000	6.9	198 780 600	5.1	208 980 600
OCEANIA									
Australia	10 500 000	7.6	11 300 000	15.0	13 000 000	9.2	14 190 000	7.8	15 300 000
New Zealand	1 908 000	10.6	2 110 000	11.4	2 350 000	17.2	2 754 000	16.2	3 200 000
DEVELOPING ECONOMIES									
AFRICA									
Algeria	500 000	30.0	650 000	130.8	1 500 000	28.0	1 920 000	28.1	2 460 000
Angola	41 000	0.0	41 000	82.9	75 000	13.3	85 000	..	..
Benin	50 000	40.0	70 000	42.9	100 000	325.0	425 000	64.7	700 000
Botswana	60 000	0.0	60 000	0.0	60 000	0.0	60 000	..	..
Burkina Faso	25 000	92.0	48 000	10.8	53 200	21.4	64 600	23.8	80 000
Burundi	8 000	75.0	14 000	78.6	25 000	60.0	40 000	50.0	60 000
Cameroon	60 000	66.7	100 000	70.0	170 000	47.1	250 000	48.0	370 000
Cape Verde	16 000	25.0	20 000	25.0	25 000	0.0	25 000	16.0	29 000
Central African Republic	5 000	20.0	6 000	50.0	9 000	22.2	11 000	18.2	13 000
Chad	15 000	100.0	30 000	16.7	35 000	14.3	40 000	50.0	60 000
Comoros	3 200	56.3	5 000	60.0	8 000	150.0	20 000	5.0	21 000
Congo	5 000	200.0	15 000	140.0	36 000	38.9	50 000	40.0	70 000
Côte d'Ivoire	90 000	55.6	140 000	14.3	160 000	25.0	200 000	50.0	300 000
Democratic Republic of the Congo	50 000	50.0	75 000	50.0	112 500	25.0	140 625	28.0	180 000
Djibouti	4 500	44.4	6 500	38.5	9 000	11.1	10 000	10.0	11 000
Egypt	1 900 000	57.9	3 000 000	30.0	3 900 000	28.2	5 000 000	20.0	6 000 000
Equatorial Guinea	1 800	66.7	3 000	66.7	5 000	40.0	7 000	14.3	8 000
Eritrea	9 000	5.6	9 500	426.3	50 000	40.0	70 000	42.9	100 000
Ethiopia	50 000	50.0	75 000	50.7	113 000	45.1	164 000	..	..
Gabon	25 000	40.0	35 000	14.3	40 000	67.5	67 000	20.9	81 000
Gambia	25 000	40.0	35 000	40.0	49 000	18.4	58 000	..	..
Ghana	170 000	47.1	250 000	47.2	368 000	9.1	401 310	52.0	609 800
Guinea	35 000	14.3	40 000	15.0	46 000	8.7	50 000	0.0	50 000
Guinea-Bissau	14 000	35.7	19 000	36.8	26 000	19.2	31 000	19.4	37 000
Kenya	400 000	150.0	1 000 000	5.5	1 054 920	5.3	1 111 000	149.4	2 770 300
Lesotho	21 000	42.9	30 000	43.3	43 000	19.8	51 500	..	..
Liberia	..	..	..	..	..	..	..	..	..
Libyan Arab Jamahiriya	125 000	28.0	160 000	28.1	205 000	13.2	232 000	..	..
Madagascar	55 000	28.2	70 500	27.7	90 000	11.1	100 000	10.0	110 000
Malawi	27 000	33.3	36 000	28.2	46 140	13.8	52 500	13.7	59 700
Mali	25 000	40.0	35 000	42.9	50 000	20.0	60 000	16.7	70 000
Mauritania	10 000	20.0	12 000	16.7	14 000	42.9	20 000	400.0	100 000
Mauritius	125 000	20.0	150 000	20.0	180 000	0.0	180 000	1.1	182 000
Mayotte	..	..	..	..	..	..	..	..	..
Morocco	700 000	42.9	1 000 000	250.0	3 500 000	31.4	4 600 000	32.6	6 100 000
Mozambique	50 000	66.0	83 000	66.3	138 000	29.0	178 000	..	..
Namibia	50 000	30.0	65 000	15.4	75 000	7.5	80 600	..	..
Niger	15 000	26.7	19 000	26.3	24 000	20.8	29 000	37.9	40 000

Table 1.14 *(continued)*

	2002	% change 2002–2003	2003	% change 2003–2004	2004	% change 2004–2005	2005	% change 2005–2006	2006
Nigeria	420 000	78.6	750 000	136.0	1 769 661	182.5	5 000 000	60.0	8 000 000
Rwanda	25 000	24.0	31 000	22.6	38 000	31.6	50 000	30.0	65 000
Sao Tome and Principe	11 000	36.4	15 000	33.3	20 000	15.0	23 000	..	..
Senegal	105 000	114.3	225 000	114.2	482 000	12.0	540 000	20.4	650 000
Seychelles	11 736	2.2	12 000	66.7	20 000	5.0	21 000	38.1	29 000
Sierra Leone	8 000	12.5	9 000	11.1	10 000	0.0	10 000	..	..
Somalia	86 000	4.7	90 000	0.0	90 000	0.0	90 000	4.4	94 000
South Africa	3 100 000	7.3	3 325 000	7.2	3 566 000	43.0	5 100 000	..	..
Sudan	300 000	212.3	937 000	21.7	1 140 000	145.6	2 800 000	25.0	3 500 000
Swaziland	20 000	35.0	27 000	33.3	36 000	15.6	41 600	..	..
Togo	200 000	5.0	210 000	5.2	221 000	35.7	300 000	6.7	320 000
Tunisia	505 500	24.6	630 000	32.5	835 000	14.2	953 770	35.8	1 294 900
Uganda	100 000	25.0	125 000	60.0	200 000	150.0	500 000	50.0	750 000
United Republic of Tanzania	80 000	212.5	250 000	33.2	333 000	15.4	384 300	..	..
Zambia	52 420	109.8	110 000	110.0	231 000	44.9	334 800	..	..
Zimbabwe	500 000	60.0	800 000	2.5	820 000	22.0	1 000 000	22.0	1 220 000
ASIA									
Afghanistan	1 000	1900.0	20 000	25.0	25 000	1100.0	300 000	78.3	535 000
Bahrain	122 794	22.2	150 000	1.8	152 721	1.5	155 000	1.5	157 300
Bangladesh	204 000	19.1	243 000	23.5	300 000	23.3	370 000	21.6	450 000
Bhutan	10 000	50.0	15 000	33.3	20 000	25.0	25 000	20.0	30 000
Brunei Darussalam	..	..	48 000	16.7	56 000	16.1	65 000	154.8	165 600
Cambodia	30 000	16.7	35 000	17.1	41 000	7.3	44 000	..	..
China	59 100 000	34.5	79 500 000	18.2	94 000 000	18.1	111 000 000	23.4	137 000 000
Dem. People's Republic of Korea	..	..	..	..	..	..	..	..	..
Hong Kong (China)	2 918 800	10.1	3 212 800	8.3	3 479 700	1.3	3 526 200	6.9	3 770 400
India	16 580 000	11.5	18 481 044	89.4	35 000 000	71.4	60 000 000	..	..
Indonesia	4 500 000	79.6	8 080 000	38.9	11 226 143	42.5	16 000 000	..	..
Iran (Islamic Republic of)	3 168 000	51.5	4 800 000	14.6	5 500 000	27.3	7 000 000	157.1	18 000 000
Iraq	25 000	20.0	30 000	20.0	36 000	..	..	..	..
Jordan	307 469	44.4	444 000	41.8	629 524	14.3	719 800	10.7	796 900
Kuwait	250 000	126.8	567 000	5.8	600 000	16.7	700 000	16.7	816 700
Lao PDR	15 000	26.7	19 000	10.0	20 900	19.6	25 000	..	..
Lebanon	400 000	25.0	500 000	20.0	600 000	16.7	700 000	35.7	950 000
Macao (China)	115 000	4.3	120 000	25.0	150 000	13.3	170 000	17.6	200 000
Malaysia	7 842 000	10.2	8 643 000	14.3	9 879 000	11.5	11 016 000	2.5	11 292 000
Maldives	15 000	13.3	17 000	11.8	19 000	5.8	20 100	..	..
Mongolia	50 000	185.6	142 800	40.1	200 000	34.2	268 300	..	..
Myanmar	200	5650.0	11 500	1.7	11 700	169.2	31 500	..	..
Nepal	80 000	25.0	100 000	20.0	120 000	-6.3	112 500	121.7	249 400
Oman	180 000	16.7	210 000	16.7	245 000	16.3	285 000	12.0	319 200
Pakistan	1 000 000	700.0	8 000 000	25.0	10 000 000	5.0	10 500 000	14.3	12 000 000
Palestine	105 000	38.1	145 000	10.3	160 000	51.9	243 000	..	..
Philippines	3 500 000	14.3	4 000 000	10.0	4 400 000	4.9	4 614 800	..	..
Qatar	70 000	101.1	140 760	17.2	165 000	32.7	219 000	32.4	289 900

Table 1.14 *(continued)*

	2002	% change 2002–2003	2003	% change 2003–2004	2004	% change 2004–2005	2005	% change 2005–2006	2006
Republic of Korea	26 270 000	11.2	29 220 000	8.1	31 580 000	4.5	33 010 000	3.4	34 120 000
Saudi Arabia	1 418 880	5.7	1 500 000	5.7	1 586 000	89.2	3 000 000	56.7	4 700 000
Singapore[a]	2 100 000	1.7	2 135 034	13.4	2 421 782	-28.5	1 731 600	-0.8	1 717 100
Sri Lanka	200 000	25.0	250 000	12.0	280 000	25.0	350 000	22.3	428 000
Syrian Arab Republic	365 000	67.1	610 000	31.1	800 000	37.5	1 100 000	36.4	1 500 000
Taiwan Province of China	10 720 000	9.5	11 740 000	4.0	12 210 000	8.2	13 210 000	..	..
Thailand	4 800 000	25.6	6 030 000	15.6	6 971 500	1.6	7 084 200	19.5	8 465 800
Turkey	4 300 000	39.5	6 000 000	70.3	10 220 000	9.6	11 204 300	9.6	12 283 500
United Arab Emirates	1 175 516	-5.6	1 110 207	11.6	1 238 464	12.8	1 397 207	22.3	1 708 500
Viet Nam	1 500 000	106.5	3 098 007	104.8	6 345 049	68.8	10 710 980	37.1	14 683 800
Yemen	100 000	20.0	120 000	50.0	180 000	22.5	220 500	22.4	270 000
LATIN AMERICA AND THE CARIBBEAN									
Antigua and Barbuda	10 000	40.0	14 000	42.9	20 000	45.0	29 000	10.3	32 000
Argentina	4 100 000	10.5	4 530 000	35.8	6 153 603	11.5	6 863 466	19.2	8 183 700
Aruba	24 000	0.0	24 000	0.0	24 000	0.0	24 000	..	..
Bahamas	60 000	40.0	84 000	10.7	93 000	10.8	103 000	..	..
Barbados	30 000	233.3	100 000	50.0	150 000	6.7	160 000	..	..
Belize	..	..	..	..	16 000	62.5	26 000	30.8	34 000
Bolivia	270 000	14.8	310 000	29.0	400 000	20.0	480 000	20.8	580 000
Brazil	14 300 000	25.9	18 000 000	22.2	22 000 000	65.3	36 356 000	17.2	42 600 000
Cayman Islands	..	..	..	..	..	..	..	..	..
Chile	3 575 000	11.9	4 000 000	7.5	4 300 000	4.9	4 510 900	-7.9	4 155 600
Colombia	2 000 113	36.6	2 732 201	41.5	3 865 860	22.6	4 738 544	41.5	6 705 000
Costa Rica	815 745	10.3	900 000	11.1	1 000 000	10.0	1 100 000	10.4	1 214 400
Cuba	160 000	-38.8	98 000	53.1	150 000	26.7	190 000	26.3	240 000
Dominica	12 500	36.0	17 000	20.6	20 500	26.8	26 000	0.0	26 000
Dominican Republic	500 000	30.0	650 000	23.1	800 000	87.5	1 500 000	33.3	2 000 000
Ecuador	537 881	5.9	569 727	9.6	624 579	55.0	968 000	60.0	1 549 000
El Salvador	300 000	83.3	550 000	6.8	587 475	8.4	637 050	..	..
Grenada	15 000	26.7	19 000	..	..	..	..	..	..
Guatemala	400 000	37.5	550 000	38.2	760 000	31.6	1 000 000	32.0	1 320 000
Guyana	125 000	12.0	140 000	3.6	145 000	10.3	160 000	..	..
Haiti	80 000	87.5	150 000	233.3	500 000	20.0	600 000	8.3	650 000
Honduras	168 560	10.1	185 510	21.3	225 000	15.6	260 000	29.7	337 300
Jamaica	600 000	33.3	800 000	33.4	1 067 000	15.5	1 232 300	..	..
Mexico	10 764 715	13.5	12 218 830	14.9	14 036 475	32.7	18 622 509	18.1	22 000 000
Netherlands Antilles	..	..	..	..	..	..	..	..	..
Nicaragua	90 000	11.1	100 000	25.0	125 000	12.0	140 000	10.7	155 000
Panama	144 963	19.4	173 085	13.6	196 548	4.9	206 178	6.7	220 000
Paraguay	100 000	20.0	120 000	66.7	200 000	0.0	200 000	30.0	260 000
Peru	2 400 000	18.8	2 850 000	13.0	3 220 000	42.9	4 600 000	32.6	6 100 000
Saint Kitts and Nevis	10 000	..	..	..	..	..	..	..	..
Saint Lucia	..	..	34 000	61.8	55 000	..	..	..	..
Saint Vincent and the Grenadines	6 000	16.7	7 000	14.3	8 000	25.0	10 000	..	..
Suriname	20 000	15.0	23 000	30.4	30 000	6.7	32 000	..	..

Table 1.14 *(continued)*

	2002	% change 2002–2003	2003	% change 2003–2004	2004	% change 2004–2005	2005	% change 2005–2006	2006
Trinidad and Tobago	138 000	10.9	153 000	4.6	160 000	1.9	163 000	..	..
Uruguay	380 000	39.5	530 000	7.0	567 175	17.8	668 000	13.2	756 000
Venezuela	1 244 000	55.5	1 935 000	14.1	2 207 000	50.1	3 313 000	25.0	4 139 800
Virgin Islands (United States)	30 000	0.0	30 000	0.0	30 000	0.0	30 000	..	..
OCEANIA									
American Samoa	..	..	..	..	..	..	..	..	..
Fiji	50 000	10.0	55 000	10.9	61 000	6.6	65 000	23.1	80 000
French Polynesia	20 000	75.0	35 000	28.6	45 000	22.2	55 000	18.2	65 000
Kiribati	2 000	0.0	2 000	0.0	2 000	0.0	2 000	..	..
Marshall Islands	1 250	12.0	1 400	42.9	2 000	10.0	2 200	..	..
Micronesia (Fed. States of)	6 000	66.7	10 000	20.0	12 000	16.7	14 000	14.3	16 000
Nauru	..	..	..	..	..	..	..	..	..
New Caledonia	50 000	20.0	60 000	16.7	70 000	8.6	76 000	5.3	80 000
Northern Mariana Islands	..	..	..	..	..	..	..	..	..
Palau	4 000	..	..	..	..	..	..	..	..
Papua New Guinea	75 000	6.7	80 000	12.5	90 000	16.7	105 000	4.8	110 000
Samoa	4 000	25.0	5 000	10.0	5 500	9.1	6 000	33.3	8 000
Solomon Islands	2 200	13.6	2 500	20.0	3 000	33.3	4 000	100.0	8 000
Tonga	2 900	3.4	3 000	0.0	3 000	0.0	3 000	3.3	3 100
Tuvalu	1 250	44.0	1 800	66.7	3 000	..	..	..	..
Vanuatu	7 000	7.1	7 500	0.0	7 500	0.0	7 500	..	..
TRANSITION ECONOMIES									
Albania	12 000	150.0	30 000	150.0	75 000	150.7	188 000	150.6	471 200
Armenia	60 000	133.3	140 000	7.1	150 000	7.3	161 000	7.3	172 800
Azerbaijan	300 000	16.7	350 000	16.6	408 000	66.4	678 800	22.1	829 100
Belarus	808 481	72.2	1 391 903	76.8	2 461 093	37.9	3 394 421	61.4	5 477 500
Bosnia and Herzegovina	100 000	50.0	150 000	50.0	225 000	258.4	806 421	17.8	950 000
Bulgaria	630 000	..	..	..	1 234 000	29.0	1 591 705	17.5	1 870 000
Croatia	789 000	28.5	1 014 000	30.9	1 327 700	9.3	1 451 100	8.6	1 576 400
Georgia	73 500	59.2	117 020	50.1	175 600	54.6	271 400	22.3	332 000
Kazakhstan	250 000	20.0	300 000	33.3	400 000	52.3	609 200	104.7	1 247 000
Kyrgyzstan	152 000	31.6	200 000	31.5	263 000	6.5	280 000	6.5	298 100
Moldova	150 000	92.0	288 000	41.0	406 000	35.5	550 000	32.3	727 700
Montenegro	..	..	..	..	..	..	..	..	266 000
Romania	2 200 000	81.8	4 000 000	12.5	4 500 000	6.1	4 773 000	6.1	5 062 500
Russian Federation	6 000 000	100.0	12 000 000	54.2	18 500 000	17.8	21 800 000	17.8	25 688 600
Serbia	640 000	32.3	847 000	..	..	..	..	..	1 400 000
Tajikistan	3 500	17.7	4 120	21.4	5 000	290.0	19 500	..	..
TFYR Macedonia	100 000	26.0	126 000	26.2	159 000	0.6	159 889	67.6	268 000
Turkmenistan	..	..	20 000	80.0	36 000	34.2	48 300	34.2	64 800
Ukraine	900 000	177.8	2 500 000	50.0	3 750 000	21.6	4 560 000	21.6	5 545 000
Uzbekistan	275 000	78.9	492 000	78.9	880 000	..	..	..	1 700 000

Source: UNCTAD calculations based on the ITU World Telecommunication/ICT Indicators database, 2007.

[a] Singapore: Data from the IDA Household Survey. From 2003-2004 data refer to internet users out of the total population, and from 2005-2006 to Internet users out of the resident population (Singapore Citizens and Permanent residents only) aged 15+.

Table 1.15

Internet penetration: economies by level of development and by region

Internet users per 100 inhabitants

	2002	% change 2002–2003	2003	% change 2003–2004	2004	% change 2004–2005	2005	% change 2005–2006	2006
DEVELOPED ECONOMIES									
ASIA									
Israel	16.9	10.2	18.7	16.8	21.8	12.0	24.4	8.7	26.6
Japan	46.5	3.9	48.3	28.9	62.2	7.2	66.8	2.6	68.5
EUROPE									
Andorra	10.4	33.3	13.9	3.0	14.3	95.0	27.9	2.3	28.6
Austria	41.2	11.1	45.8	3.7	47.5	1.5	48.2	4.9	50.6
Belgium	32.8	17.2	38.5	4.5	40.2	14.2	45.9	..	..
Cyprus	26.6	17.6	31.3	16.3	36.3	7.3	39.0	7.0	41.7
Czech Republic	25.5	-7.9	23.5	7.5	25.2	7.0	27.0	27.6	34.4
Denmark (incl. Faroe Islands)	44.6	3.3	46.0	9.7	50.5	4.3	52.7	10.6	58.3
Estonia	32.6	35.7	44.2	12.1	49.6	3.3	51.2	10.4	56.5
Finland	48.5	1.0	49.0	4.7	51.3	3.9	53.3	4.3	55.6
France (incl. Guadeloupe. French Guiana. Martinique and Réunion)	30.8	20.1	36.9	8.3	40.0	10.1	44.0	12.6	49.6
Germany	33.9	17.9	40.0	6.7	42.7	6.3	45.4	2.9	46.7
Greece	13.5	11.3	15.0	17.5	17.6	2.2	18.0	2.2	18.4
Guernsey	50.0	10.0	55.0	9.1	60.0	..	..	..	..
Hungary	15.8	50.3	23.7	12.7	26.7	11.1	29.7	17.0	34.8
Iceland	51.7	10.7	57.2	1.1	57.9	9.1	63.1	2.5	64.7
Ireland	28.0	12.9	31.7	-6.3	29.7	13.8	33.7	1.2	34.1
Italy	35.1	12.7	39.5	18.5	46.8	2.9	48.2	3.0	49.6
Jersey	..	..	22.8	34.8	30.8	..	..	..	..
Latvia	13.3	81.4	24.1	46.5	35.4	26.1	44.6	4.4	46.6
Liechtenstein	62.5	-3.0	60.6	6.8	64.7	-2.9	62.9	0.0	62.9
Lithuania	14.4	40.3	20.2	10.2	22.2	15.8	25.7	23.0	31.7
Luxembourg	36.7	3.0	37.8	55.8	58.9	16.3	68.5	5.3	72.1
Malta	20.8	18.8	24.7	15.7	28.6	13.6	32.5	..	..
Monaco	47.8	2.6	49.1	5.6	51.8	5.2	54.5	11.1	60.6
Netherlands	50.6	3.1	52.2	18.1	61.6	20.1	74.0	20.1	88.8
Norway	30.7	12.5	34.6	12.7	39.0	88.9	73.6	19.3	87.8
Poland	23.2	1.1	23.5	.4	23.6	11.2	26.2	10.1	28.9
Portugal	21.8	17.3	25.5	-4.2	24.5	14.5	28.0	8.8	30.5
San Marino	51.2	-0.8	50.8	1.8	51.7	-3.3	50.0	-0.6	49.7
Slovakia	16.0	59.4	25.6	20.0	30.7	15.3	35.4	18.3	41.8
Slovenia	37.6	6.6	40.1	18.7	47.6	14.4	54.4	13.2	61.6
Spain	19.1	20.0	22.9	44.7	33.2	1.9	33.8	21.8	41.2
Sweden	57.3	9.8	63.0	19.8	75.5	1.0	76.2	1.0	77.0
Switzerland	41.2	12.1	46.2	2.1	47.2	5.1	49.6	17.5	58.3
United Kingdom (incl. Gibraltar)	42.3	3.4	43.7	7.4	47.0	1.7	47.8	17.2	56.0
NORTH AMERICA									
Bermuda	..	..	58.1	6.6	61.9	6.0	65.6	..	..
Canada	48.3	14.7	55.4	12.5	62.3	8.9	67.9	..	..

Table 1.15 *(continued)*

	2002	% change 2002–2003	2003	% change 2003–2004	2004	% change 2004–2005	2005	% change 2005–2006	2006
Greenland	44.3	23.6	54.8	22.1	66.9	-0.4	66.7	..	..
United States (incl. Puerto Rico and Guam)	54.7	0.7	55.1	13.3	62.5	5.3	65.7	4.2	68.5
OCEANIA									
Australia	53.5	6.3	56.8	14.9	65.3	7.8	70.4	6.7	75.1
New Zealand	48.4	10.3	53.4	10.3	58.9	16.0	68.3	15.3	78.8
DEVELOPING ECONOMIES									
AFRICA									
Algeria	1.6	28.0	2.0	127.1	4.6	25.9	5.8	26.4	7.4
Angola	0.3	-2.9	0.3	86.6	0.5	2.5	0.5	..	..
Benin	0.7	35.4	1.0	38.3	1.4	311.4	5.7	42.0	8.0
Botswana	3.4	-1.7	3.4	-0.6	3.3	-2.0	3.3	..	..
Burkina Faso	0.2	87.3	0.4	1.5	0.4	22.9	0.5	20.2	0.6
Burundi	0.1	70.6	0.2	82.1	0.4	49.8	0.5	44.6	0.8
Cameroon	0.4	62.3	0.6	69.6	1.0	46.9	1.5	45.5	2.2
Cape Verde	3.6	22.3	4.3	22.3	5.3	-2.1	5.2	7.1	5.6
Central African Republic	0.1	14.8	0.1	58.8	0.2	18.3	0.3	16.7	0.3
Chad	0.2	94.8	0.4	6.5	0.4	3.7	0.4	45.8	0.6
Comoros	0.4	..	0.6	58.0	1.0	146.9	2.5	2.4	2.6
Congo	0.2	182.9	0.4	119.9	0.9	32.6	1.3	35.9	1.7
Côte d'Ivoire	0.5	54.2	0.8	12.5	0.9	16.4	1.1	47.6	1.6
Democratic Republic of the Congo	0.1	45.8	0.1	45.6	0.2	21.3	0.2	24.2	0.3
Djibouti	0.6	40.6	0.9	34.0	1.2	7.1	1.2	6.2	1.3
Egypt	2.8	54.8	4.4	27.5	5.6	25.8	7.0	13.5	8.0
Equatorial Guinea	0.4	63.1	0.6	63.2	1.1	37.1	1.4	9.1	1.6
Eritrea	0.2	63.1	0.2	63.2	1.2	34.3	1.6	37.8	2.2
Ethiopia	0.1	45.7	0.1	44.3	0.2	35.7	0.2	..	..
Gabon	1.9	35.8	2.6	13.4	3.0	63.9	4.9	18.3	5.7
Gambia	1.9	36.9	2.6	30.4	3.4	13.7	3.8	..	..
Ghana	0.8	43.2	1.2	46.5	1.7	5.5	1.8	48.9	2.7
Guinea	0.5	13.0	0.5	14.3	0.6	6.0	0.6	-16.7	0.5
Guinea-Bissau	1.1	13.0	1.5	33.7	2.0	16.6	2.3	-1.9	2.3
Kenya	1.3	148.3	3.2	2.0	3.2	0.9	3.2	143.3	7.9
Lesotho	1.1	42.1	1.6	40.4	2.2	17.3	2.6	..	..
Liberia	..	..	..	..	..	..	..	..	..
Libyan Arab Jamahiriya	2.2	28.7	2.9	25.2	3.6	9.5	4.0	..	..
Madagascar	0.3	24.6	0.4	24.3	0.5	8.1	0.5	7.2	0.6
Malawi	0.3	32.7	0.3	9.0	0.4	9.0	0.4	11.2	0.5
Mali	0.2	37.0	0.3	39.8	0.5	17.5	0.5	-5.0	0.5
Mauritania	0.4	16.9	0.4	7.7	0.5	38.7	0.7	385.8	3.2
Mauritius	10.3	19.0	12.3	19.0	14.6	-1.6	14.4	0.3	14.4
Mayotte	..	..	..	..	..	..	..	..	..
Morocco	2.4	40.9	3.4	245.3	11.7	29.6	15.2	30.8	19.9
Mozambique	0.3	62.1	0.4	62.3	0.7	23.6	0.9	..	..
Namibia	2.7	27.3	3.4	10.2	3.7	6.4	4.0	..	..

Table 1.15 *(continued)*

	2002	% change 2002–2003	2003	% change 2003–2004	2004	% change 2004–2005	2005	% change 2005–2006	2006
Niger	0.1	21.1	0.2	25.0	0.2	7.5	0.2	33.4	0.3
Nigeria	0.3	73.9	0.6	128.9	1.4	173.1	3.8	56.6	6.0
Rwanda	0.3	20.6	0.4	21.4	0.4	23.4	0.6	27.3	0.7
Sao Tome and Principe	7.6	33.6	10.2	30.7	13.3	12.7	15.0	..	..
Senegal	1.0	108.5	2.2	114.6	4.7	-0.6	4.6	17.5	5.4
Seychelles	14.1	1.0	14.3	64.7	23.5	3.8	24.4	36.5	33.3
Sierra Leone	0.2	7.4	0.2	6.5	0.2	..	0.2	..	..
Somalia	1.2	1.3	1.2	-3.1	1.1	-3.3	1.1	1.1	1.1
South Africa	6.7	6.4	7.1	6.6	7.6	42.4	10.8	..	..
Sudan	0.9	208.3	2.8	17.4	3.3	141.6	8.0	18.5	9.5
Swaziland	1.9	33.7	2.6	28.4	3.3	10.9	3.7	..	..
Togo	3.7	1.3	3.8	-0.1	3.7	28.4	4.8	1.1	4.9
Tunisia	5.2	23.4	6.4	31.2	8.4	13.0	9.5	34.2	12.7
Uganda	0.4	20.8	0.5	54.5	0.7	141.3	1.7	44.8	2.5
United Republic of Tanzania	0.2	206.4	0.7	30.7	0.9	13.3	1.0	..	..
Zambia	0.5	106.3	1.0	106.5	2.0	45.0	2.9	..	..
Zimbabwe	4.3	58.2	6.8	1.4	6.9	10.5	7.6	19.9	9.1
ASIA									
Afghanistan	0.0	1803.0	0.1	19.1	0.1	1048.2	1.0	71.3	1.7
Bahrain	17.5	20.4	21.1	0.4	21.2	0.1	21.2	0.1	21.3
Bangladesh	0.2	16.9	0.2	21.2	0.2	21.1	0.3	19.4	0.3
Bhutan	1.8	..	2.5	25.2	3.1	18.4	3.7	18.0	4.4
Brunei Darussalam	..	..	13.7	13.4	15.6	12.9	17.6	148.1	43.6
Cambodia	0.2	14.0	0.3	14.6	0.3	4.6	0.3	..	..
China	4.6	33.7	6.1	17.5	7.2	17.4	8.4	23.0	10.4
Dem. People's Republic of Korea	..	..	..	..	..	..	..	..	..
Hong Kong (China)	43.0	9.7	47.2	6.6	50.3	-0.6	50.0	4.7	52.3
India	1.6	9.8	1.7	86.5	3.2	68.9	5.4	..	..
Indonesia	2.1	77.1	3.8	34.2	5.0	42.4	7.2	..	..
Iran (Islamic Republic of)	4.8	49.3	7.2	11.8	8.1	24.7	10.1	150.7	25.3
Iraq	0.1	16.7	0.1	15.6	0.1	..	..	..	..
Jordan	5.8	40.2	8.1	38.5	11.2	12.5	12.6	8.1	13.6
Kuwait	10.6	115.8	22.9	2.9	23.5	10.6	26.0	13.3	29.5
Lao PDR	0.3	23.3	0.3	7.9	0.4	17.0	0.4	..	..
Lebanon	11.7	22.1	14.3	18.3	16.9	15.7	19.6	34.6	26.3
Macao (China)	26.1	2.0	26.7	20.9	32.2	7.6	34.7	10.8	38.4
Malaysia	32.0	7.9	34.5	11.9	38.6	11.2	42.9	1.9	43.8
Maldives	5.4	9.4	5.9	8.0	6.3	4.7	6.6	..	..
Mongolia	2.1	182.1	5.8	31.0	7.6	33.1	10.1	..	..
Myanmar	0.0	5635.5	0.0	-0.4	0.0	163.8	0.1	..	..
Nepal	0.3	22.5	0.4	14.8	0.5	-14.5	0.4	117.3	0.9
Oman	7.2	15.7	8.4	15.7	9.7	14.5	11.1	10.3	12.2
Pakistan	0.7	684.5	5.3	22.6	6.6	4.0	6.8	12.1	7.6
Palestine	3.0	32.0	4.0	8.5	4.3	51.1	6.6	..	..
Philippines	4.4	12.1	4.9	7.9	5.3	2.9	5.5	..	..

Table 1.15 *(continued)*

	2002	% change 2002–2003	2003	% change 2003–2004	2004	% change 2004–2005	2005	% change 2005–2006	2006
Qatar	10.4	89.8	19.8	12.5	22.3	25.9	28.1	22.9	34.5
Republic of Korea	55.2	10.7	61.1	7.6	65.7	4.1	68.4	3.0	70.4
Saudi Arabia	6.3	2.9	6.4	3.0	6.6	84.4	12.2	52.8	18.7
Singaporeª	50.5	1.4	51.2	13.2	57.9	-31.3	39.8	-1.5	39.2
Sri Lanka	1.1	23.1	1.3	10.8	1.4	17.3	1.7	21.3	2.0
Syrian Arab Republic	2.1	63.1	3.5	26.3	4.4	31.6	5.8	33.1	7.7
Taiwan Province of China	47.6	9.1	51.9	3.6	53.8	7.8	58.0	..	..
Thailand	7.7	24.5	9.6	14.6	10.9	0.8	11.0	18.5	13.1
Turkey	6.1	37.6	8.4	68.0	14.2	8.2	15.3	8.2	16.6
United Arab Emirates	31.3	-12.3	27.5	5.3	28.9	7.3	31.0	18.1	36.7
Viet Nam	1.9	103.7	3.8	102.1	7.6	66.6	12.7	35.3	17.2
Yemen	0.5	16.1	0.6	45.8	0.9	21.0	1.1	18.7	1.2
LATIN AMERICA AND THE CARIBBEAN									
Antigua and Barbuda	12.5	36.6	17.1	41.1	24.1	43.3	34.5	9.0	37.6
Argentina	10.9	9.5	12.0	34.6	16.1	10.5	17.8	17.6	20.9
Aruba	24.0	-1.0	23.8	-1.0	23.5	-1.0	23.3	..	..
Bahamas	19.4	40.0	27.1	7.3	29.1	10.8	32.2	..	..
Barbados	10.5	231.0	34.7	49.0	51.7	5.9	54.8	..	..
Belize	..	..	..	..	6.2	56.5	9.6	26.1	12.1
Bolivia	3.1	12.5	3.5	26.6	4.4	17.8	5.2	18.6	6.2
Brazil	8.0	24.1	9.9	20.6	12.0	63.0	19.5	15.6	22.6
Cayman Islands	..	..	..	..	..	..	..	..	..
Chile	23.8	10.6	26.3	6.2	27.9	3.7	28.9	-12.8	25.2
Colombia	4.6	35.1	6.2	36.7	8.5	21.8	10.4	39.4	14.5
Costa Rica	20.3	6.4	21.6	9.0	23.5	8.0	25.4	8.6	27.6
Cuba	1.4	-38.9	0.9	52.9	1.3	26.3	1.7	26.1	2.1
Dominica	18.4	36.0	25.0	20.6	30.1	26.8	38.2	1.5	38.8
Dominican Republic	5.8	28.2	7.5	21.5	9.1	85.0	16.8	31.7	22.2
Ecuador	4.1	5.1	4.3	8.9	4.7	54.5	7.3	57.8	11.5
El Salvador	4.6	78.4	8.3	7.3	8.9	4.2	9.3	..	..
Grenada	14.7	25.4	18.4	..	..	..	..	..	..
Guatemala	3.3	33.9	4.5	34.5	6.0	32.2	7.9	28.8	10.2
Guyana	17.0	11.8	19.0	3.4	19.6	10.2	21.7	..	..
Haiti	1.0	96.2	1.9	222.0	6.1	15.5	7.0	6.8	7.5
Honduras	2.5	7.1	2.7	17.8	3.2	12.3	3.6	26.9	4.6
Jamaica	22.9	32.3	30.3	31.4	39.8	16.8	46.5	..	..
Mexico	10.7	11.9	12.0	11.8	13.4	30.1	17.4	16.7	20.3
Netherlands Antilles	..	..	..	..	..	..	..	..	..
Nicaragua	1.7	9.2	1.9	22.7	2.3	10.7	2.6	7.9	2.8
Panama	4.8	15.2	5.5	11.8	6.2	3.0	6.4	4.8	6.7
Paraguay	1.7	17.0	2.0	64.2	3.3	-2.3	3.2	27.1	4.1
Peru	9.0	15.8	10.4	11.6	11.6	41.7	16.4	30.7	21.5
Saint Kitts and Nevis	21.7	..	..	..	..	..	..	..	..
Saint Lucia	..	..	21.7	59.7	34.6	..	..	..	..
Saint Vincent and the Grenadines	5.2	15.7	6.0	13.3	6.8	23.9	8.4	..	..

Table 1.15 *(continued)*

	2002	% change 2002–2003	2003	% change 2003–2004	2004	% change 2004–2005	2005	% change 2005–2006	2006
Suriname	4.5	14.0	5.2	29.3	6.7	5.7	7.1	..	..
Trinidad and Tobago	10.6	10.4	11.7	4.2	12.1	1.3	12.3	..	..
Uruguay	11.8	39.0	16.4	6.7	17.5	17.4	20.6	5.4	21.7
Venezuela	4.9	52.6	7.5	11.9	8.4	46.9	12.4	22.8	15.2
Virgin Islands (United States)	28.3	-0.9	28.0	-1.5	27.6	-1.3	27.3	..	..
OCEANIA									
American Samoa	..	..	..	..	..	..	..	..	..
Fiji	6.2	8.9	6.8	9.8	7.4	5.5	7.9	21.5	9.5
French Polynesia	8.3	71.4	14.3	26.0	18.0	19.4	21.5	16.5	25.0
Kiribati	2.3	-2.3	2.3	-2.2	2.2	-2.2	2.2	..	..
Marshall Islands	2.4	9.9	2.6	37.7	3.6	6.1	3.9	..	..
Micronesia (Fed. States of)	5.6	0 65.1	9.3	18.9	11.0	15.6	12.7	13.3	14.4
Nauru	..	..	..	..	..	..	..	..	..
New Caledonia	22.4	17.7	26.3	14.5	30.1	6.6	32.1	3.6	33.3
Northern Mariana Islands	..	..	..	..	..	..	..	..	..
Palau	..	..	..	..	..	..	..	..	..
Papua New Guinea	1.4	3.3	1.4	8.6	1.5	15.7	1.8	2.8	1.8
Samoa	2.2	23.6	2.8	8.8	3.0	7.9	3.3	31.2	4.3
Solomon Islands	0.5	11.1	0.6	17.4	0.6	30.5	0.8	90.3	1.6
Tonga	3.0	3.2	3.0	-0.2	3.0	-0.4	3.0	2.3	3.1
Tuvalu	..	..	..	..	..	..	..	..	..
Vanuatu	3.5	4.5	3.7	-2.4	3.6	-2.3	3.5	..	..
TRANSITION ECONOMIES									
Albania	0.4	150.8	1.0	140.6	2.4	155.5	6.0	149.0	15.0
Armenia	2.0	134.1	4.6	7.5	5.0	7.7	5.3	7.7	5.7
Azerbaijan	3.6	16.0	4.2	15.9	4.9	65.2	8.1	21.3	9.8
Belarus	8.2	73.0	14.1	77.9	25.1	38.6	34.8	62.4	56.5
Bosnia and Herzegovina	2.6	49.6	3.9	48.4	5.8	254.7	20.6	17.8	24.3
Bulgaria	8.0	..	..	..	15.9	29.5	20.6	18.4	24.4
Croatia	18.1	28.5	23.2	29.5	30.0	6.2	31.9	8.4	34.6
Georgia	1.6	61.0	2.6	51.4	3.9	56.3	6.1	23.4	7.5
Kazakhstan	1.7	20.2	2.0	33.5	2.7	52.5	4.1	104.8	8.4
Kyrgyzstan	3.0	30.8	4.0	30.2	5.2	3.0	5.3	5.3	5.6
Moldova	3.5	92.9	6.8	40.0	9.5	37.4	13.1	32.3	17.3
Montenegro	..	..	..	..	..	..	..	..	..
Romania	10.1	82.7	18.5	12.6	20.8	6.4	22.1	6.0	23.4
Russian Federation	4.1	101.0	8.3	54.9	12.9	17.8	15.1	19.0	18.0
Serbia	8.5	32.3	11.3	..	..	..	..	..	13.8
Tajikistan	0.1	14.8	0.1	26.0	0.1	277.4	0.3	..	..
TFYR Macedonia	5.0	25.4	6.2	26.2	7.8	0.6	7.9	66.8	13.1
Turkmenistan	..	..	0.4	77.4	0.7	37.2	1.0	32.2	1.3
Ukraine	1.9	180.9	5.3	50.6	7.9	22.6	9.7	22.3	11.9
Uzbekistan	1.1	76.7	1.9	73.0	3.3	..	..	..	6.3

Source: UNCTAD calculations based on the ITU World Telecommunication/ICT Indicators database, 2007.

[a] Singapore: Data from the IDA Household Survey. From 2003-2004 data refer to internet users out of the total population, and from 2005-2006 to Internet users out of the resident population (Singapore Citizens and Permanent residents only) aged 15+.

Table 1.16

Economies ranked by 2006 Internet penetration

Internet users per 100 inhabitants

		2002	% change 2002–2003	2003	% change 2003–2004	2004	% change 2004–2005	2005	% change 2005–2006	2006
1	Netherlands	50.6	3.1	52.2	18.1	61.6	20.1	74.0	20.1	88.8
2	Norway	30.7	12.5	34.6	12.7	39.0	88.9	73.6	19.3	87.8
3	New Zealand	48.4	10.3	53.4	10.3	58.9	16.0	68.3	15.3	78.8
4	Sweden	57.3	9.8	63.0	19.8	75.5	1.0	76.2	1.0	77.0
5	Australia	53.5	6.3	56.8	14.9	65.3	7.8	70.4	6.7	75.1
6	Luxembourg	36.7	3.0	37.8	55.8	58.9	16.3	68.5	5.3	72.1
7	Republic of Korea	55.2	10.7	61.1	7.6	65.7	4.1	68.4	3.0	70.4
8	Japan	46.5	3.9	48.3	28.9	62.2	7.2	66.8	2.6	68.5
9	United States (incl. Puerto Rico and Guam)	54.7	0.7	55.1	13.3	62.5	5.3	65.7	4.2	68.5
10	Canada	48.3	14.7	55.4	12.5	62.3	8.9	67.9	-0.5	67.5
11	Greenland	44.3	23.6	54.8	22.1	66.9	-0.4	66.7	-1.7	65.5
12	Iceland	51.7	10.7	57.2	1.1	57.9	9.1	63.1	2.5	64.7
13	Bermuda	..	..	58.1	6.6	61.9	6.0	65.6	-1.5	64.6
14	Liechtenstein	62.5	-3.0	60.6	6.8	64.7	-2.9	62.9	0.0	62.9
15	Slovenia	37.6	6.6	40.1	18.7	47.6	14.4	54.4	13.2	61.6
16	Monaco	47.8	2.6	49.1	5.6	51.8	5.2	54.5	11.1	60.6
17	Guernsey	50.0	10.0	55.0	9.1	60.0	0.0	60.0	0.0	60.0
18	Denmark (incl. Faroe Islands)	44.6	3.3	46.0	9.7	50.5	4.3	52.7	10.6	58.3
19	Switzerland	41.2	12.1	46.2	2.1	47.2	5.1	49.6	17.5	58.3
20	Taiwan Province of China	47.6	9.1	51.9	3.6	53.8	7.8	58.0	-0.1	57.9
21	Estonia	32.6	35.7	44.2	12.1	49.6	3.3	51.2	10.4	56.5
22	Belarus	8.2	73.0	14.1	77.9	25.1	38.6	34.8	62.4	56.5
23	United Kingdom (incl. Gibraltar)	42.3	3.4	43.7	7.4	47.0	1.7	47.8	17.2	56.0
24	Finland	48.5	1.0	49.0	4.7	51.3	3.9	53.3	4.3	55.6
25	Barbados	10.5	231.0	34.7	49.0	51.7	5.9	54.8	-0.7	54.4
26	Hong Kong (China)	43.0	9.7	47.2	6.6	50.3	-0.6	50.0	4.7	52.3
27	Austria	41.2	11.1	45.8	3.7	47.5	1.5	48.2	4.9	50.6
28	San Marino	51.2	-0.8	50.8	1.8	51.7	-3.3	50.0	-.6	49.7
29	Italy	35.1	12.7	39.5	18.5	46.8	2.9	48.2	3.0	49.6
30	France (incl. Guadeloupe. French Guiana. Martinique and Réunion)	30.8	20.1	36.9	8.3	40.0	10.1	44.0	12.6	49.6
31	Germany	33.9	17.9	40.0	6.7	42.7	6.3	45.4	2.9	46.7
32	Latvia	13.3	81.4	24.1	46.5	35.4	26.1	44.6	4.4	46.6
33	Jamaica	22.9	32.3	30.3	31.4	39.8	16.8	46.5	-0.4	46.3
34	Belgium	32.8	17.2	38.5	4.5	40.2	14.2	45.9	-.2	45.8
35	Malaysia	32.0	7.9	34.5	11.9	38.6	11.2	42.9	1.9	43.8
36	Brunei Darussalam	..	..	13.7	13.4	15.6	12.9	17.6	148.1	43.6
37	Slovakia	16.0	59.4	25.6	20.0	30.7	15.3	35.4	18.3	41.8
38	Cyprus	26.6	17.6	31.3	16.3	36.3	7.3	39.0	7.0	41.7
39	Spain	19.1	20.0	22.9	44.7	33.2	1.9	33.8	21.8	41.2
40	Singapore[a]	50.5	1.4	51.2	13.2	57.9	-31.3	39.8	-1.5	39.2
41	Dominica	18.4	36.0	25.0	20.6	30.1	26.8	38.2	1.5	38.8
42	Macao (China)	26.1	2.0	26.7	20.9	32.2	7.6	34.7	10.8	38.4

Table 1.16 *(continued)*

		2002	% change 2002–2003	2003	% change 2003–2004	2004	% change 2004–2005	2005	% change 2005–2006	2006
43	Antigua and Barbuda	12.5	36.6	17.1	41.1	24.1	43.3	34.5	9.0	37.6
44	United Arab Emirates	31.3	-12.3	27.5	5.3	28.9	7.3	31.0	18.1	36.7
45	Hungary	15.8	50.3	23.7	12.7	26.7	11.1	29.7	17.0	34.8
46	Croatia	18.1	28.5	23.2	29.5	30.0	6.2	31.9	8.4	34.6
47	Qatar	10.4	89.8	19.8	12.5	22.3	25.9	28.1	22.9	34.5
48	Czech Republic	25.5	-7.9	23.5	7.5	25.2	7.0	27.0	27.6	34.4
49	Ireland	28.0	12.9	31.7	-6.3	29.7	13.8	33.7	1.2	34.1
50	Seychelles	14.1	1.0	14.3	64.7	23.5	3.8	24.4	36.5	33.3
51	Saint Lucia	..	..	21.7	59.7	34.6	-1.2	34.2	-2.4	33.3
52	New Caledonia	22.4	17.7	26.3	14.5	30.1	6.6	32.1	3.6	33.3
53	Lithuania	14.4	40.3	20.2	10.2	22.2	15.8	25.7	23.0	31.7
54	Malta	20.8	18.8	24.7	15.7	28.6	13.6	32.5	-3.7	31.3
55	Bahamas	19.4	40.0	27.1	7.3	29.1	10.8	32.2	-3.0	31.2
56	Jersey	..	..	22.8	34.8	30.8	-0.6	30.6	0.0	30.6
57	Portugal	21.8	17.3	25.5	-4.2	24.5	14.5	28.0	8.8	30.5
58	Kuwait	10.6	115.8	22.9	2.9	23.5	10.6	26.0	13.3	29.5
59	Poland	23.2	1.1	23.5	.4	23.6	11.2	26.2	10.1	28.9
60	Andorra	10.4	33.3	13.9	3.0	14.3	95.0	27.9	2.3	28.6
61	Costa Rica	20.3	6.4	21.6	9.0	23.5	8.0	25.4	8.6	27.6
62	Virgin Islands (United States)	28.3	-0.9	28.0	-1.5	27.6	-1.3	27.3	-0.9	27.0
63	Israel	16.9	10.2	18.7	16.8	21.8	12.0	24.4	8.7	26.6
64	Lebanon	11.7	22.1	14.3	18.3	16.9	15.7	19.6	34.6	26.3
65	Iran (Islamic Republic of)	4.8	49.3	7.2	11.8	8.1	24.7	10.1	150.7	25.3
66	Chile	23.8	10.6	26.3	6.2	27.9	3.7	28.9	-12.8	25.2
67	French Polynesia	8.3	71.4	14.3	26.0	18.0	19.4	21.5	16.5	25.0
68	Bulgaria	8.0	..	..	..	15.9	29.5	20.6	18.4	24.4
69	Bosnia and Herzegovina	2.6	49.6	3.9	48.4	5.8	254.7	20.6	17.8	24.3
70	Romania	10.1	82.7	18.5	12.6	20.8	6.4	22.1	6.0	23.4
71	Aruba	24.0	-1.0	23.8	-1.0	23.5	-1.0	23.3	-1.0	23.1
72	Brazil	8.0	24.1	9.9	20.6	12.0	63.0	19.5	15.6	22.6
73	Dominican Republic	5.8	28.2	7.5	21.5	9.1	85.0	16.8	31.7	22.2
74	Guyana	17.0	11.8	19.0	3.4	19.6	10.2	21.7	0.1	21.7
75	Uruguay	11.8	39.0	16.4	6.7	17.5	17.4	20.6	5.4	21.7
76	Peru	9.0	15.8	10.4	11.6	11.6	41.7	16.4	30.7	21.5
77	Bahrain	17.5	20.4	21.1	0.4	21.2	0.1	21.2	0.1	21.3
78	Argentina	10.9	9.5	12.0	34.6	16.1	10.5	17.8	17.6	20.9
79	Mexico	10.7	11.9	12.0	11.8	13.4	30.1	17.4	16.7	20.3
80	Morocco	2.4	40.9	3.4	245.3	11.7	29.6	15.2	30.8	19.9
81	Saudi Arabia	6.3	2.9	6.4	3.0	6.6	84.4	12.2	52.8	18.7
82	Greece	13.5	11.3	15.0	17.5	17.6	2.2	18.0	2.2	18.4
83	Russian Federation	4.1	101.0	8.3	54.9	12.9	17.8	15.1	19.0	18.0
84	Moldova	3.5	92.9	6.8	40.0	9.5	37.4	13.1	32.3	17.3
85	Viet Nam	1.9	103.7	3.8	102.1	7.6	66.6	12.7	35.3	17.2
86	Turkey	6.1	37.6	8.4	68.0	14.2	8.2	15.3	8.2	16.6
87	Venezuela	4.9	52.6	7.5	11.9	8.4	46.9	12.4	22.8	15.2
88	Albania	0.4	150.8	1.0	140.6	2.4	155.5	6.0	149.0	15.0

Table 1.16 *(continued)*

		2002	% change 2002–2003	2003	% change 2003–2004	2004	% change 2004–2005	2005	% change 2005–2006	2006
89	Sao Tome and Principe	7.6	33.6	10.2	30.7	13.3	12.7	15.0	-3.2	14.6
90	Colombia	4.6	35.1	6.2	36.7	8.5	21.8	10.4	39.4	14.5
91	Mauritius	10.3	19.0	12.3	19.0	14.6	-1.6	14.4	0.3	14.4
92	Micronesia (Fed. States of)	5.6	0 65.1	9.3	18.9	11.0	15.6	12.7	13.3	14.4
93	Serbia	8.5	32.3	11.3	..	..	..	..	..	13.8
94	Jordan	5.8	40.2	8.1	38.5	11.2	12.5	12.6	8.1	13.6
95	TFYR Macedonia	5.0	25.4	6.2	26.2	7.8	0.6	7.9	66.8	13.1
96	Thailand	7.7	24.5	9.6	14.6	10.9	0.8	11.0	18.5	13.1
97	Tunisia	5.2	23.4	6.4	31.2	8.4	13.0	9.5	34.2	12.7
98	Oman	7.2	15.7	8.4	15.7	9.7	14.5	11.1	10.3	12.2
99	Trinidad and Tobago	10.6	10.4	11.7	4.2	12.1	1.3	12.3	-0.7	12.2
100	Belize	..	..	..	..	6.2	56.5	9.6	26.1	12.1
101	Ukraine	1.9	180.9	5.3	50.6	7.9	22.6	9.7	22.3	11.9
102	Ecuador	4.1	5.1	4.3	8.9	4.7	54.5	7.3	57.8	11.5
103	South Africa	6.7	6.4	7.1	6.6	7.6	42.4	10.8	-0.3	10.7
104	China	4.6	33.7	6.1	17.5	7.2	17.4	8.4	23.0	10.4
105	Guatemala	3.3	33.9	4.5	34.5	6.0	32.2	7.9	28.8	10.2
106	Mongolia	2.1	182.1	5.8	31.0	7.6	33.1	10.1	-1.1	10.0
107	Azerbaijan	3.6	16.0	4.2	15.9	4.9	65.2	8.1	21.3	9.8
108	Fiji	6.2	8.9	6.8	9.8	7.4	5.5	7.9	21.5	9.5
109	Sudan	0.9	208.3	2.8	17.4	3.3	141.6	8.0	18.5	9.5
110	Zimbabwe	4.3	58.2	6.8	1.4	6.9	10.5	7.6	19.9	9.1
111	El Salvador	4.6	78.4	8.3	7.3	8.9	4.2	9.3	-1.7	9.1
112	Kazakhstan	1.7	20.2	2.0	33.5	2.7	52.5	4.1	104.8	8.4
113	Saint Vincent and the Grenadines	5.2	15.7	6.0	13.3	6.8	23.9	8.4	-0.8	8.3
114	Benin	0.7	35.4	1.0	38.3	1.4	311.4	5.7	42.0	8.0
115	Egypt	2.8	54.8	4.4	27.5	5.6	25.8	7.0	13.5	8.0
116	Kenya	1.3	148.3	3.2	2.0	3.2	0.9	3.2	143.3	7.9
117	Syrian Arab Republic	2.1	63.1	3.5	26.3	4.4	31.6	5.8	33.1	7.7
118	Pakistan	0.7	684.5	5.3	22.6	6.6	4.0	6.8	12.1	7.6
119	Haiti	1.0	96.2	1.9	222.0	6.1	15.5	7.0	6.8	7.5
120	Georgia	1.6	61.0	2.6	51.4	3.9	56.3	6.1	23.4	7.5
121	Algeria	1.6	28.0	2.0	127.1	4.6	25.9	5.8	26.4	7.4
122	Indonesia	2.1	77.1	3.8	34.2	5.0	42.4	7.2	-1.2	7.1
123	Suriname	4.5	14.0	5.2	29.3	6.7	5.7	7.1	-1.3	7.0
124	Panama	4.8	15.2	5.5	11.8	6.2	3.0	6.4	4.8	6.7
125	Maldives	5.4	9.4	5.9	8.0	6.3	4.7	6.6	-1.0	6.6
126	Palestine	3.0	32.0	4.0	8.5	4.3	51.1	6.6	0.0	6.6
127	Uzbekistan	1.1	76.7	1.9	73.0	3.3	..	..	..	6.3
128	Bolivia	3.1	12.5	3.5	26.6	4.4	17.8	5.2	18.6	6.2
129	Nigeria	0.3	73.9	0.6	128.9	1.4	173.1	3.8	56.6	6.0
130	Gabon	1.9	35.8	2.6	13.4	3.0	63.9	4.9	18.3	5.7
131	Armenia	2.0	134.1	4.6	7.5	5.0	7.7	5.3	7.7	5.7
132	Kyrgyzstan	3.0	30.8	4.0	30.2	5.2	3.0	5.3	5.3	5.6
133	Cape Verde	3.6	22.3	4.3	22.3	5.3	-2.1	5.2	7.1	5.6
134	Philippines	4.4	12.1	4.9	7.9	5.3	2.9	5.5	-0.3	5.5

Table 1.16 *(continued)*

		2002	% change 2002–2003	2003	% change 2003–2004	2004	% change 2004–2005	2005	% change 2005–2006	2006
135	Senegal	1.0	108.5	2.2	114.6	4.7	-0.6	4.6	17.5	5.4
136	India	1.6	9.8	1.7	86.5	3.2	68.9	5.4	-1.4	5.4
137	Togo	3.7	1.3	3.8	-0.1	3.7	28.4	4.8	1.1	4.9
138	Honduras	2.5	7.1	2.7	17.8	3.2	12.3	3.6	26.9	4.6
139	Bhutan	1.8	..	2.5	25.2	3.1	18.4	3.7	18.0	4.4
140	Samoa	2.2	23.6	2.8	8.8	3.0	7.9	3.3	31.2	4.3
141	Paraguay	1.7	17.0	2.0	64.2	3.3	-2.3	3.2	27.1	4.1
142	Namibia	2.7	27.3	3.4	10.2	3.7	6.4	4.0	-1.0	3.9
143	Libyan Arab Jamahiriya	2.2	28.7	2.9	25.2	3.6	9.5	4.0	-2.0	3.9
144	Marshall Islands	2.4	9.9	2.6	37.7	3.6	6.1	3.9	-3.4	3.7
145	Gambia	1.9	36.9	2.6	30.4	3.4	13.7	3.8	-2.6	3.7
146	Swaziland	1.9	33.7	2.6	28.4	3.3	10.9	3.7	-1.4	3.6
147	Vanuatu	3.5	4.5	3.7	-2.4	3.6	-2.3	3.5	-4.9	3.3
148	Botswana	3.4	-1.7	3.4	-0.6	3.3	-2.0	3.3	-2.4	3.2
149	Mauritania	0.4	16.9	0.4	7.7	0.5	38.7	0.7	385.8	3.2
150	Tonga	3.0	3.2	3.0	-0.2	3.0	-0.4	3.0	2.3	3.1
151	Zambia	0.5	106.3	1.0	106.5	2.0	45.0	2.9	-3.2	2.8
152	Nicaragua	1.7	9.2	1.9	22.7	2.3	10.7	2.6	7.9	2.8
153	Ghana	0.8	43.2	1.2	46.5	1.7	5.5	1.8	48.9	2.7
154	Lesotho	1.1	42.1	1.6	40.4	2.2	17.3	2.6	-1.3	2.6
155	Comoros	0.4	..	0.6	58.0	1.0	146.9	2.5	2.4	2.6
156	Uganda	0.4	20.8	0.5	54.5	0.7	141.3	1.7	44.8	2.5
157	Guinea-Bissau	1.1	..	1.5	33.7	2.0	16.6	2.3	-1.9	2.3
158	Cameroon	0.4	62.3	0.6	69.6	1.0	46.9	1.5	45.5	2.2
159	Eritrea	0.2	..	0.2	..	1.2	34.3	1.6	37.8	2.2
160	Cuba	1.4	-38.9	0.9	52.9	1.3	26.3	1.7	26.1	2.1
161	Kiribati	2.3	-2.3	2.3	-2.2	2.2	-2.2	2.2	-3.2	2.1
162	Sri Lanka	1.1	23.1	1.3	10.8	1.4	17.3	1.7	21.3	2.0
163	Papua New Guinea	1.4	3.3	1.4	8.6	1.5	15.7	1.8	2.8	1.8
164	Afghanistan	0.0	1803.0	0.1	19.1	0.1	1048.2	1.0	71.3	1.7
165	Congo	0.2	182.9	0.4	119.9	0.9	32.6	1.3	35.9	1.7
166	Côte d'Ivoire	0.5	54.2	0.8	12.5	0.9	16.4	1.1	47.6	1.6
167	Solomon Islands	0.5	11.1	0.6	17.4	0.6	30.5	0.8	90.3	1.6
168	Equatorial Guinea	0.4	63.1	0.6	63.2	1.1	37.1	1.4	9.1	1.6
169	Turkmenistan	..	..	0.4	77.4	0.7	37.2	1.0	32.2	1.3
170	Djibouti	0.6	40.6	0.9	34.0	1.2	7.1	1.2	6.2	1.3
171	Yemen	0.5	16.1	0.6	45.8	0.9	21.0	1.1	18.7	1.2
172	Somalia	1.2	1.3	1.2	-3.1	1.1	-3.3	1.1	1.1	1.1
173	United Republic of Tanzania	0.2	206.4	0.7	30.7	0.9	13.3	1.0	-1.8	1.0
174	Nepal	0.3	22.5	0.4	14.8	0.5	-14.5	0.4	117.3	0.9
175	Mozambique	0.3	62.1	0.4	62.3	0.7	23.6	0.9	-1.8	0.9
176	Burundi	0.1	70.6	0.2	82.1	0.4	49.8	0.5	44.6	0.8
177	Rwanda	0.3	20.6	0.4	21.4	0.4	23.4	0.6	27.3	0.7
178	Chad	0.2	94.8	0.4	6.5	0.4	3.7	0.4	45.8	0.6
179	Burkina Faso	0.2	87.3	0.4	1.5	0.4	22.9	0.5	20.2	0.6
180	Madagascar	0.3	24.6	0.4	24.3	0.5	8.1	0.5	7.2	0.6

Table 1.16 *(continued)*

		2002	% change 2002–2003	2003	% change 2003–2004	2004	% change 2004–2005	2005	% change 2005–2006	2006
181	Angola	0.3	-2.9	0.3	86.6	0.5	2.5	0.5	-1.5	0.5
182	Guinea	0.5	13.0	0.5	14.3	0.6	6.0	0.6	-16.7	0.5
183	Mali	0.2	37.0	0.3	39.8	0.5	17.5	0.5	-5.0	0.5
184	Malawi	0.3	32.7	0.3	9.0	0.4	9.0	0.4	11.2	0.5
185	Lao PDR	0.3	23.3	0.3	7.9	0.4	17.0	0.4	-2.3	0.4
186	Central African Republic	0.1	14.8	0.1	58.8	0.2	18.3	0.3	16.7	0.3
187	Bangladesh	0.2	16.9	0.2	21.2	0.2	21.1	0.3	19.4	0.3
188	Cambodia	0.2	14.0	0.3	14.6	0.3	4.6	0.3	-3.4	0.3
189	Democratic Republic of the Congo	0.1	45.8	0.1	45.6	0.2	21.3	0.2	24.2	0.3
190	Tajikistan	0.1	14.8	0.1	26.0	0.1	277.4	0.3	-1.2	0.3
191	Niger	0.1	21.1	0.2	25.0	0.2	7.5	0.2	33.4	0.3
192	Ethiopia	0.1	45.7	0.1	44.3	0.2	35.7	0.2	-2.3	0.2
193	Sierra Leone	0.2	7.4	0.2	6.5	0.2	..	0.2	..	0.2
194	Iraq	0.1	16.7	0.1	15.6	0.1	-0.8	0.1	-11.8	0.1
195	Myanmar	0.0	5635.5	0.0	-0.4	0.0	163.8	0.1	-0.7	0.1

Developed economy.

Developing economy.

Transition economy.

Source: UNCTAD calculations based on the ITU World Telecommunication/ICT Indicators database, 2007.

[a] Singapore: Data from the IDA Household Survey. From 2003-2004 data refer to internet users out of the total population, and from 2005-2006 to Internet users out of the resident population (Singapore Citizens and Permanent residents only) aged 15+.

Table 1.17

Broadband subscribers: economies by level of development and by region

	2002	% change 2002–2003	2003	% change 2003–2004	2004	% change 2004–2005	2005	% change 2005–2006	2006
DEVELOPED ECONOMIES									
ASIA									
Israel	216 163	192.9	633 100	54.8	980 000	25.5	1 229 626	15.6	1 421 000
Japan	9 397 426	58.7	14 917 165	31.1	19 557 146	14.4	22 365 148	15.2	25 755 100
EUROPE									
Andorra	1 148	213.7	3 601	74.5	6 282	64.6	10 341	..	..
Austria	539 500	11.4	601 000	36.4	820 000	43.4	1 176 000	21.4	1 428 000
Belgium	815 418	52.4	1 242 928	30.1	1 617 185	24.0	2 004 859	0.3	2 010 600
Cyprus	5 879	70.7	10 033	33.2	13 368	99.6	26 684	85.9	49 600
Czech Republic	15 300	126.7	34 690	580.3	235 996	89.7	447 682	142.7	1 086 600
Denmark (incl. Faroe Islands)	451 297	59.2	718 299	42.1	1 020 893	32.9	1 356 283	28.2	1 738 500
Estonia	45 700	97.6	90 300	23.7	111 699	60.4	179 200	27.3	228 100
Finland	273 500	79.6	491 100	62.9	800 000	46.8	1 174 200	21.6	1 428 000
France (incl. Guadeloupe. French Guiana. Martinique and Réunion)	1 682 992	112.4	3 575 381	85.1	6 619 077	43.0	9 465 600	33.8	12 669 000
Germany	3 205 000	40.4	4 500 000	53.3	6 900 000	55.1	10 700 000	31.6	14 085 200
Greece	..	..	10 476	391.2	51 455	211.2	160 113	204.7	487 900
Guernsey	..	..	..	..	..	..	..	..	..
Hungary	111 458	137.1	264 311	55.6	411 171	58.5	651 689	49.9	976 700
Iceland	24 270	66.5	40 419	36.4	55 112	41.6	78 017	12.4	87 700
Ireland	10 600	294.3	41 800	263.9	152 100	78.0	270 700	91.1	517 300
Italy	850 000	164.7	2 250 000	97.8	4 450 000	52.4	6 780 000	27.4	8 638 900
Jersey	..	..	..	..	..	..	..	..	..
Latvia	10 000	95.3	19 533	151.6	49 147	23.7	60 800	80.4	109 700
Liechtenstein	1 400	7.1	1 500	213.3	4 700	83.0	8 600	..	..
Lithuania	20 000	234.0	66 790	93.2	129 051	81.4	234 081	57.5	368 700
Luxembourg	5 697	169.5	15 351	137.8	36 500	92.1	70 100	33.0	93 200
Malta	17 679	28.6	22 736	65.6	37 642	18.7	44 672	-5.8	42 100
Monaco	4 900	32.7	6 500	16.9	7 600	23.7	9 400	..	..
Netherlands	1 068 966	86.0	1 988 000	61.3	3 206 000	27.9	4 100 000	26.6	5 192 200
Norway	205 307	94.2	398 758	68.4	671 666	47.6	991 352	28.9	1 278 300
Poland	121 684	60.9	195 752	314.7	811 796	53.2	1 243 949	112.2	2 640 000
Portugal	262 789	91.5	503 128	70.6	858 419	41.2	1 212 034	20.5	1 460 300
San Marino	..	..	600	..	..	..	..	..	1 500
Slovakia	..	..	7 708	538.1	49 188	181.7	138 569	128.8	317 000
Slovenia	56 735	2.2	57 992	98.4	115 069	47.7	169 950	55.2	263 700
Spain	1 247 496	76.5	2 202 000	56.3	3 441 630	45.1	4 994 274	33.3	6 654 900
Sweden	716 085	10.7	793 000	56.0	1 237 000	56.1	1 931 000	21.5	2 346 300
Switzerland	455 220	72.2	783 874	55.3	1 217 000	41.8	1 725 446	24.0	2 140 300
United Kingdom (incl. Gibraltar)	1 821 225	110.0	3 824 500	86.4	7 130 500	33.8	9 539 900	36.2	12 995 100
NORTH AMERICA									
Bermuda	..	..	..	..	..	..	18 500	27.6	23 600
Canada	3 515 000	28.4	4 513 000	20.0	5 416 000	23.8	6 706 699	14.4	7 675 500
Greenland	..	..	..	..	..	..	..	..	..

Table 1.17 *(continued)*

	2002	% change 2002–2003	2003	% change 2003–2004	2004	% change 2004–2005	2005	% change 2005–2006	2006
United States (incl. Puerto Rico and Guam)	19 904 281	41.8	28 230 149	34.2	37 890 646	30.4	49 391 060	17.9	58 254 900
OCEANIA									
Australia	258 100	100.2	516 800	98.4	1 025 500	105.1	2 102 800	85.5	3 900 000
New Zealand	43 500	90.8	83 000	131.0	191 695	72.7	331 000	74.0	576 100
DEVELOPING ECONOMIES									
AFRICA									
Algeria	..	..	18 000	100.0	36 000	441.7	195 000	..	..
Angola	..	..	..	..	..	..	..	..	..
Benin	21	0.0	21	285.7	81	142.0	196	2.0	200
Botswana	..	..	..	..	..	..	1 600	..	..
Burkina Faso	50	190.0	145	6.2	154	68.8	260	553.8	1700
Burundi	..	..	..	..	..	..	..	..	..
Cameroon	..	..	..	..	..	..	..	..	200
Cape Verde	..	..	..	..	283	231.1	937	92.1	1800
Central African Republic	..	..	..	..	..	..	..	..	..
Chad	..	..	..	..	..	..	..	..	..
Comoros	..	..	..	..	1	300.0	4	..	..
Congo	..	..	..	..	..	..	..	..	..
Côte d'Ivoire	..	..	1 000	-17.4	826	45.3	1 200	..	..
Democratic Republic of the Congo	..	..	..	..	1 450	3.4	1 500	..	..
Djibouti	..	..	..	..	..	..	42	..	..
Egypt	937	417.6	4 850	879.5	47 504	139.0	113 526	81.0	205 500
Equatorial Guinea	..	..	..	..	..	..	..	..	200
Eritrea	..	..	..	..	..	..	..	..	..
Ethiopia	..	..	57	..	..	..	..	..	200
Gabon	..	..	170	282.4	650	133.1	1 515	-20.8	1 200
Gambia	..	..	..	..	..	..	71	40.8	100
Ghana	..	..	..	..	..	..	1 904	567.0	12 700
Guinea	..	..	..	..	..	..	..	..	..
Guinea-Bissau	..	..	..	..	..	..	..	..	..
Kenya	..	..	..	..	..	..	..	..	..
Lesotho	..	..	..	..	..	..	45	..	..
Liberia	..	..	..	..	..	..	..	..	..
Libyan Arab Jamahiriya	..	..	..	..	..	..	..	..	..
Madagascar	..	..	..	..	..	..	..	..	..
Malawi	..	..	69	100.0	138	192.8	404	..	..
Mali	..	..	..	..	..	..	..	..	..
Mauritania	..	..	..	..	..	..	164	326.8	700
Mauritius	285	315.8	1 185	128.5	2 708	14.5	3 100	606.5	21 900
Mayotte	..	..	..	..	..	..	..	..	..
Morocco	2 000	35.6	2 712	2284.2	64 660	285.3	249 138	56.9	390 800
Mozambique	..	..	..	..	..	..	..	..	..
Namibia	..	..	..	..	..	..	..	..	..
Niger	..	..	..	..	77	175.3	212	..	..

Table 1.17 *(continued)*

	2002	% change 2002–2003	2003	% change 2003–2004	2004	% change 2004–2005	2005	% change 2005–2006	2006
Nigeria	..	..	..	..	..	..	500	..	..
Rwanda	..	..	..	..	..	..	..	..	..
Sao Tome and Principe	..	..	..	..	..	..	..	..	..
Senegal	1 200	75.0	2 100	264.9	7 663	140.1	18 396	57.1	28 900
Seychelles	..	..	..	..	349	64.8	575	126.1	1 300
Sierra Leone	..	..	..	..	..	..	..	..	..
Somalia	..	..	..	..	..	..	..	..	..
South Africa	2 669	661.1	20 313	195.4	60 000	175.5	165 290	..	..
Sudan	..	..	..	..	1 400	28.6	1 800	16.7	2 100
Swaziland	..	..	..	..	..	..	..	..	..
Togo	..	..	..	..	..	..	..	..	..
Tunisia	..	..	2 590	9.6	2 839	480.9	16 491	165.6	43 800
Uganda	..	..	..	..	..	..	..	..	1 200
United Republic of Tanzania	..	..	..	..	..	..	..	..	..
Zambia	48	89.6	91	119.8	200	0.0	200	..	..
Zimbabwe	..	..	4 618	94.2	8 967	13.6	10 185	0.1	10 200
ASIA									
Afghanistan	..	..	..	..	200	10.0	220	127.3	500
Bahrain	4 980	95.5	9 737	54.1	15 000	42.7	21 400	80.4	38 600
Bangladesh	..	..	..	..	..	..	..	..	..
Bhutan	..	..	..	..	..	..	..	..	..
Brunei Darussalam	2 800	35.7	3 800	65.8	6 300	28.6	8 100	29.6	10 500
Cambodia	50	738.0	419	90.9	800	25.0	1 000	..	..
China	5 367 000	107.7	11 147 000	123.2	24 875 000	50.8	37 504 000	35.8	50 916 000
Dem. People's Republic of Korea	..	..	..	..	..	..	..	..	..
Hong Kong (China)	1 038 995	22.0	1 267 966	19.9	1 519 837	9.2	1 659 098	8.3	1 796 200
India	82 409	70.3	140 362	67.4	235 000	453.2	1 300 000	76.9	2 300 000
Indonesia	38 300	60.8	61 600	37.8	84 900	27.4	108 200	..	..
Iran (Islamic Republic of)	16 171	9.5	17 700	8.5	19 200	8.3	20 800	2136.1	465 100
Iraq	..	..	..	..	..	..	..	..	..
Jordan	3 177	57.3	4 996	108.6	10 424	132.2	24 200	100.8	48 600
Kuwait	10 500	23.8	13 000	53.8	20 000	25.0	25 000	..	..
Lao PDR	..	..	..	..	..	..	100	..	..
Lebanon	35 000	100.0	70 000	14.3	80 000	62.5	130 000	30.8	170 000
Macao (China)	16 954	63.6	27 744	63.0	45 218	50.4	68 030	35.5	92 200
Malaysia	19 302	472.0	110 406	128.9	252 701	94.2	490 630	82.9	897 300
Maldives	190	164.7	503	42.5	717	191.6	2 091	124.8	4 700
Mongolia	90	455.6	500	80.0	900	100.0	1 800	..	..
Myanmar	..	..	..	..	..	..	119	68.1	200
Nepal	..	..	..	..	..	..	..	..	..
Oman	97	40.2	136	391.2	668	1154.2	8 378	81.4	15 200
Pakistan	..	..	..	..	22 300	100.0	44 600	26.9	56 600
Palestine	..	..	..	..	..	..	7 665	..	..
Philippines	21 000	161.9	55 000	61.8	89 000	38.2	123 000	..	..
Qatar	228	1211.8	2 991	256.1	10 652	136.3	25 168	86.0	46 800

Table 1.17 *(continued)*

	2002	% change 2002–2003	2003	% change 2003–2004	2004	% change 2004–2005	2005	% change 2005–2006	2006
Republic of Korea	10 405 486	7.4	11 178 499	6.6	11 921 440	2.3	12 190 711	15.2	14 042 700
Saudi Arabia	34 800	32.2	46 000	49.3	68 700	-1.3	67 800	221.8	218 200
Singapore	270 000	56.2	421 700	21.5	512 400	29.9	665 500	19.7	796 500
Sri Lanka	592	477.2	3 417	497.0	20 400	27.9	26 100	11.5	29 100
Syrian Arab Republic	..	..	..	..	600	100.0	1 200	366.7	5600
Taiwan Province of China	2 100 000	44.9	3 043 273	23.3	3 751 214	15.7	4 340 900	3.8	4 505 800
Thailand	15 000	200.0	45 000	66.7	75 000	40.0	105 000	..	..
Turkey	21 205	840.0	199 324	189.9	577 931	175.1	1 589 768	74.5	2 773 700
United Arab Emirates	16 177	84.4	29 831	86.2	55 541	131.3	128 493	87.2	240 600
Viet Nam	1 076	753.2	9 180	474.2	52 709	298.5	210 024	146.0	516 600
Yemen	..	..			..		..	..	..
LATIN AMERICA AND THE CARIBBEAN									
Antigua and Barbuda	..	..	..	..	1 600	256.3	5 700	..	..
Argentina	115 000	104.0	234 625	112.0	497 513	69.0	841 000	86.4	1 567 700
Aruba	..	..	1 400	400.0	7 000	75.7	12 300	..	..
Bahamas	7 540	45.1	10 941	17.0	12 803	4.7	13 400	..	..
Barbados	..	..	27 319	8.3	29 600	7.9	31 942	..	..
Belize	..	..	940	200.7	2 827	51.4	4 280	30.8	5 600
Bolivia	3 330	72.4	5 740	52.0	8 723	23.7	10 788	..	..
Brazil	731 000	64.0	1 199 000	88.2	2 256 000	46.5	3 304 000	79.2	5 921 900
Cayman Islands	..	..	..	..	..	..	..	..	..
Chile	188 454	86.9	352 234	36.0	478 883	47.9	708 358	38.1	978 100
Colombia	34 888	84.7	64 436	97.3	127 113	150.7	318 683	97.0	627 800
Costa Rica	363	3998.6	14 878	87.7	27 931	67.2	46 700	26.6	59 100
Cuba	..	..	..	..	..	..	..	..	..
Dominica	2 238	18.5	2 651	22.7	3 253	1.4	3 300	..	..
Dominican Republic	..	..	..	..	37 257	76.8	65 856	1.0	66 500
Ecuador	..	..	..	..	11 620	130.5	26 786	..	..
El Salvador	..	..	..	..	29 321	44.3	42 314	..	..
Grenada	563	6.6	600	1.5	609	425.5	3 200	71.9	5 500
Guatemala	..	..	..	..	..	..	27 106	..	..
Guyana	..	..	..	..	..	..	2 000	..	..
Haiti	..	..	..	..	..	..	..	..	..
Honduras	..	..	..	..	..	..	..	..	..
Jamaica	9 000	0.0	9 000	200.0	27 000	66.7	45 000	..	..
Mexico	231 486	85.1	428 378	142.2	1 037 455	122.1	2 304 520	61.8	3 728 200
Netherlands Antilles	..	..	..	..	..	..	..	..	..
Nicaragua	2 319	89.9	4 403	13.6	5 001	110.6	10 534	80.4	19 000
Panama	12 235	22.9	15 039	11.4	16 746	4.9	17 567	..	..
Paraguay	500	0.0	500	520.0	3 100	80.6	5 600	185.7	16 000
Peru	34 400	172.4	93 695	47.6	138 277	152.8	349 582	38.7	484 900
Saint Kitts and Nevis	500	..	..	..	..	..	..	..	..
Saint Lucia	..	..	..	..	..	..	..	..	..
Saint Vincent and the Grenadines	1 086	5.5	1 146	15.2	1 320	176.3	3 647	53.6	5 600
Suriname	94	129.8	216	94.4	420	138.3	1 001	169.7	2 700

Table 1.17 *(continued)*

	2002	% change 2002–2003	2003	% change 2003–2004	2004	% change 2004–2005	2005	% change 2005–2006	2006
Trinidad and Tobago	95	830.5	884	378.8	4 233	155.2	10 803	90.7	20 600
Uruguay	..	..	..	..	27 000	126.6	61 186	74.9	107 000
Venezuela	78 151	49.7	116 997	79.8	210 303	69.7	356 898	50.6	537 500
Virgin Islands (United States)	..	..			..	1 500	100.0	3 000	..	..
OCEANIA									
American Samoa	..	..	..	..	..	..	..	..	..
Fiji	..	..	..	..	..	..	7 000	..	..
French Polynesia	..	..	946	359.9	4 351	152.8	11 000	65.5	18 200
Kiribati	..	..	..	..	..	..	..	..	..
Marshall Islands	..	..	..	..	..	..	..	..	..
Micronesia (Fed. States of)	..	..	..	..	..	..	..	..	..
Nauru	..	..	..	..	..	..	..	..	..
New Caledonia	700	138.3	1 668	208.5	5 146	86.6	9 600	..	..
Northern Mariana Islands	..	..	..	..	..	..	..	..	..
Palau	..	..	..	..	..	..	..	..	..
Papua New Guinea	..	..	..	..	..	..	..	..	..
Samoa	..	..	..	..	..	..	100	..	..
Solomon Islands	108	89.8	205	0.0	205	95.1	400	..	..
Tonga	..	..	..	..	300	100.0	600	..	..
Tuvalu	..	..	..	..	..	..	..	..	..
Vanuatu	..	..	15	53.3	23	334.8	100	..	..
TRANSITION ECONOMIES									
Albania	..	..	..	..	..	..	..	..	300
Armenia	8	25.0	10	9 900.0	1 000	100.0	2 000	..	..
Azerbaijan	..	..	..	..	900	142.7	2 184	..	..
Belarus	20	515.0	123	509.8	750	108.5	1 564	628.9	11 400
Bosnia and Herzegovina	213	604.2	1 500	342.5	6 637	106.4	13 702	191.9	40 000
Bulgaria	..	..	..	..	..	..	165 469	132.3	384 310
Croatia	12 000	..	..	..	26 800	235.1	89 800	180.4	251 800
Georgia	920	53.3	1 410	34.8	1 900	26.3	2 400	1025.0	27 000
Kazakhstan	..	..	..	..	1 997	50.2	3 000	916.7	30 500
Kyrgyzstan	36	286.1	139	1271.9	1 907	28.9	2 459	..	..
Moldova	418	42.8	597	306.9	2 429	328.0	10 395	109.7	21 800
Montenegro	..	..	..	..	..	..	..	..	25 800
Romania	15 800	1141.2	196 106	95.2	382 783	96.2	751 060	..	..
Russian Federation	11 000	..	..	..	675 000	135.4	1 589 000	82.5	2 900 000
Serbia	..	..	..	..	..	..	..	..	121 700
Tajikistan	..	..	..	..	..	..	..	..	..
TFYR Macedonia	..	..	..	..	..	..	12 436	193.5	36 500
Turkmenistan	..	..	..	..	..	..	..	..	..
Ukraine	..	..	..	..	..	..	..	..	..
Uzbekistan	..	..	2 757	99.5	5 500	50.9	8 300	..	..

Source: UNCTAD calculations based on the ITU World Telecommunication/ICT Indicators database, 2007.

Table 1.18

Broadband penetration: economies by level of development and by region

Broadband subscribers per 100 inhabitants

	2002	% change 2002–2003	2003	% change 2003–2004	2004	% change 2004–2005	2005	% change 2005–2006	2006
DEVELOPED ECONOMIES									
ASIA									
Israel	3.3	187.3	9.4	52.8	14.3	24.7	17.8	11.5	19.9
Japan	7.4	58.5	11.7	31.0	15.3	14.3	17.5	15.2	20.2
EUROPE									
Andorra	1.7	191.3	5.0	64.1	8.2	61.1	13.2	..	..
Austria	6.7	10.9	7.4	35.3	10.0	42.0	14.2	21.3	17.2
Belgium	7.9	51.8	12.0	29.5	15.5	23.9	19.2	0.1	19.2
Cyprus	0.7	68.5	1.3	30.0	1.6	95.8	3.2	81.7	5.8
Czech Republic	0.2	126.7	0.3	579.6	2.3	89.5	4.4	141.1	10.6
Denmark (incl. Faroe Islands)	8.3	58.3	13.2	41.9	18.7	32.4	24.7	27.7	31.6
Estonia	3.4	98.4	6.7	24.2	8.3	60.9	13.3	27.6	17.0
Finland	5.2	79.2	9.4	62.9	15.3	45.9	22.4	21.4	27.1
France (incl. Guadeloupe. French Guiana. Martinique and Réunion)	2.8	111.5	6.0	83.5	11.0	42.8	15.6	33.4	20.9
Germany	3.9	40.4	5.5	53.4	8.4	54.7	12.9	31.6	17.0
Greece	..	..	0.1	407.1	0.5	210.6	1.4	204.2	4.4
Guernsey	..	..	..	..	..	..	..	..	..
Hungary	1.1	137.6	2.6	55.9	4.1	58.5	6.5	50.3	9.7
Iceland	8.4	66.5	13.9	36.4	19.0	41.6	26.9	8.7	29.2
Ireland	0.3	289.4	1.1	258.5	3.8	73.3	6.5	88.4	12.3
Italy	1.5	158.2	3.9	97.4	7.7	52.1	11.7	27.3	14.9
Jersey	..	..	..	..	..	..	..	..	..
Latvia	0.4	96.2	0.8	154.9	2.1	22.6	2.6	81.2	4.8
Liechtenstein	4.4	3.9	4.5	204.1	13.8	77.8	24.6	..	..
Lithuania	0.6	236.9	1.9	93.2	3.7	82.4	6.8	58.0	10.8
Luxembourg	1.3	169.5	3.4	132.6	7.9	92.1	15.2	30.1	19.8
Malta	4.6	28.0	5.8	64.8	9.6	18.3	11.4	-9.3	10.3
Monaco	15.1	31.8	19.9	16.2	23.2	22.9	28.5	..	..
Netherlands	6.6	84.9	12.2	61.9	19.8	27.3	25.2	26.1	31.7
Norway	4.5	93.0	8.7	67.7	14.6	47.0	21.5	28.4	27.5
Poland	0.3	61.0	0.5	314.9	2.1	53.3	3.3	112.4	6.9
Portugal	2.5	90.4	4.8	69.6	8.2	41.7	11.6	19.9	13.9
San Marino	..	..	2.1	..	..	..	..	..	4.8
Slovakia	..	..	0.1	537.8	0.9	181.6	2.6	128.6	5.9
Slovenia	2.8	2.2	2.9	98.3	5.8	47.3	8.5	53.1	13.0
Spain	3.0	69.9	5.2	54.5	8.0	40.2	11.2	32.0	14.8
Sweden	8.0	10.2	8.8	55.5	13.7	55.6	21.4	21.1	25.9
Switzerland	6.3	70.3	10.7	54.0	16.4	41.0	23.1	23.6	28.6
United Kingdom (incl. Gibraltar)	3.1	108.5	6.4	85.5	11.9	34.1	16.0	35.8	21.7
NORTH AMERICA									
Bermuda	..	..	..	..	..	..	28.9	25.6	36.3
Canada	11.2	27.2	14.2	18.8	16.9	22.6	20.7	13.9	23.6

Table 1.18 *(continued)*

	2002	% change 2002–2003	2003	% change 2003–2004	2004	% change 2004–2005	2005	% change 2005–2006	2006
Greenland	..	..	..	..	..	..	..	..	..
United States (incl. Puerto Rico and Guam)	6.8	40.5	9.6	32.9	12.7	28.4	16.3	16.8	19.1
OCEANIA									
Australia	1.3	97.8	2.6	98.1	5.2	102.5	10.4	83.6	19.1
New Zealand	1.1	90.3	2.1	128.6	4.8	71.0	8.2	72.8	14.2
DEVELOPING ECONOMIES									
AFRICA									
Algeria	..	..	0.1	96.8	0.1	432.6	0.6	..	..
Angola	..	..	..	..	..	..	..	..	..
Benin	0.0	-3.3	0.0	273.5	0.0	134.2	0.0	-12.0	0.0
Botswana	..	..	..	..	..	..	0.1	..	..
Burkina Faso	0.0	182.9	0.0	-2.8	0.0	70.9	0.0	534.7	0.0
Burundi	..	..	..	..	..	..	..	..	..
Cameroon	..	..	..	..	..	..	..	..	0.0
Cape Verde	..	..	..	..	0.1	224.2	0.2	77.3	0.3
Central African Republic	..	..	..	..	..	..	..	..	..
Chad	..	..	..	..	..	..	..	..	..
Comoros	..	..	..	..	0.0	295.0	0.0	..	..
Congo	..	..	..	..	..	..	..	..	..
Côte d'Ivoire	..	..	0.0	-18.7	0.0	35.3	0.0	..	..
Democratic Republic of the Congo	..	..	..	..	0.0	0.4	0.0	..	..
Djibouti	..	..	..	..	..	..	..	..	..
Egypt	0.0	407.5	0.0	860.6	0.1	134.5	0.2	71.2	0.3
Equatorial Guinea	..	..	..	..	..	..	..	..	0.0
Eritrea	..	..	..	..	..	..	..	..	..
Ethiopia	..	..	..	..	..	..	..	..	0.0
Gabon	..	..	0.0	279.5	0.0	128.0	0.1	-22.5	0.1
Gambia	..	..	..	..	..	..	0.0	37.2	0.0
Ghana	..	..	..	..	..	..	0.0	553.7	0.1
Guinea	..	..	..	..	..	..	..	..	..
Guinea-Bissau	..	..	..	..	..	..	..	..	..
Kenya	..	..	..	..	..	..	..	..	..
Lesotho	..	..	..	..	..	..	..	..	..
Liberia	..	..	..	..	..	..	..	..	..
Libyan Arab Jamahiriya	..	..	..	..	..	..	..	..	..
Madagascar	..	..	..	..	..	..	..	..	..
Malawi	..	..	0.0	70.0	0.0	180.5	0.0	..	..
Mali	..	..	..	..	..	..	..	..	..
Mauritania	..	..	..	..	..	..	0.0	314.7	0.0
Mauritius	0.0	312.4	0.1	126.7	0.2	12.6	0.2	600.8	1.7
Mayotte	..	..	..	..	..	..	..	..	..
Morocco	0.0	33.8	0.0	2252.3	0.2	280.0	0.8	54.7	1.3
Mozambique	..	..	..	..	..	..	..	..	..
Namibia	..	..	..	..	..	..	..	..	..
Niger	..	..	..	..	0.0	145.0	0.0	..	..

Table 1.18 *(continued)*

	2002	% change 2002–2003	2003	% change 2003–2004	2004	% change 2004–2005	2005	% change 2005–2006	2006
Nigeria	..	..	..	..	..	..	0.0	..	..
Rwanda	..	..	..	..	..	..	..	..	..
Sao Tome and Principe	..	..	..	..	..	..	..	..	..
Senegal	0.0	70.3	0.0	265.6	0.1	112.9	0.2	53.4	0.2
Seychelles	..	..	..	..	0.4	62.8	0.7	123.5	1.5
Sierra Leone	..	..	..	..	..	..	..	..	..
Somalia	..	..	..	..	..	..	..	..	..
South Africa	0.0	655.2	0.0	193.6	0.1	174.2	0.3	..	..
Sudan	..	..	..	..	0.0	26.5	0.0	10.6	0.0
Swaziland	..	..	..	..	..	..	..	..	..
Togo	..	..	..	..	..	..	..	..	..
Tunisia	..	..	0.0	8.5	0.0	474.5	0.2	162.5	0.4
Uganda	..	..	..	..	..	..	..	..	0.0
United Republic of Tanzania	..	..	..	..	..	..	..	..	..
Zambia	0.0	86.4	0.0	116.1	0.0	0.0	0.0	..	..
Zimbabwe	..	..	0.0	92.1	0.1	2.9	0.1	-1.6	0.1
ASIA									
Afghanistan	..	..	..	..	0.0	5.2	0.0	118.4	0.0
Bahrain	0.7	92.8	1.4	51.9	2.1	40.7	2.9	77.9	5.2
Bangladesh	..	..	..	..	..	..	..	..	..
Bhutan	..	..	..	..	..	..	..	..	..
Brunei Darussalam	0.8	31.8	1.1	61.2	1.8	25.1	2.2	26.2	2.8
Cambodia	0.0	719.1	0.0	86.7	0.0	21.8	0.0	..	..
China	0.4	106.4	0.9	121.8	1.9	49.9	2.9	35.3	3.9
Dem. People's Republic of Korea	..	..	..	..	..	..	..	..	..
Hong Kong (China)	15.3	21.7	18.6	18.0	22.0	7.0	23.5	6.0	24.9
India	0.0	67.7	0.0	64.9	0.0	445.0	0.1	74.4	0.2
Indonesia	0.0	58.6	0.0	33.2	0.0	27.3	0.0	..	..
Iran (Islamic Republic of)	0.0	7.9	0.0	5.8	0.0	6.1	0.0	2079.9	0.7
Iraq	..	..	..	..	..	..	..	..	..
Jordan	0.1	52.7	0.1	103.8	0.2	128.5	0.4	96.0	0.8
Kuwait	0.4	17.8	0.5	49.6	0.8	18.5	0.9	..	..
Lao PDR	..	..	..	..	..	..	0.0	..	..
Lebanon	1.0	95.4	2.0	12.7	2.3	61.1	3.6	29.7	4.7
Macao (China)	3.9	60.0	6.2	57.6	9.7	42.9	13.9	27.6	17.7
Malaysia	0.1	460.1	0.4	124.1	1.0	93.6	1.9	81.8	3.5
Maldives	0.1	155.6	0.2	37.8	0.2	188.7	0.7	122.6	1.5
Mongolia	0.0	448.8	0.0	68.4	0.0	98.5	0.1	..	..
Myanmar	..	..	..	..	..	..	0.0	66.8	0.0
Nepal	..	..	..	..	..	..	..	..	..
Oman	0.0	39.1	0.0	387.3	0.0	1134.7	0.3	78.6	0.6
Pakistan	..	..	..	..	0.0	98.1	0.0	24.4	0.0
Palestine	..	..	..	..	..	..	0.2	..	..
Philippines	0.0	156.8	0.1	58.7	0.1	35.6	0.1	..	..
Qatar	0.0	1137.9	0.4	241.7	1.4	124.2	3.2	72.7	5.6
Republic of Korea	21.9	6.9	23.4	6.1	24.8	1.8	25.2	14.8	29.0

Table 1.18 *(continued)*

	2002	% change 2002–2003	2003	% change 2003–2004	2004	% change 2004–2005	2005	% change 2005–2006	2006
Saudi Arabia	0.2	28.6	0.2	45.5	0.3	-3.8	0.3	213.9	0.9
Singapore	6.5	55.8	10.1	21.2	12.3	24.8	15.3	18.9	18.2
Sri Lanka	0.0	468.2	0.0	490.6	0.1	20.0	0.1	10.6	0.1
Syrian Arab Republic	..	..	..	..	0.0	91.4	0.0	355.4	0.0
Taiwan Province of China	9.3	44.4	13.5	22.8	16.5	15.3	19.1	3.7	19.8
Thailand	0.0	197.4	0.1	65.2	0.1	38.8	0.2	..	..
Turkey	0.0	827.2	0.3	186.1	0.8	171.4	2.2	72.2	3.7
United Arab Emirates	0.4	71.2	0.7	75.7	1.3	120.0	2.9	80.8	5.2
Viet Nam	0.0	741.5	0.0	466.4	0.1	293.2	0.2	142.8	0.6
Yemen	..	..	..	..	..	..	..	..	..
LATIN AMERICA AND THE CARIBBEAN									
Antigua and Barbuda	..	..	..	..	1.9	252.0	6.8	..	..
Argentina	0.3	102.1	0.6	110.0	1.3	67.5	2.2	83.8	4.0
Aruba	..	..	1.4	395.1	6.9	74.0	11.9	..	..
Bahamas	2.4	45.1	3.5	13.4	4.0	4.7	4.2	..	..
Barbados	..	..	9.5	7.6	10.2	7.2	10.9	..	..
Belize	..	..	0.4	200.7	1.1	45.8	1.6	26.2	2.0
Bolivia	0.0	68.9	0.1	49.1	0.1	21.4	0.1	..	..
Brazil	0.4	61.7	0.7	85.6	1.2	44.5	1.8	76.9	3.1
Cayman Islands	..	..	..	..	..	..	..	..	..
Chile	1.3	84.7	2.3	34.4	3.1	46.2	4.5	30.7	5.9
Colombia	0.1	82.6	0.1	90.5	0.3	149.2	0.7	94.1	1.4
Costa Rica	0.0	3851.2	0.4	84.2	0.7	64.1	1.1	24.5	1.3
Cuba	..	..	..	..	..	..	..	..	..
Dominica	3.3	18.5	3.9	22.7	4.8	1.4	4.9	..	..
Dominican Republic	..	..	..	..	0.4	74.4	0.7	-0.3	0.7
Ecuador	..	..	..	..	0.1	129.8	0.2	..	..
El Salvador	..	..	..	..	0.4	38.6	0.6	..	..
Grenada	0.6	5.5	0.6	0.5	0.6	420.4	3.0	70.3	5.2
Guatemala	..	..	..	..	..	..	0.2	..	..
Guyana	..	..	..	..	..	..	0.3	..	..
Haiti	..	..	..	..	..	..	..	..	..
Honduras	..	..	..	..	..	..	..	..	..
Jamaica	0.3	-0.8	0.3	195.5	1.0	68.6	1.7	..	..
Mexico	0.2	82.4	0.4	135.7	1.0	117.8	2.2	59.8	3.4
Netherlands Antilles	..	..	..	..	..	..	..	..	..
Nicaragua	0.0	86.6	0.1	11.5	0.1	108.2	0.2	75.9	0.3
Panama	0.4	18.6	0.5	9.6	0.5	3.0	0.5	..	..
Paraguay	0.0	-2.5	0.0	510.7	0.1	76.5	0.1	179.4	0.3
Peru	0.1	165.7	0.3	45.8	0.5	150.8	1.2	36.7	1.7
Saint Kitts and Nevis	1.1	..	..	..	..	..	..	..	..
Saint Lucia	..	..	..	..	..	..	..	..	..
Saint Vincent and the Grenadines	0.9	4.6	1.0	14.2	1.1	174.0	3.1	52.3	4.7
Suriname	0.0	127.7	0.0	92.7	0.1	136.2	0.2	166.2	0.6
Trinidad and Tobago	0.0	827.0	0.1	377.0	0.3	153.9	0.8	89.4	1.5

Table 1.18 (continued)

	2002	% change 2002–2003	2003	% change 2003–2004	2004	% change 2004–2005	2005	% change 2005–2006	2006
Uruguay	..	..	..	..	0.8	125.9	1.9	62.9	3.1
Venezuela	0.3	46.9	0.5	76.4	0.8	66.1	1.3	48.0	2.0
Virgin Islands (United States)	..	..	..	..	1.4	97.5	2.7	..	..
OCEANIA									
American Samoa	..	..	..	..	..	..	..	..	..
Fiji	..	..	..	..	..	..	0.8	..	..
French Polynesia	..	..	0.4	350.7	1.7	146.9	4.3	63.0	7.0
Kiribati	..	..	..	..	..	..	..	..	..
Marshall Islands	..	..	..	..	..	..	..	..	..
Micronesia (Fed. States of)	..	..	..	..	..	..	..	..	..
Nauru	..	..	..	..	..	..	..	..	..
New Caledonia	0.3	133.8	0.7	202.7	2.2	83.2	4.1	..	..
Northern Mariana Islands	..	..	..	..	..	..	..	..	..
Palau	..	..	..	..	..	..	..	..	..
Papua New Guinea	..	..	..	..	..	..	..	..	..
Samoa	..	..	..	..	..	..	0.1	..	..
Solomon Islands	0.0	85.6	0.0	-2.2	0.0	91.0	0.1	..	..
Tonga	..	..	..	..	0.3	99.2	0.6	..	..
Tuvalu	..	..	..	..	..	..	..	..	..
Vanuatu	..	..	0.0	49.7	0.0	324.7	0.0	..	..
TRANSITION ECONOMIES									
Albania	..	..	..	..	..	..	..	..	0.0
Armenia	0.0	25.4	0.0	9933.0	0.0	100.7	0.1	..	..
Azerbaijan	..	..	..	..	0.0	140.9	0.0	..	..
Belarus	0.0	518.1	0.0	513.5	0.0	109.6	0.0	633.4	0.1
Bosnia and Herzegovina	0.0	602.4	0.0	337.9	0.2	104.3	0.4	191.9	1.0
Bulgaria	..	..	..	..	..	..	2.1	134.1	5.0
Croatia	0.3	..	..	..	0.6	225.5	2.0	179.8	5.5
Georgia	0.0	54.9	0.0	35.9	0.0	27.7	0.1	1035.2	0.6
Kazakhstan	..	..	..	..	0.0	50.4	0.0	917.4	0.2
Kyrgyzstan	0.0	283.8	0.0	1258.5	0.0	24.8	0.0	..	..
Moldova	0.0	43.5	0.0	304.0	0.1	334.1	0.2	109.7	0.5
Montenegro	..	..	..	..	..	..	..	..	4.2
Romania	0.1	1147.5	0.9	95.3	1.8	96.8	3.5	..	..
Russian Federation	0.0	..	..	..	0.5	135.3	1.1	84.3	2.0
Serbia	..	..	..	..	..	..	..	..	1.2
Tajikistan	..	..	..	..	..	..	..	..	..
TFYR Macedonia	..	..	..	..	..	..	0.6	192.1	1.8
Turkmenistan	..	..	..	..	..	..	..	..	..
Ukraine	..	..	..	..	..	..	..	..	..
Uzbekistan	..	..	0.0	92.9	0.0	50.3	0.0	..	..

Source: UNCTAD calculations based on the ITU World Telecommunication/ICT Indicators database, 2007.

Table 1.19

Economies ranked by 2006 broadband penetration

Broadband subscribers per 100 inhabitants

		2002	% change 2002–2003	2003	% change 2003–2004	2004	% change 2004–2005	2005	% change 2005–2006	2006
1	Bermuda	..	..	..	..	..	..	28.9	25.6	36.3
2	Netherlands	6.6	84.9	12.2	61.9	19.8	27.3	25.2	26.1	31.7
3	Denmark (incl. Faroe Islands)	8.3	58.3	13.2	41.9	18.7	32.4	24.7	27.7	31.6
4	Iceland	8.4	66.5	13.9	36.4	19.0	41.6	26.9	8.7	29.2
5	Republic of Korea	21.9	6.9	23.4	6.1	24.8	1.8	25.2	14.8	29.0
6	Switzerland	6.3	70.3	10.7	54.0	16.4	41.0	23.1	23.6	28.6
7	Norway	4.5	93.0	8.7	67.7	14.6	47.0	21.5	28.4	27.5
8	Finland	5.2	79.2	9.4	62.9	15.3	45.9	22.4	21.4	27.1
9	Sweden	8.0	10.2	8.8	55.5	13.7	55.6	21.4	21.1	25.9
10	Hong Kong (China)	15.3	21.7	18.6	18.0	22.0	7.0	23.5	6.0	24.9
11	Canada	11.2	27.2	14.2	18.8	16.9	22.6	20.7	13.9	23.6
12	United Kingdom (incl. Gibraltar)	3.1	108.5	6.4	85.5	11.9	34.1	16.0	35.8	21.7
13	France (incl. Guadeloupe. French Guiana. Martinique and Réunion)	2.8	111.5	6.0	83.5	11.0	42.8	15.6	33.4	20.9
14	Japan	7.4	58.5	11.7	31.0	15.3	14.3	17.5	15.2	20.2
15	Israel	3.3	187.3	9.4	52.8	14.3	24.7	17.8	11.5	19.9
16	Luxembourg	1.3	169.5	3.4	132.6	7.9	92.1	15.2	30.1	19.8
17	Taiwan Province of China	9.3	44.4	13.5	22.8	16.5	15.3	19.1	3.7	19.8
18	Belgium	7.9	51.8	12.0	29.5	15.5	23.9	19.2	0.1	19.2
19	Australia	1.3	97.8	2.6	98.1	5.2	102.5	10.4	83.6	19.1
20	United States (incl. Puerto Rico and Guam)	6.8	40.5	9.6	32.9	12.7	28.4	16.3	16.8	19.1
21	Singapore	6.5	55.8	10.1	21.2	12.3	24.8	15.3	18.9	18.2
22	Macao (China)	3.9	60.0	6.2	57.6	9.7	42.9	13.9	27.6	17.7
23	Austria	6.7	10.9	7.4	35.3	10.0	42.0	14.2	21.3	17.2
24	Germany	3.9	40.4	5.5	53.4	8.4	54.7	12.9	31.6	17.0
25	Estonia	3.4	98.4	6.7	24.2	8.3	60.9	13.3	27.6	17.0
26	Italy	1.5	158.2	3.9	97.4	7.7	52.1	11.7	27.3	14.9
27	Spain	3.0	69.9	5.2	54.5	8.0	40.2	11.2	32.0	14.8
28	New Zealand	1.1	90.3	2.1	128.6	4.8	71.0	8.2	72.8	14.2
29	Portugal	2.5	90.4	4.8	69.6	8.2	41.7	11.6	19.9	13.9
30	Slovenia	2.8	2.2	2.9	98.3	5.8	47.3	8.5	53.1	13.0
31	Ireland	0.3	289.4	1.1	258.5	3.8	73.3	6.5	88.4	12.3
32	Lithuania	0.6	236.9	1.9	93.2	3.7	82.4	6.8	58.0	10.8
33	Czech Republic	0.2	126.7	0.3	579.6	2.3	89.5	4.4	141.1	10.6
34	Malta	4.6	28.0	5.8	64.8	9.6	18.3	11.4	-9.3	10.3
35	Hungary	1.1	137.6	2.6	55.9	4.1	58.5	6.5	50.3	9.7
36	French Polynesia	..	..	0.4	350.7	1.7	146.9	4.3	63.0	7.0
37	Poland	0.3	61.0	0.5	314.9	2.1	53.3	3.3	112.4	6.9
38	Chile	1.3	84.7	2.3	34.4	3.1	46.2	4.5	30.7	5.9
39	Slovakia	..	..	0.1	537.8	0.9	181.6	2.6	128.6	5.9
40	Cyprus	0.7	68.5	1.3	30.0	1.6	95.8	3.2	81.7	5.8
41	Qatar	0.0	1137.9	0.4	241.7	1.4	124.2	3.2	72.7	5.6
42	Croatia	0.3	..	..	..	0.6	225.5	2.0	179.8	5.5

Table 1.19 *(continued)*

		2002	% change 2002–2003	2003	% change 2003–2004	2004	% change 2004–2005	2005	% change 2005–2006	2006
43	Bahrain	0.7	92.8	1.4	51.9	2.1	40.7	2.9	77.9	5.2
44	Grenada	0.6	5.5	0.6	0.5	0.6	420.4	3.0	70.3	5.2
45	United Arab Emirates	0.4	71.2	0.7	75.7	1.3	120.0	2.9	80.8	5.2
46	Bulgaria	..	..	..	..	..	..	2.1	134.1	5.0
47	San Marino	..	..	2.1	..	..	..	..	..	4.8
48	Latvia	0.4	96.2	0.8	154.9	2.1	22.6	2.6	81.2	4.8
49	Lebanon	1.0	95.4	2.0	12.7	2.3	61.1	3.6	29.7	4.7
50	Saint Vincent and the Grenadines	0.9	4.6	1.0	14.2	1.1	174.0	3.1	52.3	4.7
51	Greece	..	..	0.1	407.1	0.5	210.6	1.4	204.2	4.4
52	Montenegro	..	..	..	..	..	..	..	..	4.2
53	Argentina	0.3	102.1	0.6	110.0	1.3	67.5	2.2	83.8	4.0
54	China	0.4	106.4	0.9	121.8	1.9	49.9	2.9	35.3	3.9
55	Turkey	0.0	827.2	0.3	186.1	0.8	171.4	2.2	72.2	3.7
56	Malaysia	0.1	460.1	0.4	124.1	1.0	93.6	1.9	81.8	3.5
57	Mexico	0.2	82.4	0.4	135.7	1.0	117.8	2.2	59.8	3.4
58	Brazil	0.4	61.7	0.7	85.6	1.2	44.5	1.8	76.9	3.1
59	Uruguay	..	..	..	..	0.8	125.9	1.9	62.9	3.1
60	Brunei Darussalam	0.8	31.8	1.1	61.2	1.8	25.1	2.2	26.2	2.8
61	Russian Federation	0.0	..	..	..	0.5	135.3	1.1	84.3	2.0
62	Belize	..	..	0.4	200.7	1.1	45.8	1.6	26.2	2.0
63	Venezuela	0.3	46.9	0.5	76.4	0.8	66.1	1.3	48.0	2.0
64	TFYR Macedonia	..	..	..	..	..	..	0.6	192.1	1.8
65	Mauritius	0.0	312.4	0.1	126.7	0.2	12.6	0.2	600.8	1.7
66	Peru	0.1	165.7	0.3	45.8	0.5	150.8	1.2	36.7	1.7
67	Trinidad and Tobago	0.0	827.0	0.1	377.0	0.3	153.9	0.8	89.4	1.5
68	Maldives	0.1	155.6	0.2	37.8	0.2	188.7	0.7	122.6	1.5
69	Seychelles	..	..	..	..	0.4	62.8	0.7	123.5	1.5
70	Colombia	0.1	82.6	0.1	90.5	0.3	149.2	0.7	94.1	1.4
71	Costa Rica	0.0	3851.2	0.4	84.2	0.7	64.1	1.1	24.5	1.3
72	Morocco	0.0	33.8	0.0	2252.3	0.2	280.0	0.8	54.7	1.3
73	Serbia	..	..	..	..	..	..	..	..	1.2
74	Bosnia and Herzegovina	0.0	602.4	0.0	337.9	0.2	104.3	0.4	191.9	1.0
75	Saudi Arabia	0.2	28.6	0.2	45.5	0.3	-3.8	0.3	213.9	0.9
76	Jordan	0.1	52.7	0.1	103.8	0.2	128.5	0.4	96.0	0.8
77	Dominican Republic	..	..	..	..	0.4	74.4	0.7	-0.3	0.7
78	Iran (Islamic Republic of)	0.0	7.9	0.0	5.8	0.0	6.1	0.0	2079.9	0.7
79	Georgia	0.0	54.9	0.0	35.9	0.0	27.7	0.1	1035.2	0.6
80	Viet Nam	0.0	741.5	0.0	466.4	0.1	293.2	0.2	142.8	0.6
81	Suriname	0.0	127.7	0.0	92.7	0.1	136.2	0.2	166.2	0.6
82	Oman	0.0	39.1	0.0	387.3	0.0	1134.7	0.3	78.6	0.6
83	Moldova	0.0	43.5	0.0	304.0	0.1	334.1	0.2	109.7	0.5
84	Tunisia	..	..	0.0	8.5	0.0	474.5	0.2	162.5	0.4
85	Cape Verde	..	..	..	..	0.1	224.2	0.2	77.3	0.3
86	Nicaragua	0.0	86.6	0.1	11.5	0.1	108.2	0.2	75.9	0.3
87	Egypt	0.0	407.5	0.0	860.6	0.1	134.5	0.2	71.2	0.3
88	Paraguay	0.0	-2.5	0.0	510.7	0.1	76.5	0.1	179.4	0.3

Table 1.19 (continued)

		2002	% change 2002–2003	2003	% change 2003–2004	2004	% change 2004–2005	2005	% change 2005–2006	2006
89	Senegal	0.0	70.3	0.0	265.6	0.1	112.9	0.2	53.4	0.2
90	Kazakhstan	..	..	..	..	0.0	50.4	0.0	917.4	0.2
91	India	0.0	67.7	0.0	64.9	0.0	445.0	0.1	74.4	0.2
92	Sri Lanka	0.0	468.2	0.0	490.6	0.1	20.0	0.1	10.6	0.1
93	Belarus	0.0	518.1	0.0	513.5	0.0	109.6	0.0	633.4	0.1
94	Gabon	..	..	0.0	279.5	0.0	128.0	0.1	-22.5	0.1
95	Zimbabwe	..	..	0.0	92.1	0.1	2.9	0.1	-1.6	0.1
96	Ghana	..	..	..	..	..	..	0.0	553.7	0.1
97	Equatorial Guinea	..	..	..	..	..	..	..	..	0.0
98	Pakistan	..	..	..	..	0.0	98.1	0.0	24.4	0.0
99	Syrian Arab Republic	..	..	..	..	0.0	91.4	0.0	355.4	0.0
100	Mauritania	..	..	..	..	..	..	0.0	314.7	0.0
101	Burkina Faso	0.0	182.9	0.0	-2.8	0.0	70.9	0.0	534.7	0.0
102	Albania	..	..	..	..	..	..	..	..	0.0
103	Gambia	..	..	..	..	..	..	0.0	37.2	0.0
104	Sudan	..	..	..	..	0.0	26.5	0.0	10.6	0.0
105	Uganda	..	..	..	..	..	..	..	..	0.0
106	Benin	0.0	-3.3	0.0	273.5	0.0	134.2	0.0	-12.0	0.0
107	Afghanistan	..	..	..	..	0.0	5.2	0.0	118.4	0.0
108	Cameroon	..	..	..	..	..	..	..	..	0.0
109	Myanmar	..	..	..	..	..	..	0.0	66.8	0.0
110	Ethiopia	..	..	..	..	..	..	..	..	0.0

Developed economy.

Developing economy.

Transition economy.

Source: UNCTAD calculations based on the ITU World Telecommunication/ICT Indicators database, 2007.

Table 1.20 Core indicators on use of ICT by businesses and on the ICT sector

Enterprises with 10 or more employees, latest available year

	Reference year	Proportion of:				Proportion of enterprises:				Proportion of enterprises accesing the Internet by:				
		Enterprises using computers	Employees using computers	Enterprises using Internet	Employees using Internet	With a website (or those using Internet)	With an intranet	Receiving orders online (or those using Internet)	Placing orders online (or those using Internet)	Analogue modem	ISDN	Fixed line connection under 2 Mbps	Fixed line connection of 2 Mbps or more	Other modes of access
		B1	B2	B3	B4	B5	B6	B7	B8	B9.a	B9.b	B9.c	B9.d	B9.e
Argentina	2005	100.0	39.7	95.8	24.8	74.4	47.0	45.6	44.6	15.4	..	..	..	9.6
Australia	2005	95.5	..	87.3	..	51.8	..	20.6	54.7	24.2	7.1	..	68.7	..
Austria	2006	98.0	53.0	98.0	..	80.0	41.0	18.0	51.0	5.0	34.0	29.0	71.0	16.0
Azerbaijan	2006	38.4	9.1	8.7	1.7	32.6	..	..	..	67.3	4.8	12.1	5.1	17.8
Belarus	2005	83.6	..	37.6	..	27.2	..	..	..	..	..	..	..	..
Belgium	2006	97.0	57.0	95.0	41.0	72.0	45.0	15.0	44.0	17.0	28.0	26.0	89.0	12.0
Bermuda	2005	80.0	62.0	71.0	37.0	56.3	34.0	14.1	40.8	..	..	..	..	..
Brazil	2006	99.4	47.6	94.3	36.5	49.6	39.0	50.3	52.2	13.9	..	68.7	4.7	4.8
Bulgaria	2006	89.6	21.2	74.8	14.5	43.8	35.0	4.7	8.4	26.4	10.9	26.5	6.0	19.1
Canada	2006	..	..	94.9	..	71.2	..	13.2	65.0	..	..	..	21.9	..
Chile	2005	60.3	..	48.8	..	38.8	..	4.2	6.7	16.4	..	19.4	80.6	..
China	2005	..	..	47.4	..	23.7	..	12.4	9.6	13.6	6.0	32.9	0.2	..
Cuba	2006	94.9	58.7	70.9	29.6	23.7	34.0	1.0	3.7	51.1	..	..	..	0.1
Cyprus	2006	95.0	43.0	86.0	31.0	50.0	21.0	6.0	21.0	33.0	26.0	11.0	63.0	5.0
Czech Republic	2006	97.0	39.0	95.0	29.0	74.0	23.0	9.0	27.0	12.0	27.0	41.0	73.0	31.0
Denmark	2006	98.0	68.0	98.0	61.0	85.0	35.0	35.0	59.0	4.0	17.0	17.0	84.0	10.0
Egypt	2006	100.0	17.9	53.2	9.9	71.0	34.0	34.8	21.0	6.3	1.8	71.9	22.8	0.9
Estonia	2006	94.0	38.0	92.0	33.0	63.0	35.0	14.0	25.0	10.0	17.0	20.0	82.0	9.0
Finland	2006	99.0	67.0	99.0	59.0	81.0	39.0	12.0	56.0	19.0	22.0	21.0	90.0	22.0
France	2006	99.0	63.0	94.0	34.0	65.0	40.0	16.0	26.0	..	22.0	22.0	92.0	..
Germany	2006	96.0	56.0	95.0	39.0	77.0	41.0	19.0	54.0	4.0	38.0	13.0	77.0	7.0
Greece	2006	97.0	37.0	94.0	26.0	64.0	39.0	8.0	14.0	28.0	43.0	10.0	62.0	5.0
Hong Kong (China)	2006	87.8	58.0	82.8	45.9	51.5	28.9	2.9	21.6	3.6	..	93.4	10.2	10.6
Hungary	2006	89.0	31.0	80.0	21.0	53.0	17.0	11.0	12.0	14.0	26.0	26.0	77.0	13.0
Iceland	2006	98.8	57.9	97.0	46.0	71.7	36.1	7.1	12.6	10.2	13.4	63.9	21.5	5.3
Ireland	2006	97.0	54.0	94.0	37.0	67.0	46.0	23.0	56.0	27.0	33.0	30.0	64.0	9.0
Italy	2006	96.0	40.0	93.0	28.0	61.0	33.0	3.0	27.0	23.0	30.0	8.0	75.0	8.0

Table 1.20 *(continued)*

	Reference year	Proportion of:				Proportion of enterprises:				Proportion of enterprises accesing the Internet by:				
		Enterprises using computers	Employees using computers	Enterprises using Internet	Employees using Internet	With a website (or those using Internet)	With an intranet	Receiving orders online (or those using Internet)	Placing orders online (or those using Internet)	Analogue modem	ISDN	Fixed line connection under 2 Mbps	Fixed line connection of 2 Mbps or more	Other modes of access
		B1	B2	B3	B4	B5	B6	B7	B8	B9.a	B9.b	B9.c	B9.d	B9.e
Japan	2005	..	..	97.7	..	85.6	89.5	15.6	20.6	9.6	16.1	16.0	63.5	68.1
Latvia	2005	89.7	26.5	75.0	20.9	43.5	21.6	3.7	15.3	10.3	20.6	..	73.7	12.6
Lithuania	2006	92.0	27.0	88.0	23.0	47.0	57.0	15.0	22.0	33.0	14.0	21.0	65.0	17.0
Luxembourg	2006	98.0	49.0	93.0	32.0	65.0	44.0	..	40.0	18.0	42.0	16.0	81.0	14.0
Macao (China)	2003	75.6	..	53.3	..	26.3	..	15.8	21.0	9.9	..	8.1	78.8	::
Malta	2005	93.0	..	90.0	..	61.0	43.0	14.0	47.0	20.0	7.0	21.0	87.0	7.0
Mauritius	2006	94.4	..	87.4	..	46.1	37.3	32.9	34.8	..	..	..	..	::
Netherlands	2006	100.0	61.0	97.0	45.0	81.0	36.0	28.0	45.0	6.0	23.0	15.0	84.0	5.0
New Zealand	2006	96.4	98.7	94.5	97.4	62.7	22.4	36.8	60.3	35.0	..	81.6	..	21.1
Norway	2006	97.0	59.0	94.0	50.0	76.0	34.0	25.0	66.0	6.0	16.0	33.0	91.0	15.0
Panama	2006	89.5	32.2	80.1	20.3	..	28.0	39.0	43.5	8.2	4.0	61.7	..	36.8
Poland	2006	93.0	38.0	89.0	28.0	60.0	30.0	7.0	23.0	39.0	34.0	16.0	52.0	14.0
Portugal	2006	95.0	35.0	83.0	25.0	42.0	33.0	5.0	20.0	25.0	18.0	24.0	79.0	5.0
Qatar	2005	84.4	97.1	68.4	90.0	99.0	38.2	50.9	41.3	..	..	..	..	::
Republic of Korea	2005	96.6	..	95.9	..	58.9	37.3	7.9	33.9	0.7	0.8	..	98.2	0.2
Romania	2005	77.3	22.4	58.4	15.9	41.2	23.2	4.1	11.0	33.8	11.8	86.4	7.8	42.4
Russian Federation	2005	91.1	29.8	53.3	12.4	27.8	..	23.6	30.7	..	..	..	..	::
Singapore	2006	92.8	..	91.0	..	75.0	74.1	14.8	33.9	27.0	24.7	73.0	19.2	15.0
Slovakia	2006	97.0	39.0	93.0	29.0	65.0	31.0	7.0	22.0	19.0	32.0	18.0	65.0	29.0
Slovenia	2006	97.0	48.0	96.0	35.0	65.0	27.0	12.0	22.0	10.0	23.0	14.0	78.0	9.0
Spain	2006	98.0	49.0	93.0	35.0	50.0	28.0	8.0	16.0	16.0	19.0	8.0	94.0	9.0
Sweden	2006	96.0	66.0	96.0	53.0	90.0	43.0	23.0	70.0	18.0	17.0	30.0	92.0	28.0
Switzerland	2005	99.1	56.5	98.2	47.6	91.6	61.4	23.2	58.0	..	..	54.8	42.2	::
Thailand	2006	88.4	..	69.6	..	50.5	..	11.1	13.9	52.2	5.3	..	39.4	19.4
Turkey	2005	87.8	40.6	80.4	34.1	59.9	38.9	..	..	35.3	6.8	52.3	27.4	14.1
United Kingdom	2006	96.0	51.0	93.0	42.0	81.0	34.0	19.0	62.0	37.0	33.0	16.0	83.0	2.0

Table 1.20 (continued)

	Reference year	Proportion of enterprises with:		Proportion of enterprises using the Internet for:								
		LAN	An extranet	Sending and receiving e-mail	Information about goods or services	Information from public authorities	Other information searches or research	Internet banking or financial services	Transacting with public authorities	Providing customer services	Delivering products online	Other types of activity
		B10	B11	B12.a	B12.b.i	B12.b.ii	B12.b.iii	B12.c	B12.d	B12.e	B12.f	B12.g
Argentina	2005	81.9	20.5	97.2	88.1	74.9	40.2	83.7	56.6	43.0	6.1	54.9
Australia	2005	..	..	..	..	..	..	..	50.4	..	..	..
Austria	2006	..	19.0	..	..	..	47.0	88.0	81.0	..	..	..
Azerbaijan	2006	10.5	..	..	..	25.8	..	25.5	26.4	..	..	..
Belarus	2005	41.1	..	..	..	..	..	..	..	..	..	..
Belgium	2006	..	28.0	..	..	..	62.0	88.0	59.0	..	..	..
Bermuda	2005	..	..	..	..	..	..	..	..	..	..	..
Brazil	2006	94.5	21.6	97.7	78.4	59.4	82.4	80.1	84.1	30.9	13.6	..
Bulgaria	2006	52.8	4.1	..	..	57.3	46.9	53.4	61.4	..	..	..
Canada	2006	..	16.7	97.6	..	..	..	..	..	..	..	..
Chile	2005	12.6	3.4	99.2	..	..	..	..	..	..	..	..
China	2005	16.3	..	80.4	65.0	46.1	38.9	..	37.4	35.2	11.0	..
Cuba	2006	..	..	..	..	..	..	..	..	..	..	..
Cyprus	2006	..	7.0	..	..	..	67.0	57.0	44.0	..	..	..
Czech Republic	2006	..	7.0	..	..	..	71.0	91.0	76.0	..	..	..
Denmark	2006	..	22.0	..	..	..	53.0	94.0	87.0	..	..	..
Egypt	2006	78.9	2.4	93.3	58.9	..	58.9	26.8	5.8	36.2	..	0.4
Estonia	2006	..	12.0	..	..	..	49.0	98.0	69.0	..	..	..
Finland	2006	..	25.0	..	..	..	71.0	93.0	93.0	..	..	..
France	2006	..	22.0	..	..	..	..	77.0	66.0	..	..	..
Germany	2006	..	24.0	..	..	..	67.0	77.0	49.0	..	..	..
Greece	2006	..	11.0	..	..	..	62.0	74.0	84.0	..	..	..
Hong Kong (China)	2006	60.7	10.1	96.9	96.0	72.6	..	42.2	..	23.4	43.2	53.0
Hungary	2006	..	4.0	..	..	..	52.0	68.0	45.0	..	..	..
Iceland	2006	50.3	30.1	..	90.0	87.3	..	94.3	64.7	..	..	..
Ireland	2006	..	18.0	..	..	..	46.0	86.0	84.0	..	..	..
Italy	2006	..	13.0	..	..	..	66.0	81.0	87.0	..	..	..

Table 1.20 (continued)

	Reference year	Proportion of enterprises with:		Proportion of enterprises using the Internet for:								
		LAN	An extranet	Sending and receiving e-mail	Information about goods or services	Information from public authorities	Other information searches or research	Internet banking or financial services	Transacting with public authorities	Providing customer services	Delivering products online	Other types of activity
		B10	B11	B12.a	B12.b.i	B12.b.ii	B12.b.iii	B12.c	B12.d	B12.e	B12.f	B12.g
Japan	2005	39.6	60.1	..	..	..	..	..	..	..	..	..
Latvia	2005	65.9	7.6	..	..	46.3	..	89.2	50.1	..	..	..
Lithuania	2006	..	8.0	..	..	..	61.0	94.0	76.0	..	..	..
Luxembourg	2006	..	25.0	..	..	..	53.0	76.0	83.0	..	..	..
Macao (China)	2003	..	..	88.8	..	20.1	68.5	..	..	14.8	..	3.8
Malta	2005	..	23.0	..	..	..	56.0	67.0	68.0	..	..	..
Mauritius	2006	..	..	..	..	..	..	..	..	..	..	..
Netherlands	2006	..	15.0	..	..	..	52.0	76.0	70.0	..	..	..
New Zealand	2006	61.6	7.5	..	..	68.0	..	87.2	76.8	29.9	..	..
Norway	2006	..	16.0	..	..	..	55.0	92.0	74.0	..	..	..
Panama	2006	53.3	13.7	97.3	80.6	67.9	60.7	70.1	35.8	39.1	..	69.6
Poland	2006	..	7.0	..	..	..	56.0	75.0	61.0	..	..	..
Portugal	2006	..	20.0	..	..	..	40.0	75.0	60.0	..	..	..
Qatar	2005	..	..	..	..	..	..	..	..	..	..	..
Republic of Korea	2005	66.5	..	88.6	60.9	53.5	77.5	67.4	43.4	34.5	13.2	2.4
Romania	2005	45.1	19.1	93.9	..	64.9	65.3	51.7	10.2	8.7	3.6	..
Russian Federation	2005	52.4	..	91.6	54.7	42.5	..	14.9	..	4.5	5.3	..
Singapore	2006	74.1	35.8	92.7	93.4	..	..	64.1	..	..	41.5	..
Slovakia	2006	..	12.0	..	..	..	64.0	84.0	77.0	..	..	..
Slovenia	2006	..	13.0	..	..	..	77.0	93.0	75.0	..	..	..
Spain	2006	..	13.0	..	..	..	33.0	85.0	58.0	..	..	..
Sweden	2006	..	20.0	..	..	..	72.0	92.0	80.0	..	..	..
Switzerland	2005	79.6	33.1	..	97.7	..	59.5	85.0	56.5	21.3	22.4	..
Thailand	2006	..	..	80.7	..	..	65.0	9.5	..	24.0	20.5	14.2
Turkey	2005	64.6	7.6	..	..	56.3	..	75.4	62.5	15.5	38.0	..
United Kingdom	2006	..	10.0	..	..	..	53.0	73.0	52.0	..	..	..

Note: See after table 1.20. in the Statistical Annex for notes to data.

Source: UNCTAD information economy database, 2007.

Notes to data for tables 1.5–1.10 and 1.20

Argentina: Data from the "Encuesta Nacional a Empresas sobre Innovación, Investigación y Desarrollo y Tecnologías de la Información y de las Comunicaciones" conducted by the Instituto Nacional de Estadística y Censos de Argentina (INDEC). The survey refers only to the manufacturing sector. Enterprise size is defined by revenue, not number of employees. Results refer to the sample and have not been extrapolated to the target population, but INDEC considers them representative.

Australia: Data from the "Business Use of Information Technology Survey" conducted by the Australian Bureau of Statistics (ABS). The survey is not designed to provide high-quality estimates of numbers of businesses for any of the output classifications (for example, state and territory or industry); a more robust source of counts of Australian businesses is available from the ABS Business Register, Counts of Businesses (cat. no. 8161.0.55.001). Data refer to enterprises with five or more employees. Enterprises "transacting with Government organisations/public authorities" (B12.d) refer to enterprises making "electronic lodgements with government organisations".

Austria: Data from the Eurostat database. "Proportion of employees using computers" (B2) refers to "percentage of employees using any computer (at least once a week)". "Proportion of enterprises using the Internet" (B3) refers to "percentage of enterprises with Internet access". "Proportion of employees using the Internet" (B4) refers to "percentage of persons employed using computers connected to the Internet in their normal work routine at least once a week". "Proportion of enterprises using the Internet that receive orders online" (B7) refers to "percentage of enterprises which have received orders via Internet over the last calendar year (excluding manually typed e-mails)". "Proportion of enterprises using the Internet that place orders online" (B8) refers to "percentage of enterprises which have ordered via Internet over the last calendar year (excluding manually typed e-mails)". "Proportion of enterprises accessing the Internet by fixed line connection of 2 Mbps or more" (B9.d) refers to "percentage of enterprises with broadband access". "Proportion of enterprises accessing the Internet by fixed line connection under 2 Mbps" (B9.c) refers to "percentage of enterprises connecting to the Internet via another fixed connection (e.g. cable etc.)". "Proportion of enterprises accessing the Internet by other modes of access" (B9.e) refers to "percentage of enterprises connecting to the Internet via wireless connection (satellite, mobile phones etc.)" and refers to 2005. "Proportion of enterprises using the Internet for other information searches or research" (B12.b.iii) refers to "percentage of enterprises that use the Internet for market monitoring (e.g. prices)". "Proportion of enterprises using the Internet for transacting with public authorities" (B12.d) refers to "percentage of enterprises which use the Internet for interaction with public authorities".

Azerbaijan: Data from the Azerbaijan "Census of enterprises" conducted by the State Statistical Committee.

Belarus: Data from the survey on "Usage of global information networks" conducted by the Ministry of Statistics and Analysis.

Belgium: See notes for Austria.

Bermuda: Data from the annual "Economic Activity Survey" (EAS) and "Employment Survey" (ES) conducted by the Department of Statistics of Bermuda.

Brazil: Data from the "Survey on the use of ICT" conducted by the Brazilian Internet Steering Committee. There were methodological changes in the sampling of the survey from 2005 to 2006, and so caution should be used when comparing results for both years. For example, in 2005 the "proportion of enterprises using the Internet for other types of activity" (B12.g) included enterprises that ordered travel and accommodation services via the Internet in the previous 12 months, paid online for any product/service ordered via the Internet in the previous 12 months, or sold any product to another enterprise via a specialized Internet marketplace in the previous 12 months. This response category was not included for 2006. The sampling frame is the RAIS (Relação Anual de Informações Sociais, or Social Information Annual List), which covers approximately 97 per cent of the formal sector.

Bulgaria: Data from the annual "Survey on ICT usage in enterprises" conducted by the National Statistical Institute.

For 2005, "other modes of access" to the Internet (B9.e) included wireless connection (e.g. satellite, mobile phone) and other broadband connection (e.g. cable).

Canada: Data from the annual "Survey of Electronic Commerce and Technology" conducted by Statistics Canada. Enterprises using the Internet for "delivering products online" (B12.f) refer to enterprises "delivering digitised products (via Web site or other Internet)".

Chile: Data from the 2006 Structural Surveys on trade (commerce), services, and hotels and restaurants, conducted by the Instituto Nacional de Estadística.

China: Data from the "Specialized Survey of ICT Usage of China in 2005" conducted by the National Bureau of Statistics of China.

Cuba: Data from the "Censo Económico TIC" (ICT Economic Census) conducted by the Dirección de Turismo, Comercio y Servicios (Tourism, Trade, and Services Directorate). Results refer to the sample and have not been extrapolated to the target population.

Cyprus: See notes for Austria.

Czech Republic: See notes for Austria.

Denmark: See notes for Austria.

Egypt: Data from the 2007 survey on "Use of ICT in businesses" conducted by the Central Agency for Public Mobilization and Statistics (CAPMAS). The results of the survey refer to the sample and have not been extrapolated to the target population, but CAPMAS indicates that the results can be considered representative at the national level.

Estonia: See notes for Austria.

Finland: See notes for Austria.

France: See notes for Austria.

Germany: See notes for Austria.

Greece: See notes for Austria.

Hong Kong (China): Data from the "Annual Survey on Information Technology Usage and Penetration in the Business Sector" conducted by the Census and Statistics Department. "Fixed line connection under 2 Mbps" (B9. c) refers to Internet and Internet-related services with a transmission speed from hundreds of Kbps to several Mbps (cable modems, asynchronous transfer mode, Ethernet, ADSL (asymmetric digital subscriber line) and other types of DSL (digital subscriber line) are commonly used technologies). "Fixed line connection over 2 Mbps" (B9.d) refers to dedicated circuits (not fixed-line) or unswitched connections with capacity dedicated to the users. "Other modes of access" refers to the public cellular telephone network or Wireless Fidelity. Enterprises using the Internet for "getting information from Government organisations/public authorities" (B12.b.ii) include enterprises engaging in transactions with authorities. Enterprises using the Internet "for other types of activity" (B12.g) include online purchase/ordering and sales of goods, services or information, software download and miscellaneous activities.

Hungary: See notes for Austria.

Iceland: Data from "Enterprises' use of ICT" survey conducted by Statistics Iceland.

Ireland: See notes for Austria.

Italy: See notes for Austria.

Japan: Data from the annual "Communications Usage Trend Survey" conducted by the Ministry of Internal Affairs and Communications. The results of this survey refer to the sample and have not been extrapolated to the target population. Data refer to enterprises with 100 or more employees. "Fixed line connection under 2 Mbps" and "fixed line connection of 2 Mbps or more" (B9.c and B9.d) refer, respectively, to under 1.5Mbps and 1.5Mbps or more. "Other modes of access" (B9.e) refer to accessing by broadband (CATV, FTTH, FWA, DSL etc.).

Latvia: Data from the annual "Community Survey on ICT Usage and e-Commerce in Enterprises" conducted by the Central Statistical Bureau. "Fixed line connection of 2 Mbps or more" (B9.d) refers to all types of fixed broadband (DSL and other fixed Internet connection, e.g. cable, leased line) and includes connection speeds under 2 Mbps.

Lithuania: See notes for Austria.

Luxembourg: "Proportion of enterprises using the Internet that place orders online" (B8) refers to "percentage of enterprises which have ordered via the Internet over the last calendar year (excluding manually typed e-mails)" and refers to 2005. For other indicators, see notes for Austria.

Macao (China): Data from the survey on "Usage of Information Technology in Business Sector" conducted by the Statistics and Census Service.

Malta: See notes for Austria.

Mauritius: Data from the "Survey of Employment and Earnings, March 2006" conducted by the Central Statistics Office. The results of this survey refer to the sample and have not been extrapolated to the target population.

Netherlands: See notes for Austria.

New Zealand: Data from the "Business Operations Survey 2006" conducted by Statistics New Zealand. Enterprises "accessing the Internet by analogue modem" (B9.a) include enterprises accessing the Internet by ISDN (B9.b). Enterprises "accessing the Internet by fixed line connection under 2 Mbps" (B9.c) refer to any fixed-line connection regardless of speed (including of 2 Mbps or more). "Enterprises using the Internet for providing customer services" (B.12e) includes delivery of products of online (B.12f) and other types of activity (B.12g).

Norway: See notes for Austria.

Panama: Data from the "Survey of Non-Financial Enterprises" (Encuesta Entre Empresas No Financieras) conducted by the Statistics and Census Directorate (Dirección de Estadística y Censo) of the Contraloría General de la República. Preliminary figures.

Poland: See notes for Austria.

Portugal: See notes for Austria.

Qatar: Data from an economy-wide ICT survey conducted by the Statistics Department of the Planning Council.

Republic of Korea: Data from the "Survey on Information Society" conducted by the National Information Society Agency (NIA).

Romania: Data from annual survey on "ICT Usage and e-Commerce in Enterprises" conducted by the National Statistical Institute (Structural Business Statistics Department).

Russian Federation: Data from the annual survey on "Statistics on use of ICT and production of ICT goods and services" conducted by the National Statistical Office and the State University Institute of Stastical Studies and the Knowledge Economy. Enterprises using the Internet for "banking or accessing other financial services" (B12.c) include enterprises using the Internet to pay for supply products (procurement).

Singapore: Data from the annual "Survey on Infocomm Usage by Companies" conducted by the Infocomm Development Authority (IDA). "Fixed line connection" (B9.c and B9.d) refers to frame relay/dedicated leased line, XDSL and cable modem. Owing to changes in methodology, results for 2005 and 2006 are not directly comparable.

Slovakia: "Proportion of enterprises using Internet that receive orders online" (B7) refers to "percentage of enterprises which have received orders via the Internet over the last calendar year (excluding manually typed e-mails)" and refers to 2005. "Proportion of enterprises using the Internet that place orders online" (B8) refers to "percentage of enterprises which have ordered via the Internet over the last calendar year (excluding manually typed e-mails)" and refers to 2005. For other indicators, see notes for Austria.

Slovenia: See notes for Austria.

Spain: See notes for Austria.

Sweden: See notes for Austria.

Switzerland: Data from the "Swiss Innovation Survey 2005" conducted by ETH Zurich, KOF Swiss Economic Institute.

Thailand: Data from the annual "Information and Communication Technology Survey (Establishment)" conducted by the National Statistical Office. Data refer to enterprises with 16 or more employees. Enterprises accessing the Internet "by fixed line connection of 2 Mbps or more" (B9.d) refer to xDSL. Enterprises accessing the Internet "by other modes of access" (B9.e) refer to cable modem, leased line, wireless and others.

Turkey: Data from the "2005 ICT Usage Survey in Enterprises" conducted by TURKSTAT.

United Kingdom: See notes for Austria.

References

Bigné E, Ruiz C, and Sanz S (2007). Key drivers of mobile commerce adoption: an exploratory study of Spanish mobile users", *Journal of Theoretical and Applied Electronic Commerce Research*, Universidad de Talca, Chile, vol. 2(2), 48–60, http://www.jtaer.com/.

Brazilian Internet Steering Committee (2006). *Survey on the Use of Information and Communication Technologies in Brazil 2006*. São Paulo.

Chinn M and Fairlie R (2004). The determinants of the global digital divide: a cross-country analysis of computer and Internet penetration, Santa Cruz Center for International Economics, Working Paper Series 1022, Center for International Economics, University of California Santa Cruz, http://repositories.cdlib.org/cgi/viewcontent.cgi?article=1022&context=sccie.

Chinn M and Fairlie R (2006). ICT use in the developing world: an analysis of differences in computer and Internet penetration, NET Institute Working Paper No. 06-03, University of Wisconsin, Madison, Robert M. La Follette School of Public Affairs and Department of Economics, and University of California Santa Cruz, http://www.ssc.wisc.edu/~mchinn/developing6.pdf.

Curry J and Kenney M (2006). Digital divide or digital development? The Internet in Mexico, *First Monday*, vol. 11, no. 3 (March 2006), http://firstmonday.org/issues/issue11_3/curry/index.html.

Davison E and Cotten S (2003). Connection discrepancies: unmasking further layers of the digital divide, *First Monday*, vol. 8, no. 3 (March 2003), http://firstmonday.org/issues/issue8_3/davison/index.html.

Economist Intelligence Unit (2007). The 2007 e-readiness rankings: raising the bar. White paper written in cooperation with the IBM Institute for Business Value, http://www.eiu.com/site_info.asp?info_name=eiu_2007_e_readiness_rankings.

Hinson R, Sorensen O and Buatsi S (2007). Internet use patterns amongst internationalizing Ghanaian exporters, *Electronic Journal on Information Systems in Developing Countries* 29, 3, pp. 1–14, http://www.ejisdc.org/.

ITU and UNCTAD (2007). *World Information Society Report 2007: Beyond WSIS*, http://www.itu.int/osg/spu/publications/worldinformationsociety/2007/index.html.

Jensen R (2007). The digital provide: information (technology), market performance, and welfare in the South Indian fisheries sector, *Quarterly Journal of Economics*, MIT Press, vol. 122(3), pp. 879–924, http://www.mitpressjournals.org/doi/pdfplus/10.1162/qjec.122.3.879.

Kotelnikov V (2007). *Small and Medium Enterprises and ICT*. Asia-Pacific Development Information Programme – e-Primers for the Information Economy, Society and Polity.

National Internet Development Agency of Korea (NIDA) (2007). *2007 Korea Internet White Paper*.

O'Donnell J, Jackson M, Shelly M, and Ligertwood J (2007). Australian case studies in mobile commerce, *Journal of Theoretical and Applied Electronic Commerce Research*, Universidad de Talca, Chile, vol. 2(2), pp. 1-18, http://www.jtaer.com/.

OECD (2004). ICT, e-business and SMEs. Paper for the OECD Conference of Ministers Responsible for SMEs, Istanbul, 3–5 June 2004.

OECD (2006). ICT diffusion to business: peer review country report Mexico. Report for the Working Party on the Information Economy (DSTI/ICCP/IE(2006)4/FINAL).

OECD (2007a). *The OECD Communications Outlook 2007*, http://213.253.134.43/oecd/pdfs/browseit/9307021E. PDF.

OECD (2007b). Fixed-mobile convergence: market developments and policy issues. Report for the Working Party on Communication Infrastructures and Services Policy (DSTI/ICCP/CISP(2006)4/FINAL).

Research ICT Africa (RIA!) (2006). *Towards and African e-Index: SME e-Access and Usage across 14 African Countries*, http://www.researchictafrica.net/images/upload/SME_book-Web.pdf.

World Bank (2006). *Information and Communications for Development: Global Trends and Policies.* Washington, DC: World Bank. http://www-wds.worldbank.org/external/default/ WDSContentServer/WDSP/IB/2006/04/20/000012009_20060420105118/Rendered/PDF/ 359240PAPER0In101OFFICIAL0USE0ONLY1.pdf.

Yannopoulos D (2006). Greece hits EU Internet-use bottom, *Athens News*, 24 April 2006, http://www. helleniccomserve.com/percentofgreeks.html.

Notes

1. There is no established convention for the designation of «developed» and «developing» countries or areas in the United Nations system. In common practice, Japan in Asia, Canada and the United States in North America, Australia and New Zealand in Oceania, and Europe are considered «developed» regions or areas. In international trade statistics, the Southern African Customs Union is also treated as a developed region and Israel as a developed country; countries emerging from the former Yugoslavia are treated as developing countries. Countries of Eastern Europe and of the Commonwealth of Independent States in Europe are not included under either developed or developing regions, although Bulgaria and Romania have recently joined the European Union; «countries in transition from centrally planned to market economies», or transition economies, is a grouping used for economic analysis. Source: United Nations Statistics Division, http://unstats.un.org/unsd/methods/m49/m49regin.htm.

2. See http://measuring-ict.unctad.org for information on the UNCTAD XII Partnership on Measuring ICT for Development, which works to improve the availability of ICT data in developing countries.

3. On this, the ITU website says that «Examples include differences in availability of infrastructure or advanced technologies by area or region, or urban/rural differences in income and hence in the affordability of ICTs (as a proportion of income). Highlighting internal disparities along these lines can raise national debate in many countries and promote action to address inequality.» http://www.itu.int/osg/spu/publications/worldinformationsociety/2006/faq.html.

4. This includes wireless application protocol (WAP), general packet radio service (GPRS), and universal mobile telecommunications system (UMTS).

5. Statement by Viviane Reding, member of the European Commission responsible for Information Society and Media, Connecting up the global village: a European view on telecommunications policy, at the Conference of the ITU, Telecom World 2006, Hong Kong (China), 4 December 2006. Document ref: SPEECH/06/772.

6. See http://www.fibreforafrica.net/ and http://mybroadband.co.za/news/Telecoms/827.html; http://www.scidev.net/News/index.cfm?fuseaction=readNews&itemid=3825&language=1.

7. O'Donnell et al. (2007) state as follows: «mServices encompass a wide range of wireless applications that involve individuals or enterprises… While mCommerce encompasses B2B and B2C transactions, mEnterprise concerns the use of mobile devices in inter- and intra-business operations. MEnterprise projects generally entail some aspect of business process reengineering. These areas overlap and the definitions are not precise.»

8. Mobile devices that can support up to 14.4 Mbps download speed are expected to become available in 2008 in the Republic of Korea.

9. World Bank Investment Climate Surveys, 2000–2003.

10. WiBro is high-speed wireless Internet and data network access over a wide area.

11. Eurostat, Statistics in Focus, 28/2006.

12. A *maquiladora* or *maquila* is a factory that imports materials and equipment on a duty-free and tariff-free basis for assembly or manufacturing and then re-exports the assembled product, usually back to the originating country.

13. To explain the difference between online sales and purchases, the EU states that «adoption of online sales can be more complex than purchases, as it can entail a new business model for the enterprise». Statistics in focus 28/2006, http://epp.eurostat.ec.europa.eu/cache/ITY_OFFPUB/KS-NP-06-028/EN/KS-NP-06-028-EN.PDF.

14. B2C trade volume for 2006 was 3.6 trillion Won (KRW) and B2B trade volume was 366.2 trillion KRW, at an exchange rate of 1 KRW = $0.00109292. Live rate of 28 September 2007 at http://www.xe.com/.

15. As at 30 June 2007. See http://www.alibaba.com/.

16. http://www.financialexpress.com/print.php?content_id=167102.

17. Interview with Mr. Walid Kooli, focal point for e-commerce at the Tunisian Ministry for Trade and Handicrafts, in the ITC e@work newsletter, September 2007.

18. See http://www.strategyanalytics.net/ and http://www.ifpi.org/.

Chapter 2

THE ICT PRODUCING SECTOR AND THE EMERGING SOUTH

A. Introduction

The role of the ICT sector in economic development and employment creation in developing countries has been widely recognized and has been discussed in previous editions of the *Information Economy Report* (UNCTAD, 2003, 2005a, 2006a). Only a few developing countries have a significant ICT goods industry. At the same time, the development of efficient and cost-effective ICT services (particularly telecommunication services) is critical for creating an enabling environment for domestic ICT uptake. Moreover, the ICT sector contributes to productivity and economic growth. As has been shown in other parts of this Report (see introductory chapter and chapter 3), the rapid technological progress in the production of ICT goods and services increases total factor productivity in the ICT-producing sector and contributes to GDP growth. ICTs used as a capital input in other sectors and services also contribute to labour productivity and total factor productivity.

As this chapter will show, the strong growth in ICT production, trade and investment since the mid-1990s has continued over the past few years, in particular in developing and transition countries, where ICT sector production and consumption have grown much faster and have higher shares than in OECD countries. The (sharp) decline in global ICT production, trade and investment following the Nasdaq crash in 2000 has been completely reversed. With the "second generation" of globalization emerging, developing countries provide an important contribution to the output of the ICT sector.[1]

ICT production and use impact on globalization in several ways. First, they are an enabler of globalization. ICTs play a critical role in the fragmentation of the global value-added chain and in shifting parts of production to different geographic at locations. This process has now been extended to also take into account the service industry and delocalization to lower-cost markets. By using ICTs, firms are able to exchange knowledge and information online from anywhere in the world, communicate just-in-time with clients and suppliers, and deliver services efficiently and promptly. Through outsourcing and offshoring, services can be provided by more cost-effective suppliers. This has contributed to a major shift in production from high- to low-cost locations, as well as a shift in investment and trade flows. At the international level, this has resulted in deeper specialization, with more productivity-driven gains on the outsourcer's side and more employment-driven gains in the host country.

Second, the ICT-producing sector itself is highly globalized. ICT components and parts, and telecom and computer equipment, are manufactured in very different locations, and there are significant shifts in production, from developed to developing countries. In 2004, the ICT sector had a lead role in international trade (13 per cent), mergers and acquisitions (20 per cent) and business value-added (9 per cent) (OECD, 2006a). In some Asian developing countries, the ICT sector contributes up to 17 per cent (Malaysia) and 21 per cent (Philippines) of business-sector value-added (UNCTAD, 2006a). The ICT sector (supply and demand) will continue to grow strongly in the next decade, given the increasing role of ICT products and use in the global economy.

FDI in the ICT sector is growing strongly, especially in ICT manufacturing (electronic and electrical equipment) and ICT services (communication services) (UNCTAD, 2005b). Developing countries are increasingly a destination for FDI flows in the ICT sector, especially for ICT services (UNCTAD, 2004). While most of those flows are targeted to Asian emerging economies, they account for larger shares of GDP in smaller developing countries. This corresponds to a general increase in FDI flows to developing countries, primarily in services: in 2005, in developing regions, more than half of all cross-border mergers and acquisitions (M&As), amounting to more than $167 billion, took place in the services sector.[2]

This chapter will review major developments in the ICT industry during the past 10 years. In particular, it will highlight the role of the ICT sector in global economic trends and shifts, such as those related to production, employment, trade and investment, the growth of China and India and South–South economic relations, and the increasing attention that the ICT industry receives from developing country policymakers who are developing their knowledge-based economies. The chapter will argue that the ICT-producing sector has played a major role in recent global economic developments that have led to an emerging second generation of globalization, and that this trend will continue in the years ahead. It will provide evidence demonstrating the increasing importance of the ICT sector for the developing countries and South-South relations.

The Information Economy Reports 2005 and 2006 have focused on the latest developments in the trade of ICT goods and services respectively. This chapter will provide up-to-date information on some of those trends and present concrete examples of the role of the ICT sector in developing countries' efforts to develop their information economies. In that context, the UNCTAD secretariat collects from member States data on ICT sector value-added and employment. As in previous Reports, the results of the 2006 data collection will be presented in this chapter. In addition, a special project to survey the ICT industry was carried out jointly with the Government of Egypt (box 2.1). The chapter will feature some of the key findings of the survey throughout the various sections. Given the

scarcity of comparable data from developing countries, the chapter relies on a number of examples to illustrate the role of the ICT sector in selected developing countries.

The ICT industry plays an important role in the growth of international sourcing. While international sourcing in ICT manufacturing has been a common feature in the past two decades, the outsourcing of ICT services took off only a few years ago and has prompted many developing countries to try to develop a competitive advantage in this field. International sourcing in the ICT industry is reflected in investment and trade flows, employment figures and national ICT strategies. Therefore, reference to offshoring and outsourcing will be made throughout this chapter. A more detailed discussion on the link between outsourcing and employment and outsourcing in the ICT services sector can be found in the IER 2006.

Lastly, the chapter will highlight the role of national and international policymaking for the development of the ICT sector in developing countries. In particular, it will consider the WTO's Information Technology Agreement (ITA), which is the key international trade agreement relevant to ICT goods trade and which entered into force 10 years ago, and will make suggestions on possible further work related to the review of the ITA.

Box 2.1

Joint UNCTAD – Government of Egypt project on the ICT sector

The Ministry of Communication and Information Technology (MCIT) of Egypt and its Information Technology Industry Development Agency (ITIDA) have formulated an ICT strategy aimed at transforming their economy into a regional hub of ICT production and service provision. The policies put in place have been successful in attracting foreign investment and outsourcing contracts in the Egyptian ICT sector. A large part of the existing ICT enterprises are concentrated in technology parks and incubators. There is strong government support for providing the necessary ICT infrastructure at accessible prices.

The ICT and E-Business Branch of UNCTAD and MCIT worked together to conduct a survey on Egypt's ICT sector with the objective of obtaining detailed firm-level information. UNCTAD assisted ITIDA in the design of a questionnaire. The main aim of the questionnaire was to identify the product portfolio of the Egyptian ICT sector, its imports and exports as well as key trading partners, the workforce, key financial indicators, participation in public support programmes and remaining obstacles to growth. The data were collected, verified and entered into a database by ITIDA, in cooperation with a local counterpart. The subsequent data analysis was carried out by both UNCTAD and MCIT.

The survey was sent to a sample of 250 ICT companies in Egypt, 151 of which responded. The results presented in this chapter are based on those companies' responses and are thus not representative of the entire ICT sector in Egypt.

The results of the survey will be used to fine-tune ICT policymaking in Egypt. In the first stage, the data will help to benchmark the performance of the ICT sector against policy targets. In the second stage, research will aim at revealing some of the factors responsible for success or failure in meeting certain policy targets.

B. ICT market, value added and employment

1. Overview of the ICT market

The production of ICTs is undergoing a shift from technology-oriented products, which dominated in the late 1990s, to commercial, often user-driven, and new applications of ICTs. There are also some structural changes, with the emergence of niche products, software and services, rather than the traditional commodities or IT equipment. The role of individuals and their demands, via use of ICTs, global communication and access to information, is growing and impacting on the ICT supply side, influencing production and firms' decisions.

According to the World Information Technology and Services Alliance (WITSA), which is monitoring the global ICT market, global ICT spending grew at 8.9 per cent annually between 2001 and 2005, representing 6.8 per cent of global GDP. In 2006, ICT spending slowed down to 6 per cent annual growth, with a value of over $ 3 trillion (WITSA, 2006). The highest growth can be observed in software, but the largest amount spent is in the area of communications and ICT services.

Between 2000 and 2005, revenues of the top 250 ICT firms grew by an annual 4.3 per cent and reached $3 trillion in 2005. With regard to specific sectors, revenues grew by 8.3 per cent in the services and software sectors, 6.1 per cent in the telecommunications sector, 5.1 per cent in the IT equipment sector, and 3.3 per cent in the electronics and components sector. In the area of communications equipment, revenues decreased by 4.1 per cent (OECD, 2006a).

The top 50 ICT firms are primarily from Europe (Germany, the United Kingdom, France, Netherlands),

Chart 2.1

Electronics industry output by location, 1995-2005

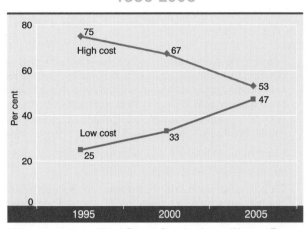

High-cost locations: United States, Canada, Japan, Western Europe.
Low-cost locations: Asia/Pacific, China, Eastern Europe, rest of the world.
Source: Reed Research, presented by Ernie Santiago, SEIPI, WTO ITA Symposium, 28 March 2007, Geneva.

North America (United States) and Asia-Pacific (Japan, Republic of Korea, Hong Kong (China), China, Singapore). In 2005, the revenues of the top 10 ICT firms ranged from $60 billion to $100 billion. However, an increasing number of the top 50 ICT firms are from developing economies, such as Samsung Electronics and LG Electronics (Republic of Korea), China Mobile (Hong Kong, China), China Telecom (China), Hon Hai Precision (Taiwan Province of China) and Flextronics (Singapore), with 2005 revenues ranging from $15 billion (Flextronics) to $48 billion (Samsung Electronics) (OECD, 2006a).

During the past decade, there has been a clear tendency to migrate the production of electronics from high-cost to low-cost locations (see chart 2.1). Asia-Pacific continues to be the fastest-growing region for electronics production (table 2.1). Asia's electronics industry grew from 48 per cent of world production in

Table 2.1

Global production of electronics, 2002-2005, $ billion

Region	2002	2003	2004	2005	CAGR 2002 – 2005 (per cent)
Europe	220.4	247.5	279.1	285.8	9.0
Americas	317.6	314.1	334.3	341.9	2.5
Japan	162.4	180.2	197.8	202.3	7.6
Asia/Pacific	343.1	386.9	448.8	492.7	12.8
Rest of the world	13.2	14.3	15.7	16.2	7.2
World	1 056.8	1 143.0	1 275.6	1 338.9	8.2

Source: Reed Research, presented by Ernie Santiago, SEIPI, WTO ITA Symposium, 28 March 2007, Geneva.

Box 2.2

Costa Rica: the Intel factor

When in 1996 Intel decided to choose Costa Rica as one of its main production sites, this was the beginning of a considerable structural change in the country's economy. In 1985, 60 per cent of Costa Rica's exports were perishable products and 3 per cent were electrical and electronic products. In 2005, the share of perishable products exports had decreased to 24 per cent, and the share of electrical and electronic products increased to 30 per cent. Exports of ICT products decreased sharply in 2000, following the global slowdown in the IT industry, but started to increase again as of 2001. Despite a 15 per cent increase since 2000, ICT goods exports have not yet reached the pre-2000 levels (chart 2.2), although at the global level, ICT goods trade fully recovered in 2005.

Even though FDI from other companies followed, Intel remained the largest ICT-producing firm in the country. Its impact on the country's economy has been estimated in various areas.

The contribution of Intel to GDP is estimated to be 5 per cent (2000—2005), with a further 0.8 per cent indirect impact through income spent by employees and local purchases. In 2005, the direct and indirect effects of Intel reached 25 per cent of GDP of the entire manufacturing industry (see chart 2.3). Intel accounts for 20 per cent of total exports. In addition, there are a number of qualitative benefits, such as publicizing the country as an IT location (attracting other investors), technology and knowledge transfer and the creation of linkages with academia. For example, Intel supported the curriculum and teaching teams in engineering courses at the Instituto Tecnológico de Costa Rica and the Universidad de Costa Rica, which led to a considerable increase in the numbers of students enrolled in those careers (from 577 in 1997 to 874 in 2000). Only 60 electrical engineers graduated in 1997, while by 2005 there were over 200 graduates. Because Intel absorbs only 10 per cent of those new graduates, other multinational enterprises and local firms can benefit from the increase in the number of those professionals by employing them. A study by the International Labour Organization (2007) found that Intel pays its employees an average monthly wage of $836, while employees in the manufacturing sector earn only an average monthly wage of $491 (as of December 2005).

The dominant role of Intel has, however, its downsides. The country is highly dependent on the ICT sector and was hit hard by the 2001 global slump, from which it still has not fully recovered (see above). In particular, most of its ICT exports are limited to a small number of electrical and electronic products, whereas exports of higher-value-added services are still limited, albeit growing.

Chart 2.2	Chart 2.3
ICT goods exports in Costa Rica, 1997-2006	**Share of Intel GDP in total industrial GDP, 2002-2005**

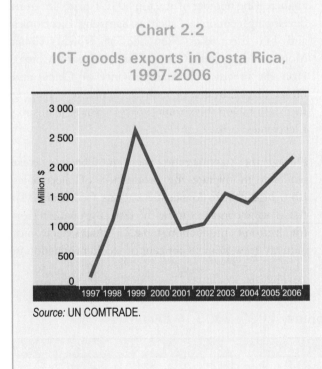

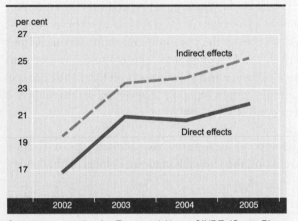

Source: UN COMTRADE.

Source: presentation by Emmanuel Hess, CINDE (Costa Rican Investment Promotion Agency), WTO ITA symposium, 28 March 2007, Geneva.

2002 to 52 per cent in 2005, mainly in China and Japan. ICT production also has an important or increasing role in countries in other regions, such as Costa Rica (box 2.2), Chile, Finland, Israel and Mauritius.

In recent years, the ICT services industry has experienced strong growth, including in economies that are important suppliers of ICT goods (see box 2.3, Philippines). Those developments have been fostered

Box 2.3

Philippines: contact centres on the increase

Traditionally, the Philippine ICT industry has been dominated by manufacturing, in particular of components and devices (e.g. chips), which accounted for 74 per cent of the electronics industry in 2005, and of computer-related products (20 per cent). In 2005, the ICT industry accounted for 66 per cent of total manufacturing exports, a fact that illustrates the transformation of domestic production in only 30 years: in 1975 the Philippine economy was 49 per cent agro-based and only 3 per cent electronics-based. As in other Asian countries, the ICT sector in the Philippines is heavily dominated by foreign firms, which account for 72 per cent of firms in the sector (30 per cent are from Japan, 10 per cent from the Republic of Korea and 9 per cent from the United States).

Recently, the ICT service sector has been growing, including with regard to contact (or call) centres, business processing, animation, medical and legal transcription, engineering and software design. Contact centres are now the fastest-growing industry in the ICT services sector. In 2006, the Philippines had 146 such centres, employing 150, 000 staff and generating $2.7 billion revenue.

The industry estimates that the ICT services workforce will increase from 99,300 (in 2004) to 920,764 (in 2010), and revenues from $1.5 billion (in 2004) to $12 billion (in 2010), with the largest growth expected in customer care (contact centres) and back-office operations (business processing).

Table 2.2

The Philippines e-services industry

	Players	Employees	Seats	Revenues (million $)	Growth rate (per cent)	Performance level
Contact Centre (2007)	146	150 000	93 750	2 688	90	..
Business Processing (2006)	62	22 500	..	180	80	..
Animation (2006)	40	4 500	..	40	25	..
Medical and legal transcription (2007)	66	9 675	..	126	80	98–99 accuracy rate
Engineering design (2007)	14	4 000	..	48	30	..
Software design (2007)	300	16 000	..	272	40	..

Source: Based on the presentation by Ernie Santiago, SEIPI, WTO ITA Symposium, 28 March 2007, Geneva, and updated with information received from the Philippine Government in July 2007.
Notes: Data provided for 2005 as of first quarter 2006 and for 2006 as of first quarter 2007.

by the growth in outsourcing of ICT and ICT-enabled services, which has prompted many developing countries to develop their ICT services industry. A number of developing country ICT services producers have grown strongly, notably India, but also China, Egypt (box 2.4) and the Philippines, as well as several of the transition economies, such as Croatia, Poland and Romania.

The telecommunications sector in particular continues to grow strongly. In the OECD countries, revenues increased each year during the past decade and exceeded $ 1 billion in 2005, contributing more than 3 per cent to GDP (OECD, 2007a). Similarly, in some developing countries, the telecommunications sector

has experienced significant growth, for example in Brazil, Chile, Malaysia, Mexico and South Africa.

2. ICT sector value added

In the OECD countries, ICT sector value added as a share of total business-sector value added continues to increase overall. After the 2000 downturn, the strongest recovery can be observed in the Republic of Korea and Ireland. ICT services account for more than two thirds of ICT sector value added in the OECD countries, with growth sectors being communications services and software services (OECD, 2006a). Between 2003 and 2005, in the EU countries, Finland and the United

---

Box 2.4

Egypt's growing ICT sector

According to the MCIT/ITIDA–UNCTAD survey of 151 businesses in the Egyptian ICT sector, average firm revenue growth increased gradually from 16 per cent in 2004 to 26 per cent in 2006, with a greater variation in the revenue growth rate of firms. In 2006, several small and medium-sized enterprises achieved higher sales growth than in the previous two years, when most growth occurred in the largest firms. At the same time, findings suggest that revenues of foreign-owned firms also grew faster.

The increase in ICT sector sales went hand in hand with growth in ICT exports. Results from the survey indicate that while domestic sales show a more equal composition between ICT goods and services, export sales of the ICT industry concentrated mainly on services. In 2006, Egypt exported $548 million worth of communications and computer and information services (based on IMF BOP data) and $18.2 million in terms of ICT goods (based on United Nations COMTRADE). In terms of ICT services, Egypt was the second largest African exporter in 2005, for both communications and computer and information services, competing strongly with Morocco and South Africa. In 2006, the value of ICT-enabled services exports was $2.3 billion. Egypt has the potential to become a competitive international location for offshore services owing to its large skilled workforce, existing business infrastructure and geographical location. In 2005 A. T. Kearney's Offshore Location Attractiveness Index ranked Egypt among the 12 destinations most attractive to offshore businesses, before other regional competitors such as Jordan, Ghana and Tunisia, and after India, China, Malaysia and the Philippines.

Even though Egyptian ICT goods exports have a considerably lower value than ICT services exports, they expanded faster at 22 per cent (CAGR) annually from 2000 to 2006, while services increased more slowly (9 per cent CAGR). Telecommunications equipment and audio and video equipment account for more than three quarters of exports and experienced the most dynamic evolution since 2000. The main importers of Egyptian ICT goods were neighbouring countries in North Africa (Algeria and Sudan) and the Middle East (Kuwait and Jordan).

Kingdom continued to have high ICT sector value-added shares, whereas those shares have been falling in Ireland and Austria. At the same time, new member countries, such as Hungary, Slovakia, Romania and Estonia, had increasing shares of ICT sector value added.

Chart 2.5 presents available comparable data at the global level, collected by UNCTAD and based on the OECD ICT sector definition (OECD, 2007b). Among the developing countries, ICT sector value added as a share of total business-sector value added is still small, with the exception of some Asian countries. Growing shares can be observed in smaller economies such as Mauritius and Cuba. In Chile, ICT sector value added reached 3 per cent of total value added in 2004, primarily from telecommunications services (60.8 per cent), followed by IT services and trade, whereas ICT manufacturing had an insignificant share. The ICT sector represented 3.4 per cent of GDP in 2004.[3]

In India, which is one of the largest producers of ICT services, the contribution of the IT software and services sector to GDP increased from 1.2 per cent (1999) to 5.4 per cent (2006). The figures do not include the telecommunications sector, the third largest in the world and which has grown at an average rate of 40-45 per cent during the last two years (WTO, 2007). The total services contribution to GDP increased

from 50 per cent (2000–2001) to 54 per cent (2005–2006), compared with a stable contribution of the manufacturing sector and a declining contribution by the agricultural sector.

Chart 2.4

Value added to sales ratio for US MNC computers and electronic products sector, 1999-2004

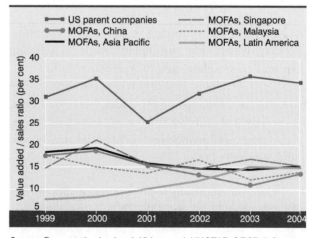

Source: Presentation by Jacob Kirkegaard, UNCTAD-OECD-ILO expert meeting, Geneva, 5 December 2006.

Note: MOFAs refers to majority-owned foreign affiliates.

Chart 2.5

Share of ICT sector value added in total business-sector value added

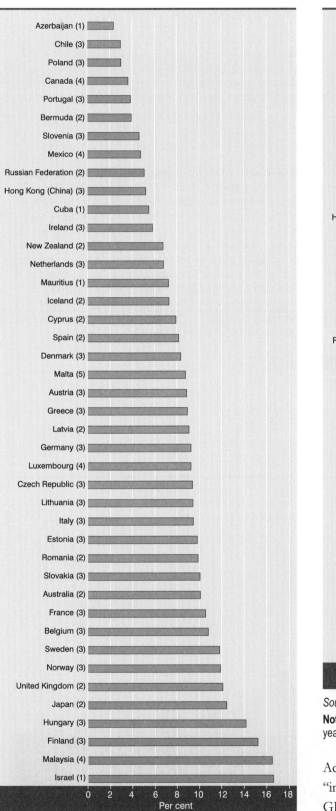

Source: UNCTAD Information Economy database (2007).
Notes: *(1)* reference year 2006; *(2)* reference year 2005; *(3)* reference year 2004; *(4)* reference year 2003; *(5)* reference year 2002.

Chart 2.6

Share of ICT sector workforce in total business-sector workforce

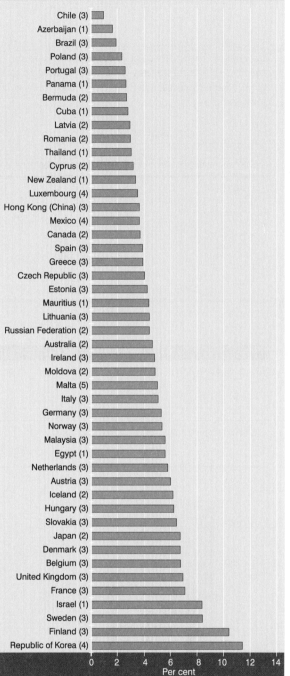

Source: UNCTAD Information Economy database (2007).
Notes: *(1)* reference year 2006; *(2)* reference year 2005; *(3)* reference year 2004; *(4)* reference year 2003; *(5)* reference year 2002.

According to Chinese sources, the value added of the "information industry" in China reached 7.5 per cent of GDP in 2004, an increase in value by 30 per cent from 2003. The value added of post and telecommunication services accounted for 8 per cent of services industry value added (2003). This compares with 32 per cent

Box 2.5

Sri Lanka's ICT employment: rising demand

In Sri Lanka, ICT employment has grown steadily over the past few years and is expected to reach 44,660 by 2008 (chart 2.7)[4] In 2006, the overall IT workforce stood at 30,120 of whom 46 per cent are in the IT sector, 47 per cent in the non-IT sector and 6.6 per cent in government. The highest shares of the overall IT workforce are in software engineering and programming, technical support, and testing and quality assurance (see chart 2.8). Even though the IT workforce represents less than 1 per cent of total employment, it is an important growth sector.

Chart 2.7

IT workforce in Sri Lanka, 2003-2008

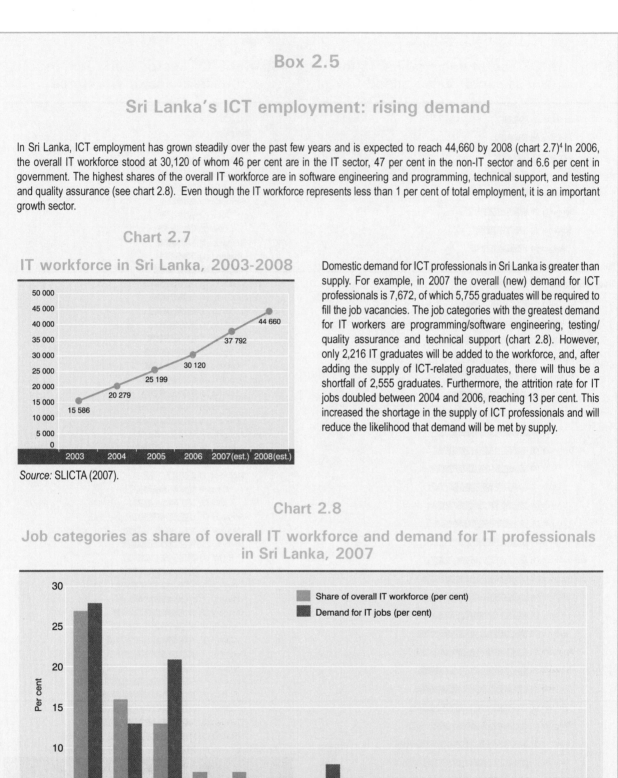

Source: SLICTA (2007).

Domestic demand for ICT professionals in Sri Lanka is greater than supply. For example, in 2007 the overall (new) demand for ICT professionals is 7,672, of which 5,755 graduates will be required to fill the job vacancies. The job categories with the greatest demand for IT workers are programming/software engineering, testing/quality assurance and technical support (chart 2.8). However, only 2,216 IT graduates will be added to the workforce, and, after adding the supply of ICT-related graduates, there will thus be a shortfall of 2,555 graduates. Furthermore, the attrition rate for IT jobs doubled between 2004 and 2006, reaching 13 per cent. This increased the shortage in the supply of ICT professionals and will reduce the likelihood that demand will be met by supply.

Chart 2.8

Job categories as share of overall IT workforce and demand for IT professionals in Sri Lanka, 2007

Source: Based on SLICTA (2007).

Box 2.5 *(continued)*

A reduction in the attrition rate and the industry's willingness to recruit IT-related degree holders and advanced diploma holders, as well non-IT graduates, may help bridge the gap. Both an increase in the number of IT graduates and an improvement in their quality are necessary. It will be a challenge to meet this goal in view of the insufficient specialization in IT courses at degree and postgraduate levels, and because courses are not tailored to particular job categories. For instance, "testing and quality assurance" is a job category that has seen rapid growth in the number of employees over the past two years, but there are no academic courses tailored to that job category. Furthermore, IT training institutions lack qualified teaching staff that could impart essential and complementary skills to graduates and improve their qualifications. A further explanation for the reduced supply of IT professionals is related to "brain drain", mainly to India, where companies find it an advantage to tap the large pool of skilled people in Sri Lanka.

About half of all IT jobs are in large organizations (with more than 300 employees). In the non-IT and government sectors, large organizations employ about three quarters of the IT workforce. By contrast, in the IT industry, almost 80 per cent of the workforce is employed by small and medium-sized enterprises (under 300 employees), a significant increase from 2004, when they employed 53 per cent of the IT workforce. A relatively small share (21 per cent) of the IT workforce is female. The Government employs most female IT professionals (36 per cent), while the non-IT sector employs 15 per cent and the IT sector 20 per cent (charts 2.9 and 2.10). No information on types of activities by gender is available.

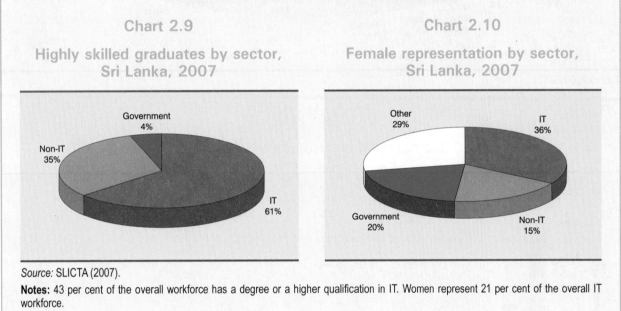

Chart 2.9

Highly skilled graduates by sector, Sri Lanka, 2007

Government 4%
Non-IT 35%
IT 61%

Chart 2.10

Female representation by sector, Sri Lanka, 2007

Other 29%
IT 36%
Government 20%
Non-IT 15%

Source: SLICTA (2007).

Notes: 43 per cent of the overall workforce has a degree or a higher qualification in IT. Women represent 21 per cent of the overall IT workforce.

of total value added in 2004 for total services, which is still relatively low compared with OECD countries. Available data from United States foreign affiliates show that the value added to sales ratios in ICT goods production in China and other developing country producers remain below that in the United States (chart 2.4).

3. ICT sector employment

In the OECD countries, between 1995 and 2003 ICT sector employment increased by 8 per cent overall, but growth is stable in most countries. In the United States, ICT sector employment shifted from primarily manufacturing (66.6 per cent in 1996) to primarily services (54.6 per cent in 2004), a trend which is likely to be observed in other OECD countries as

well (Bednarzik, 2005). The share of the ICT sector workforce in the total business-sector workforce is highest in the Republic of Korea, accounting for more than 10 per cent in 2003 (chart 2.6). As in the case of the ICT sector value added, ICT-workforce shares are increasing in some EU countries, such as Finland and the United Kingdom, but are decreasing in others, such as Ireland, Sweden and the Netherlands. Available figures from developing countries point to small but increasing shares in countries such as Cuba and Egypt. In Chile, ICT sector activities generated 50,769 jobs in 2004, equivalent to 1 per cent of all jobs in the economy. More than 50 per cent of the persons in those jobs are considered to be highly qualified professionals and technicians. The creation of high-skilled jobs in the ICT sector has also been confirmed by a recent study carried out in Sri Lanka (box 2.5).

Chart 2.11

Chart 2.11

Global distribution of employment in electrical and electronic products manufacturing, 2004

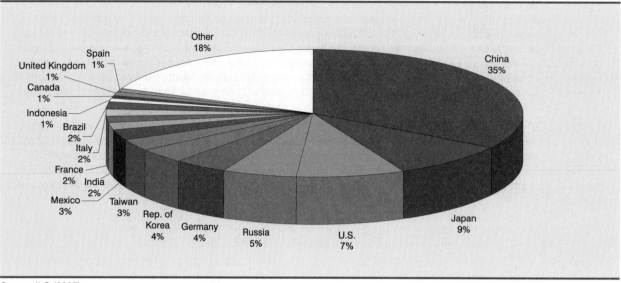

Source: ILO (2007).

Chart 2.12

Employment in electrical and electronic products manufacturing, 2002

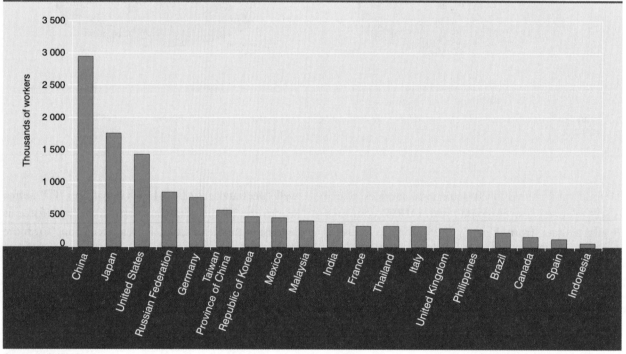

Source: ILO (2007).

In 2004, total employment in the manufacture of electrical and electronic products worldwide, which constitutes a major part of the ICT industry, was over 18 million, with a relatively high share of female employment in some developing countries (box 2.6) (ILO, 2007). Employment in this sector declined in the 1990s and until 2002, when the sector started booming, with China as the engine of growth (see section D.2.). Employment in the sector is highly concentrated, with 20 countries accounting for 87 per cent of world total employment. Among the largest employer countries are China, Japan, the United States, the Russian

Table 2.3

Employment trends in electrical and electronic products manufacturing, 1997–2004

Countries	Office, computing and accounting machinery	Electrical machinery and apparatus	Radio, TV and communication equipment	Employment trend
United States	Declining	Declining	Declining	Loss of 550,000 jobs
Japan	Declining but steady more recently	Declining	Slightly declining	Loss of 400,000 jobs
Germany	Declining	Steady	Slightly declining	Loss of 100,000 jobs
The Russian Federation	Increasing since 2002	Increasing	Declining	N/A
Thailand	Loss of 18,000 jobs	Increase of 101,000 jobs	Increase of 84,000 jobs	Increase of 167,000 jobs (1997 and 2000)
India	N/A	Increasing	Increasing since 2001	N/A
Republic of Korea	Increasing	Increasing	Increasing	N/A
Philippines	N/A	Declining	Gain of 29,000 jobs (2003 and 2004)	N/A
Mexico	Increasing since 2003	Increasing since 2003	Declining	N/A
Indonesia	N/A	N/A	Increasing	Increase of 65,000 jobs (1998 and 2001)

Source: UNCTAD, based on ILO (2007).
N/A = not available.

Box 2.6

Women in ICT manufacturing

Women account on average for almost 40 per cent of the workforce in the electrical and electronics production, with large differences between countries and industries (table 2.4) (ILO, 2007). In the manufacture of office, accounting and computing machinery, women's share of employment stands at 50 per cent in the emerging economies of Asia, such as Malaysia, the Philippines, Thailand and Viet Nam, whereas in India it is very low. In the manufacture of electrical machinery and apparatus, the average shares of women in the workforce in developing economies such as Indonesia, the Philippines, Thailand and Viet Nam are much higher than in developed economies, whereas the shares in India and Malaysia are very low. In the manufacture of radio, television and communication equipment, women's share of total employment is higher than in the other two segments, and shares in Malaysia, the Philippines, Thailand and Mexico are particularly high. High rates can also be found in Eastern Europe – in Bulgaria, the Czech Republic, Latvia, the Russian Federation, Slovakia and Ukraine. In addition, there are differences between women's and men's wages, with women earning less than men, although they work almost the same number of hours, and in all countries (except Finland) where data are available the wage discrepancy is increasing over time.

Table 2.4

Employment by gender in electrical and electronic products manufacturing

Job category	2004	2003
Office, computing and accounting machinery	40 per cent	10 per cent – 64 per cent in developing economies 25 per cent – 40 per cent in developed economies
Electrical machinery and apparatus	37 per cent	13 per cent – 60 per cent in developing economies 16 per cent – 50 per cent in developed economies
Radio, TV and communication equipment	N/A	20 per cent – 72 per cent in developing economies 17 per cent – 62 per cent in developed economies

Source: UNCTAD calculations, based on ILO (2007).
N/A = not available.

Box 2.7

Employment in the ICT sector in Egypt

The growing global demand for ICT products and services has resulted in job creation in Africa as well. Egypt in particular has seen a notable expansion of 700 per cent in revenues from its ICT industry,[5] and a fourfold increase in the total number of ICT companies[6] during the last seven years (2000–2007). According to the World Bank (2006) the share of telecommunications revenue increased from 2.8 per cent in 2000 to 3.5 per cent in 2004, above the average performance in the Middle East and North African region. Currently, approximately 2,262 ICT companies[7] are present in the Egyptian market and they employ some 50,000 people.[8] This amount corresponds to a small share of the total active workforce of 18 million,[9] but it represents a dynamic sector with higher than average participation of women in the labour market.

A 2007 survey conducted jointly by the Ministry of Communication and Information Technology (MCIT), the Egyptian Information Technology Industry Development Agency (ITIDA) and UNCTAD interviewed 151 Egyptian ICT companies representative of the ICT sector and assessed in greater detail the structure of the workforce by gender and occupations.[10] Within the sample of companies surveyed, it was found that 70 per cent of jobs are specialized ICT occupations (against only 30 per cent other occupations) (chart 2.13). Among ICT occupations, most jobs were for applications development and testing technicians and for network and hardware professionals, while there were fewer IT trainers. Even though only 5 per cent of the staff employed were in temporary positions, the survey showed that there were relatively more temporary workers employed as software and multimedia developers and analysts compared with other occupations. This finding reveals the potential for high staff turnover for this specific type of occupation.

Chart 2.13

Permanent and tempory workforce in a sample of ICT companies in Egypt

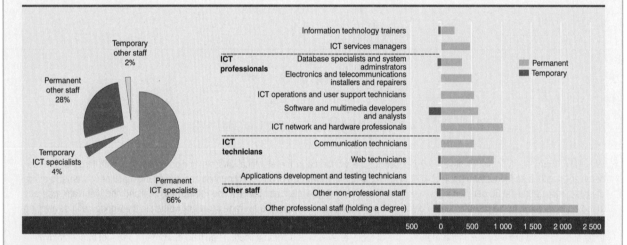

Egypt is recognized as one of the few African countries where women's participation in ICT is more pronounced.[11] Even though the results of the MCIT/ITIDA UNCTAD survey show that female employees occupied only 23 per cent of the positions in the ICT businesses surveyed (chart 2.14), this share is greater than the 19 per cent participation of women in the total workforce estimated by ILO for 2003. In 2001, the International Telecommunication Union found that women represented 22 per cent of the staff employed in the telecom sector in Egypt.[12]

Additionally, the MCIT/ITIDA UNCTAD survey revealed that in the ICT sector there are relatively more women employed in temporary positions and in non-ICT occupations. On average, for all types of ICT occupations there is a ratio of 3 male to 1 female employee, while among the other staff members the ratio is closer to 2 to 1. Also, for every 4 male permanent staff members, there is 1 permanent female staff member, while among temporary staff members this ratio is 3 to 1. There were more women working as ICT professionals than as ICT technicians. Women are more commonly employed as ICT network and hardware professionals and as applications development and testing technicians. As seen in chart 2.13, these same jobs with a high participation of women coincide with positions most in demand among specialized ICT occupations in Egypt. This indicates that women can make an important contribution to increasing the numbers of specialized ICT personnel to match demand from the ICT-producing sector.

According to the MCIT/ITIDA UNCTAD study, domestically owned firms employ relatively more women in the ICT sector compared with other businesses. Foreign-owned or joint ventures employed less than a fifth of the total surveyed workforce, corresponding to a reduced share of foreign ownership in the Egyptian ICT sector. Also, a larger female participation is found in firms with a certain experience in the market, founded some 10 years ago (both absolute and relative). Younger ICT firms, founded in 2002 or before, tend to employ relatively fewer women.

Box 2.7 *(continued)*

If demand conditions remain favourable to growth in the ICT sector in Egypt, there is scope for encouraging women's participation in the ICT workforce through targeted training and by supporting a gender-balanced use of ICTs. As other studies (Standing, 1999) have shown, hiring women when there is great demand can result in superior business competitiveness.

Chart 2.14

Male and female occupations in a sample of ICT companies in Egypt

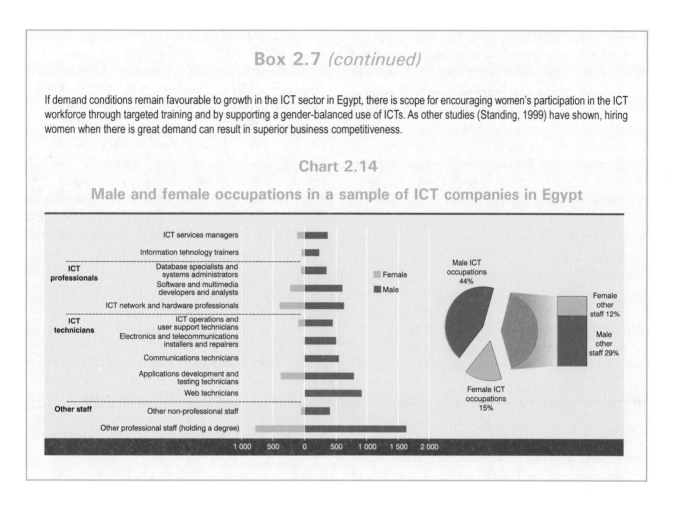

Federation, Germany, and the Republic of Korea (charts 2.11 and 2.12). The Republic of Korea, where the electronics industry has been an important part of the economy since the 1980s, there is a shift from high- to low labour-intensive production, as well as an improvement in labour productivity. This has led to a deceleration of employment in the sector, which grew annually by only 3.5 per cent (between 1997 and 2004), whereas total production grew by 9.4 per cent. The most significant growth in employment (13.7 per cent CAGR) occurred in the area of consumer products, such as television and radio transmission devices, and telephones.

The three main sub sectors are office, computing and accounting machinery, electrical machinery and apparatus, and radio, television and communication equipment. Between 1997 and 2004, most of the largest employer countries in all three sectors had flat or declining employment. In the United States 550,000 jobs (over 30 per cent) were lost from the workforce, in Japan 400,000 jobs (20 per cent) and in Germany 100,000 jobs (14 per cent). On the other hand, countries that increased their employment levels in one or several of the industry segments include China,

the Russian Federation, Thailand, Indonesia and the Philippines.

Chart 2.15

Potentially offshorable occupations in total employment: EU 15, United States (estimate), Canada and Australia, 1995-2003

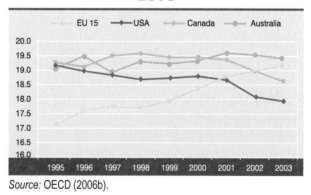

Source: OECD (2006b).

ICT sector employment and international sourcing

Given the strong growth of international sourcing during the past few years, the question of how many jobs are lost/created as result of outsourcing has received much attention. Unfortunately, there are few reliable data on the impact of international sourcing in the ICT industry on employment (in both developed and developing countries). Several studies on ICT services offshoring and job losses in the United States conclude that the impact is rather limited overall and concentrated in a few industries and occupations (Bednarzik, 2005). In the European Union, job losses in information technology due to international sourcing have been estimated at 1,570 (in 2005), or 14 per cent of all job losses due to international sourcing.[13] These figures are insignificant compared with the more than 2 million jobs lost in manufacturing in the EU 15 during the period 1996–2004. However, the types of jobs in all industries that could be outsourced have been increasing in number in the EU, whereas they seem to be stable in other markets where outsourcing could originate (see chart 2.15). This is largely explained by the fact that employment in occupations potentially affected by outsourcing has been increasing in the EU countries over the past decade. The overall increase in international sourcing and its related impact on employment have caused considerable political debate in countries such as France, Germany, the United Kingdom and the United States.[14]

C. Trade and investment in ICT goods and services

1. Overview

The past decade has witnessed strong growth in ICT-related trade and investment flows. The slowdown after the NASDAQ crash in 2000 was fully reversed by 2005, with growth rates in ICT goods trade equal to those in overall manufacturing trade and above-average growth in ICT services trade. Global ICT trade has grown faster than ICT production and more strongly compared with total exports. The global market has witnessed a general shift from developed to developing countries in the export of ICT goods and services, with the two largest developing economies, China and India, dominating the two sectors respectively. This has led to newly emerging South–South trade relationships in this dynamic sector. In 2005, the value of ICT trade among developing countries almost caught up with that among developed countries and is likely to have exceeded it as this Report goes to press.

Similarly, FDI in the ICT sector grew strongly, especially in ICT manufacturing (electronic and electrical equipment) and ICT services (communication services and software businesses) (UNCTAD, 2005b). Developing countries are increasingly becoming a destination for FDI flows, especially for ICT-related services (UNCTAD, 2004). Over the last decade, 14,566 completed cross-border M&A deals targeted the ICT sector and ICT sector companies concluded 11,634 deals. Available data on ICT sector cross-border M&As show an increase in value for target countries from $22 billion (1995) to $134 billion (2005) and in the number of deals from 386 (1995) to 1,475 (2005) (OECD, 2006a). For non-OECD countries, the value has increased from $2.9 billion (1995) to $106 billion (2005). In 2005, most cross-border M&A deals were in ICT services (610), followed by telecommunications (323) and electronics (190), which, combined, accounted for 76 per cent of all deals. The fastest growth has been in ICT services and telecommunications. In particular, South–South investment flows in the telecommunications sector are on the rise, driven by large TNCs from such countries as South Africa, Malaysia and Mexico (World Bank, 2006).

Chart 2.16

Growth of world merchandise exports and ICT goods exports, 1996-2005

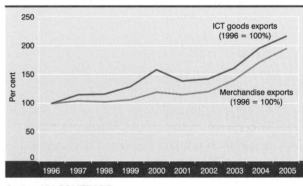

Source: UN COMTRADE.

South Asia, East Asia and South-East Asia are the main magnet for FDI inflows into developing countries, which reached $165 billion in 2005, representing 18 per cent of world inflows. Manufacturing FDI has been increasingly attracted to South, East and South-East Asia, although specific locations have changed as countries have moved up the value chain. This has included large inflows into the electronics industry (UNCTAD, 2006b).

Chart 2.17

ICT goods export by level of development, 1996 and 2005

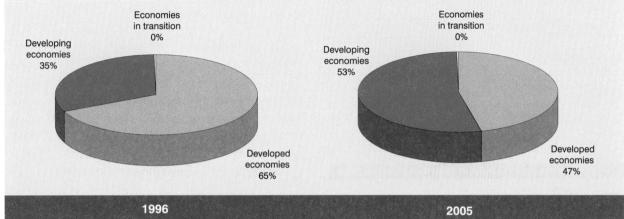

Source: UN COMTRADE.

Chart 2.18

ICT goods export by level of development, 1996-2005

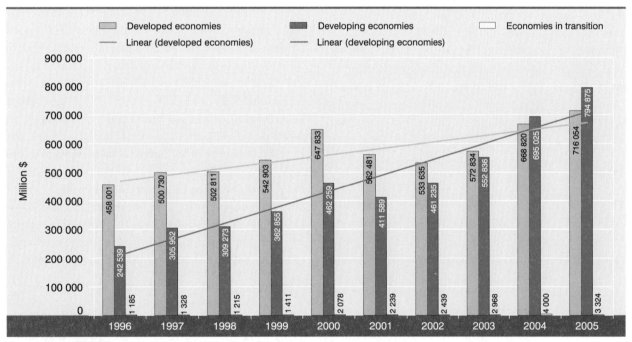

Source: UN COMTRADE.

2. ICT goods

Overall, there has been strong growth of trade in ICT goods, and full recovery from the 2000 decline. The share of world exports of ICT goods in world exports of all merchandise goods increased from 13 per cent in 1996 to 15 per cent in 2005. Between 1996 and 2005, world exports of ICT goods increased at a higher CAGR than total world merchandise exports – at an annual rate of 9 per cent compared with 8 per cent. However, this is largely based on high growth in the late 1990s, whereas growth rates have been stable since 2001.

Since 1996, the value of world exports of ICT goods has more than doubled, reaching $1.5 trillion in 2005. Between 1996 and 2005 the value of merchandise exports increased by 94 per cent, while ICT goods exports increased by over 116 per cent (chart 2.16). This is despite high commodity prices in 2005 and

Chart 2.19

World exports of ICT goods, 1996-2005

Chart 2.19

World exports of ICT goods, 1996-2005

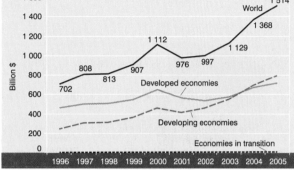

Source: UN COMTRADE.

2006, which, coupled with falls in the prices of certain ICT goods, disguise the relative performance of trade in ICT goods in volume terms and the emergence of new markets.

The strong growth in ICT goods trade coincides with the lowering of tariffs on ICT products of countries that are signatories to the WTO's Information Technology Agreement (ITA). Today, 93 per cent of WTO ITA products are imported duty-free, with larger tariff cuts by developing countries than by developed countries. An assessment of ITA's 10 years of existence is provided in section E.

Exports

The global market in ICT goods trade has witnessed a general shift from developed economies to emerging economies in the developing world. Until 2003, the United States was the largest exporter of ICT goods;

in 2004 China overtook it, and in 2006 its exports totalled $299 billion (see section D). China's exports value increased by an annual 36 per cent between 1996 and 2006, up from $19 billion in 1996, when it was the eleventh largest exporter of ICT.

Other large exporters of ICT goods include developed economies such as Germany, the United Kingdom, France, Netherlands and Japan and developing economies, such as the Republic of Korea, Singapore, Taiwan Province of China, Malaysia, Thailand, the Philippines and Mexico. While in 1996 the largest exporters of ICT goods were mainly developed countries, by 2005 developing countries had emerged as large exporters of ICT goods and such countries as France, Italy and Spain lost their place among the world's largest ICT goods exporters. Annex 1 provides country-level data on ICT exports during the past decade. Chart 2.17 shows the shift in ICT goods exporters from developed to developing countries during the same decade.

Among the emerging developing countries that have increased their share in the global ICT goods market, the cases of Mexico and Indonesia should be mentioned. In both countries, ICT goods exports increased more than overall merchandise exports (see charts 2.20 and 2.21). Since 2005, Mexico has specialized in exports of audio and video equipment, which together with telecommunications equipment increased fastest. The United States remained the main export destination for ICT goods produced in Mexico, but exports to certain South and Central American countries (such as Nicaragua, Honduras and Colombia) increased rapidly. In Indonesia, exports of electronic components picked up in the aftermath of 2000, with the main destination being Singapore.

Chart 2.20

Mexican merchandise exports and ICT goods exports (1996 = 100 per cent)

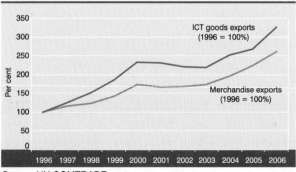

Source: UN COMTRADE.

Chart 2.21

Indonesia merchandise exports and ICT goods exports (1996 = 100 per cent)

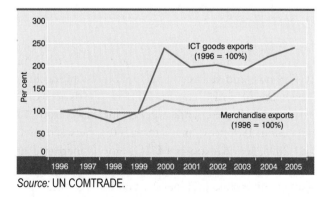

Source: UN COMTRADE.

However, computer and related equipment remained the Indonesian ICT export category with the highest value in 2006. Countries in South-East and East Asia were the main trading partners, with an exceptionally large increase in exports of computers and related equipment from Indonesia to China.

Regional trends

- Among developing countries in Latin America, the Caribbean and other America, Mexico had a 2.9 per cent share of world exports of ICT goods in 2005 and registered a CAGR of 13 per cent between 1996 and 2006 (compared with 7 per cent for total exports). Other developing countries in that region such as Guatemala, Colombia and Honduras also had higher growth in ICT goods exports between 1997 and 2005 compared with overall exports, and above-average growth compared with all developing countries. Although they do not currently have significant shares in world exports, they grew significantly from very low initial export values in 1997. Countries such as Brazil and Costa Rica, which started with higher export values in 1997, also had higher CAGRs than the overall exports and higher than all developing countries' export.

- In Africa, developing countries' ICT export values are small. However, Mauritius, Uganda and the United Republic of Tanzania recorded substantial growth (over 50 per cent each) between 1997 and 2005, which is well above their total export growth and above the average of all developing countries' ICT goods exports CAGRs. Other countries such as Cape Verde, Egypt and Senegal registered lower growth rates but still above their overall export growth rates or above the developing countries' exports growth rate (starting from very low initial values).

- Transition economies' export values are small, as well as their shares in world exports of ICT goods. Exceptions are Romania and Croatia, which have some of the highest exports values in the region and also had high CAGRs of 50 per cent and 18 per cent respectively between 1997 and 2005, which are well above the overall export growth rates or those of the developing countries' exports.

- East Asia dominates trade in ICT goods. In Asia, a number of developing economies such as China, Hong Kong (China), the Republic of Korea, Singapore, Malaysia, Taiwan Province of China, Thailand and the Philippines are either among the top 10 exporters of ICT goods or have important export shares, comparable to those of developed countries. China, Hong Kong (China), the Republic of Korea, Indonesia, Turkey and India, which have the highest export values (in billions of dollars) in the region, also had higher CAGRs than the overall export CAGRs between 1997 and 2005.

The least developed countries (LDCs) account for only a small share of global ICT trade (less than 1 per cent, which is slightly higher than their overall share of world merchandise exports). However, their overall CAGR in ICT goods exports between 1997 and 2004 was high, at 25 per cent. This was largely influenced by strong growth in some of the LDCs' exports. For example, Uganda and the United Republic of Tanzania were the two countries that registered very high CAGRs (57 per cent and 54 per cent respectively), and exports from Bangladesh, Mozambique, Senegal and Madagascar also experienced an upward trend (table 2.5). In the case of Madagascar, Senegal and Uganda, exports of ICT goods picked up only recently (2004–2005) and are concentrated mainly in telecommunication equipment, which is also the main category of ICT goods imported into those countries. Uganda and Senegal export mainly to developed European countries, while Madagascar, besides exporting to Europe, registers high values in terms of exports to Singapore, China and Hong Kong (China). Imports of ICT goods from LDCs into the European Union enjoy duty-free market access either under the ITA agreement or under the EU's Everything But Arms initiative (since 2001). In 2006, as much as 75 per cent of LDC exports of ICT goods to developed countries went to the EU. Developing economies in Asia and Africa are the other main buyers of ICT goods from LDCs.

Imports

On the importing side, the major importers of ICT goods since 1996 include developed countries such as the United States, Germany, the United Kingdom, France, the Netherlands and Japan. Developing economies such as China, Hong Kong (China), Malaysia, Thailand, the Philippines and Mexico gradually increased their imports of ICT goods, overtaking

Table 2.5

Exports of ICT goods in selected least developed countries (thousand $)

	1997	2004	2006	CAGR (1997–2004)	CAGR (2004– 2006)
Madagascar	553	545	4 656	-0.2	192.3
Uganda	508	12 103	58 805	57.3	120.4
Senegal	1 219	6 161	24 066	26	97.6
United Republic of Tanzania	105	2 161	2 179	54	0.4
Bangladesh	139	8 650	N/A	80.4	N/A
Mozambique	N/A	7 244	1 914	N/A	48.6

Source: UN COMTRADE.
N/A = not available.

other developed economies such as Italy and Canada, which were among the 10 largest importers in 1996, or Sweden, Switzerland and Austria, which were among the 20 largest importers in 1996. In fact, in 2006, China became the second largest importer of ICT goods, with an import value of $226 billion, after the United States with an import value of $280 billion. At the same time, the import shares of the United States, Japan and Singapore have decreased.

See Annex 1 for ICT goods imports by country.

Developed countries accounted for 55 per cent of world imports of ICT goods in 2005, and developing countries accounted for 44 per cent. As shown in the section on exports, there was a general shift from developing countries being the main importers of ICT goods in the mid-1990s to developed countries being the main importers in 2005. Transition countries in South-East Europe, CIS countries and countries in Oceania import only a small share of ICT goods.

Overall imports of ICT goods have grown faster than imports of other goods. The share of world imports of ICT goods in world imports of all merchandise goods increased from 13 per cent in 1996 to 15 per cent in 2005. Between 1996 and 2005, world imports of ICT goods increased at a higher CAGR than total world merchandise imports – at an annual 9 per cent CAGR compared with 8 per cent.

Regional trends

- Among developing countries in Latin America, the Caribbean and other America, Mexico (which is among the top 20 importers) had a 2.8 per cent share of imports of ICT goods in 2005 and registered a CAGR of 12 per cent between 1996 and 2006 (compared with a 7 per cent annual growth of total imports). Other countries such as Ecuador and Costa Rica, which have some of the highest import values in the region, registered higher annual growth between 1997 and 2005 (17 per cent and 25 per cent respectively) than total annual import growth (10 per cent).

Chart 2.22

Selected LDC exports and imports of ICT goods

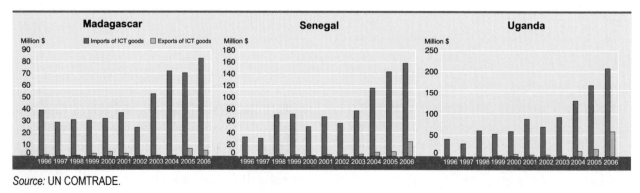

Source: UN COMTRADE.

Chart 2.23

Direction of ICT goods trade originating in developed and developing economies, 1996-2005

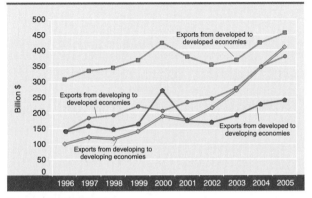

Source: UN COMTRADE.

- Developing countries in Africa have very small import values and do not command significant shares of world imports of ICT goods. Nevertheless, some countries' imports, for example those of Mauritius, Uganda and Senegal, grew at CAGRs of over 20 per cent between 1997 and 2005, which is well above the overall CAGRs for imports (albeit starting from low levels).

- A number of developing economies in Asia such as China, Hong Kong (China), Singapore, Malaysia, the Republic of Korea, the Philippines, Thailand and India have significant shares of

world imports. The share of China's imports of ICT goods in its imports of all merchandise goods increased from 12 per cent in 1996 to 31 per cent in 2006. Between 1996 and 2006, China's imports of ICT goods increased at a higher CAGR than its total merchandise imports–at an annual 32 per cent CAGR compared with 20 per cent.

- Countries in South-East Europe and the CIS do not have significant shares of world imports of ICT goods. However, Romania and Croatia experienced high CAGRs between 1997 and 2005 (19 per cent and 11 per cent respectively), which were above the overall 10 per cent CAGR for imports.

Direction of trade: the emerging South-South trade

In the LDCs, imports of ICT goods by far outnumber exports. All of the LDCs are in fact net importers of ICT goods. In addition, the share of LDCs' ICT goods imports in total ICT goods imports is small (1.1 per cent in 2005). Their overall CAGR for ICT goods imports between 1997 and 2004 was 17 per cent. In some LDCs (Madagascar, Uganda and Senegal) import growth rates are below average, while exports are increasing at above-average rates (chart 2.22). This points to the growth of the domestic ICT sector, which increasingly serves the domestic market with ICT goods.

Table 2.6

Direction of ICT goods trade and growth, 2000–2005

Exports from	2005 value of ICT goods exports (million $)	CAGR (per cent) 2000 –2005	CAGR (per cent) 2000 –2003	CAGR (per cent) 2003 –2005	Share (per cent) of world exports of ICT goods in 2005
Developed to developed economies	455.118	1.3	-2.9	11.1	30.1
Developed to developing economies	239.541	-2.5	-6.8	11.8	15.8
Developed to transition economies	20.300	70.7	52.3	33.0	1.3
Developing to developed economies	379.856	13.0	6.2	16.7	25.1
Developing to developing economies	410.554	16.8	7.5	23.0	27.1
Developing to transition economies	4.291	70.6	52.4	32.6	0.3
Transition to developed economies	1.710	-22.5	-23.2	2.5	0.1
Transition to developing economies	889	0.8	-1.6	6.3	0.1
Transition to transition economies	716	13.6	7.8	14.2	0.0

Source: UN COMTRADE.

In 2004, exports from developing to developing countries (i.e. South–South) exceeded those from developing to developed countries (chart 2.23). The $410 billion value of South–South trade in ICT goods had almost equalled the $450 billion value of North–North trade, and is likely to have exceeded it in 2006, given the strong growth of South–South ICT trade and the relatively weaker growth of North–North trade (table 2.6). This confirms the increasing importance of trade among developing countries, and the overall shift of ICT production and trade from developed to developing countries. It also demonstrates the growth of the ICT market in developing countries, where the potential for ICT uptake is considerable and hence the demand for ICT goods high. Unlike in the past, when ICT trade largely took place between developed countries, ICT goods trade is becoming trade among developing countries.

Remarkable is also the flow of ICT goods to transition economies. Exports from developed and developing economies directed to transition economies grew significantly between 2000 and 2005, although they accounted for only a small share in total world exports. As transition economies are quickly catching up in the development of their information economies, there is a great demand for ICT products.

In the developed world, most countries import from other developed countries. However, imports from the developing world increased at a CAGR of 13 per cent between 2000 and 2005, compared with a CAGR of 1.3 per cent only for imports from developed economies. It is likely that this trend will continue and

therefore developing economies will soon become the main source of imports of ICT goods for developed economies.

3. ICT services

Unlike trade in ICT goods, trade in ICT services is more difficult to capture. An OECD definition of ICT services based on the Central Product Classification (CPC) Ver. 2 (2007) was agreed in 2006 (OECD, 2007b). However, the CPC classification is not used to capture trade in services statistics, which are mainly estimated using the IMF's BOP classification. The latter is rather broad and does not identify ICT services. Therefore, UNCTAD has been using the concept of ICT-enabled services to analyse trade and investment flows. ICT-enabled services go beyond the economic activities described in the ICT sector classification and include such BOP services categories as communication services, insurance services, financial services, computer and information services, royalties and licence fees, other business services, and personal, cultural and recreational services (UNCTAD, 2002).

Key trends concerning trade and investment in ICT-enabled services were featured in detail in the IER 2006. Since then, ICTs have continued to facilitate trade in services and increase the tradability of services. Exports of ICT-enabled services grew faster than total services exports during 2000–2005, at 11 per cent compared with 8 per cent. In 2005, the $1.1 trillion value of ICT-enabled services represented about 50

Chart 2.24

ICT-enabled services share in total worldwide services exports

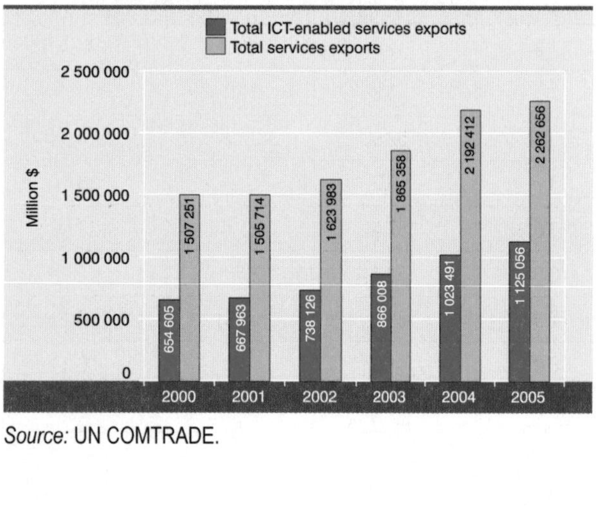

Chart 2.25

ICT-enabled services exports, by broad development categories, 2000-2005

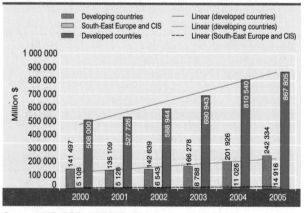

Source: UN COMTRADE.

Source: IMF BOP, UNCTAD Globstat and United Nations Statistics Division.

Box 2.8

Telecommunications sector – main target of FDI in ICT services

In the past two decades the telecommunications sector in all regions of the world has witnessed liberalization and modernization, which have increasingly attracted FDI. FDI came in the first wave in the early to mid-1990s and was characterized by privatization of State-owned telecommunications companies, and then in a second wave after the revolution in mobile communications technology in the mid-1990s, characterized by the introduction of cost-effective digital cellular services. Telecommunications FDI flows to developing countries were substantial during the mid- and late 1990s, but then decreased after 2000 following the bursting of the telecommunications bubble in developed countries, the compromised balance sheet of leading global operators and pessimism about emerging markets after the East Asian, the Russian Federation and Argentine economic crises (World Bank, 2006).

Evidence regarding large-scale foreign entry in the telecom services in 85 developing countries in the period 1985–1999 points to an improved and more competitive supply of services as a result of better firm performance (Fink et al., 2002). Research has shown that FDI has led to improved telecom services and contributed to higher economic growth (Norton, 1992; Roeller and Waverman, 1996). As the telephone system improves, the business-related costs fall, and firm output increases in individual sectors of the economy. Thus, telecommunications infrastructure investment and the derived services provide significant benefits because of productivity gains.

From a study carried out by the World Bank, it emerges that 122 out of 154 developing countries financed telecommunications infrastructure projects with foreign investment between 1990 and 2003 (World Bank, 2006). The foreign investment commitments amounted to $194 billion, which corresponds to 11.5 per cent of total FDI inflows to developing countries during that period. The same study shows that between 1990 and 2003 the 10 largest foreign direct investors in telecommunications were multinational corporations from Europe and the United States and they accounted for 57 per cent of the total FDI in telecommunications in developing countries. In 2002, the top-30 list of the largest telecommunications multinational corporations included four companies from developing countries: Datatec (South Africa), America Movil (Mexico), MTN Group (South Africa) and Telekom Malaysia.

Teléfonos de México (Telmex), the largest telecommunications operator in Mexico, expanded regionally in Latin America, including in Argentina, Brazil, Chile, Colombia, Mexico, Peru and Puerto Rico, as well as in the United States. In 2004, it acquired AT&T Latin America (with operations in Argentina, Brazil, Chile, Colombia and Peru) and paid $113.5 million for an 80 per cent stake in Techtel, Argentina's data and voice provider. In 2005, Telmex acquired MCI's equity stake in Brazilian long-distance operator Embratel. In addition, América Móvil, a Telmex spin off company that operates Mexico's largest mobile phone company – Telcel – has expanded aggressively in Latin America in recent years, and has subsidiaries in Argentina, Brazil, Colombia, Ecuador, El Salvador, Guatemala, Honduras, Nicaragua, Paraguay, Peru and Uruguay, as well as in the United States.

Table 2.7

Intraregional South-South telecommunications FDI, 1990–2003
(percentage of total South–South FDI)

Region of investor	Destination region					
	East Asia and Pacific	Europe and Central Asia	Latin America and the Caribbean	Middle East and North Africa	South Asia	Sub-Saharan Africa
North to South	72	93	90	52	75	51
South to South	28	7	10	48	25	49
East Asia and Pacific	100				24	50
Europe and Central Asia		100				
Latin America and the Caribbean			100			
Middle East and North Africa				100	36	5
South Asia					40	
Sub-Saharan Africa						45

Source: World Bank (2006).
Note: Based on the largest 75 investors in telecommunications, accounting for 95 per cent of total telecommunications-related FDI in developing countries between 1990 and 2003.

Box 2.8 (continued)

Maturing domestic markets and increased competition, geographical proximity and the withdrawal of some developed-country investors led to a surge in South–South FDI. From 2001 to 2003, South–South FDI accounted for over 36 per cent of total inflows and close to 20 per cent of the total number of telecommunications projects, compared with only 23 per cent and 11 per cent respectively in 1990–1999. This is primarily attributable to the high cost of acquiring reliable information about foreign markets for relatively small companies. They tend to invest in neighbouring countries where they are more knowledgeable about local conditions than multinationals because they have been developing trade, cultural or family links. They thus understand the complexities of investing in those markets and are more tolerant of political risk.

The surge in South–South FDI has been significant also because multinational companies either gradually phased themselves out, to be replaced by regional players, or considered markets too small or marginal to invest in directly and as a consequence invested through their subsidiaries. Investments by Vodacom of South Africa (partly owned by Vodafone of the United Kingdom) and Sonatel of Senegal (a subsidiary of France Télécom) are relevant examples.

---

per cent of total services exports, compared with only 37 per cent in 1995 (see chart 2.24). This has created new export opportunities for developing countries.

Until 2004 the top 10 exporters of ICT-enabled services were all from developed countries, but in 2005 India joined the top 10, as the first developing economy (replacing Hong Kong (China) and overtaking Italy and Luxembourg), with exports worth $41 billion and a market share of 3.8 per cent. Its annual growth rate

Chart 2.26

Developing countries' exports of ICT-enabled services, 2000-2005

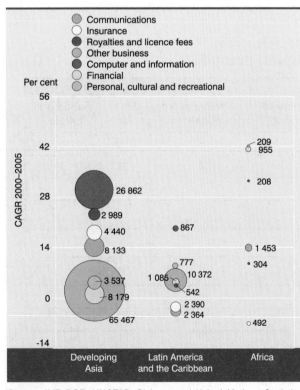

Source: IMF BOP, UNCTAD Globstat and United Nations Statistics Division.

between 2000 and 2005 of 37 per cent was higher than the growth rate of overall world ICT-enabled services exports. China is catching up quickly and is among the top 20 exporters and not far behind the top 10 in terms of export value ($26 billion in 2005). It has a share of 2.4 per cent of the world market and had annual growth of 22 per cent between 2000 and 2005, which is well above the growth rate of overall world ICT-enabled services exports. Both developed and developing countries have seen their ICT-enabled services exports increase since 2000 (chart 2.25). Computer and information services exports grew six times faster than total services exports between 1995 and 2004, and the share of developing countries in this export sector increased from 4 per cent in 1995 to 28 per cent in 2005 (although it was largely dominated by India).

In developing Asia and Latin American and the Caribbean, computer and information services exports grew at an annual rate of 30 per cent and 19 per cent respectively between 2000 and 2005, faster than all the other ICT-enabled services exports in those regions and faster also than the overall world ICT-enabled services exports. In Africa, personal, cultural and recreational services, financial services and computer and information services grew annually at higher levels than other ICT-enabled services exports and at higher levels than the overall world ICT-enabled services exports (chart 2.26).

Investment flows and outsourcing

International sourcing[15] of business activities/processes has become an integral element in the discussion on trade and investment in ICT-enabled services, and the related shifts from developed to developing countries.

Available data sometimes distinguish between outsourcing in the ICT industry and outsourcing in other industries (BPO), but there is no internationally agreed definition. Therefore, the data presented below have to be interpreted carefully.

According to the Everest Research Institute, the worldwide outsourcing market size in 2005 was $362 billion, of which IT outsourcing accounted for $233 billion (64 per cent) and BPO accounted for $129 billion (36 per cent). [16]

Rising labour costs in the most popular locations, competitive pressures and improving host-country environments, have led to a broadening of the geographical scope of locations for FDI in services and to offshoring of service activities such as IT services, business processes and call centres. Most offshored services are concentrated in India, China, Malaysia or the Czech Republic, but Canada and the United States also rank highly because of their favourable business environment and skilled labour force. A more detailed discussion of China and India as popular international sourcing destinations will be found in section D.

According to the OECD (2006a), there were 632 export-oriented FDI projects in IT services worldwide during 2002–2003, 513 call-centre projects and 139 projects related to shared services centres. The number of IT services projects in developing countries more than doubled. Developing Asia accounted for 265 (42 per cent) of the IT services projects, with India accounting for 118 (19 per cent) of the worldwide total. A total of 33 per cent of the call-centre projects and 47 per cent of the shared services centre projects were directed to Asia. The telecommunication sector in developing countries has been a preferred recipient of FDI during the past decade (see box 2.8). For example, in South Africa, FDI in telecommunications and information technology has overtaken that in mining and extraction (UNCTAD, 2004).

D. The role of the ICT sector in the growth of China and India

1. Overview

The two largest developing economies, China and India, have been growing strongly during the past decade. Both countries accounted for an increasing world GDP share and growth between 1995 and 2003; the figures are as follows: China 11 per cent (average

GDP share) and 22 per cent (growth share); India 6 per cent (average GDP share) and 11 per cent (growth share) (Jorgenson and Vu, 2005).

This has impacted on global economic developments ranging from important shifts in trade and FDI flows from developed countries to China and India, to shifts in production and employment from West to East, the growth of outsourcing in services, and an increase in South–South trade and investment, and the related emerging new geography of trade.[17]

ICTs have played a critical role in the expansion of the two economies. Both economies have benefited from the strong growth of their ICT sectors. As shown in the previous section, China overtook the United States as the world's number one producer and exporter of ICT goods in 2004. India is the world's largest exporter of ICT and ICT-enabled services and the main market for business process outsourcing (BPO).

Economic growth in China and India is driven by foreign investment and international sourcing. For example, the two countries have been described as the most attractive business locations for foreign investors (UNCTAD, 2005b), often because of the growth of the ICT industry, growth in China being in ICT manufacturing, and in India in ICT and related services. In the next few years, not only will China and India continue to be major recipients of FDI and international sourcing, but also international sourcing by those countries to other locations, for example in developing countries, may increase. Both countries are in the process of shifting from labour-intensive to knowledge-intensive goods and services. It is possible that China and India will generate a large pool of knowledge in the future, and also develop new technologies, which could be available to developed countries and hence could further contribute to global shifts in production, trade and employment.

2. China's ICT manufacturing

China's ICT sector is primarily driven by the production of ICT goods. ICT production in China grew dramatically between 2000 and 2005 (OECD, 2006a). According to Chinese sources, the value added of the "information industry" reached 7.5 per cent of GDP in 2004, a 30 per cent an increase in value from 2003. Value added of post and telecommunication services accounted for 8 per cent of services industry value added (2003). This compares with total services accounting for 32 per cent of total value added in 2004,

Chart 2.27

China's share of employment in world total employment in the electrical and electronic industry, 1997 and 2004

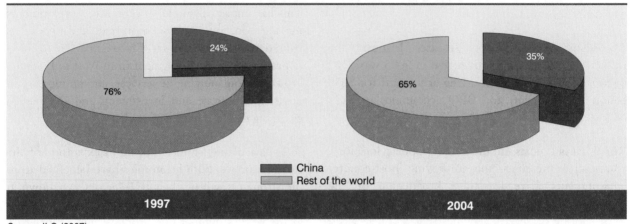

Source: ILO (2007).

which is still relatively low compared with the figures for OECD countries. The ICT sector is the largest trade sector, accounting for 34.4 per cent of total trade in 2006. [18] Employment in the electrical and electronics industry, a major part of the Chinese ICT sector, has increased sharply: in 1997 China accounted for 24 per cent of global employment in that sector, whereas that figure was 35 per cent in 2004 (ILO, 2007) (chart 2.27).

China imports mainly high-value-added electronic components and exports computer and related equipment, telecom equipment (the area of highest growth), and audio/video equipment. There is a tendency to increasingly export final products. Export

Chart 2.28

China: merchandise exports and ICT goods exports (1996 = 100 per cent)

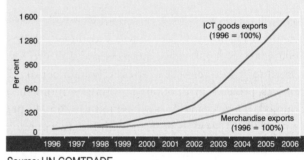

Source: UN COMTRADE.

Chart 2.29

China: share of ICT goods exports in all merchandise exports, 1996 and 2006

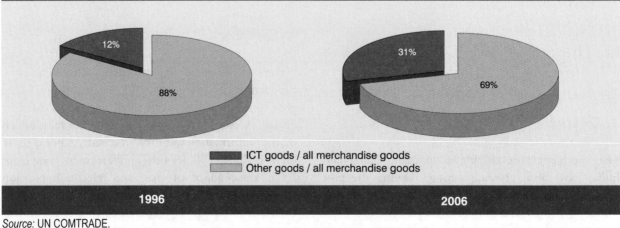

Source: UN COMTRADE.

destinations have somewhat shifted to developing Asia, but are still mainly the United States, Hong Kong (China), the EU and Japan.

In 1996, ICT goods exports accounted for 12 per cent of total merchandise exports, a figure that increased to 31 per cent in 2005. Between 1996 and 2006, exports of ICT goods increased at a higher rate than total merchandise exports – at an annual 32 per cent compared with 20 per cent for total merchandise exports (charts 2.28 and 2.29). ICT goods exports have thus been a driving force in Chinese exports over the past decade. In 2004, China overtook the United States as the world's largest exporter of ICT goods, which had a value of $299 billion in 2006.

On the import side, total merchandise imports growth slowed down between 2004 and 2005, but ICT goods imports continued to grow strongly. The share of ICT goods imports in total merchandise goods imports increased from 12 per cent in 1996 to 31 per cent in 2006. Remarkably, in only a year (between 2005 and 2006), the share of ICT goods imports in total merchandise imports increased from 26 per cent to 31 per cent. This is primarily due to a sharp increase in the imports of ICT and chips, and wired telecom equipment. Like exports, imports of ICT goods increased at a higher rate than total merchandise imports between 1996 and 2006 – at an annual 32 per cent CAGR compared with 20 per cent.

A significant share of FDI in China is driven by ICT-related investments. For example, in 2005, China received 3,000 "instances" of FDI inflows for a contractual value of $21 billion in telecom equipment, computers and other electronic equipment, which accounted for almost 30 per cent of all FDI inflows (OECD, 2006a). Most of the investment goes towards labour-intensive, low-value-added assembly and production of, for example, television sets, computers (e.g. laptops) and telephone handsets.

Leading firms from the United States, such as Dell, Hewlett Packard, Motorola and Nokia, as well as from Taiwan Province of China, have made ICT-related FDI. In 2004, the 3,384 ICT manufacturing firms from abroad accounted for 21 per cent of total assets in this sector, 30 per cent of total revenue, 20 per cent of profits and 16 per cent of employees. Most of the FDI inflows are directed towards manufacturing and less towards services. ICT and ICT-enabled services (mainly leasing and business services, followed by computer services and scientific research services) have accounted for a relatively small share of total

Chinese FDI – 6 per cent of inward FDI in 2004 (OECD, 2007a).[19] Foreign affiliates from Europe, the United States, and Taiwan Province of China have also opened R&D units in the computer, communications and electronics industry.[20]

China will continue to grow in ICT supply; its ICT industry is still low-cost manufacturing often relying on imports of intermediate inputs, but it will gradually shift to ICT-enabled services and move up the value chain. The Chinese Government is encouraging the development of capacity in ICT services, which have so far had a small share in the economy, especially compared with India. China's offshoring is more focused on industry-specific R&D activities and, unlike India, less on language-based back-office services or call centres.

China has the necessary infrastructure and IT workforce to attract investments in the ICT services area. A recent study by the International Data Corporation (IDC) claims that Shanghai and Beijing could overtake Indian top outsourcing destinations by 2011.[21] But research carried out by the OECD (2007a) concludes that China is unlikely to compete with India in ICT-enabled services unless it improves language, cultural and corporate culture skills. It also needs to strengthen its intellectual property legislative system and its regulatory system in order to create a level playing field in the supply of services, especially computer and information services. For the time being, China has the advantage of receiving BPO contracts from Japan since Japanese is spoken widely in one of its regions (North-East China's Liaoning Province – the north-east cities of Shenyang and Dalian); hence it has become the major offshoring destination for Japan. Large Indian companies are already investing in China as a springboard for entering markets in Japan and the Republic of Korea.

3. India's ICT and ICT-related services

As in the case of China, the ICT industry has been an important driver of India's economic growth in recent years and will continue to be so. Already, the ICT industry[22] contributed 5.4 per cent of GDP in 2006, up from 4.8 per cent in 2005 (agriculture contributed 18 per cent to GDP). The value of software exports alone exceeded that of foreign investment (in the same year) in a country which is also a major destination for FDI. The ICT sector is also an important source of employment in India: the number of professionals increased from 284,000 professionals in 1999 to 1.3

million in 2005, and it is estimated that it will reach 1.6 million in 2007 (table 2.8). While those figures are small compared with, for example, agriculture (which employs 60 per cent of the working population, including the non-organized sector), it is an important job growth market for the country.

The ICT sector has also played an important role in India's trade performance over the past decade. Both total merchandise and services exports increased significantly. Manufacturing exports increased from $30 billion in 1995 to $120 billion in 2006, while services exports (based on the IMF BOP classification) increased from $6.7 billion in 1995 to $48 billion in 2005. In particular, the share of services in total exports increased from 18 per cent in 1995 to 37 per cent in 2006, and this has been primarily ICT-driven. For example, the ICT-enabled services share in total services exports increased from 33.8 per cent in 1995 to 86 per cent in 2005; computer and information services alone accounted for 56 per cent of India's ICT-enabled services exports in 2005.

Table 2.8

Indian employment in the software and services sector

Sector	FY 2004	FY 2005	FY 2006	FY 2007E
IT services	215 000	297 000	398 000	562 000
ITES-BPO	216 000	316 000	415 000	545 000
Engineering services and R&D and software products	81 000	93 000	115 000	144 000
Domestic market (including user organizations)	318 000	352 000	365 000	378 000
TOTAL*	830 000	1 058 000	1 293 000	1 630 000

Source: NASSCOM, Indian IT industry – Fact Sheet, February 2007 (www.nasscom.in).
Notes: *Figures do not include employees in the hardware sector.

Table 2.9

The Indian IT industry

$ billion	FY 2004	FY 2005	FY 2006	FY 2007E
IT services	**10.4**	**13.5**	**17.8**	**23.7**
-Exports	7.3	10.0	13.3	18.1
-Domestic	3.1	3.5	4.5	5.6
ITES-BPO	**3.4**	**5.2**	**7.2**	**9.5**
-Exports	3.1	4.6	6.3	8.3
-Domestic	0.3	0.6	0.9	1.2
Engineering services and R&D, software products	**2.9**	**3.9**	**5.3**	**6.5**
-Exports	2.5	3.1	4.0	4.9
-Domestic	0.4	0.8	1.3	1.6
Total software and services revenues	**16.7**	**22.6**	**30.3**	**39.7**
Of which, exports are	12.9	17.7	23.6	31.3
Hardware	5.0	5.9	7.0	8.2
Total IT industry (including hardware)	**21.6**	**28.4**	**37.4**	**47.8**

Source: NASSCOM, Indian IT industry – Fact Sheet, February 2007 (www.nasscom.in).
Notes: Total may not match because of rounding off.
　　　*NASSCOM estimates have been reclassified to provide greater granularity.
　　　Historical values for a few segments have changed because of availability of updated information.

According to NASSCOM, ICT-related services exports grew by 34.6 per cent (CAGR) annually between 2000 and 2006, from $4 billion to $23.6 billion (table 2.9). It is estimated that the export value could reach $60 billion in 2010. NASSCOM's definition of ICT-related services includes BPO, IT software and services, and engineering services and products, and is not based on the IMF BOP classification. The main export markets are the United States and the United Kingdom, but revenues from other European markets are growing.

In addition, India has become the main global hub for BPO and international sourcing of services. According to the NASSCOM – McKinsey 2005 study, India accounts for 65 per cent of the global market in ICT offshoring and 46 per cent of global BPO.[23] The latter figure is expected to increase to 50 per cent by 2010; this could mean that by 2010 India's IT-related services exports will account for 5 per cent of Indian GDP (compared with all exports accounting for 19 per cent of GDP) and 25 per cent of total exports of goods and services. NASSCOM estimates that one United States dollar now offshored to India will increase its value to $1.45, including the delivery of value to India ($0.33), savings for the United States ($0.67) and the creation of new value through the re-employment of US labour ($0.45).[24]

The global market for international sourcing in ICT and other services is forecast to increase further, with a huge growth potential for the Indian ICT services industry (see chart 2.30).

A research report by the Everest Research Institute found that India's labour arbitrage with the United States is likely to be sustained for another two decades.[25] Currently, more than 50 per cent of the Fortune 500 companies source to India, mostly because of the success of the BPO model. The BPO segment will grow, but at a rate of 30 per cent. This sector accounted for 4 per cent of India's GDP and 29 per cent of exports in 2004–2005 and is projected to grow to 7 per cent of GDP and 35 per cent of exports by 2008–2009. India's software and services export sales are well on track to meet a target of $60 billion for 2010.[26]

While Indian companies clearly dominate the ICT services industry, business processing in other sectors is largely in the hands of foreign companies (chart 2.31), although, with more domestic firms gaining ground (Rowthorne, 2006).

India has been one of the main destinations for FDI: TNC investments announced in 2006, and to be made over next few years, total $10 billion. The Indian Government aims to attract $150 billion in FDI in the next decade by setting up special economic zones, science parks, and free-trade and warehousing zones (UNCTAD, 2005b).

Chart 2.30

Potential market for ICT-related international sourcing

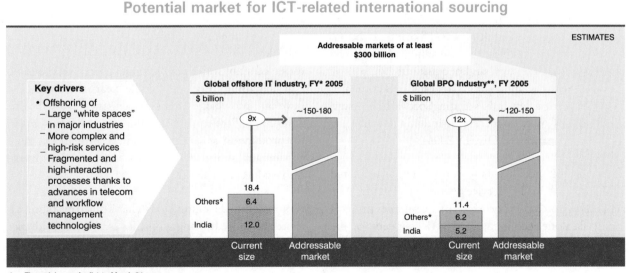

* Financial year, April 1 to March 31
** Includes addressable markets in currently offshoring industries
*** Includes Philippines, China, Russia, Eastern Europe, Ireland, Mexico

Source: NASSCOM, presentation at UNCTAD-OECD-ILO expert meeting "In support of the implementation and follow-up of WSIS: using ICTs to achieve growth and development", Geneva, 4-5 December 2006.

Chart 2.31

India-based IT services providers, export revenue by corporate characteristic (2005), in billion $

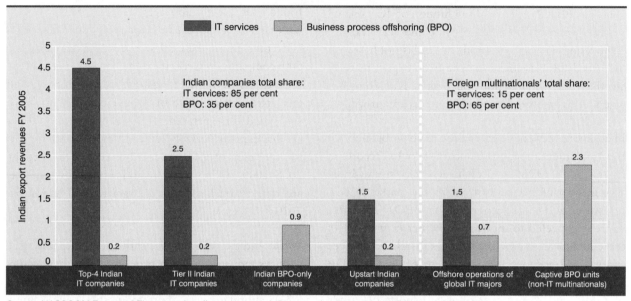

Source: NASSCOM Facts and Figures, at http://www.nasscom.in/Nasscom/templates/NormalPage.aspx?id=28487.

Approximately one third of Indian exports of ICT services and two thirds of ICT-enabled services are estimated to be generated by foreign-owned companies.[27] Export-oriented affiliates that locate in India's software technology parks to serve foreign markets, joint ventures that have expanded in the Indian market by buying local companies, and units of Indian ICT companies that are established, managed and expanded by a foreign company and then later taken over by the Indian companies are examples of ways in which TNCs have gained a foothold in the country. Joint ventures such as Mahindra British Telecom Ltd, India's eighth largest ITS company, NEC-HCL Infosystems, Deloitte Consulting-Mastek, and Microsoft-TCS-Uniware, and acquisitions by IBM of Dakesh eServices, India's third largest ITES company, or by Oracle of i-flex, India's leading software company, are examples of how successful India is in attracting foreign multinationals (OECD, 2006a).

Between 2002 and 2003, there were 632 export-oriented FDI projects in IT services worldwide, with Asia accounting for 265 (42 per cent) of IT services projects and India alone accounting for 118 (19 per cent) of the world total.[28] In 2005, total inflows into India stood at a record level of $5.6 billion. Cross-border M&As in India increased in 2004 in the telecommunications, BPO and pharmaceutical industries. FDI inflows have been encouraged by an improving economic situation and a more open FDI climate. For example,

the Government allows 100 per cent FDI under the automatic route in software and related services, as well as in electronics and ICT hardware manufacturing (WTO, 2007). In telecommunications, the FDI limit was increased from 49 per cent to 74 per cent in 2005, and this resulted in the telecom sector being one of the largest recipients of FDI, after the electronics and electrical equipment sector.

A key factor in India's performance in the ICT services sector has been its relatively strong ICT-skilled labour force. These strong ICT skills will also help increase productivity in other manufacturing sectors and develop new high-value-added products. It is expected that India will develop more sophisticated manufacturing products in the years to come and develop knowledge-intensive activities (Rowthorne, 2006). The Government is making an effort to support the production of quality manpower through a number of initiatives and reforms to improve the overall tertiary-level education system.

A more recent development is what is called nearshoring: more and more Indian companies establish subsidiaries in other Asian developing countries (e.g. in the field of software development) in order to be closer to their clients. This also helps to overcome restrictions on the movement of persons, which are often important for the ICT industry since staff have to move between home and client locations.

Promoting and supporting the domestic ICT sector have been important elements in the Government's national ICT plan, and it has put in place a number of initiatives in this regard. For example, the number of Software Technology Parks of India has increased to 47, spread all over the country. As India is a member of the WTO ITA, tariffs on all ICT goods were eliminated in 2005; and infrastructure facilities for the ICT sector, in particular to promote the BPO industry, are planned for all major cities in the country (WTO, 2007). The telecom sector has experienced a continuous opening up to competition (in particular, the mobile service), which has contributed significantly to a reduction in communication tariffs.

E. International and national environment and policymaking

1. Government policies on the ICT sector

Government policies can favourably affect ICT sector growth. Therefore, the development of the domestic ICT industry, including ICT production, trade and investment, has been included in many national ICT policies in developing countries. The nature of such policies differs substantially among countries, but a few common features can be identified.

The telecommunication sector is crucial for any ICT-related strategy and thus telecommunications-related policies are included in virtually all national ICT master plans. Attracting new investments in telecommunication, building ICT-relevant infrastructure and developing a wider range of services are usually the basic elements of such policies.

The ICT manufacturing industry is well developed in a number of countries, mainly in South-East Asia, but also in smaller economies such as Costa Rica, Mauritius and Romania. Following their successful example, a number of developing countries have implemented proactive measures to attract domestic and foreign investment in the ICT industry through the establishment of technology parks, the development of new infrastructure projects, the provision of special incentives and the creation of public–private partnerships.

The ICT services industry is a more recent growth sector. The availability of domestic ICT services is essential for supporting ICT uptake by businesses in other sectors. Moreover, ICT services create new growth and employment opportunities, such as those related to contact centres and internationally sourced contracts. As this chapter has shown, some countries that were traditional ICT manufacturers (Republic of Korea, Philippines and China) are now shifting their focus to developing ICT services, aiming to create higher-skilled jobs and develop higher-value-added products. Other countries that are starting to develop their ICT industry are also often focusing on ICT services, taking advantage of local language skills, geographical location and a well-educated workforce. As in the case of the ICT manufacturing industries, policymakers in those countries are including the development of ICT services in their national ICT strategies and plans (see boxes 2.9 and 2.10).

Government policy can be instrumental in the development of the ICT sector. In particular, in the area of telecommunications infrastructure and services, it can contribute to creating a more competitive market, which will lower prices and improve the quality of services. Furthermore, governments can play a critical role by ensuring a substantial commitment to technical education and the creation of a high-skilled workforce for the IT industry, and by providing a stable regulatory and enabling environment to attract BPO contracts and promote call centres.

Often, government policies are best designed and implemented in close dialogue with other stakeholders, particularly the businesses concerned. For example, close cooperation between public and private sectors has been crucial to successes in the ICT industry in such countries as the Republic of Korea and India. As mentioned in the previous section, the Indian Government has made an effort to involve private industry associations in its policy formulation related to the development of the domestic ICT sector. Following the success of India, the Government of Egypt is working with the local private sector to develop a competitive ICT industry (box 2.11). It is also developing a close relationship with India's NASSCOM to exchange know-how and learn from best practice. Egypt envisages becoming a hub in the Middle East for Indian ICT companies.

2. Multilateral trade agreements relevant to the ICT sector: the ITA ten years later

International trade in ICT goods and services is subject to multilateral trade agreements, which include the ITA,

Box 2.9

Ghana BPO sector found to have great potential - focus of government policy

Ghana is one of the sub-Saharan African countries that decided to prioritize the development of ICT as a key strategy in its national development plan and its efforts to achieve middle- income status. Thus, the Ghana ICT Policy for Accelerated Development has been integrated as a driver and enabler within the national Growth and Poverty Reduction Strategy II. In addition to the building of infrastructure and skills, and the establishment of a legal framework, the development of the ICT sector (particularly services) is a central component of the national ICT plan. The objective of the "e-Ghana project" is to increase employment and income generated by the sector, to increase its competitiveness and human capacity, and to promote investment in it.

The Government has sought the support of a private company (Hewitt Associates) to develop the country's ICT services, including BPO. It wants to build on its comparative advantage related to language skills, geographical location and related time zone, and low-cost labour force. It is also developing a technology park in a free trade zone in Tema, the port city of Ghana. Initial research by Hewitt estimates that the Ghana BPO sector has the potential to create 37,000 direct jobs and 150,000 indirect jobs within five years, and revenues of $750 million. In 2005, another private company (A.T. Kearney) ranked Ghana as the number one BPO destination in sub-Saharan Africa, ahead of South Africa. To realize its potential, the country still needs to improve the quality and cost of communication services and strengthen the regulatory environment related to the ICT services sector.

Sources: Paper presented by Mike Aaron, Minister of ICT, Ghana, at UNCTAD-OECD-ILO expert meeting, December 2006; ICT Provides Additional Growth for Ghana, World Bank News, 8 March 2007, available at http://web.worldbank.org/WBSITE/EXTERNAL/TOPICS/ EXTINFORMATIONANDCOMMUNICATIONANDTECHNOLOGIES/0,,contentMDK:21249466~menuPK:2643833~pagePK:64020865~ piPK:149114~theSitePK:282823,00.html

Box 2.10

Morocco: offshoring as a growth strategy

Morocco, like other African and Arab countries, is a new supplier in the global ICT sector market. As part of its ICT development strategy, the Government decided to promote the ICT industry and put in place a number of projects, including "Casashore", a domestic IT park, which was established in 2005.

While the ICT industry is still small, it has been growing strongly: in 2001-2004 at 6.5 per cent (annually), in 2004-2006 at 10 per cent (annually) and in 2006-2007 at 11 per cent. ICT sector revenue increased from 2.6 billion euros (2004) to 3.5 billion euros (2006), and the sector's contribution to GDP increased from 5 per cent in 2004 to 6.5 per cent in 2006. The industry is heavily concentrated in the two urban centres, with 85 per cent of activities located in Casablanca and Rabat, which employ 0.5 per cent of the actively employed population.

In 2006, 80 per cent of companies were in the area of telecom (operators, distributors, network installation). But offshoring activities have been growing, mainly with French companies, and they now constitute one of the most dynamic sectors in the economy with high growth potential. Therefore, the Government decided to make offshoring a central strategy in its "emergency" programme (a programme aimed at prioritizing the support of certain economic sectors); it estimates that, by 2013, offshoring could generate 91,000 jobs (currently 25,000) and contribute $1.7 billion to GDP.

Source: Fédération Marocaine des Technologies de l'Information, des Télécommunications et de l'Offshoring, 2007.

the General Agreement on Trade in Services (GATS) and the agreements related to telecommunications. A detailed analysis of each of the agreements and its implications for ICT sector trade would go beyond the scope of this chapter. The IER 2006 analysed the trade in ICT services and related GATS commitments. This section will review the latest developments regarding the ITA, which entered into force ten years ago, and which covers more than 90 per cent of the global trade in ICT goods.

Box 2.11

Developing the Egyptian ICT industry: need for training and finance

Egypt's geographical location makes it an ideal place for developing regional links with Europe, the Mediterranean, Africa and the Arab region. Since Egypt has a large population under the age of 25 and strong foreign-language skills, the Government has decided to make the development of the Egyptian ICT industry, and in particular ICT services, a priority in its national development plan.

The Ministry of Communication and Information Technology (MCIT), created in 1999, drew up its first National Plan to develop and expand the telecommunications infrastructure, establishing hundreds of information technology clubs to expand the pool of IT-skilled labour, and creating national information systems and databases. The plan was followed by the Egyptian Information Society Initiative in 2003, designed to help bridge the digital divide and facilitate Egypt's becoming an information society. In 2004, the Information Technology Industry Development Agency (ITIDA) was established to support the Egyptian ICT industry and exports.

In May 2007, the MCIT announced Egypt's 2010 ICT strategy, which consolidates and builds on the progress made by the Government in partnership with the private sector and civil society. The strategy has three main pillars: restructuring Egypt's ICT sector, maximizing the benefits of ICT for development, and nurturing innovation and supporting industry development. In line with the third pillar, ITIDA adopted the IT Industry National Development Strategy - targeting export growth, whose aim is achieve specific goals by 2010, such as increasing IT exports to reach more than 1 billion Egyptian pounds and creating 50,000 to 60,000 jobs in the IT sector.

In the joint MCIT/ITIDA–UNCTAD survey of 151 ICT companies[29] in 2007, respondents mentioned "difficulties in attracting new customers" and "lack of access to financial markets" as the main factors inhibiting their business development. This was followed by "high costs of office rent" and "slow Internet connections". A large majority of the respondents were domestically owned SMEs. Only 13 per cent of them had broadband, while the others had low-speed Internet connections. In comparison, an A.T. Kearney study (2005) focusing on services offshore business in Egypt showed that from a financial perspective, including office-renting costs, Egypt is an attractive location for foreign companies. When taking into account the spending on specialized personnel, infrastructure costs and tax regulations, A.T. Kearney ranked Egypt among the top five most financially attractive offshoring locations, before countries such as Indonesia, India, Thailand and China.

In this context, the Government of Egypt took a number of steps to foster the growth of small and medium-sized ICT enterprises through ITIDA, which assists such enterprises through programmes aimed at:

• Facilitating access to international markets through trade missions to international exhibitions;

• Facilitating the opportunities for winning tenders for joint projects with ITIDA;

• Offering training and quality certifications;

• Providing financial support to reduce business costs.

In addition, ITIDA helps NGOs working to support small and medium-sized ICT enterprises by providing office space in the Smart Village in order to facilitate contact with both the MCIT and ITIDA and enable them to provide the best services to SMEs in the ICT sector.

The survey also revealed the need to reinforce ICT education and training, as well as general business skills, including in management. Government policy can be helpful in a number of those areas, in particular in the area of education and training, and improving communications infrastructure. While significant progress has already been made in deregulating the telecommunications sector, especially with regard to mobile telephony, further improvements are to be expected in voice telephony and broadband connectivity. As part of its national ICT master plan, the Government has put in place a number of policies and initiatives to create a supportive environment for the ICT industry to improve its competitiveness, the results of which should become visible in the next couple of years.

In December 1996, the Ministerial Declaration on Trade in Information Technology Products was adopted by 29 countries at the first WTO Ministerial Conference, held in Singapore. The Declaration was adopted in preparation for the ITA, which entered into force in April 1997, with an additional 11 members. This made a total of 40 members, which accounted for 90 per cent of trade in IT products at that time. Under the ITA, duty-free imports are traded on a most-favoured-nation (MFN) basis; in other words, all WTO members benefit from the concessions made by members of the agreement. The ITA is a sectoral initiative and has therefore not been part of the overall negotiations (e.g. the Doha Round). Since tariffs in developing countries were higher, the tariff cuts for members from the developing world were larger than those for developed countries. For example, India had not only a very high applied rate but also an even higher bound rate, and therefore significantly reduced tariffs upon entering the agreement. Developing countries were granted a longer period of time in which to implement the ITA (at the latest by 2005) and new members still have time to implement the commitments.

Ten years later, in 2007, there were 70 members of the ITA membership, accounting for 97 per cent of IT trade; and developing country members are now in the majority (see annex 2 for a list of current members).

During the past ten years, world exports of ITA products have more than doubled, reaching in 2005 a value of $1.45 billion, which represents an annual average growth of 8.5 per cent. ITA trade accounts for 19 per cent of world exports, compared with agricultural products (8.4 per cent) and automotive products (7.2 per cent). Research on the ITA's impact on trade indicates that participation in the agreement increases bilateral trade, particularly if both countries are WTO members (Bora and Liu, 2006; Mann and Liu, 2007).

Product coverage and classification: keeping pace with innovation

The ICT market is typically characterized by rapid product development and technological change, as well as by a high level of innovation, with the result that new products enter the market on a continuous basis and more frequently than in many other manufacturing industries. Repeated calls for an expansion of the ITA's product coverage are therefore justified. For example, the distinction between intermediate and final goods is increasingly artificial and needs to be reconsidered. Moreover, in some countries, such as the Republic of Korea, the export growth of non-ITA products is now higher than that of ITA products, because of the exports of products such as colour TVs, CDs, DVDs and monitors, which are not included in the ITA. ICT goods that have changed their functionality and thus need to be reconsidered include set top boxes, flat panel computer monitors, laser multifunctional printers and some "parts" of ITA products.[30]

Attempts to revise the ITA product coverage have failed so far. Discussions in the WTO Committee of Participants on the Expansion of Trade in Information Technology Products did not result in any agreement on the number of consumer electronics included in the package, which for some member countries is too large, and for others too small. Some members are hesitant about revising the product coverage because they fear that existing products may be traded for new ones and thus eliminated from the ITA list. The impact of possible product erosion on trade flows and tariff levels needs to be further researched, from a development perspective.

Product classification is also being discussed. The ITA product list is largely based on the Harmonized System (HS) 1996 (ITA Attachment A), but includes in Attachment B a description of 13 specific items with no corresponding HS codes. The original intention was that descriptions could override HS codes and thus ensure the coverage of converging technologies.

This classification divergence between ITA products and the HS applied in Customs can lead to discrepancies in classifications among member states. Most countries have made ratifications in their WTO schedules by including ITA commitments at the national tariff line level, which is more detailed than the HS. As a result, some countries include certain products under the ones covered by the ITA - and hence apply duty-free treatment - while others do not. This is especially the case with products included in ITA Attachment B.

The HS nomenclature is currently being amended from the current version, HS02, to HS07. While only few modifications occurred when it changed from HS96 to HS02, the move to HS07 implies major changes for ITA products, with 158 out of 241 subheadings affected. As a result, the WTO schedules (bound and applied duties) and ITA product lists will need to be updated.

It should also be noted that the ITA product coverage is not identical to the OECD·ICT goods classification. This makes it difficult to compare international trade flows of ICT goods, which are usually based on the OECD classification. Further research is needed in order to examine the types of ICT goods not included in the ITA (and vice versa), and identify the related trade and tariff implications, in particular for developing countries. In addition, products that are critical to enhancing productivity and contributing to growth should be identified.

Geographical coverage: free riders?

Although the current 70 ITA members represent 97 per cent of trade in ITA products (in 2007), two regions – South America and Africa – remain virtually absent from the ITA. These are important exporters in South America, for example Brazil, Chile, Mexico and Venezuela. Mexico, for instance, is among the top 20 exporters of ITA products. At the same time, because of the MFN nature of the ITA, all countries enjoy full duty-free market access. Therefore, some members, in particular the EU, have stated that they will not accept any further discussion on product expansion unless some advanced developing countries become members.

In the case of some of the African countries joining the ITA, one possibility could be to introduce special and differential treatment with extended implementation periods on certain products. Also, countries focusing on the development of their ICT sector may consider the positive impact that becoming an ITA signatory has on attracting foreign investment, which was the case of Costa Rica. Further research needs to investigate the pattern of trade in ITA products by non-ITA members compared with members.

F. Conclusions and policy recommendations

The ICT sector is a dynamic and fast-changing market, with an important growth potential in developing countries. As an ICT supplier, it plays an important role in the development of a competitive information economy in developing countries. As a key technology producer, it contributes to total factor productivity and GDP growth. This chapter has provided a comprehensive overview of latest trends in the sector, including ICT production, employment, value added, trade and investment, from a development perspective. The information presented is summarized below, and relevant conclusions and suggestions for policy and research are provided.

Production, trade and investment

- ICT production, trade and investment will continue to increase, with continued shifts from developed to developing countries. The ICT sector thus plays a major role in the second generation of globalization and the emerging South and South–South trade. The impressive growth of some large developing countries, including China, India and Mexico, is having a significant impact on ICT sector performance in other countries in the South.

- South–South trade in ICT goods has grown substantially over the past five years. In terms of value, it overtook South–North trade in 2004, and is expected to have overtaken North–North trade in 2007. This corresponds not only to the overall trend of an emerging South and South–South trade, but also to the rapid catching up

of many developing countries in terms of ICT uptake and the development of their information economies.

- The developing world ICT market is concentrated in a few Asian economies, but a number of small economies (including some LDCs) have succeeded in building competitive advantage and increasing their export shares in ICT goods and services. China and India are the world's largest players in the export of ICT goods and services, respectively.

- Computer and information services exports continue to grow exponentially and the developing countries' shares rose from 4 per cent in 1995 to 28 per cent in 2005.

Services and international sourcing

- In 2005, India joined the top 10 group of ICT-enabled services exporters (replacing Hong Kong, China). The ICT services industry will continue to increase its contribution to domestic value-added, employment and export earnings, and to attracting foreign investment.

- The international sourcing of ICT production and ICT-enabled services will continue, with a huge potential for developing countries, while the impact on employment in developed countries is insignificant overall, although more noticeable in certain sectors. In the developing world, an increasing demand for ICT-skilled labour will have to be addressed.

- The expanding ICT industry and international sourcing offer a huge potential for developing countries. At the same time, competition will increase and countries wishing to attract FDI and BPO contracts will need to invest in their domestic labour skills, telecommunication infrastructure and improving the investment climate.

National and international policy responses

- The dynamic nature of the ICT industry has led a number of developing countries to include the promotion of the ICT sector (in particular, ICT-related services) in their national ICT plans and overall development strategies. This may be sustainable in the long term in the context of increased global competition only if global demand remains strong and sustained.

- At the international level, the WTO ITA has contributed to facilitating trade in ICT goods, 93 per cent of which are now imported duty-free. Some member countries indicated the need for a revision of the ITA to harmonize the product coverage based on international classifications and to take into consideration the fast-changing nature of the ICT market. Such revision should fully take into consideration the implications for developing countries. A careful analysis at the tariff-line level is needed in order to assess the implications for developing economies not yet signatories to the Agreement.

Annex 2.1

Trade in ICT goods

Table 2.10

Exports of ICT goods by level of development and by region, 1996– 2005 ($ million)

	1996	2000	2005	CAGR 1996–2005 (per cent)	CAGR 2000–2005 (per cent)
World	**701 724**	**1 112 170**	**1 514 254**	**8.9**	**6.4**
Developed economies	**458 001**	**647 833**	**716 054**	**5.1**	**2.0**
Asia	106 797	131 470	125 876	1.8	-0.9
Europe	211 115	309 122	415 890	7.8	6.1
North America	137 677	204 888	171 532	2.5	-3.5
Oceania	2 412	2 353	2 757	1.5	3.2
Developing economies	**242 539**	**462 259**	**794 875**	**14.1**	**11.5**
Africa	513	1 391	2 311	18.2	10.7
Asia	224 315	417 964	742 332	14.2	12.2
Latin America and the Caribbean	17 701	42 900	50 224	12.3	3.2
Oceania	10	5	8	-3.0	10.4
Transition economies	**1 185**	**2 078**	**3 324**	**12.1**	**9.9**

Source: UN COMTRADE.

Table 2.11

Exports of ICT goods, 1996– 2005: economies ranked by 2005 export values ($ million)

Economy	1996	2000	2005	CAGR 1996–2005 (per cent)	CAGR 2000–2005 (per cent)
China	18 584	46 996	235 167	32.6	38.0
United States	123 802	182 262	154 917	2.5	-3.2
Japan	103 213	123 548	121 474	1.8	-0.3
Hong Kong (China)	37 643	55 313	118 237	13.6	16.4
Singapore	67 742	77 345	106 576	5.2	6.6
Germany	42 812	57 452	99 127	9.8	11.5
Republic of Korea	34 316	61 525	87 163	10.9	7.2
Taiwan, Province of China	..	64 409	66 506	..	0.6
Netherlands	24 899	41 218	64 748	11.2	9.5
Malaysia	36 987	55 582	64 472	6.4	3.0
United Kingdom	43 116	55 865	59 755	3.7	1.4
Mexico	16 422	38 267	43 870	11.5	2.8
France	25 892	35 689	33 187	2.8	-1.4
Thailand	14 208	20 318	26 169	7.0	5.2
Ireland	13 265	26 349	24 931	7.3	-1.1
Philippines	10 294	26 422	24 418	10.1	-1.6
Canada	13 875	22 626	16 615	2.0	-6.0
Hungary	663	7 776	16 537	43.0	16.3
Sweden	11 407	16 883	15 818	3.7	-1.3
Italy	13 047	12 842	15 162	1.7	3.4
Belgium	..	..	14 620	..	..
Finland	5 935	11 555	14 557	10.5	4.7
Czech Republic	894	2 128	9 778	30.4	35.7
Spain	4 969	6 137	8 280	5.8	6.2
Austria	3 270	4 883	8 134	10.7	10.7
Indonesia	3 287	7 844	7 911	10.2	0.2
Denmark	3 154	4 177	7 102	9.4	11.2
Switzerland	4 143	4 712	5 554	3.3	3.3
Israel	3 584	7 921	4 402	2.3	-11.1
Poland	648	1 424	4 123	22.8	23.7
Brazil	1 010	2 513	4 038	16.6	10.0
Turkey	496	1 103	3 395	23.8	25.2
Slovakia	154	464	3 200	40.0	47.2
Portugal	1 371	1 893	3 184	9.8	11.0
Australia	2 180	2 068	2 262	0.4	1.8
Norway	1 301	1 430	1 858	4.0	5.4
Costa Rica	9	1 740	1 758	78.7	0.2
India	659	714	1 424	8.9	14.8
Estonia	161	996	1 403	27.2	7.1
Luxembourg	..	..	1 390	..	..
Russian Federation	794	799	1 157	4.3	7.7
Malta	908	1 565	1 121	2.4	-6.4
South Africa	333	521	798	10.2	8.9
Romania	37	552	770	39.9	6.9

Table 2.11 (continued)

Economy	1996	2000	2005	CAGR 1996–2005 (per cent)	CAGR 2000–2005 (per cent)
Morocco	4	552	705	77.5	5.0
Lithuania	192	199	661	14.7	27.2
Slovenia	368	397	588	5.4	8.2
Croatia	129	161	548	17.4	27.8
Greece	182	481	525	12.5	1.8
New Zealand	232	286	494	8.8	11.6
Tunisia	144	199	385	11.5	14.1
Cyprus	41	16	381	28.3	88.1
Saudi Arabia	28	114	369	33.2	26.4
Ukraine	144	283	302	8.6	1.3
Mauritius	7	11	295	52.5	91.5
Bulgaria	47	50	232	19.4	35.7
Argentina	123	181	205	5.9	2.6
Belarus	..	138	199	..	7.6
Jordan	..	47	172	..	29.4
Latvia	49	31	139	12.4	34.9
Oman	42	70	121	12.5	11.3
Sri Lanka	..	..	102	..	..
Pakistan	..	..	81	..	..
Colombia	12	20	74	22.4	29.9
Chile	25	34	52	8.7	8.7
Côte d'Ivoire	3	3	47	36.4	74.8
Peru	7	25	42	22.1	11.4
Serbia and Montenegro	9	9	36	17.0	30.9
Kazakhstan	..	55	34	..	-8.9
Venezuela (Bolivarian Republic of)	11	22	34	13.5	9.4
Barbados	..	25	32	..	5.0
Iceland	2	12	25	36.0	17.0
Trinidad and Tobago	12	3	25	9.0	51.4
Ecuador	2	2	24	29.5	71.7
Iran (Islamic Republic of)	..	6	23	..	29.6
Honduras	7	0	21	12.4	166.0
Namibia	..	13	18	..	7.4
Uganda	0	5	17	72.8	30.1
Guatemala	2	2	16	28.0	50.2
Egypt	0	6	14	47.1	21.2
Bosnia and Herzegovina	..	..	13	..	..
Qatar	..	12	11	..	-1.2
Moldova	12	8	11	-1.8	6.9
Bahrain	..	3	10	..	31.5
TFYR, Macedonia	2	4	8	20.2	15.4
Antigua and Barbuda	..	9	8	..	-3.8
Senegal	0	2	7	34.9	34.4
Madagascar	1	4	6	22.6	10.4
Uruguay	2	4	6	14.9	8.8

Table 2.11 (continued)

Economy	1996	2000	2005	CAGR 1996–2005 (per cent)	CAGR 2000–2005 (per cent)
Albania	0	2	5	33.5	23.0
Armenia	..	6	5	..	-5.6
Saint Lucia	3	4	4	2.8	-0.8
Bolivia	1	11	4	13.6	-17.5
New Caledonia	..	2	3	..	17.1
Paraguay	1	2	3	11.4	11.8
Fiji	..	0	3	..	65.6
Cape Verde	..	1	3	..	21.4
Jamaica	5	4	3	-6.6	-5.7
United Republic of Tanzania	..	3	3	..	1.2
Timor-Leste	..	..	2	..	..
Yemen	..	..	2	..	..
Cameroon	1	0	2	17.4	55.2
Azerbaijan	3	5	2	-5.2	-15.8
Gabon	6	2	2	-11.5	0.1
Mozambique	..	1	2	..	19.8
Ghana	0	3	2	16.6	-8.6
Kyrgyzstan	5	4	2	-11.7	-15.6
Nicaragua	43	1	1	-32.1	8.1
Malawi	1	1	1	2.0	12.0
Georgia	1	1	1	7.9	5.8
Syrian Arab Republic	..	..	1	..	..
Niger	1	1	1	-0.6	13.9
French Polynesia	10	3	1	-21.5	-15.3
Zambia	2	1	1	-5.1	7.0
Saint Kitts and Nevis	1	0	1	5.7	18.6
Guyana	..	0	1	..	35.3
Panama	0	..	1	51.1	..
Saint Vincent and the Grenadines	0	0	1	4.5	14.9
Mayotte	..	0	0	..	51.6
Togo	2	0	0	-19.6	36.5
Maldives	..	..	0	..	..
Seychelles	2	..	0	-19.7	..
Burundi	..	0	0	..	165.9
Dominica	0	0	0	0.9	-17.8
Benin	..	0	0	..	30.0
Belize	0	0	0	-13.8	-18.8
Mongolia	0	1	0	-17.0	-29.8
Montserrat	..	0	0	..	-30.7
Gambia	0	0	0	-3.3	-4.2
Central African Republic	0	0	0	-14.4	-7.8
Macao (China)	28	110	..	..	..
Lesotho	..	25	..	..	..
Botswana	..	20	..	..	..
Grenada	0	14	..	..	..

Table 2.11 (continued)

Economy	1996	2000	2005	CAGR 1996–2005 (per cent)	CAGR 2000–2005 (per cent)
Lebanon	..	11	..	..	..
Kuwait	..	9	..	..	..
El Salvador	0	8	..	..	..
Algeria	2	8	..	..	..
Bangladesh	0	7	..	..	..
Swaziland	..	3	..	..	..
Zimbabwe	..	3	..	..	..
Bahamas	..	3	..	..	..
Cuba	..	3	..	..	..
Burkina Faso	2	2	..	..	..
Andorra	1	2	..	..	..
Kenya	..	1	..	..	..
Sudan	0	1	..	..	..
Mali	0	1	..	..	..
Nepal	..	1	..	..	..
Suriname	0	1	..	..	..
Cambodia	..	1	..	..	..
Turkmenistan	..	0	..	..	..
Anguilla	..	0	..	..	..
Turks and Caicos Islands	..	0	..	..	..
Guinea	0	0	..	..	..
Greenland	0	0	..	..	..
Papua New Guinea	..	0	..	..	..
Nigeria	1	0	..	..	..
Comoros	0	0	..	..	..
Cook Islands	..	0	..	..	..
Dominican Republic	1	..	..	..	..
Faroe Islands	0	..	..	..	..
Rwanda	0	..	..	..	..
Aruba	..	..	..	..	..
Bermuda	..	..	..	..	..
Bhutan	..	..	..	..	..
Brunei Darussalam	..	..	..	..	..
Eritrea	..	..	..	..	..
Ethiopia	..	..	..	..	..
Haiti	..	..	..	..	..
Kiribati	..	..	..	..	..
Mauritania	..	..	..	..	..
Samoa	..	..	..	..	..
Sao Tome and Principe	..	..	..	..	..
Sierra Leone	..	..	..	..	..
Tuvalu	..	..	..	..	..
Wallis and Futuna Islands	..	..	..	..	..

Source: UN COMTRADE.

Table 2.12

Imports of ICT goods by level of development and by region, 1996–2005
($ million)

	1996	2000	2005	CAGR 1996–2005 (per cent)	CAGR 2000–2005 (per cent)
World	**718 213**	**1 128 748**	**1 574 158**	**9.1**	**6.9**
Developed economies	**480 808**	**716 614**	**863 035**	**6.7**	**3.8**
Asia	51 492	72 745	81 634	5.3	2.3
Europe	244 288	356 555	473 638	7.6	5.8
North America	174 028	273 933	289 576	5.8	1.1
Oceania	10 999	13 381	18 187	5.7	6.3
Developing economies	**232 073**	**406 137**	**691 373**	**12.9**	**11.2**
Africa	6 275	8 494	13 197	8.6	9.2
Asia	194 344	338 885	603 901	13.4	12.2
Latin America and the Caribbean	31 397	58 588	73 997	10.0	4.8
Oceania	57	170	279	19.2	10.5
Transition economies	**5 332**	**5 996**	**19 750**	**15.7**	**26.9**

Source: UN COMTRADE.

Table 2.13

Imports of ICT goods, 1996– 2005:
economies ranked by 2005 import values ($ million)

Economy	1996	2000	2005	CAGR 1996–2005 (per cent)	CAGR 2000–2005 (per cent)
United States	150 475	237 943	256 770	9.1	6.9
China	16 850	50 597	183 025	6.7	3.8
Hong Kong (China)	44 831	64 403	119 967	12.9	11.2
Germany	48 736	65 268	99 100	15.7	26.9
Singapore	50 429	59 769	80 415	5.8	1.1
Japan	47 858	66 871	76 454	5.3	2.3
United Kingdom	47 144	67 727	68 735	7.6	5.8
Netherlands	23 938	42 118	60 430	5.7	6.3
Republic of Korea	23 482	39 086	47 037	8.6	9.2
Malaysia	27 024	37 249	46 105	10.0	4.8
France	28 458	39 571	45 835	13.4	12.2
Taiwan, Province of China	..	44 851	45 068	19.2	10.5
Mexico	14 968	36 332	43 354	6.1	1.5
Canada	23 526	35 970	32 806	30.3	29.3
Italy	18 452	23 515	30 183	11.6	13.2
Philippines	9 911	12 621	23 333	8.2	8.7
Thailand	13 160	15 660	23 213	5.3	6.1
Spain	10 565	14 238	22 571	5.3	2.7
Belgium	..	..	17 719	4.3	0.3
Ireland	9 297	17 232	17 319	10.8	7.5
Australia	9 380	11 626	15 499	8.0	3.8
Sweden	9 094	12 254	13 690	6.1	4.4
Hungary	1 483	7 612	13 535	5.4	3.0
India	1 368	3 300	12 516	..	0.1
Austria	5 454	7 058	10 745	12.5	3.6
Brazil	7 318	9 133	10 634	3.8	-1.8
Switzerland	7 267	9 225	10 587	5.6	5.1
Denmark	4 651	5 909	9 831	10.0	13.1
Czech Republic	2 761	3 900	9 723	6.5	8.2
Poland	2 989	5 107	9 070	8.8	9.7
Finland	4 214	6 293	9 063	..	..
Russian Federation	2 979	1 883	8 859	7.2	0.1
Turkey	2 567	6 035	8 240	5.7	5.9
South Africa	3 514	3 648	6 741	4.7	2.2
Portugal	2 616	3 588	5 564	27.8	12.2
Norway	3 206	3 642	5 381	27.9	30.6

Table 2.13 (continued)

Economy	1996	2000	2005	CAGR 1996–2005 (per cent)	CAGR 2000–2005 (per cent)
Israel	3 635	5 874	5 180	7.8	8.8
Saudi Arabia	1 399	1 556	4 492	4.2	3.1
Argentina	2 491	3 869	3 726	4.3	2.8
Slovakia	472	1 002	3 292	8.7	10.7
Greece	1 593	2 465	3 281	15.0	20.0
Romania	688	1 563	3 230	13.1	12.2
Colombia	1 503	1 208	3 014	8.9	7.6
Venezuela (Bolivarian Republic of)	698	1 369	2 843	12.9	36.3
New Zealand	1 620	1 755	2 688	13.8	6.4
Iran (Islamic Republic of)	..	747	2 639	7.5	13.1
Chile	1 415	1 834	2 459	8.7	9.2
Indonesia	2 851	1 001	2 426	5.9	8.1
Pakistan	..	..	2 324	4.0	-2.5
Costa Rica	225	1 112	2 288	13.8	23.6
Ukraine	418	416	1 739	4.6	-0.7
Morocco	324	1 311	1 676	24.1	26.9
Luxembourg	..	..	1 633	8.4	5.9
Croatia	494	483	1 469	18.7	15.6
Estonia	355	1 028	1 469	8.0	20.1
Ecuador	195	220	1 153	16.9	15.7
Bulgaria	203	336	1 137	5.8	8.9
Peru	857	748	1 132	..	28.7
Slovenia	537	694	1 106	6.3	6.0
Lithuania	287	306	1 077	-1.8	19.4
Malta	888	1 525	1 045	..	..
Egypt	504	656	965	29.4	15.5
Kazakhstan	..	348	900	17.2	33.1
Tunisia	411	468	803	20.0	5.0
Jordan	..	237	782	..	..
Qatar	..	167	724	12.9	24.9
Cyprus	207	278	715	17.1	7.4
Serbia and Montenegro	274	165	659	21.9	39.3
Guatemala	137	450	657	21.1	27.6
Paraguay	415	244	646	3.1	8.6
Belarus	..	249	555	8.4	9.8
Latvia	150	245	551	15.8	28.6
Ghana	135	113	526	1.8	-7.3
Oman	231	188	459	7.5	8.0

Table 2.13 (continued)

Economy	1996	2000	2005	CAGR 1996–2005 (per cent)	CAGR 2000–2005 (per cent)
Mauritius	94	127	416	..	20.9
Sudan	24	108	413	7.7	11.4
Sri Lanka	..	..	411	..	27.0
Panama	177	277	389	..	34.1
Iceland	177	262	352	14.7	20.8
Bosnia and Herzegovina	..	..	325	10.2	31.9
Syrian Arab Republic	..	..	291	19.0	7.9
Honduras	98	13	277	5.0	21.5
Uruguay	253	257	268	..	17.4
Trinidad and Tobago	97	135	264	15.6	17.6
Jamaica	142	182	259	16.3	36.1
Côte d>Ivoire	91	105	250	7.9	19.5
Azerbaijan	52	124	242	17.9	26.8
United Republic of Tanzania	58	96	207	37.5	30.7
Nicaragua	58	70	179	..	..
Bahrain	..	134	169	9.1	7.0
Uganda	39	58	169	7.9	6.1
TFYR, Macedonia	88	91	153	..	..
Senegal	32	50	144	..	..
Georgia	24	52	142	12.2	85.3
Namibia	..	119	142	0.6	0.8
Barbados	..	101	137	11.7	14.3
Zambia	37	71	130	6.9	7.3
Mozambique	..	68	124	11.9	19.0
Albania	35	47	122	18.5	14.3
Bolivia	138	140	118	15.2	16.7
French Polynesia	56	78	105	13.3	20.8
Yemen	..	..	104	..	4.8
Moldova	29	43	98	17.5	23.8
Cameroon	35	45	95	6.4	10.9
New Caledonia	..	69	93	18.1	23.7
Maldives	16	24	89	21.9	22.5
Gabon	63	75	82	..	3.5
Fiji	..	..	80	..	6.3
Armenia	..	59	72	15.1	12.8
Madagascar	39	32	71	..	12.6
Mongolia	18	42	69	14.7	20.9
Malawi	..	25	52	-1.7	-3.3

Table 2.13 (continued)

Economy	1996	2000	2005	CAGR 1996–2005 (per cent)	CAGR 2000–2005 (per cent)
Suriname	19	35	49	7.2	6.0
Kyrgyzstan	47	25	46	..	..
Faroe Islands	22	28	36	14.3	17.8
Niger	7	8	30	11.8	16.0
Benin	..	18	30	..	6.1
Belize	9	20	29	20.8	30.6
Saint Lucia	15	25	26	2.8	1.7
Togo	13	11	24	..	..
Antigua and Barbuda	..	34	22	..	4.3
Mauritania	..	13	22	6.9	17.4
Saint Kitts and Nevis	11	19	20	15.7	10.5
Burundi	5	17	19	..	16.0
Cape Verde	..	17	19	11.0	7.0
Guyana	..	21	18	-0.2	12.9
Mayotte	..	6	17	5.6	5.2
Saint Vincent and the Grenadines	..	9	17	18.1	31.3
Dominica	7	11	15	..	10.7
Seychelles	23	..	14	14.2	7.7
Gambia	7	6	13	6.0	0.4
Central African Republic	7	1	5	7.2	17.9
Timor-Leste	..	..	4	..	-8.3
Montserrat	..	2	2	..	10.7
Kiribati	1	..	1	7.1	1.0
Algeria	480	545	..	15.5	2.0
Kuwait	..	448	..	..	2.2
El Salvador	141	355	..	..	-2.9
Lebanon	..	279	..	..	23.0
Cuba	..	258	..	..	12.3
Bangladesh	100	245	..	7.9	5.9
Nigeria	259	196	..	-5.3	..
Macao (China)	107	162	..	7.1	16.4
Kenya	..	154	..	-3.7	27.9
Botswana	..	125	..	..	..
Turkmenistan	..	112	..	..	-3.6
Andorra	73	89	..	0.9	..
Bahamas	..	78	..	..	..
Ethiopia	..	75	..	..	..
Nepal	..	50	..	..	..

Table 2.13 (continued)

Economy	1996	2000	2005	CAGR 1996–2005 (per cent)	CAGR 2000–2005 (per cent)
Burkina Faso	25	41	..	..	..
Cambodia	..	34	..	..	..
Swaziland	..	31	..	..	..
Mali	18	28	..	..	..
Papua New Guinea	..	22	..	..	..
Greenland	27	19	..	..	..
Lesotho	..	17	..	..	..
Grenada	9	16	..	..	..
Guinea	18	10	..	..	..
Turks and Caicos Islands	..	7	..	..	..
Anguilla	..	4	..	..	..
Comoros	4	1	..	..	..
Sao Tome and Principe	..	0	..	..	..
Rwanda	9	..	..	..	..
Aruba	..	..	..	..	..
Bermuda	..	..	..	..	..
Bhutan	..	..	..	..	..
Brunei Darussalam	..	..	..	..	..
Cook Islands	..	..	..	..	..
Dominican Republic	..	..	..	..	..
Eritrea	..	..	..	..	..
Haiti	..	..	..	..	..
Samoa	..	..	..	..	..
Sierra Leone	..	..	..	..	..
Tuvalu	..	..	..	..	..
Wallis and Futuna Islands	..	..	..	..	..
Zimbabwe	..	..	..	..	..

Source: UN COMTRADE.

Table 2.14

Top 50 exporters of ICT-enabled services, 1996– 2005: ranked by 2005 export values ($ million)

Rank	Economy	1996	2000	2005	CAGR 1996– 2005 (per cent)	CAGR 2000– 2005 (per cent)
1	United States	84 793	127 234	184 691	9.0	7.7
2	United Kingdom	49 896	77 418	132 848	11.5	11.4
3	Germany	34 934	36 849	73 836	8.7	14.9
4	Japan	34 757	33 483	52 469	4.7	9.4
5	Ireland	1 962	14 331	46 574	42.2	26.6
6	Netherlands	17 884	21 796	42 683	10.1	14.4
7	France[2]	30 282	27 933	42 032	3.7	8.5
8	India[3]	2 359	8 490	41 659	37.6	37.5
9	Italy	20 853	17 867	35 639	6.1	14.8
10	Luxembourg	..	..	34 183	..	..
11	Hong Kong (China)	..	21 346	32 776	..	9.0
12	Canada	14 094	20 736	28 857	8.3	6.8
13	Switzerland[4]	12 645	16 296	28 739	9.6	12.0
14	Belgium	..	..	28 098	..	..
15	Spain	9 221	13 291	28 016	13.1	16.1
16	Singapore	12 466	12 308	26 994	9.0	17.0
17	China	7 297	9 642	26 594	15.5	22.5
18	Sweden	5 118	10 913	26 023	19.8	19.0
19	Austria	15 848	16 123	25 411	5.4	9.5
20	Republic of Korea	9 003	9 196	14 307	5.3	9.2
21	Norway[5]	3 511	5 556	10 548	13.0	13.7
22	Israel	2 977	7 869	9 933	14.3	4.8
23	Brazil	2 307	5 514	7 845	14.6	7.3
24	Russian Federation	2 456	2 410	7 549	13.3	25.7
25	Australia	3 873	5 385	6 471	5.9	3.7
26	Hungary	1 839	1 486	6 026	14.1	32.3
27	Saudi Arabia	2 769	4 779	5 916	8.8	4.4
28	Malaysia	7 667	5 684	5 690	-3.3	0.0
29	Finland[6]	3 302	2 596	5 518	5.9	16.3
30	Thailand	4 972	2 822	5 510	1.1	14.3
31	Lebanon	..	..	4 870	..	..
32	Indonesia	278	86	4 729	37.0	122.9
33	Poland	2 256	1 977	3 614	5.4	12.8
34	Nigeria	621	1 512	3 415	20.9	17.7
35	Portugal	1 371	2 045	3 380	10.5	10.6
36	Greece	5 163	1 805	2 902	-6.2	10.0

Table 2.14 (continued)

Rank	Economy	1996	2000	2005	CAGR 1996– 2005 (per cent)	CAGR 2000– 2005 (per cent)
37	Czech Republic	2 192	2 221	2 677	2.2	3.8
38	Mexico	2 348	3 903	2 541	0.9	-8.2
39	Turkey	3 554	7 643	2 491	-3.9	20.1
40	Egypt	3 174	2 604	2 350	-3.3	-2.0
41	Romania	399	699	2 290	21.4	26.8
42	Cyprus	714	1 090	2 128	12.9	14.3
43	Argentina	541	719	1 842	14.6	20.7
44	South Africa	1 229	1 029	1 786	4.2	11.7
45	Morocco	342	330	1 659	19.2	38.2
46	Chile	913	988	1 628	6.6	10.5
47	Croatia	394	543	1 414	15.3	21.1
48	Kuwait	94	91	1 412	35.2	72.9
49	Philippines	11 007	660	1 225	-21.6	13.2
50	Ukraine	494	448	1 192	10.3	21.6

Source: IMF BOP.

Notes: 1. Including Puerto Rico,
2. Including Monaco, Guadeloupe, Martinique, French Guiana and Réunion,
3. Including Sikkim,
4. Including Liechtenstein,
5. Including Svalbard and Jan Mayen Islands, excluding Bouvet Island,
6. Including Åland Islands.

Annex 2.2

ITA members as of September 2007

Albania	Kyrgyzstan
Australia	Macao (China)
Bahrain	Malaysia
Bulgaria	Mauritius
Canada	Moldova
China	Morocco
Costa Rica	New Zealand
Croatia	Nicaragua
Dominican Republic	Norway
Egypt	Oman
El Salvador	Panama
European Communities	Philippines
Georgia	Republic of Korea
Guatemala	Romania
Honduras	Saudi Arabia
Hong Kong (China)	Singapore
Iceland	Switzerland (incl. Liechtenstein)
India	Taiwan, Province of China
Indonesia	Thailand
Israel	Turkey
Japan	United States
Jordan	

References

American Chamber of Commerce in Egypt (2007). Information and communications technology developments in Egypt. Business Studies Series.

A.T. Kearney (2005). Assessment of Egypt's potential as a competitive location for offshore services: findings and recommendations.

Bednarzik RW (2005). Restructuring information technology: is offshoring a concern? *Monthly Labor Review*, August, pp. 11–21.

Bora B and Liu X (2006). Evaluating the impact of the Information Technology Agreement. Draft, WTO.

Fink C et al. (2002). An assessment of telecommunications reform in developing countries, volume 1. World Bank Policy Research Working Paper No. WPS2909.

ILO (2007). *The Production of Electronic Components for the IT industries: Changing Labour Force Requirements in a Global Economy*. Geneva: ILO.

ILO (2006). Occupations in information and communications technology. Options for updating the International Standard Classification of Occupations. Discussion Paper, ILO.

Jorgenson DW and Vu K (2005). Information technology and the world economy, *Scandinavian Journal of Economics*, vol. 107, no. 4, pp. 631–650.

Mann C and Liu (2007). *The information technology agreement: sui generis or model stepping stone?* Paper presented at the Conference on Multilateralising Regionalism, WTO-HEI, 10–12 September 2007, Geneva, Switzerland.

Norton S (1992). Transactions costs, telecommunications, and the microeconomics of macroeconomic growth, *Economic Development and Cultural Change,* vol. 41, no. 1, pp. 175–196.

OECD (2006a). *Information Technology Outlook 2006*. Paris: OECD.

OECD (2006b). *Potential Impacts of International Sourcing on Different Occupations.* DSTI/ICCP/IE(2006)1/FINAL.

OECD (2007a). *Communications Outlook 2005*. Paris: OECD.

OECD (2007b). *Guide to Measuring the Information Society.* Forthcoming.

Roeller LH and Waverman L (2001). Telecommunications infrastructure and economic development: a simultaneous approach. *American Economic Review*, vol. 19, no. 4, September.

Rowthorne R (2006). The renaissance of China and India: implications for advanced economies. UNCTAD Discussion Paper No. 182.

Sri Lanka Information and Communication Technology Association (SLICTA) (2007). Rising demand: the increasing demand for IT workers spells a challenging opportunity for the IT industry. SLICTA.

Standing G (1999). *Global Labour Flexibility: Seeking Distributive Justice.* London: Macmillan Press.

Statistics Denmark (2007). Measuring International Sourcing. Presentation of a harmonized survey in EU member States on international sourcing, Joint OECD WPIIS-WPIE Workshop, 22 May 2007, London.

UNCTAD (2002). *E-Commerce and Development Report 2002.* New York and Geneva: United Nations.

UNCTAD (2003). *Information Economy Report 2003*. New York and Geneva: United Nations.

UNCTAD (2004). *World Investment Report 2004*. New York and Geneva: United Nations.

UNCTAD (2005a). *Information Economy Report 2005*. New York and Geneva: United Nations.

UNCTAD (2005b). *World Investment Report 2005*. New York and Geneva: United Nations.

UNCTAD (2006a). *Information Economy Report 2006*. New York and Geneva: United Nations.

UNCTAD (2006b). *World Investment Report 2006*. New York and Geneva: United Nations.

UNCTAD (2007). *Report of the Secretary-General of UNCTAD to UNCTAD XII: Globalization for Development: Opportunities and Challenges*. TD/413.

World Bank (2006). *Information and Communications for Development: Global Trends and Policies,* Chapter 2, Foreign direct investment in telecommunications in developing countries.

World Information Technology and Services Alliance (2006). *Digital Planet 2006: The Global Information Economy*.

WTO (2007). *Trade Policy Review India*.

Notes

1. The «second generation» of globalization refers to a new phase of globalization in which the developing countries play a significant role. It stands in comparison with the wave of globalization during the 1990s, when developing countries were mainly providers of raw materials. It is based on the expansion of some large emerging markets, in particular China and India, and their import demands, which have created significant economic opportunities for other developing countries. In particular, South–South trade has increased significantly during the past few years (UNCTAD, 2007).

2. UNCTAD (2006b, p. 39).

3. See Chile Satellite Account 2006, available at http://www.economia.cl/aws00/servlet/ aawsconver?2,,500798.

4. SLICTA (2007). The data mainly cover ICT-related occupations in both the ICT sector and other industries.

5. Ministry of Communications and Information Technology Egypt, http://www.mcit.gov.eg consulted 13 September 2007.

6. American Chamber of Commerce in Egypt (2007).

7. American Chamber of Commerce in Egypt (2007).

8. Ministry of Communication and Information Technology Egypt, http://www.mcit.gov.eg consulted 13 September 2007.

9. Figure refers to 2003, www.laborsta.ilo.org, consulted in September 2007.

10. The UNCTAD MCIT/ITIDA 2007 survey used as a basis the classification of occupations in ICT proposed by ILO (2006).

11. ILO–ITU–ECA study on the impact of ICTs on employment and poverty alleviation in Africa, presentation in 2007, http://www.uneca.org/CODI/codi5/content/AOM_Joint_ITU_ILO_ECA_Study-Opoku-Mensah-EN.ppt.

12. See http://www.itu.int/ITU-D/ict/statistics/at_glance/f_staff.html.

13. Presentation by Peter Borg Nielsen, Statistics Denmark, at the OECD WPIIS-WPIE Workshop, London, May 2007, based on European Restructuring Monitor information.

14. A discussion on the political debate of international sourcing goes beyond the scope of this Report. For further reading, see, for example, Alan S. Blinder, Fear of offshoring, Princeton University, CEPS Working Paper No. 119, December 2005. There are hundreds of press articles on this subject, see, for example, Bush, adviser assailed for stance on offshoring jobs, *Washington Post*, 11 February , 2004, http:// www.washingtonpost.com/ac2/wp-dyn/A30194-2004Feb10?language=printer, Outsourcing made in Germany, *Deutsche Welle*, 22.07.2004, http://www.dw-world.de/dw/article/0,,1273178,00.html; India says outsourcing jobs good for U.S., UPI, 31 December 2004, http://www.upi.com/International_Intelligence/ Analysis/2004/12/31/india_says_outsourcing_job_good_for_us/4941/; Indian company creates 600 UK jobs, *The Register*, 8 September 2005, http://www.the register.co.uk/2005/09/08/india_jobs_to_uk/.

15. Definition of international sourcing: the total or partial movement of business functions (core or support business functions) currently performed in-house or currently domestically sourced by the resident

enterprise to either non-affiliated (external suppliers) or affiliated enterprises located abroad. Exemptions: movement of business functions (core or support business functions) abroad without reducing activity and/or jobs in the enterprise concerned; temporary subcontracting abroad (one-year limit could be used) (Statistics Denmark, 2007).

16. See http://www.everestresearchinstitute.com/.

17. UNCTAD (2004) and UNCTAD XI São Paulo documents.

18. Presentation by Zhongzou Li, UNCTAD-OECD-ILO expert meeting "In support of the implementation and follow-up of WSIS: using ICTs to achieve growth and development", Geneva, 4-5 December 2006.

19. OECD, 29 March, 2007.

20. OECD (2006a, p. 151).

21. IDC press release, 3 July, 2007, available at http://www.idc.com/getdoc.jsp?containerId=prSG20768607.

22. The definition used by the country is not the same as the OECD ICT sector definition; for example, India does not include telecommunication services, but includes business process outsourcing services (such as those in accounting, medical services and financial services).

23. See also WTO (2007, p.21).

24. NASSCOM presentation, UNCTAD-OECD-ILO expert meeting, December 2006.

25. *The Hindu Business Line*, 27 February, 2006, in "demographic complementarities and outsourcing: implications and challenges for India", Mukul G. Asher and Amarendu Nandy, Research and information system for developing countries, July 2006.

26. *Financial Times*, February 10, 2006.

27. Business Standard (2005).

28. OECD (2006a, p. 78).

29. It is worth mentioning that the majority of these companies are small and medium-sized enterprises.

30. Product coverage issues have been discussed at, for example, a WTO workshop on 18 January, 2007. For presentations see http://www.wto.org/English/tratop_e/inftec_e/inftec_e.htm.

Chapter 3

MEASURING THE IMPACT OF ICT ON PRODUCTION EFFICIENCY

A. Introduction

The economic impact of information and communication technologies is a complex subject of great interest to both developing and developed countries, particularly in the context of globalization. Previous editions of UNCTAD's *Information Economy Report*, formerly known as the E-Commerce and Development Report, have addressed this issue from a variety of angles in order to provide a comprehensive view of how and how much ICTs matter to economies and development. The first edition of the ECDR (2000) showed through its computable general equilibrium estimates that developing countries would be at a disadvantage in terms of income and welfare if they fell too much behind as regards ICT use. In the meantime, ICTs have evolved, improved and become more widespread and constitute a very dynamic sector of economic activity. Developing countries' exports of ICT goods and services have grown spectacularly, establishing new international specialization patterns.[1] At this stage it is important to establish to what extent ICTs have impacted on economic growth. The analysis of differences in economic gains from ICT between countries, industries and firms aims at revealing where most positive effects have occured and where new efforts should focus.

The Introduction to this Report clarifies the role played by ICTs in the wider context of science and technology and the latter's contribution to growth and development. This chapter presents empirical results regarding the economic impact of ICT use from the productivity perspective. More specifically, it reviews quantitative results of empirical studies on the impact of ICT use on production efficiency.

The focus on productivity and the supply side is of interest to developing countries. In recent years, developing countries have continued to grow extensively by absorbing new investment and labour. In particular, the policy choice to build up and support an ICT-producing sector has generated jobs and attracted additional foreign investment and offshoring in several emerging economies. This chapter shows that in addition to recognizing the benefits of the ICT-producing sector, it is important to recognize the development potential of other sectors of economic activity where ICTs could be used intensively (ICT-enabled activities).

An important result of the intellectual debate initiated by Gordon (1999) was that in developed countries, while initial productivity gains from ICT accrued mainly from a fast growing ICT-producing sector, in the second stage there were additional gains deriving from increasingly widespread use of ICTs in many economic sectors. In developing countries, by contrast, productivity gains from ICT are still being largely generated by the ICT-producing industry.[2] This is because developing countries with tighter budget constraints have been spending less on ICT. In that context, developing countries wishing to increase their gains from ICT should take measures to extend the use of ICT from the ICT-producing sector to other economic sectors and to domestic consumers.

Researchers studying the impact of ICTs on productivity used different economic models of interpretation, depending on the type of ICT data made available and the issue researched. The analysis depended very much on the quality of measurements and the availability of data on ICT uptake. At the firm level, the specialized literature tended to move away from evaluating the impact of all-inclusive ICT investment and instead focused on the way in which specific ICT use resulted in higher production efficiency. As ICTs become more complex and widespread, more information is needed on how ICT use determines competitiveness, in addition to information on the amount spent on ICT investment. In developing countries as well, more information on how firms use ICT would help in refining policy decisions to scale up the most efficient enterprise practices.

In keeping with the conceptual structure adopted in past issues of this Report,[3] the present chapter proceeds with a literature review at the macroeconomic and microeconomic levels of the main empirical findings on ICT and production efficiency. The review of studies applied to developed country data is aimed at facilitating the interpretation of similar empirical exercises conducted with more recent data from developing countries. Also included in the review are a number of research papers that use developing country data. The literature review links three defining elements: the economic model of interpretation, the ICT or IT[4] variable used and the significance of results for developing countries.

Section D of this chapter draws on a commonly used theoretical framework (Cobb–Douglas production function) for studying the relationship between specific ICT use and labour productivity in firms. It provides an illustration of how information on ICT use by firms can be linked with economic data in order to quantify productivity gains associated with ICTs.[5] The study described is the outcome of a joint research project of UNCTAD and the Thai National Statistical Office. It is an application of a verified econometric model to a large sample of Thai manufacturing firms. The project is part of a broader international initiative to improve ICT measurement and the quality of data on ICT uptake in order to enable cross-country comparisons and allow the measuring of ICT impact. This initiative is promoted by UNCTAD through the Partnership on Measuring ICT for Development.[6]

B. Economic impact of ICT at the macroeconomic level

The macroeconomic impact of ICT on production efficiency is assessed in the context of growth accounting models. The introduction of this Report presents the theoretical assumptions underlying both neoclassical and endogenous growth models. Past issues of the UNCTAD IER and ECDR[7] have also reviewed the channels through which ICTs impact on economic growth. This section reviews the recent literature on the impact of ICT on growth and productivity in order to highlight the main factors that apply in the context of developing countries.

Chart 3.1 summarizes the way in which ICT use leads to productivity growth: first it increases the efficiency of factor inputs (capital and labour) and, second, it fosters technological innovation as a source of total factor productivity growth. Labour productivity in particular grows as a result of capital deepening through

incorporating ICT capital inputs into the production process. In this case, ICT investment results in improved labour efficiency without changing the technology of production. When, in addition to capital deepening, economic agents are able to relocate resources in a way that improves technological efficiency and better incorporates ICTs into their production processes, ICT use can result in total factor productivity gains.

This section reviews the findings of the specialized literature on several research questions at the macroeconomic level related to ICT and growth in both developed and developing countries. One of the questions addressed in the specialized literature is whether ICTs have contributed to GDP growth as an ordinary production input or whether they generated, besides capital deepening, additional spillover effects, other positive externalities and technological innovations. This allows a comparison of the magnitude of gains from ICT production and from ICT use. The main purpose of the review is to show instances where results regarding ICT and productivity differed between developed and developing countries, in order to better understand how developing economies can maximize gains from ICT.

1. ICT capital, ICT use and growth

Measuring the amount of capital invested in ICT goods and services is one way of measuring the intensity of ICT uptake in an economy. Within the neoclassical growth accounting setting, ICT investment is assumed to contribute to labour productivity through capital deepening, this assumption being consistent with the logic that workers who have more computing and communication equipment are more efficient than workers who have only ordinary capital. Thus ICT capital contributes to GDP per capita growth as an additional input to the production function. This theoretical framework does not account for the contribution of ICT to total factor productivity arising from an ICT-enabled innovative reallocation and reorganization of production (such as the substitution of ICT capital for ordinary capital). Several such growth accounting exercises were conducted by a number of authors over different periods of time to estimate the contribution of ICT capital to GDP growth.

Gains to productivity from ICT have low values when ICTs are nascent in the economy. When ICT inputs have a very small share in total production, growth accounting models can detect only a feeble effect on productivity from those inputs. For example, early

Chart 3.1

Channels through which ICT contributes to productivity growth

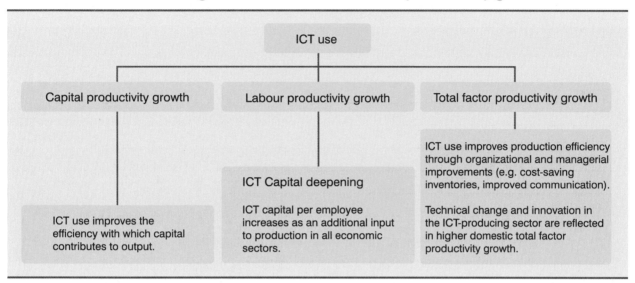

results in developed countries derived for the 1980s and early 1990s showed that investment in computer equipment had a minimal impact on GDP growth. Research[8] estimating the impact of investment in computing equipment on GDP growth in the United States in the 1980s found a rather small contribution of just 0.2 per cent compared with an average GDP growth rate of 2.3 per cent. The ICT variable used in that research was the yearly variation in the stock of computer capital. The broader concept of ICT investment was developed later, in parallel with research and has been used only recently. It was suggested by several authors that inputs with a very small share in total output could not have a great impact on GDP even if investment in that input was growing at high rates (Triplett, 1999). Although computer investment had reached a sizeable amount in certain specialized sectors, at the national level investment in computing equipment represented only 0.5 per cent of the total capital stock in 1993 (Jorgenson and Stiroh, 1995[9]). This measurement problem was subsequently addressed by broadening the concept of computer capital to also incorporate other ICT goods and services, including software and communication equipment. Other measurement problems related to accurately capturing price changes in IT and composite IT goods and services since the rather insubstantial changes in nominal ICT investment were hiding much larger variations in real ICT investment (IMF, 2001).

Conversely, when specific ICTs such as computers become ubiquitous, especially in developed countries, they can no longer contribute to explaining variations in output between economic actors (Carr, 2003). In that

case, it is necessary to find other measurements relating to a particular use of ICT (for example, establishing web presence) where significant differences in ICT use explain changes in output.

An additional insight into growth accounting for ICT was provided by the structural change in the mid-1990s and the shift in the trend of ICT investment. Table 3.1 shows the evolution of ICT investment as a share in total capital investment in OECD countries before and after the mid-1990s.[10] Some associated this structural change with the beginning of the Internet revolution, arguing that connected computers allowed a larger variety of combined ICT use among firms. Others attributed the structural change to an unusually rapid technological improvement in the semiconductor industry (Oliner and Sichel, 2002). Starting in 1995, there was a marked decline in global ICT prices and a considerable acceleration of investment in ICT. At first, this trend was apparent only in developed countries. With a time lag of almost a decade, developing countries also experienced a similar increase in their ICT capital investments.[11] In the mid-1990s developing countries started to invest in ICT at faster rates and some, notably from Asia,[12] have acquired a considerable stock of ICT capital that matches ICT capital intensity in the United States in the early 1990s. In that connection, the analysis of more recent data on ICT investment in developing countries benefits from comparisons with

Table 3.1

Share of ICT investment in non-residential fixed capital formation in selected OECD countries (5-year averages), 1980–2004

	1980–1984 (per cent)	1985–1989 (per cent)	1990–1994 (per cent)	1995–1999 (per cent)	2000–2004 (per cent)
United States	15.1	19.8	23.1	27.0	29.9
Sweden	11.9	16.1	20.1	24.9	27.6
Finland	8.3	11.6	18.9	24.5	26.6
Australia	8.4	11.4	17.4	20.7	23.8
United Kingdom	7.9	12.5	16.2	22.7	23.1
Belgium	13.1	18.1	17.4	19.8	21.9
New Zealand	8.6	12.7	16.8	16.7	20.0
Canada	9.6	12.0	15.4	18.2	19.4
Denmark	12.7	14.6	17.6	19.4	19.3
Rep. of Korea	N/A	N/A	N/A	14.3	18.3
France	8.7	11.4	10.8	15.5	17.6
Netherlands	9.3	11.9	13.1	15.5	17.5
Germany	N/A	N/A	13.3	14.8	16.5
Italy	10.5	13.8	14.4	15.5	16.1
Japan	N/A	N/A	9.0	13.0	15.5
Austria	8.9	12.0	12.0	12.8	14.0
Portugal	12.3	13.2	10.7	12.6	12.8
Spain	11.0	14.7	12.7	14.2	12.7
Greece	4.5	7.5	10.2	11.2	12.4
Norway	8.3	9.4	8.5	10.0	12.3
Ireland	6.8	7.5	5.6	9.4	10.8

Notes: ICT investment has three components: information technology equipment (computers and related hardware), communications equipment and software. Software includes acquisition of prepackaged software, customized software and software developed in house. The investment shares shown in the table are percentages of each country's gross fixed capital formation, excluding residential construction. Also, the figures shown here do not account for price falls since they are ratios of nominal investment values. Since the price of ICT goods and services fell considerably it is reasonable to assume that the share of ICT investment increased faster than this table shows.

N/A = not available.

Source: OECD (2007).

ICT investment trends in developed countries in the 1980s and 1990s.

Using the new data from after 1995, Jorgenson and Vu (2005) found that the contribution of ICT capital to world GDP had more than doubled and now accounts for 0.53 per cent of the world average GDP growth of 3.45 per cent. The percentage was higher for the group of G7 countries, where ICT investments contributed with 0.69 per cent to a GDP growth of 2.56 per cent during 1995–2003. The Economic Commission for Latin America and the Caribbean has recently conducted a series of growth accounting exercises for

Latin American countries. Their estimates are reflected below for a comparison with certain Asian developing countries.

Chart 3.2 shows the results of several growth accounting exercises in a number of developing countries in the period after the mid-1990s, benchmarked against estimates for the G7 countries before and after 1995. For each country, the various inputs broken down between ICT capital, non-ICT capital, labour and residual total factor productivity growth add up to average GDP growth. Economies appear in increasing order of the contribution of ICT capital to growth,

Chart 3.2

Sources of GDP growth, 1990–2003

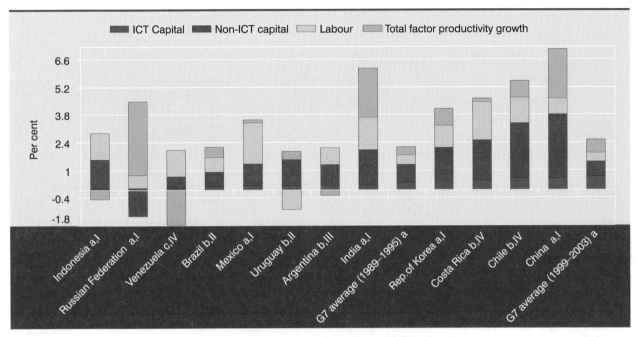

Notes: Time period considered in analysis I: 1995-2003, II: 1995-2004, III: 1992-2004, IV: 1990-2004.

Source of estimates – a: Jorgenson & Vu (2005); b: de Vries et al. (2007), c: de Vries et al. (2006).

ranging from 0.09 per cent in Indonesia to 0.69 per cent in the G7 countries and reaching 0.38 per cent in G7 countries prior to 1995. In a number of developing countries, accelerated investment in ICT made a significant and positive contribution to GDP growth after the mid-1990s. However, owing to the low starting levels of ICT capital, this type of input could make only a reduced contribution, below 0.25 per cent in many cases. It was necessary for ICT capital stock to reach a certain level before greater benefits from investment in those technologies could be achieved. There were exceptions to that general finding, however. In the Republic of Korea, Costa Rica, Chile and China benefits from ICT investment were about 0.5 per cent, closer to the G7 average. In three out of those four developing countries, the ICT-producing sector had a large share in economic output.[13]

As shown in chart 3.2, several emerging economies have experienced strong growth in the last 10 years. In fast-growing developing countries ICT investment made only a small contribution to GDP growth. Other inputs such as ordinary capital investment and labour remained major contributors to the rapid GDP growth in emerging economies.[14] China and India stood out with some of the largest output growth in recent times, but their performance was led only partly by greater investment in ICT, as the other inputs distinguished by

the model had a considerably higher share. Previous editions of the ECDR[15] have shown that the effect of ICT use on other fundamental variables of the economy such as labour was also significant. It is therefore possible that in emerging economies ICT use made a larger contribution to economic growth than suggested by the capital-deepening effect. Indeed, ICT use has also contributed to improving labour productivity in emerging economies (see section 3). However, as growth accounting estimates show, developing countries in which ICT investments had a greater impact on growth also managed to leverage more capital and to improve the skill mix of their labour force. Benefits from ICT capital did not occur in isolation from a general upturn in the economic evolution of emerging economies.

As mentioned in the Introduction to this Report, ICT use is also reflected in multifactor productivity increases, which account for an important part of GDP growth in emerging economies.

2.   The ICT-producing sector and the ICT using sectors

In a debate initiated by Gordon (1999), researchers asked whether there was proof of gains from the widespread use of ICTs in addition to gains to GDP from rapid technological advances in the ICT-producing sector. For the United States economy, Stiroh (2002) showed that gains from ICT investment in a broad range of economic sectors were positive and complementary to gains from the ICT-producing industry itself. Furthermore, in a comprehensive empirical study at the sectoral level comparing productivity growth in the United States with that in the European Union, van Ark, Inklaar and McGuckin (2003) found that the faster productivity growth in the United States is explained by the larger employment share in the ICT-producing sector and productivity growth in services industries with a highly ICT-intensive profile. Accordingly, wholesale and retail trade and the financial securities industry account for most of the difference in aggregate productivity growth between the EU and the United States.

This debate is of relevance to development because in many developing countries ICT investment has remained largely concentrated in the ICT-producing sector. Several studies[16] have shown that while in developed countries, the impact of ICT use was an important element accounting for growth, in most developing countries productivity gains were generated predominantly by the ICT-producing industry itself. On a global scale, developing countries managed to

attract investment and started to produce a large share of global ICT goods and services. However, demand for ICTs originated predominantly from developed countries. This pattern has recently changed in the case of certain developing countries. For example, after a rapid development, China became in 2006 the world's second largest importer of ICT goods, with a large share of those imports originating from developing economies of Asia (see chapter 2).

Bayoumi and Haacker (2002) compared gains to real GDP from the IT-producing sector with gains to total consumption from domestic IT spending (1996–2000).[17] They found that several developing economies received a smaller demand boost from IT spending even if they were relatively specialized in IT production (chart 3.3). Because ICT use in several developing economies is limited to a few economic sectors those economies gained less from ICT. A comparison between developed and developing ICT producers in chart 3.3 shows that IT spending had a proportionately greater impact on demand in Ireland and Finland as compared with the effect in the Philippines and Thailand, for example. Among developing countries specialized in the production of ICTs, the Republic of Korea, Singapore and to a certain extent also Malaysia had a larger contribution of IT spending to domestic demand growth. However, as the authors suggest,

Chart 3.3

Impact of the IT sector on GDP growth and domestic demand, 1996–2000

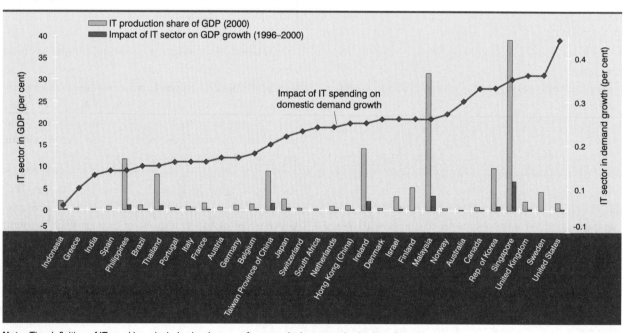

Note: The definition of IT used here includes hardware, software and telecommunications equipment.
Source: Bayoumi and Haacker (2002).

developing countries gained from IT mostly as IT producers, while developed countries gained relatively more as IT consumers.

There are several factors that can explain why firms in developing countries use ICT less. Data on ICT use by firms in Thailand (see section D), for example, show that the main factors hindering the adoption of ICT applications are high costs and lack of knowledge about how to apply ICTs for improving performance in a particular type of business. Undoubtedly, the lack of knowledge about how to apply ICT to business goes hand in hand with the reduced size of e-markets in specific sectors. Certain domestic producers continue to be confronted with lower levels of ICT penetration among domestic suppliers and consumers. Owing to lack of knowledge about how to implement ICT in order to boost sales, domestic producers find it best to refrain from adopting ICTs. Furthermore, data on Thai firms show that exporting firms connect their computers to the Internet and establish web presence more often than firms producing for domestic markets, so as to access information and reach consumers in foreign markets. This is confirmed by research findings in Clarke and Wallsten (2004), which show that the Internet affects exports from developing countries in an asymmetric manner. Their study measures ICT adoption by the number of Internet users within a country (the Internet penetration rate). They find that developing countries with higher Internet penetration rates tend to export more merchandise to partners from developed countries, while there was no significant relationship between Internet access and trade in goods between developing countries. This finding indicates that in developing countries the Internet has been used more effectively in trade relations with developed countries.

3. ICT impact on total factor productivity growth

Models reviewed in this section estimate a two-sided impact of ICT on labour productivity growth: the effect of ICT capital investment as an additional input into production, and the share of total factor productivity growth explained by technological change in the ICT-producing sector.

Atkinson and McKay (2007) describe three ways in which ICT can contribute to total factor productivity: network externalities, complementary improvements generated by ICT adoption, and improved access to knowledge. All three suggest that the positive effect

of ICTs on productivity will occur with a time lag. ICTs can generate network externalities in the same way as connecting a popular service provider to the telephone network will increase the satisfaction of all telephone subscribers. However, it goes without saying that building valuable networks of ICT users takes time and the process is not without difficulties. For example, the different technologies involved are not always compatible and interconnections may fail. The same applies to organizational change – which is costly, time-consuming and sometimes unsuccessful – and also to accessing information: not all ICTs are equally user-friendly and users may find it hard to identify the best source of information available.

Estimates of the effect of ICT on total factor productivity growth are provided either by growth accounting models or by other regression models explaining total factor productivity growth. As seen in the introduction of this Report, total factor productivity growth reflects technical change in the economy that cannot be explained by mere increases in capital and labour, but it also captures other factors such as policy and institutions. There are several ways in which ICTs could contribute to total factor productivity growth, but so far there is no agreement on the size of this contribution. More rapid technological progress in the ICT-producing sector is one way in which ICTs influence total factor productivity. For example, Oliner and Sichel (2002) ran a sectoral growth accounting model on data corresponding to the non-farm business sector in the United States and found that both the contribution of IT use to labour productivity and the total factor productivity growth arising from the IT-producing sector are positive and statistically significant. Table 3.2 provides a summary of their results. Accordingly, IT capital investment accounted for the larger share of labour productivity growth in the United States through a general IT capital deepening effect in several economic sectors. The IT-producing sector had a relatively lower contribution to labour productivity in the United States, which was above all influenced by productivity gains in the semiconductor industry.

Other growth accounting exercises were performed with European data (van Ark et al., 2003) and with East Asian data (Lee and Khatri, 2003). As the results in chart 3.4 indicate, the ICT-producing sector had a sizeable positive effect on total factor productivity growth in several East Asian economies specialized in the production of ICT, and also in Ireland. In those economies, productivity growth in the ICT sector was the main source of technological improvement and

Table 3.2

Oliner and Sichel (2002) estimates for the IT contribution to labour productivity growth in the United States (per cent)

	1974–1990	1991–1995	1996–2001
Growth of labour productivity (non-farm business sector)	1.4	1.5	2.4
Percentage contributions from:			
• IT capital (IT capital deepening), of which:	0.4	0.5	1.0
Computer hardware	0.2	0.2	0.5
Software	0.1	0.2	0.4
Communication equipment	0.1	0.1	0.1
• Other capital (capital deepening)	0.4	0.1	0.2
• Labour quality	0.2	0.4	0.3
• Total factor productivity (TFP), of which:	0.4	0.6	1.0
IT-producing sector contribution to TFP, of which:	0.3	0.4	0.8
Semiconductors	0.1	0.1	0.4
Computer hardware	0.1	0.1	0.2
Software	0.0	0.1	0.1
Communication equipment	0.0	0.1	0.1
Other sectors	0.1	0.2	0.2
Total IT contribution to labour productivity	0.7	0.9	1.8

Source: Oliner and Sichel (2002).

Chart 3.4

Contribution of ICT capital to labour productivity, and of ICT production to total factor productivity

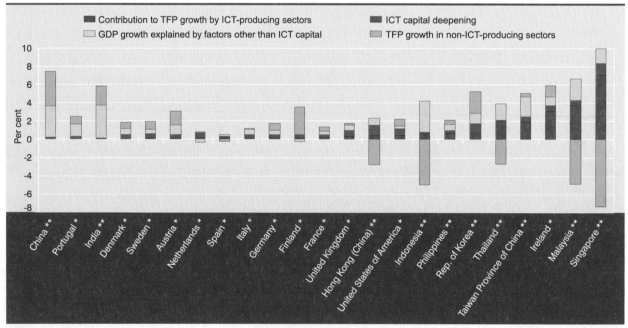

Note: Countries are in increasing order of the contribution of the ICT-producing sector to total factor productivity growth. An ICT-related TFP growth estimate was not available for China.
Source: * van Ark et al. (2003) for 1995–2000; ** Lee and Khatri (2003) for 1995–1999.

gains to GDP from ICT production exceeded gains from ICT use (capital deepening).

Research has also investigated whether ICT-using industries have contributed to total factor productivity growth. Schreyer (2000), for instance, developed a growth accounting model in which ICT capital contributed to GDP growth through both capital deepening and spillover effects as a source of technological improvement. However, to date there is no agreement about whether ICT investment has a significant effect on total factor productivity growth in developing countries. More research is needed on this topic.

There is an emerging literature attempting to estimate the contribution of ICT to total factor productivity growth in developing countries. For example, Shiu and Heshmati (2006) estimated the impact of foreign direct investment and ICT investment on total factor productivity growth in China over the last decade (1993–2003). While Chinese total factor productivity grew at an average rate of 8.86 per cent to 9.22 per cent during that decade, ICT investment was found to contribute to it to a lesser extent: a percentage increase in ICT investment resulted in 0.46 per cent total factor productivity growth, compared with foreign direct investment that generated 0.98 per cent.

Seo and Lee (2006) estimated a model of total factor productivity growth during 1992–1996 in 23 OECD and 15 non-OECD developing economies as a function of GDP growth, human capital, openness, domestic ICT capital intensity, ICT spillover effects and a linear trend. In their model, ICTs can have a double effect on total factor productivity growth – through higher domestic investments and through network effects. The ICT network variable is calculated as the ICT capital intensity of foreign countries. Their analysis compared two samples of OECD economies and developing economies.[18] Findings suggest that in developed economies total factor productivity growth is more strongly correlated with domestic ICT investments, while in developing countries it is not significantly linked to domestic investments in ICT but rather to ICT investments in OECD countries.

The results reviewed confirm that gains from ICT to GDP occurred both in developed and in developing countries, with a lower intensity in countries that started to invest in ICT later and at a slower pace. The positive macroeconomic impact of ICT coincided with a marked contribution to growth deriving from other factors such as increased ordinary capital investment,

improved skills and a better allocation of resources. A number of developing countries, particularly Asian ones, have specialized in the production and exporting of ICTs. As a result, it is estimated that ICT production increased GDP by as much as 7 per cent in Singapore and 3 per cent in Malaysia yearly (1996–2000). Developed countries gained more than developing countries from the consumption of ICTs. In terms of impact of ICT investment on total factor productivity, as a sustainable factor for growth, research results do not always confirm the existence of significant effects. However, when a positive effect is verified, it is explained by a more efficient use of ICT in specific sectors (the case of wholesale and retail trade services in the United States). It is therefore necessary to study in detail how the specific use of ICT leads to greater business efficiency.

The next section looks further at how private firms have managed to leverage gains from ICT use.

C. Firm-level impact of ICTs

This section draws on a growing number of empirical studies at the firm level conducted in a number of OECD countries. Annex 3.1 provides a summary of results in selected research on ICT and productivity. In the context of the Partnership on Measuring ICT for Development, several international institutions have pledged to work on improving the information available on ICT measurement and the quality of the data, and to make ICT statistics cross-country-comparable. As the next step, the OECD and Eurostat are promoting research to estimate the economic impact of ICTs at the macroeconomic, industry and firm levels. Initiated in 2003, the EU KLEMS project aims at building a comprehensive time series database, including measures of economic growth and productivity derived through the use of growth accounting techniques.

Through the Partnership on Measuring ICT for Development, UNCTAD has worked to improve the measurement of ICT use by businesses in developing countries. In those countries there is little research estimating the impact of ICT on production efficiency at the firm level. In order to conduct that type of study, statistical offices or research institutes need to collect data on the use of ICTs[19] and link this information with output and productivity data at the firm level. While several developing countries have made important progress in measuring and producing data on ICT use, linking this information to other economic variables often has remained a challenge.

Investing in ICTs makes a difference to firms' sales and labour productivity. Although ICT investment cannot confer long-term competitive advantage, not investing in ICT clearly leaves firms at a disadvantage (Carr, 2004).

Several recent studies[20] highlight the need to complement data on ICT investment with more information on the way in which firms use specific ICT. Particularly in a developed country context, investment in ICT has become nearly universal and spending more on acquiring additional and more complex technology may not lead to increased market shares in firms (Carr, 2004). Accordingly, it became increasingly relevant to study the way in which ICT use by firms is related to higher sales per employee. In developing countries, the percentage of companies with access to ICTs remained lower on a national scale (see chapter 1), even though for certain economic sectors or geographical regions indicators showed higher ICT penetration. Further to the overview of the modelling framework, section D presents an empirical application quantifying the relationship between ICT and labour productivity in a large sample of Thai manufacturing firms. In developing countries as well, more information on the way in which specific ICTs are used by firms can help in assessing the relationship between ICT and productivity. UNCTAD recommends that countries give priority to gathering statistical data on the core ICT indicators (United Nations, 2005).[21]

1. ICT use and firm labour productivity

The different dimensions of firm-level analyses considerably increase the complexity of research results on this topic. If previously presented models focused mostly on ICT investments, the studies reviewed here go into greater detail about specific ICT technologies or applications and interactions between them. Moreover, variables on ICT use have different productivity effects in conjunction with other control measures such as firm age, the share of foreign capital participation or access to more skilled human resources. To make it easier to keep track of the different empirical model variants, this section groups the many dimensions into three categories, by elements of the regression equation (table 3.3).

Labour productivity is commonly measured either as value added per employee or as sales per employee. Criscuolo and Waldron (2003) derive results for both measures of productivity and find that the impact of e-commerce[22] was slightly stronger on value added than on sales. Conversely, Atrostic and Nguyen (2002) rely on the findings of Baily (1986) to argue that using value added as a measure of labour productivity yields systematically biased estimates of the theoretically correct growth model. Outside the context of empirical models, value added is a more precise measure of labour productivity since it subtracts from the value of sales the costs incurred with intermediate consumption. However, as suggested by Carr (2003), firms take

Table 3.3

Key variables for measuring ICT impact on labour productivity

Labour productivity variables	ICT variables	Complementary / control variables
• Sales per employee • Gross output per employee • Value added per employee • Or recalculations of the above variables based on effective hours worked by employees	Binary (dummy) variables: these take on value 1 if the firm has access to a specific technology and 0 otherwise. Numerical variables: • Spending on specific ICTs • ICT capital stock • Share of employees using ICTs • Number of computers available in the firm	• Firm age • Ownership • Affiliation to a multi-unit firm • Skill mix (share of employees working directly in production) • Level of education • Industry sector of activity (corresponding to ISIC codes) • Geographical region • Typical factors of Cobb–Douglas production functions (ordinary capital stock, employment, spending on materials)

decisions to adopt ICTs on the basis of expectations about maintaining or increasing market share and competitiveness. For those considerations, analysing about the impact of ICT on sales per employee is more appropriate.

The empirical models presented here draw on several types of variables for describing the use of ICT. Binary variables, which distinguish between firms with and without access to, for example, Internet, are easy to collect and provide input for analyses of differences between the haves and the have nots. Empirical studies also test for the effect of intensity in ICT uptake – for example, the share of capital devoted to computer investments – and intensity of ICT use, such as the number of computers available or the share of employees using e-mail. From a theoretical point of view, findings based on numerical rather than binary variables are more powerful. Maliranta and Rouvinen (2006) proposed a slightly different modelling structure in which they estimate the net effects of several complementary features of computers: processing and storage capacity, portability and wireless and wireline connectivity. In their model, the positive labour productivity effect associated to the portability of computers is complementary to that of the basic processing and storage capacity of any computers whether portable or not.

The different variables of intensity in ICT use by enterprises analysed by the specialized literature are not the ideal measurements. As some have emphasized,[23] ICT use becomes increasingly relevant to productivity when combined with soft skills such as good management and superior marketing abilities. Unfortunately, such soft skills and soft technology inputs cannot be quantified directly and therefore their effect is hard to assess. Empirical research usually corrects for this unknown effect by accounting for different economic results in foreign-owned firms, in exporting companies, in establishments belonging to multi-unit corporations or simply in more experienced firms. Therefore, policy implications derived from such research do not directly recommend that the intensity of ICT use be scaled up (for example by increasing the number of computers per employee). Rather they recommend investigating how the combined use of ICT and superior managerial capabilities can account for variations in ICT gains between firms with different characteristics.

ICTs can generate higher market shares either by reducing input costs and thus allowing firms to produce more of the same products, or by improving the quality of products or product packages, with, as a result, additional sales or higher-priced products. Empirical results presented here cannot distinguish between those two effects. More information on the evolution of prices in different sectors is needed in order to assess which effect prevailed in defined periods of time.

2. Complementary factors explaining the ICT–productivity relationship

Control variables are additional elements in the growth equation. They also give a different dimension to results relating to ICT use and productivity when interacted with measures of ICT.

In several studies, firm age has proved to be an important element explaining productivity effects. The European Commission's Enterprise and Industry Directorate General showed in a report (Koellinger, 2006) that the dynamic evolution of new firms is a source of economic growth and employment and that new firms also contribute significantly to the diffusion of e-business applications in Europe. In terms of econometric results, Maliranta and Rouvinen (2003) estimate that young manufacturing firms in Finland, unlike older ones, have 3 per cent higher productivity gains from the use of computers. Also, young Finnish services firms appeared to be 1 per cent more productive thanks to access to the Internet. In a different study, Farooqui (2005) runs four different growth models on young and older British firms in manufacturing and services taken separately. Results show that ICT indicators such as investment in IT hardware and software and the share of ICT-equipped employment have a more pronounced impact on young manufacturing firms as compared with older ones. The same finding did not apply to young British services companies, however, on that issue, Atrostic and Nguyen (2005) draw attention to the fact that the measure of capital input used in most papers – the book value of capital – is a more accurate proxy in the case of the new firms. Older firms' capital input is not properly captured by book values because this measure is evaluated at initial prices when capital assets were acquired as opposed to current asset prices. The first best proxy to use would be the current value of the capital stock computed by means of the perpetual inventory method by using information on yearly capital investments, depreciation and current asset prices. But in many cases, data are not available on all the above-mentioned variables. Regression results using the book value of capital assets are likely to give biased results for older firms and more accurate results for younger ones.

Firms with foreign capital participation seemed to have higher labour productivity. With regard to developed countries, Bloom, Sadun and van Reenen (2005) estimated that in their large sample of UK firms from all business sectors, US-owned establishments had significantly higher productivity gains from IT capital than other foreign-owned firms or domestically owned firms. This result can be linked with macro-level findings which indicated that the United States had acquired greater labour productivity from investment in ICT than all other developed countries, especially since the mid-1990s. More productive US-owned firms appear to be better managed or have access to more efficient ICT solutions.

Similarly, firms belonging to multi-unit networks of affiliates may have greater labour productivity since they dispose of additional resources to draw from in the subsidiary–headquarters management structure. A multi-unit corporate configuration may justify benefits from network effects (a success story replicated in several subsidiary branches) and access to superior management resources.

The skill mix of production and non-production workers and the level of education in the regions where companies are located were considered by some studies to be complementary to the measures of ICT use by firms. Better-skilled workers are more likely to be able to develop, use and maintain more advanced technology. Maliranta and Rouvinen (2003) comment that growth models need to control for the human capital characteristics of employment and labour because these variables are essentially complementary to ICT uptake and omitting them would inflate the labour productivity gains from ICT.

Last but not least, when quantifying the relationship between ICTs and labour productivity one needs to control for differences in demand and supply factors. For example, in many countries businesses located in the vicinity of the capital benefit from higher demand than those located in isolated provinces simply because there is a high concentration of the population in capitals. In a similar way, different industries have distinct labour productivity averages owing to both demand and supply factors. For example, an oil-producing company is very likely to have higher sales per employee than a light industry manufacturer specialized in food and beverages of the same size (because of industry characteristics such as price, labour intensity

and type of consumer good). It is therefore necessary to take into account regional and industry-specific characteristics when accounting for the contribution of ICT to labour productivity growth.

The ICT-producing sector itself benefited from ICT use that considerably exceeded domestic industry averages. Maliranta and Rouvinen (2003) estimate that in Finland firms belonging to the ICT-producing sector had 3 to 4.5 per cent higher labour productivity gains from ICT use than the rest of the manufacturing and services companies in the sample. This may be because ICT producers have a know-how advantage over other ordinary users in terms of how to best put to work specific technology to enhance labour productivity.

3. Impact of specific ICTs on productivity

Use of computer networks (such as the Internet, intranet, LAN, EDI and Extranet) had an estimated 5 per cent positive effect on labour productivity in a large sample of American manufacturing businesses (Atrostic and Nguyen, 2002). The model considered a theoretical framework in which use of computer networks made a "disembodied" contribution to technological change other than that of capital and labour. Atrostic and Nguyen (2005) take up again the impact of computer networks in a slightly modified empirical model. The novelty of their approach consists in using two different computer-related measures in the labour productivity regression: computer capital, as distinct from ordinary capital, and the computer network binary variable used previously. In their view, having separate measures for the presence of computers (computer investment) and for how computers are used (computer networks) is crucial for estimating accurately the two effects on labour productivity. When using a sample composed only of newly registered US manufacturing firms, they find that the contribution of computer networks added 5 per cent to labour productivity while investment in computers added 12 per cent. Within the entire data set of older and younger US manufacturing firms, the contribution of computer capital dropped to 5 per cent and there was no evidence of a positive effect on computer networks any more. However, as mentioned before, most empirical studies tend to find that ICT use has less impact on older manufacturing firms, and this may be due to a measurement bias as explained in Atrostic and Nguyen (2005).

E-commerce also has a significant impact on labour productivity in firms, with a marked difference between businesses that buy and those that sell online.

Criscuolo and Waldron (2003) analyse a panel of UK manufacturing firms and find that the positive effect of placing orders online ranged between 7 and 9 per cent. On the other hand, they estimate that firm labour productivity was lower 5 per cent for those that used e-commerce for receiving orders online (online sellers). Lower labour productivity associated with selling products online is likely to be due to price effects. The prices of products sold online are considerably lower than the prices of similar goods sold through different channels. Additionally, firms which specialize in selling online may have difficulties in finding suppliers from which to buy online as much as they would want. Larger firms with a stronger position in the market may be able to better cope with balancing the extent of e-buying and e-selling. In a larger and updated UK data set of manufacturing and services firms, Farooqui (2005) finds again that e-selling negatively impacts on labour productivity in manufacturing, while e-buying has a larger and positive effect. In particular, Farooqui (2005) finds that in distribution services e-buying boosts labour productivity by 4 per cent.

Several studies show that measures of ICT use by employees are also reflected in enhanced firm productivity. Within a large panel of Finnish firms Maliranta and Rouvinen (2003) compare the impact of computer use on labour productivity in manufacturing and services sectors. A 10 per cent increase in the share of computer-equipped labour raises productivity by 1.8 per cent in manufacturing and 2.8 per cent in services. On the other hand, a higher share of employees with Internet access was found to have a significant impact only on services firms (2.9 per cent). The study considers Internet use as a proxy for external electronic communication and LAN use as a measure of internal electronic communication. Findings show that manufacturing firms benefit more from better internal communication, while services firms gain more from improved external electronic communication. A 10 per cent higher share of employees using LAN in the manufacturing sector results in 2.1 per cent higher labour productivity. A similar study on a mixed sample of Swedish manufacturing and services firms estimated that a 10 per cent higher share of computer-equipped labour boosts productivity by 1.3 per cent (Hagén and Zeed, 2005). The Swedish study also estimates productivity effects deriving from access to broadband of 3.6 per cent. With the help of a composite ICT indicator, Hagén and Zeed (2005) show that adopting an ever-increasing number of ICT solutions has positive but decreasing effects on labour productivity in Swedish firms. Each additional level of ICT complexity seems to add less to firm productivity. Farooqui (2005)

also identifies the use of computers and the Internet by employees as a proxy of work organization and skills. In the United Kingdom, a 10 per cent increase in the share of employees using computers raised productivity by 2.1 per cent in manufacturing and 1.5 per cent in services. These effects are additional to the impact of ICT investment, also accounted for in the Farooqui (2005) model. Similar estimates for Internet use by British firm employees showed 2.9 per cent for manufacturing and no significant impact for services.

Maliranta and Rouvinen (2006) estimate the impact of different complementary computer features on labour productivity in a 2001 sample of Finnish services and manufacturing firms. Their computer variables are measured in terms of share of employees using computers with one or several of the following features: processing and storage capabilities, portability and wireline or wireless connectivity. They find that a 10 per cent higher share of labour with access to basic computer attributes such as providing processing and storage capabilities increases labour productivity by 0.9 per cent. In addition, computer portability boosts output per employee by 3.2 per cent, wireline connection to the Internet adds 1.4 per cent, while wireless connectivity adds only 0.6 per cent.

There is an emerging literature estimating the impact of broadband on firm productivity. Gillett et al. (2006) were the first to quantify the economic impact of broadband and found that there were positive and significant effects on the number of workers and the number of businesses in IT-intensive sectors. Chapter 6 of this Report reviews the economic impact of mobile phones, with a focus on the African region.

4. ICT investment, soft technologies and total factor productivity gains

Brynjolfsson and Hitt (2002) explore the impact of computerization on multi-factor productivity and output growth in a panel data set of 527 large US firms over a period of eight years (1987-1994). They compare the short-run and long-run effects of computerization on total factor productivity growth by taking first and five- to seven-year differences of the log-linear output function. The aim of their analysis is to understand the mechanism through which private returns from computerization accrue, since they are the ultimate long-run determinants of decisions to invest in ICTs.

They model production as a function of computer capital and other inputs, and assume that in the presence of computers, the efficiency of employees, internal firm organization and supply-chain management systems are improved. They find that, in the short run, investments in computers generated an increase in labour productivity primarily through capital deepening, and found little evidence of an impact on total factor productivity growth. However, when the analysis is based on longer time differences, results show that computers have a positive effect on total factor productivity growth. In the long run, the contribution of computer capital to growth rises substantially above computer capital costs, and this is then reflected in terms of total factor productivity. Their interpretation is that computers create new opportunities for firms to combine input factors through business reorganization. Brynjolfsson and Yang (1999) had previously estimated that computer adoption triggers complementary investments in "organizational capital" up to 10 times as large as direct investments in computers.

More research, based on developing country data, is needed in order to ascertain how and when ICT use increased production efficiency in firms and which ICT was used. A comparison of estimation results across different countries, industry sectors and technologies can provide policymakers with additional information for fine-tuning ICT policy master plans. As an illustration, section D presents the results of a firm-level growth model for manufacturing firms in Thailand.

D. Measuring ICT use and productivity in Thai manufacturing firms

1.   Background and objectives of the project

Measuring ICT access and use is an important element in the Thai national ICT policy. ICT indicators chosen were ones that made it possible to provide the necessary data to monitor progress in the Thai ICT Master Plan, to compare with ICT developments in other countries and to help in further policy decision-making (Smutkupt and Pooparadai, 2005). To analyse and monitor ICT use by businesses, the National Statistical Office (NSO) of Thailand gathered a valuable pool of information through a series of surveys such as the ICT Survey (2004 and 2005), the Manufacturing Industry Survey (1999, 2000 and 2003) and the Business Services Survey (1999 and 2003). More than 66,000 firms were covered

in the general ICT Survey 2005, approximately 8,800 manufacturing firms in the Manufacturing Survey 2003 and some 25,000 services firms in the Business Services Survey 2004. The information provided by those surveys helped build up a thorough description of the main recent patterns of ICT use in the Thai economy at a microeconomic level.[24] On the basis of the information provided by the surveys, the Thai NSO has published yearly reports describing in detail the use of ICT by the whole economy, by sectors of economic activity and by regions.

Using a firm-level data set of combined ICT uptake and several economic variables, the NSO and UNCTAD have conducted a joint research project. A well-known empirical model (the Cobb–Douglas production function) was used to assess the relationship between ICT use and productivity of Thai manufacturers. In that context, UNCTAD has conducted a training course for NSO staff on applying econometric methods to ICT data analysis and provided additional technical assistance to facilitate the analysis throughout the research phase. On the Thai side, the impact analysis was carried out by the NSO's Economic Statistics Analysing and Forecasting Group. The study is one of the first using official data from a developing country to analyse the economic impact of ICT in the business sector.[25]

The aim of the project was to evaluate the intensity of ICT use by firms and to quantify empirically the relationship between ICT uptake and productivity in a developing country. More specifically, the project aimed at studying the link between ICT use in firms and labour productivity by applying the theoretical framework used by similar studies conducted with developed country data. The relationship between ICT use and economic efficiency in firms has different magnitudes, depending on a series of factors specific to the economic environment and time span under consideration. For example, as suggested by other studies (in section C), ICT uptake is reflected in higher productivity gains at the firm level when complemented with superior know-how and managerial skills. Therefore, the analysis also identified differences in the ICT use–productivity relationship based on geographic location, industry sector, firm size and age.

To study the relationship between ICT uptake and economic output it is necessary to bring together ICT and economic variables in a firm level-data set. This was possible only for a 2002 cross-section of Thai manufacturing firms. The empirical study was therefore limited to the available sample of data. When more data become available, research could be based on larger data sets with information on more years and sectors.

The first part of this section reviews the use of basic ICTs such as computers, the Internet and the web by Thai manufacturing firms. That is followed by a brief presentation of the theoretical framework employed in the econometric application. The study ends with a detailed presentation of empirical results, a summary and conclusions.

2. ICT use by manufacturing firms in Thailand

The Manufacturing Industry Survey 2003 covered a large sample of more than 8,800 firms representative of the Thai manufacturing sector of approximately 360,000 establishments. By using weights corresponding to a stratified sampling by regions, employment size and industry sectors it was possible to represent the characteristics of the entire population[26]. The information provided below illustrates the use of ICTs in Thai manufacturing firms in 2002[27].

The study took into account small, medium-sized and large manufacturing firms with more than 10 employees. In the Thai manufacturing sector firms with more than 10 employees account for 98 per cent of revenues and 96 per cent of value added. This characteristic is common in other economies as well[28]

and usually research is conducted separately for micro-enterprises (fewer than 10 employees).

Among firms with more than 10 employees, 60 per cent use computers, 35 per cent access the Internet and 12 per cent have a website presence (table 3.4). According to the sample information, 99 per cent of the firms with access to the Internet also had computers on their premises; similarly, all businesses present on the web also had Internet access. The analysis therefore compared the economic performance of firms not using ICT with that of firms using computers, firms using both computers and the Internet, and firms using computers, the Internet and the web[29]. Accordingly, there is a gradual increase in the share of businesses with website presence, Internet access and computers (table 3.4). Many more firms employ computers in their daily routine work and considerably fewer have a website presence.

Adoption of websites by Thai manufacturers was at an early stage in 2002. Even among Internet-connected firms relatively few had opted in favour of a website presence (slightly more than one third). A considerable proportion of large firms that had computers and Internet connection had not established a web presence (27 per cent of businesses with more than 1,000 employees).

In larger firms with computers, access to the Internet was more often available than in smaller firms using computers. This may suggest that cost factors prevented

Table 3.4

Computer, Internet and website presence in Thai manufacturing businesses

Proportion of businesses using computers	59.8
Proportion of businesses using the Internet	35.1
Proportion of businesses with a web presence	12.3
Proportion of businesses connected to the Internet among businesses using computers	58.3
Proportion of businesses with a web presence among businesses using the Internet	35.2

Source: Thai Manufacturing Survey 2003, businesses with more than 10 employees.

Table 3.5

Intensity of computer use in Thai manufacturing businesses

Average number of employees using computers per 100 employees	7.8
Average number of computers per 100 employees	6.5

Source: Thai Manufacturing Survey 2003, businesses with more than 10 employees.

smaller firms from connecting to the Internet even when they already had computers.

The intensity of computer use remains relatively low in manufacturing in Thailand. On average, for each 100 employees[30] there are approximately seven computers available. For each 100 employees there are on average eight that use computers regularly in their work (table 3.5). Moreover, while in smaller firms the ratio of computers to employees using computers is closer to 1, in larger firms there are relatively more employees using computers for each physical computer machine. In other words, on average, there are more employees using computers than available computers. This suggests that Thai firms have, to a certain extent, valuable human capital resources (i.e. computer-literate staff) that would allow them to increase the use of computers.

Larger firms have on average more computer resources available per employee. A comparison with information from the ICT survey of all economic activities in Thailand shows that in manufacturing, firms use computers less intensively than in services, where businesses have more computers available per employee.

Most businesses connect to the Internet through regular ISP subscriptions (52 per cent of businesses) or through prepaid Internet packages (51 per cent of businesses) (table 3.6). Internet cafés are not a common solution for accessing the Internet since only 3 per cent of businesses choose it. Prepaid Internet packages and Internet cafés are modalities for Internet connection that are more often used by small firms.

Table 3.6

Modality of Internet connection (in manufacturing businesses using the Internet)

Proportion of businesses accessing Internet through ISP subscriptions	51.7
Proportion of businesses accessing Internet through Internet cafés	3.3
Proportion of businesses accessing Internet through prepaid Internet packages	50.7
Proportion of businesses accessing Internet through other modalities	3.9

Note: Shares presented here do not add up to 100 per cent. A firm could access the Internet in different ways.
Source: Thai Manufacturing Survey 2003, businesses with more than 10 employees.

Table 3.7
Type of activity performed while using the Internet
(in manufacturing businesses using the Internet)

Proportion of businesses sending and receiving e-mail	84.7
Proportion of businesses obtaining information	85.2
Proportion of businesses placing orders online	47.4
Proportion of businesses carrying out business promotion	23.6
Proportion of businesses performing other activities	5.7

Source: Thai Manufacturing Survey 2003, businesses with more than 10 employees.

Table 3.8
Type of activity performed on the web
(in manufacturing businesses with web presence)

Proportion of businesses advertising own business	83.1
Proportion of businesses receiving orders online	52.3

Source: Thai Manufacturing Survey 2003, businesses with more than 10 employees.

Manufacturing firms with Internet access indicated that they used it mainly for retrieving information (85 per cent) and e-mailing (85 per cent) (table 3.7). Only a limited number of businesses engage in other Internet activities such as placing orders online (48 per cent) and business promotion (24 per cent). Information search was the most popular activity on the Internet for all size groups, but larger firms were able to engage in it more often than smaller firms. Large and middle-sized firms used the Internet more often for e-mailing compared with small firms.

Manufacturing firms with a website presence use it mostly for advertising their business (83 per cent) and less for selling goods and services online (52 per cent) (table 3.8).

Analysing the use of ICT in micro-enterprises was beyond the scope of this study. While overall in micro-firms ICT penetration rates are considerably lower, there are aspects where the use of ICT differs between micro-enterprises and the rest of the firms (for example, a larger share of micro-firms uses web presence for advertising their business). A separate study could further analyse the economic implications of ICT in micro-enterprises.

Economic performance is stronger in firms that use ICT and even more so in firms that use a combination of several ICTs. Manufacturers using computers have on average 10 times higher sales per employee than manufacturers without computers (chart 3.5). The order of magnitude is higher when comparing sales per employee in firms with and without a website presence.

In 2002, exporting firms generated three quarters of total sales in the Thai manufacturing sector, while firms with foreign capital participation accounted for 62 per cent of that total industry revenues. There were very few Thai manufacturing firms without computers that were able to export or to attract foreign capital investment. On average, of 100 exporting Thai manufacturers, 91 had at least one computer, 77 were also connected to the Internet and only 35 had a website presence.

Chart 3.5 illustrates average sales per employee in firms with computers, Internet and web presence in terms of their export position and foreign capital participation. A typical manufacturer with at least one computer receives revenues of 1,538,284 baht per employee yearly. A total of 33 per cent of businesses with computers are exporters and 19 per cent benefit from foreign capital participation. This confirms that computers are distributed through Thai manufacturing and are not used exclusively by foreign-owned firms or by exporting firms.

Among firms with Internet access and website presence, exporting and foreign-owned firms had a higher

Chart 3.5

Average sales per employee in manufacturing Thailand (baht per employee)

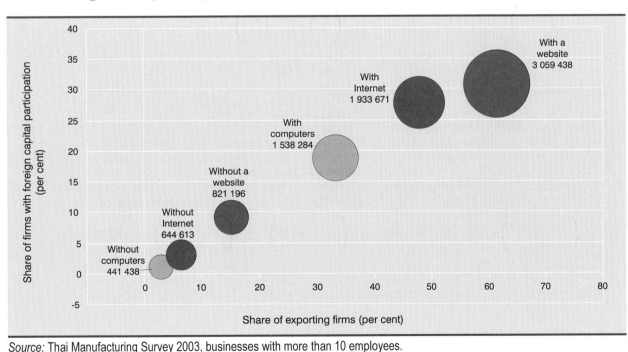

Source: Thai Manufacturing Survey 2003, businesses with more than 10 employees.

participation. A typical manufacturer with at least one computer connected to the Internet earns on average more than firms with computers only (1,933,671 baht per employee); but 48 per cent of Internet-connected businesses are exporters and 28 per cent receive foreign capital. The shares of exporting and foreign-owned firms are even higher among website-present manufacturers (61 per cent and 31 per cent). This suggests that websites are used more frequently by exporting firms, possibly for the purpose of following foreign market developments (information search).

Manufacturers with foreign capital participation do not establish web presence as frequently as exporting firms. Foreign capital participation seems to make more difference in terms of Internet access and less in terms of web presence, possibly owing to the existence of different management practices and requirements in foreign firms. The high share of exporters among manufacturers present on the web could be related to the language used or the relative scarcity of Internet content in the Thai language. However, available data provided no information on the language used by firms on the websites.

3.   The model and a few theoretical considerations

The primary goal of the analysis was to quantify the relationship between ICT use and labour productivity in Thai manufacturing firms in 2002. In the analysis, which built on methods employed in similar studies, firm productivity was modelled on the assumptions of a Cobb–Douglas production function with three input factors: capital, labour and spending on materials (see box 3.1).

The analysis also identified differences based on geographical location, industry sector, firm size and age and their influence on the ICT use–productivity relationship. This was done by estimating interactions of ICT measures with control variables within the same Cobb–Douglas theoretical framework.

Box 3.1

The empirical model

The regression equation is based on a linearized version of the Cobb–Douglas function (equation 1). Labour productivity is regressed on factor inputs (capital, labour, spending on materials), one or several ICT variables and a set of controls for industry and regional attributes of demand and supply, the presence of foreign capital participation and the activity of multi-unit firms (head offices or branches).

$$\ln\left(\frac{sales_j}{L_j}\right) = \beta_0 + \beta_1 ICTVariable_j + \beta_2 \ln\left(\frac{K_j}{L_j}\right) + \beta_3 \ln\left(\frac{M_j}{L_j}\right) + \beta_4 \ln(L_j) + \beta_5 Multi\_unit_j +$$
$$+ \beta_6 Foreign\_capital_j + \sum_r \beta_r region\_r_j + \sum_i \beta_i Industry\_i_j + u_j$$

,where K is capital, L is employment and M is spending on materials.   (equation 1)

With the data available from the Manufacturing Survey 2003 it was possible to construct two similar models based on the same Cobb–Douglas framework: one using total employment as a common denominator and the other using total effective employment. Effective employment is obtained by multiplying total employment times the declared number of hours effectively worked during 2003. Results derived from the two models could be used as a check for the robustness of the estimates. From a theoretical point of view, the effective labour productivity model is more accurate since it accounts for variations in the hours effectively worked by employees rather than assuming that all employees worked on average an equal number of hours. The model employed for effective labour productivity is described by equation 2.

$$\ln\left(\frac{sales_j}{L\_effective_j}\right) = \beta_0 + \beta_1 ICTVariable_j + \beta_2 \ln\left(\frac{K_j}{L\_effective_j}\right) + \beta_3 \ln\left(\frac{M_j}{L\_effective_j}\right) + \beta_4 \ln(L\_effective_j)$$
$$+ \beta_5 Multi\_unit_j + \beta_6 Foreign\_capital_j + \sum_r \beta_r region\_r_j + \sum_i \beta_i Industry\_i_j + u_j$$

,where K is capital, L_effective is effective employment and M is spending on materials.   (equation 2)

White (1980) heteroskedasticity consistent standard deviations were calculated.

By looking at the performance of firms with and without specific ICTs the analysis was able to quantify the extent to which during 2002 Thai manufacturers with similar characteristics had higher productivity when using ICTs. Further research with comparable data on several years could investigate the impact of past decisions to adopt ICTs on current labour productivity. The analysis could not establish whether beyond correlation, there is a causal relationship between ICT use and firm labour productivity. To deal with that shortcoming, other specialized papers applied instrumental variable estimation techniques in data sets covering several years.[31]

In the light of the discussion in Atrostic and Nguyen (2002) (see section C) the preferred dependent variable for measuring labour productivity was total sales per employee.

A set of additional variables was used in each equation to control for the effect of foreign capital participation, for the multi-unit firm organizational aspect and also for unknown demand and supply factors across industries and regions. Firm age appeared not to play a significant role as a control variable for explaining variations in labour productivity. Information on firm age, however, was used for testing if among ICT users more experienced firms had superior labour productivity as compared with younger firms. Most regression estimates showed that establishments with foreign capital participation have on average 7 to 8 per cent higher sales. Firms belonging to multi-unit organizational structures have 2 to 4 per cent higher sales, presumably because they can more easily gain access to a larger pool of resources. Also, there appeared to be decreasing returns to scale: larger

firms had on average 0.5 to 3 per cent lower labour productivity given the set of controls.

4. Results

Firstly, the study evaluated the relationship between computer, Internet and website presence and the value of sales per employee. The three measures of ICT were regarded as progressive steps adding to the complexity of ICT uptake since all firms using the Internet also had computers on their premises and all firms present on the web also had Internet access. Accordingly, estimates on the Internet factor are interpreted as additional to computer-related gains; similarly, website-related gains are complementary to those from Internet and computer use. Results are shown in table 3.9.

After controlling for a series of firm-specific economic characteristics, as well as industry and regional aspects of demand and supply, estimated results showed that firms with a combined use of computers, the Internet and the web had on average 21 per cent higher sales than firms without any of the ICTs considered. Among the three ICTs considered, computers contributed with 14 per cent, Internet access with 3 per cent and web presence with 4 per cent. Similar estimates were also obtained with the effective labour productivity variant of the model (table 3.9, second column). Atrostic and Nguyen (2005) estimated that in 1999 computer networks (such as the Internet, intranet, LAN, EDI, extranet or other) had a 5 per cent positive impact on labour productivity in a large sample of United States manufacturing firms. In comparison, the results derived in this study show that computer presence was more closely associated with labour productivity in Thailand than Internet connectivity.

Table 3.9

Estimation results for computer, the Internet and website presence

Independent variables	OLS (White heteroskedasticity – consistent standard errors)	
	Dependent variable: log (sales per employee)	Dependent variable: log (effective sales per employee)
	R^2 = 0.923877 Number of observations included: 5 877	R^2 = 0.893852 Number of observations included: 5 651
Computer presence	0.1359***	0.1346***
Internet access	0.0322*	0.0365*
Web presence	0.0418**	0.0491***

Note: The regression included controls for employment size, capital, costs incurred with materials, foreign capital participation, multi-unit firms, and industry- and region-specific characteristics. (Level of significance at *** 1 per cent, ** 5 per cent and * 10 per cent).

Studies based on developed country data rarely estimate the impact of computer presence on labour productivity because in many developed countries computer penetration rates in the business sector have already reached levels of as much as 95 per cent. However, in developing countries the share of firms that use at least one computer for business purposes has remained lower (60 per cent in manufacturing Thailand in 2002). Furthermore, data from the 2003 Manufacturing Survey in Thailand show that in absence of computers firms cannot access the Internet and establish web presence. This explains why in developing countries computer presence in firms is more closely related to economic performance than in developed countries.

Because the use of at least one computer seemed to account for a large share of the variation in sales per employee, it was interesting to estimate also the relationship between the intensity of computer use and labour productivity (tables 3.10 and 3.11). Results show that an increase of 10 per cent in the share of employees using computers is associated with 3.5 per cent higher sales per employee in Thai manufacturing firms. For the same variable, the estimated coefficient in Maliranta and Rouvinen (2003) was only 1.8 in a panel of Finnish firms (1998–2000). Since the variation in the intensity of computer use in Thailand was greater – with many firms not having computers – computer use was associated with larger differences in labour productivity in that country. Similarly, a 10 per cent improvement in the number of computers available per employee was correlated with a 4.5 per cent increase in sales per employee (table 3.11).

Computer intensity in firms, as captured by the number of physical computers per employee, can be interpreted as a measure of investment in computer capital.

Table 3.10

Estimation results for the share of employees using computers, the Internet and website presence

Independent variables	OLS (White heteroskedasticity – consistent standard errors)	
	Dependent variable: log (sales per employee)	Dependent variable: log (effective sales per employee)
	$R^2 = 0.923260$ Number of observations included: 5 863	$R^2 = 0.893054$ Number of observations included: 5 637
Share of employees using computers	0.3492***	0.3969***
Internet access	0.0561***	0.0582***
Web presence	0.0160	0.0229

Note: The regression included controls for employment size, capital, costs incurred with materials, foreign capital participation, multi-unit firms, and industry- and region-specific characteristics. (Level of significance at *** 1 per cent, ** 5 per cent and * 10 per cent).

Table 3.11

Estimation results for the share of computers per employee, the Internet and website presence

Independent variables	OLS (White heteroskedasticity – consistent standard errors)	
	Dependent variable: log (sales per employee)	Dependent variable: log (effective sales per employee)
	$R^2 = 0.923570$ Number of observations included: 5 871	$R^2 = 0.893028$ Number of observations included: 5 645
Number of computer per employees	0.4466***	0.5185***
Internet access	0.0545***	0.0562***
Web presence	0.0147	0.0213

Note: The regression included controls for employment size, capital, costs incurred with materials, foreign capital participation, multi-unit firms, and industry- and region-specific characteristics. (Level of significance at *** 1 per cent, ** 5 per cent and * 10 per cent).

Similarly, the share of employees using computers also represents a proxy for computer capital investment, including investment in human capital and training for work with computers. As estimates in tables 3.10 and 3.11 show, Internet access contributed as a significant factor additional to computer intensity in explaining differences in labour productivity among firms. At the same time, when accounting for the intensity of computer use, it is noted that web presence is no longer significantly contributing to higher sales per employee. This suggests that a greater intensity of computer use and Internet access are factors facilitating the decision to establish web presence in the businesses analysed.

Both the labour productivity and the effective labour productivity models produced similar estimates for the overall use of computers, the Internet and the web in Thai manufacturing firms.

The study also estimated coefficients relating labour productivity to the different modalities of accessing the Internet (ISP subscribers, prepaid Internet package, Internet café, etc.), as well as to the different activities carried out on the Internet (e-mailing, information search, placing orders online, business promotion, etc.) and on the web (advertising own business, receiving orders online). However, results did not show significant differences in the way in which those factors were reflected in the value of sales per employee. More analysis would be needed to assess the role of the different modalities of Internet access and of the different activities carried out online in explaining firms' economic performance. This was beyond the scope of the present analysis.

In the second stage, the analysis aimed at identifying firms' characteristics that influenced the relationship between specific ICTs and labour productivity. For that purpose, the regression equation was slightly modified to estimate the effect of ICT uptake in groups of firms with different size, age, located in different regions and belonging to different industry branches. The next subsection analysed whether firm size contributed to explaining the ICT–labour productivity relationship.

Differences between employment size groups

To analyse the implications of firm size in determining the ICT–productivity relationship, three employment size groups were considered: small firms (11 to 25 employees), medium-sized firms (26 to 80 employees) and larger firms (more than 80 employees). The groups

chosen have equal numbers of firms. Table 3.12 summarizes the results of three different empirical specifications. In the first column the estimation model took into account information on the presence of computers in firms, Internet access and web presence. In the second and third columns additional information was included on the intensity of computer use as measured by the share of employees using computers (the third column) and the number of computers per employee (the fourth column). Accordingly, results in the first column show that Internet access matters more to labour productivity in small firms and computer presence is correlated with higher productivity in middle-sized firms, while what makes a difference in large firms is web presence. The link between ICT use and labour productivity is stronger in large and middle-sized firms, with a high contribution from the presence of computers.

The third and fourth columns of table 3.12 show the estimated relationship between the intensity of computer use and productivity. Both measures of computer intensity, the share of employees using computers and the number of computers per employee yielded similar results. Most productivity gains associated with the intensity of computer use occur in large firms. A 10 per cent higher share of employees using computers leads on average to 4.2 per cent higher sales per employee in large firms and 2.9 per cent higher sales in middle-sized firms. Similarly, 10 per cent more computers per employee are correlated with 5.9 per cent higher sales per employee in large firms and 4.3 in middle-sized firms respectively. For smaller firms, variations in the intensity of computer use did not translate into significantly greater labour productivity. Also, in accounting for the intensity of computer use, the Internet and web presence no longer had a significant effect. This indicated that most middle-sized and large firms with greater intensity of computer use also had Internet access and web presence.

Differences between age groups

Several studies reviewed in the previous section found that firm age was an additional factor explaining how much enterprises gain from ICT. To analyse whether the same effect appeared in manufacturing Thailand, this study grouped businesses according to their founding year and hence experience in the market. There are three groups with an equal number of firms: young (founded between 1997 and 2002), middle-aged (founded between 1991 and 1996) and old (founded

Table 3.12

Estimation results by employment size groups

OLS (White heteroskedasticity – consistent standard errors)		
Dependent variable: log (sales per employee)		
	R2 = 0.923573 Number of observations included: 5 873	R2 = 0.922656 Number of observations included: 5 858	R2 = 0.923411 Number of observations included: 5 867
Computer presence in small firms	-0.0047	-	-
Computer presence in medium-sized firms	0.1200***	-	-
Computer presence in large firms	0.0981***	-	-
Share of employees using computers in small firms	-	0.1596	-
Share of employees using computers in medium-sized firms	-	0.2917**	-
Share of employees using computers in large firms	-	0.4139***	-
Number of computers per employee in small firms	-	-	0.1908
Number of computers per employee in medium-sized firms	-	-	0.4310**
Number of computers per employee in large firms	-	-	0.5947***
Internet access in small firms	0.0809**	0.0443	0.0427
Internet access in medium-sized firms	-0.0271	0.0427	0.0374
Internet access in large firms	-0.0339	-0.0021	-0.0055
Web presence in small firms	0.0097	-0.0005	0.0022
Web presence in medium-sized firms	0.0533	0.0368	0.0341
Web presence in large firms	0.0764*	0.0656	0.0619

Note: The regression included controls for employment size, capital, costs incurred with materials, foreign capital participation, multi-unit firms, and industry- and region-specific characteristics. (Level of significance at *** 1 per cent, ** 5 per cent and * 10 per cent).

before 1991). The applied estimation technique was the same as in the case of firm size. Results are shown in table 3.13 with the first column showing the effect of computer presence, Internet access and web in the different age groups, while in the third and fourth columns the analysis also takes into account the intensity of computer use.

In young firms computer presence is associated with the greatest value of gains in terms of sales per employee, a fact that suggests that young firms use computers more effectively. However, older firms seem to gain most from the combined use of computers, the Internet and the web. In older firms the presence of computers also matters, albeit less than in the younger ones, while there is an additional contribution to productivity from Internet access and the web.

With regard to accounting for the intensity of computer use, results confirm that for larger firms, with more experience in the market, there was a stronger correlation between ICT uptake and labour productivity. Results also show that younger firms with a lower intensity of computer use achieve higher sales per employee when they have access to the Internet.

Both firm age and employment size influence the magnitude of the relationship between ICT use and labour productivity. Larger and more experienced firms appear to gain more from the combined use of the three specific ICTs analysed here (computers, Internet and web). The presence of computers and the intensity of computer use contribute substantially to explaining those differences. However, in smaller and younger firms Internet access matters more in the sense that a lower intensity of computer use can be compensated

Table 3.13

Estimation results by firm age groups

OLS (White heteroskedasticity – consistent standard errors)			
Dependent variable: log (sales per employee)			
	R^2 = 0.924005 Number of observations included: 5 877	R^2 = 0.923348 Number of observations included: 5 863	R^2 = 0.923653 Number of observations included: 5 871
Computer presence in young firms	0.1729***	-	-
Computer presence in middle-aged firms	0.1498***	-	-
Computer presence in old firms	0.0902***	-	-
Share of employees using computers in young firms	-	0.1786	-
Share of employees using computers in middle-aged firms	-	0.4286***	-
Share of employees using computers in old firms	-	0.3713***	-
Number of computers per employee in young firms	-	-	0.2500*
Number of computers per employee in middle-aged firms	-	-	0.5188***
Number of computers per employee in old firms	-	-	0.4922***
Internet access in young firms	0.0313	0.1203***	0.1146***
Internet access in middle-aged firms	0.0144	0.0449*	0.0465*
Internet access in old firms	0.0484*	0.0260	0.0236
Web presence in young firms	0.0372	0.0206	0.0226
Web presence in middle-aged firms	0.0220	-0.0021	-0.0039
Web presence in old firms	0.0613**	0.0323	0.0292

Note: The regression included controls for employment size, capital, costs incurred with materials, foreign capital participation, multi-unit firms, and industry- and region-specific characteristics. (Level of significance at *** 1 per cent, ** 5 per cent and * 10 per cent).

for by use of the Internet. Chinn and Fairlie (2006) analysed the factors leading to higher computer and Internet penetration rates and show that income has a greater influence on decisions to acquire computers than does Internet access. Small and young firms with computers may find it easier to buy Internet access rather than additional computers. This study shows that small firms with at least one computer gain most from Internet access, while young firms use computers more effectively.

Regional and industry characteristics also have a bearing on the strength of ICT's impact on firm performance. These are further explained in the next two subsections.

Regional differences

Owing to factors such as infrastructure, labour force training and qualifications and market size, firms located in different geographical regions use ICTs differently.

Dominated by larger manufacturing firms, the south had the highest share of computer-equipped firms and a relatively high intensity of computer use. The northern region had different characteristics: more small enterprises, fewer firms with computers and a more reduced intensity of computer use (see charts 3.6 a and b).

In terms of Internet access, Bangkok and the southern region had the two highest Internet penetration rates.

Charts 3.6 a and b

Computer, Internet and website use by regions in manufacturing Thailand

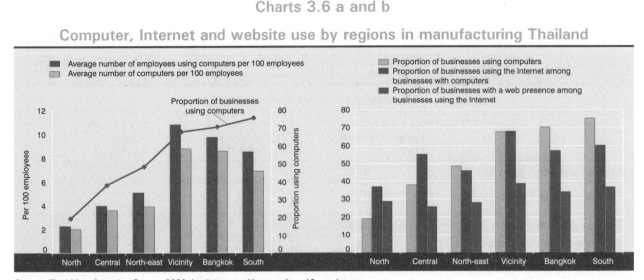

Source: Thai Manufacturing Survey 2003, businesses with more than 10 employees.

Not surprisingly, this region, together with Bangkok, has the highest proportion of ISP regular subscribers. Access through Internet cafés and prepaid Internet packages is more popular among manufacturers located in the north. However, in the north and the centre, there is a greater proportion of businesses with computers connected to the Internet. For example, in the north only 20 per cent of businesses have computers but among those with computers almost 40 per cent have also acquired Internet access. This confirms that the use of computers is an important source of differentiation among firms located in different geographical regions. Among businesses with computers, the proportion of businesses using the Internet and present on the web varied less across regions (chart 3.6 b).

Table 3.14 shows the results of two different models estimating the relationship between ICT use and labour productivity in businesses from different regions. The first one takes into account computer presence, Internet access and the web, while the second one also considers the intensity of computer use as measured by the share of employees using computers.

Table 3.14

Estimation results by regional location

OLS (White heteroskedasticity – consistent standard errors)					
Dependent variable: log (sales per employee)					
$R^2 = 0.924005$ Number of observations included: 5 877			$R^2 = 0.923348$ Number of observations included: 5 863		
	Computer presence	Internet access	Web presence	Share of employees using computers	Internet access	Web presence
Bangkok	0.1298***	0.0892*	0.0254	0.4735***	0.0844**	-0.0007
Vicinity	0.1411***	0.0511	0.0731*	0.2037**	0.0852***	0.0496
Central	0.1553***	0.0121	0.1533**	0.2851	0.0730**	0.1372**
North	0.1921***	0.0302	-0.1720**	0.5234**	0.0954	-0.2187**
North-east	0.1187***	0.0669	-0.0101	0.2029	0.0865	-0.0394
South	0.1436***	0.0236	0.0400	0.5642***	0.0285	0.0089

Note: The regression included controls for employment size, capital, costs incurred with materials, foreign capital participation, multi-unit firms, and industry- and region-specific characteristics. (Level of significance at *** 1 per cent, ** 5 per cent and * 10 per cent).

Both sets of results confirm that computers are more important among the ICTs considered here in accounting for variations in sales per employee. That is shown by the highly significant coefficients for presence of computers and share of employees using computers. For firms located in the centre, website presence matters more than in other regions. In Bangkok higher sales per employee are generated in businesses with computers and Internet access. In the vicinity of Bangkok similar results apply, but Internet access is relatively more important. In the north, sales per employee are correlated much more with computer use than with the Internet and the web. The negative coefficient estimated for web presence shows that in the northern region businesses with computers, the Internet and the web have lower sales than businesses equipped only with computers. This finding can be explained by the fact that there is a very small share of web-present firms in the north (only 2 per cent) and that the use of web presence leads to fewer efficiency gains in this region than in others. A similar problem seems to occur in the north-east for the web-present firms, but the estimated negative effect is much smaller and not significant. A more detailed study of the web-present firms in the northern and north-eastern regions is needed in order to find out why these firms have lower sales per employee than competitors that use only computers. In the south higher labour productivity is recorded in the most computer-intensive firms

Industry differences

In order to simplify the analysis, industries were classified into 14 broader categories of manufacturing activity (for reference purposes ISIC Rev. 3 codes are provided in parentheses). As expected, ICT use also varied considerably across industries (chart 3.7). Measures of ICT uptake in particular sectors of activity are a reflection of industry characteristics and the degree to which particular business types are able to integrate computers, the Internet and the web in their production process. The computing, electrical and precision instruments sector emerges as the most frequent user of the Internet and web presence. The paper and printing industry has the highest share of businesses with computers but lags behind in terms of the Internet and website presence. The chemical industry and machinery and transport equipment follow closely behind with high penetration rates for computers, the Internet and the web. In coke and petroleum an unusually high share (70 per cent) of the firms with computers and the Internet also find it useful to establish web presence. Computerized

firms in the group of other manufacturing industries (such as furniture, jewellery and musical instruments) and textiles, clothing and leather are connected to the Internet in a higher proportion than in other sectors. This indicated that Internet access is considered more relevant in those industries than for example in coke and petroleum.

Chart 3.8 shows information regarding the intensity of computer use in businesses. Coke and petroleum stands out as the industry with the highest intensity of computer use. The least computer-intensive sectors are the recycling industry but also the wood industry, textiles, clothing and leather, the processed food, beverages and tobacco industry. Typical of the last three manufacturing sectors was the very high share of small businesses.

Table 3.15 shows for comparison the results of two empirical exercises: one taking into account computer presence, the Internet and the web, and the other estimating the importance of the intensity of computer use as captured by the share of employees using computers. Estimation results identified four sectors where the use of particular ICTs was correlated with higher than average sales per employee: machinery and equipment, basic metal industry, computing, electrical and precision instruments and processed food, and beverages and tobacco.

The combined use of computers, the Internet and the web was associated with the highest gains in:

- Machinery equipment;

- Basic metal industry; and

- Computing, electrical and precision instruments.

Internet access was more significantly related to higher sales per employee in:

- Machinery equipment;

- Computing, electrical and precision instruments; and

- Processed food, beverages and tobacco.

Website presence seemed to add significantly to gains from computers and the Internet in:

- Machinery equipment;

- Basic metal industry; and

- Processed food, beverages and tobacco.

Chart 3.7

Computer, Internet and website presence across manufacturing industries in Thailand

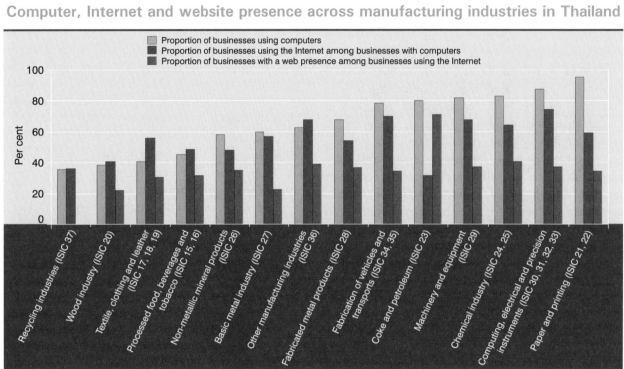

Source: Thai Manufacturing Survey 2003, businesses with more than 10 employees.

Chart 3.8

Computer use by manufacturing industries in Thailand

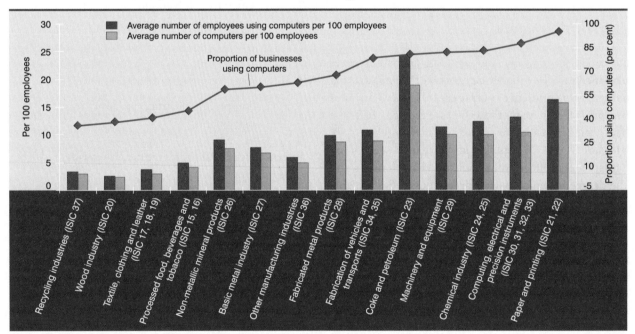

Source: Thai Manufacturing Survey 2003, businesses with more than 10 employees.

Table 3.15

Estimation results by industry affiliation

OLS (White heteroskedasticity – consistent standard errors)						
Dependent variable: log (labour productivity)						
	R² = 0.922725 Number of observations included: 5 877			R² = 0.922234 Number of observations included: 5 863		
	Computer presence	Internet access	Web presence	Share of employees using computers	Internet access	Web presence
Processed food, beverages and tobacco (ISIC 15, 16)	0.1409***	0.0492*	0.0745*	0.4735***	0.0739**	0.0486
Textile, clothing and leather (ISIC 17, 18, 19)	0.1912***	0.0222	-0.0219	0.3728**	0.0609*	-0.0369
Wood industry (ISIC 20)	0.1354***	0.0563	0.0443	1.0008**	0.1056	0.0504
Paper and printing (ISIC 21, 22)	0.1357***	0.0132	0.0635	0.3009**	0.0172	0.0044
Coke and petroleum (ISIC 23)	0.2550**	-0.0038	-0.1637	0.5608***	-0.1150	-0.3177
Chemical industry (ISIC 24, 25)	0.1148***	-0.0128	0.0189	0.3487***	0.0049	-0.0201
Non-metallic mineral products (ISIC 26)	0.1123***	0.0285	0.0331	0.1943	0.0431	-0.0222
Basic metal industry (ISIC 27)	0.1558***	0.0775	0.1672**	0.2751	0.0764	0.1202
Fabricated metal products (ISIC 28)	0.1810***	0.0638*	0.0522	0.5150***	0.0865***	0.0298
Machinery and equipment (ISIC 29)	0.2090***	0.0998**	0.1619*	0.6667***	0.1203***	0.1384
Computing, electrical and precision instruments (ISIC 30, 31, 32, 33)	0.2283***	0.0745**	0.0292	0.4350***	0.0819***	-0.0061
Fabrication of vehicles and transports (ISIC 34, 35)	0.1134***	0.0315	0.1038	0.3983**	0.0488	0.0734
Other manufacturing industries (ISIC 36)	0.0744**	-0.0057	-0.0055	0.3118	0.0448	-0.0006
Recycling industries (ISIC 37)	-	-	-	-	-	-

Note: The regression included controls for employment size, capital, costs incurred with materials, foreign capital participation, multi-unit firms, and industry- and region-specific characteristics. (Level of significance at *** 1 per cent, ** 5 per cent and * 10 per cent).

Chart 3.9 shows the contribution of the different industry branches to revenue and value added in manufacturing Thailand. The computing, electrical and precision instruments industry is the most important contributor to sales and value added in the Thai manufacturing sector. This industry is part of the Thai ICT-producing sector and has some of the highest Internet and web coverage. Estimation results indicate that in this sector the presence of computers and Internet access leads to as much as 30 per cent higher labour productivity on average.

The processed food, beverages and tobacco sector is the second highest contributor to value added in Thai manufacturing. This light industry sector is characterized by considerably low use of computers, the Internet and the web as compared to other sectors. However, estimation results show that productivity differentials within this industry were strongly correlated with the use of the Internet (4.9 per cent) and websites (7.5 per cent). If demand for food, beverages and tobacco remains favourable, there is scope for producers to increase productivity as they start using ICTs more frequently.

Last but not least, estimates showed that for the machinery equipment and the basic metal industries web presence seems to make the most difference.

Chart 3.9

Share of manufacturing industry sectors in total sales and value added in Thailand

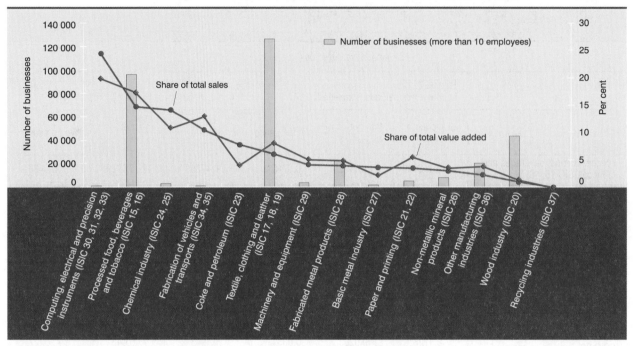

Note: The 23 ISIC Rev.3 sectors were grouped into 14 broader manufacturing industry categories.
Source: Thai Manufacturing Survey 2003, businesses with more than 10 employees.

Firms in those industries with web presence have on average 16 per cent higher sales per employee than firms with just computers and Internet access. A more detailed study of how websites are used in the machinery equipment and basic metal industries in Thailand would help further understand why these firms have higher labour productivity.

E. Concluding remarks and policy issues

Measuring the economic impact of ICT allows a thorough assessment of the extent to which ICTs contribute to development. This chapter focused on the productivity impact of ICT from a macroeconomic and a microeconomic perspective.

Studies reviewed confirm that developing countries can gain as much as developed countries from ICTs in terms of productivity. In recent years developing countries have received a significant positive contribution to GDP growth from investment in ICTs. Those gains correspond to a period of sustained demand for ICT goods and services coming from developed OECD countries, but also from fast-growing developing

economies such as China and India. Additionally, there are several developing countries where exports of ICT goods and ICT-enabled services have only recently started to grow (as shown in chapter 2).

At the macroeconomic level, growth accounting estimates reviewed in this chapter identify two principal sources of productivity gains from ICT: capital deepening through investment in ICT and technological progress in the ICT-producing industry. A comparison of the estimated impact of ICT on GDP growth shows that larger gains accrued to developing countries specialized in producing and exporting ICT goods and services. Developing countries have also benefited from ICT investment. However, in countries with a reduced level of ICT use the effects of ICT investment at a macroeconomic level remained low.

Since the mid-1990s, there has been an increase in global ICT investment. While this resulted in significantly higher gains from ICT to GDP in developed countries, in developing countries benefits had only started to become significant. Estimates show that over time technological progress in the ICT-producing sector has become an important factor contributing to GDP growth in developing countries. Furthermore, other estimates indicate that in a group of Asian developing economies total factor productivity growth is more strongly correlated with ICT investment in OECD countries rather than with ICT investment in the domestic market. This indicates that the demand for ICT goods and services coming from developed countries has had a positive effect on the productivity of developing economies. On the other hand, results from developed countries show that they gained relatively less from producing ICTs themselves, but rather from using ICTs as capital inputs in many sectors of economic activity.

Several studies reviewed point out that an innovative ICT use can lead to superior production efficiency. One of the ways to improve the efficiency of ICT use is to combine it with complementary skills such as a qualified labour force or superior management skills. An important study of the effects of computer capital investment on total factor productivity in the United States showed that significant gains occur with the time lag that is needed to effectively incorporate ICT solutions into production systems. This time lag can be reduced when "soft skills" such as superior management and marketing know-how are available.

At the firm level, this chapter reviews a number of studies using developed country statistical data to measure the contribution of specific ICTs to labour productivity in the business sector. Results show that the productivity effect of a specific information and communication technology can have different magnitudes, depending on the characteristics of the micro-environment analysed. For example, a 10 per cent increase in the share of employees using computers leads to 1.8 per cent higher labour productivity in manufacturing firms in Finland, 2.8 per cent in services firms in Finland and 1.3 per cent in a mixed sample of Swedish enterprises. Additionally, there are a series of complementary factors that come into play in determining the magnitude of the impact of ICT on labour productivity. These include firm age, foreign ownership and industry affiliation. These findings suggest that the profile of firms that gain most from the use of ICTs differs slightly from country to country, depending on the characteristics of the business environment. It is therefore necessary to further research and analyse the characteristics of firms from developing countries in which the impact of ICT use is greatest.

This chapter also presents the results of a joint research project of the Thai National Statistical Office and UNCTAD to assess the link between ICT use and labour productivity in Thai manufacturing firms.

This study is one of the first using official data from a developing country to analyse the economic impact of ICT in the business sector. The empirical analysis is based on the theoretical framework applied in similar studies in developed countries. The project is part of a broader initiative of UNCTAD to improve ICT measurement and to enable the analysis of ICT impact in the business sector in developing and transition economies. It is promoted by several international organizations and statistical offices through the Partnership on Measuring ICT for Development.

The results of the firm-level study in Thailand show that computer use, Internet access and web presence are associated with significantly higher sales per employee. In comparison with similar analyses carried out in developed countries, this project finds that the use of computers is a key ICT factor for explaining higher productivity in firms. In developed countries the penetration rates of basic ICTs such as computers are already close to saturation levels and therefore computers are present in nearly all businesses (with more than 10 employees). However, in developing countries (see chapter 1) there is a lower share of businesses that use at least one computer. Variation in the intensity of computer use in Thailand was also reflected in greater productivity gains. Estimates show that a 10 per cent increase in the share of employees using computers was correlated with 3.5 per cent higher labour productivity, more than the 1.8 per cent estimated impact in a sample of Finnish manufacturing firms. In the theoretical setting considered, computers bring value to businesses both through their intrinsic characteristics such as processing and storage capacity and as necessary means for acquiring a greater complexity of ICT use such as Internet access and web presence. Internet access and website presence are also found to be correlated with higher sales per employee in Thailand, with a coefficient similar to that estimated in other studies. For example, Atrostic and Nguyen (2005) estimated that computer networks brought a 5 per cent positive net effect to firm labour productivity in the United States after accounting for the contribution of computer capital. Similarly, results derived here suggest that Thai firms with access to the Internet had on average 4 to 6 per cent higher sales per employee, additional to the effect of computers.

This study also quantified differences in the ICT–labour productivity relationship across employment size groups, firm age, regional location and industry affiliations. This type of estimates aimed at indicating the areas where the use of specific ICTs is more strongly correlated with superior economic performance. Among Thai manufacturing businesses the groups that seem to benefit from a stronger ICT–labour productivity relationship are the larger and medium-sized more experienced firms located in the central region and also in Bangkok and its vicinity. Younger firms, however, tended to use computers more effectively, while in smaller firms Internet access made a considerable difference. Further case study evidence is needed in order to establish how specific ICTs contribute to improved economic performance in specific fields and how their efficiency can be scaled up to other groups of enterprises.

The chapter concludes that in order to maximize productivity gains from ICT in a period of economic prosperity, developing countries need to encourage the efficient use of ICT in their economies. The impact of ICT on productivity in developed countries is well researched, but only few similar studies have been carried out in developing countries. Therefore, it is necessary to measure and monitor the use of ICT in a way that enables international comparisons and allows the economic impact of ICTs in developing countries to be assessed.

Estimates of the impact of ICT uptake in firms may give different results depending on the maturity level of ICTs in a specific market. These considerations make country, regional and sectoral studies more pertinent when analysing the impact of the use of specific ICTs on firm-level performance. More research is needed in developing countries to better assess how productivity gains from ICT can be scaled up.

Annex 3.1

Summary of Literature on ICTs and Productivity at the Firm Level

Author	Dataset	Dependent variable	ICT variables used	Type of regression	Impact of ICT variables on labour productivity
Atrostic & Nguyen (2002)	US manufacturing panel, 3 years (1992, 1997 and 1999) 38 000 firms	Sales per employee	Dummy variable for presence of computer networks (such as Internet, intranet, LAN, EDI, extranet or other).	Cobb–Douglas production function with 3 factors (labour, capital and materials). Control variables: size of establishment instead of number of employees, skill mix, multi-unit firm, industry.	5 per cent from presence of computer networks.
Atrostic & Nguyen (2005)	US manufacturing, 1999	Sales per employee	Computer capital and computer networks dummy (as above).	Cobb–Douglas production function with 4 factors (labour, ordinary capital, computer capital and materials). Control variables: size of establishment instead of number of employees, firm age, skill mix, multi-unit firm, industry.	5 per cent from presence of computer networks and 12 per cent for computer capital.
Maliranta & Rouvinen (2003)	Finland panel; 3 years (1998–2000) Manufacturing and services separately with 1 500 firms each. ICT-producing sector considered separately.	Output per employee	Share of computer-equipped labour, share of Internet-using labour, share of labour using LAN at work.	Cobb–Douglas production function with 2 factors (labour and capital). Control variables: skill mix and firm age.	A 10 per cent increase in computer-equipped labour results in 1.8 per cent higher labour productivity in manufacturing and 2.8 per cent in services. A 10 per cent increase in Internet-equipped labour results in 2.9 per cent higher labour productivity in services (negative estimated impact in manufacturing). A 10 per cent increase in the share of labour using LAN results in 2.1 per cent higher labour productivity in manufacturing (no significant impact in services).
Maliranta & Rouvinen (2006)	Finland, 2001, 2 358 services and manufacturing firms	Output per employee	Share of workforce using desktops, laptops, wireline access to Internet (LAN) and wireless access to Internet (WLAN).	Cobb–Douglas production function with 2 factors (labour and capital). Control variables: skill mix, firm age and gender participation in labour force.	9 per cent labour productivity gains from a higher share of labour force using computers (desktop). Additional 32 per cent from computer portability. Additional 14 per cent from wireline Internet connectivity. Additional 6 per cent from wireless connectivity.

Annex 3.1 (continued)

Author	Dataset	Dependent variable	ICT variables used	Type of regression	Impact of ICT variables on labour productivity
Hagén & Zeed (2005)	Sweden, 2002 2 752 firms with 10 employees or more (manufacturing and services together)	Value added per employee	Share of employees using computers, dummy variable for access to broadband, a composite ICT index.	Cobb–Douglas production function with 2 factors (labour and capital). Control variables: size of establishment instead of number of employees, skill mix, ownership, perceived lack of IT competence and industry.	A 10 per cent increase in computer-equipped labour results in 1.3 per cent higher labour productivity. 3.6 per cent labour productivity gains from broadband. Positive but decreasing impact of adopting an increasing number of ICT solutions.
Crisculo & Waldron (2003)	UK manufacturing panel, 2 years (2000, 2001) 5 500 firms	Gross output per employee and value added per employee	Dummy variables for e-commerce (either placing or receiving orders online)	Cobb–Douglas production function with 3 factors (labour, capital and materials). Control variables: region, ownership, firm age, industry and year fixed effects.	Between 7 and 9 per cent labour productivity gains from placing orders online. Receiving orders online has negative effects (-5 per cent).
Farooqui (2005)	UK panel, 4 years (2000–2003); 2 277 manufacturing firms and 3 490 services firms taken separately	Value added per employee	Hardware capital stock, software capital stock, share of employees using ICT (computer and Internet), spending on telecommunication services, e-commerce (placing and receiving orders online).	Cobb–Douglas production function with 2 factors (labour and capital). Control variables: firm age, ownership, region and industry.	A 10 per cent increase in the share of employees using computers results in 2.1 per cent higher labour productivity in manufacturing and 1.5 per cent in services. A 10 per cent increase in the share of employees using Internet results in 2.9 per cent higher labour productivity in manufacturing (no significant impact on services).
Brynjolfsson & Hitt (2002)	US manufacturing panel, 8 years (1987–1994); 527 large firms	Total factor productivity growth	Computer capital	Total factor productivity growth framework.	Gain in multifactor productivity over time from 1.9 per cent for 1 year differences to 5.3 per cent for 7 year differences in computer capital investments.

References

Atkinson RD and McKay A (2007). *Digital Prosperity: Understanding the Economic Benefits of the Information Technology Revolution*. Washington, DC: Information Technology and Innovation Foundation.

Atrostic BK and Nguyen SV (2002). *Computer Networks and US Manufacturing Plant Productivity: New Evidence from the CNUS Data*. Center for Economic Studies, US Census Bureau. Washington, DC.

Atrostic BK and Nguyen SV (2005). *Computer Investment, Computer Networks, and Productivity*. Center for Economic Studies, US Census Bureau. Washington, DC.

Baily MN (1986). Productivity growth and materials use in US manufacturing. *Quarterly Journal of Economics*, Vol. 101, No. 1, pp. 185–196.

Bayoumi T and Haacker M (2002). It's not what you make, it's how you use IT: measuring the welfare benefits of the IT revolution across countries. Discussion Paper No. 3555, International Macroeconomics, Centre for Economic Policy Research. London.

Bloom N, Sadun R and van Reenen J (2005). It ain't what you do it's the way that you do IT: testing explanations of productivity growth using US affiliates. Centre for Economic Performance, London School of Economics.

Brynjolfsson E and Yang S (1999). The intangible benefits and costs of computer investments: evidence from financial markets. Working Paper, MIT Sloan School of Management.

Brynjolfsson E and Hitt LM (2002). Computing productivity: firm-level evidence. *Review of Economics and Statistics* 85(4): 793 – 808.

Carr N (2003). IT doesn't matter. *Harvard Business Review*.

Carr N (2004). *Does IT Matter? Information Technology and the Corrosion of Competitive Advantage*. Harvard Business School Press.

Chinn MD and Fairlie RW (2006). ICT use in the developing world: an analysis of differences in computer and internet penetration. National Bureau of Economic Research, NBER Working Paper 12382.

Clarke GRG and Wallsten SJ (2004). Has the Internet increased trade? Evidence from industrial and developing countries. World Bank Policy Research Working Paper 3215.

Crafts N (2002). The Solow Productivity Paradox in Historical Perspective. Centre for Economic Policy Research Discussion Paper No. 3142. London.

Criscuolo C and Waldron K (2003). *E-Commerce and Productivity*. Economic Trends 600, UK Office for National Statistics.

de Vries G, Mulder N, Dal Borgo M, Hofman A, Aravena C and Hurtado C (2006). ICT investment and its contribution to economic growth in Latin America. Presentation by Nanno Mulder to the UNCTAD expert meeting in support of the implementation and follow-up of WSIS: using ICTs to achieve growth and development, 4– 5 December 2006.

de Vries G, Mulder N, Dal Borgo M and Hofman A (2007). ICT investment in Latin America: does IT matter for economic growth?. Presentation to Seminario Crecimiento, Productividad y Tecnologías de la Información 29– 30 March 2007.

Eurostat (2003). The Observatory of European SMEs, No. 7: SMEs in Europe 2003, accessed online at http://ec.europa.eu/enterprise/enterprise_policy/analysis/doc/smes_ observatory_2003_report7_en.pdf.

Farooqui S (2005). *Information and Communication Technology Use and Productivity*. Economic Trends 625, UK Office for National Statistics.

Gillett SE et al. (2006). Measuring broadband's economic impact, accessed online at www.eda.gov/PDF/MITCMUBBImpactReport.pdf.

Gordon R (1999). *Has the "New Economy" Rendered the Productivity Paradox Obsolete?*, Northwestern University and NBER, Cambridge, MA.

Hagén H-O and Zeed J (2005). Does ICT use matter for firm productivity?. Yearbook on Productivity 2005, Statistics Sweden.

IMF (2001). World Economic Outlook.

Jorgenson DW and Stiroh KJ (1995). Computers and growth. *Economics of Innovation and New Technology 3*, 3–4, pp. 295-316.

Jorgenson DW and Vu K (2005). Information technology and the world economy. *Scandinavian Journal of Economics*, vol. 107, no. 4, pp. 631–650.

Koellinger P (2006). *Impact of ICT on Corporate Performance, Productivity and Employment Dynamics*. Special Report of the European Commission Enterprise & Industry Directorate General No. 01/2006, European e-Business Market Watch.

Lee IH and Khatri Y (2003). Information technology and productivity growth in Asia. IMF Working Paper WP/03/15.

Maliranta M and Rouvinen P (2003). Productivity effect of ICT in Finnish business. Discussion Paper No. 852, Research Institute of the Finnish Economy.

Maliranta M and Rouvinen P (2006). Informational mobility and productivity: Finnish evidence. *Economics of Innovation and New Technology*, Vol. 15(6), September.

OECD (2004). *OECD Information Technology Outlook 2004*. Paris.

OECD (2006). *OECD Information Technology Outlook* Paris.

OECD (2007). *Factbook 2007: Economic, Environmental and Social Statistics* Paris.

Oliner SD and Sichel DE (1994). Computers and output growth revisited: how big is the puzzle?. Brookings Papers on Economic Activity 2, 273–317.

Oliner SD and Sichel DE (2002). *Information Technology and Productivity: Where Are We Now and Where Are We Going?*. Federal Reserve Board, Washington DC.

Schreyer P (2000). The contribution of information and communication technology to output growth: a study of the G7 countries. OECD Science, Technology and Industry Working Papers, 2000/2, OECD Publishing.

Seo H-J and Lee SL (2006). Contribution of information and communication technology to total factor productivity and externalities effects. *Information Technology for Development*, vol. 12 (2), pp. 159-173.

Shiu A and Heshmati A (2006). Technical change and total factor productivity growth for Chinese provinces: a panel data analysis. Ratio Working Papers 98, Ratio Institute. Stockholm.

Smutkupt P and Pooparadai K (2005). ICT indicators initiatives in Thailand: progress and lessons learned. Paper distributed at "Measuring the Information Society", WSIS Thematic Meeting organized jointly by the members of the Partnership on Measuring ICT for Development: ITU, OECD, UNCTAD, UIS, UN Regional Commissions, UN ICT Task Force and World Bank, Geneva, 7–9 February 2005.

Stiroh KJ (2002). Information technology and the US productivity revival: what do the industry data say?. *American Economic Review*, 92(5). pages 1559-1576, December.

Triplett JE (1999). Solow productivity paradox: what do computers do to productivity?. *Canadian Journal of Economics*, 32 (2), April, pp. 309–334.

UNCTAD (2008, forthcoming). *The Impact of ICT Use in Manufacturing Firms in Thailand.* Joint UNCTAD–Thailand National Statistical Office project.

UNCTAD (2000). *E-Commerce and Development Report 2000.*

UNCTAD (2003). *E-Commerce and Development Report 2003.*

UNCTAD (2006). *Information Economy Report 2006.*

United Nations (2005). *Core ICT Indicators.*

van Ark B, Inklaar R and McGuckin RH (2003). ICT and productivity in Europe and the United States: where do the differences come from?. *CESifo Economic Studies*, vol. 49, 3/2003, pp. 295–318.

van Ark B, Melka J, Mulder N, Timmer M and Ypma G (2003). ICT investments and growth accounts for the European Union. Final report on "ICT and growth accounting" for the DG Economics and Finance of the European Commission. Research memorandum GD-56, Groningen Growth and Development Centre.

White H. (1980), A heteroskedasticity-consistent covariance matrix and a direct test for heteroskedasticity, *Econometrica*, Vol. 48, pp. 817–838.

Notes

1. Chapter 2 of the present report deals with the ICT producing sector in detail.

2. See, for example, Atkinson and McKay (2007) and Bayoumi and Haacker (2002).

3. Chapter 1 of IER 2006 (UNCTAD, 2006).

4. Initial studies focused on estimating the effect of IT on growth rather than that of ICT. As technology evolved, concepts also developed to better capture technology. Today the OECD definition of the ICT sector (see chapter 2) is the reference for measuring the impact of ICT on growth in a way that enables international comparisons.

5. For more detailed information see UNCTAD (2008, forthcoming).

6. For more information see http://measuring-ict.unctad.org.

7. A more comprehensive literature review of the theoretical concepts was carried out in Chapter 1 of IER 2006, and in Chapter 2 of ECDR 2003.

8. For example, Oliner and Sichel (1994) and Jorgenson and Stiroh (1995).

9. Quoted by Triplett (1999), who shows updated tables from the original authors.

10. Rather than capturing investment growth rates defined as changes in gross capital formation, this table shows the increase in the share of ICT capital inputs in total investment.

11. This time lag of one decade was considerably shorter than the time lag needed by developing countries during previous technological revolutions (Crafts, 2002).

12. For example, in the Republic of Korea venture capital investments in ICT represented 40 per cent of all venture capital investment (1999–2002) as compared with the 50 per cent OECD average for the same period (OECD, 2004).

13. More information on the ICT-producing sector can be found in chapter 2.

14. See also discussion in the introduction of this Report on the contribution of technology to growth in the newly industrialized countries in South-East Asia. Similar findings by various authors indicated that the major contribution to growth in those countries came from capital and labour accumulation, with technology playing a minor part. However, it has been counter-argued that technology enhanced labour productivity, thus indirectly contributing to growth in a significant way.

15. ECDR 2003, chapter 3.

16. For example Atkinson and McKay (2007).

17. For the purpose of that comparison they used both the demand and the production sides of the GDP national account.

18. Argentina, Brazil, Chile, China, Colombia, Egypt, India, Indonesia, Malaysia, Philippines, Singapore, South Africa, Taiwan Province of China, Turkey and Venezuela.

19. UNCTAD recommends that countries give priority to gathering data on the core ICT indicators (United Nations, 2005).

20. For example, Atrostic and Nguyen (2005), Bloom, Sadun and van Reenen (2005) and Farooqui (2005).

21. More information on the core ICT indicators can be found in chapter 1.

22. Measured as placing or receiving orders on line in Criscuolo and Waldron (2003).

23. For example, Brynjolfsson and Hitt (2002).

24. More information on this can also be found in UNCTAD (2008, forthcoming).

25. The Information for Development Programme (InfoDev) launched in 2006 a project for measuring the impact of ICT use in Poland, the Russian Federation and the Baltic States.

26. All descriptive statistics in this section were constructed with sampling weights.

27. The information presented here is not directly comparable with the data in chapter 1 on the use of ICT by businesses in Thailand since those businesses comprise only firms located in urban areas.

28. For example 93 per cent of all European enterprises have fewer than 10 employees and employ 34 per cent of the workforce (Eurostat, 2003).

29. A similar approach was used in Maliranta and Rouvinen (2006) with respect to certain complementary computer characteristics (processing and storage capacity, portability, wireless and wireline connectivity).

30. Total employment in this study refers to paid and unpaid persons engaged.

31. See for example, Atrostic and Nguyen (2002).

Chapter 4

ICT, E-BUSINESS AND INNOVATION POLICIES IN DEVELOPING COUNTRIES

A. Introduction

As economies, mostly in the developed world, but increasingly in developing countries too, move towards an even more knowledge-intensive stage of evolution, understanding the role of the technologies that facilitate the accumulation, diffusion and absorption of knowledge in innovation-driven competition becomes crucially important. In particular, knowledge has long been recognized as one of the key factors underpinning innovation and, through it, economic growth and development in market economies.

The Introduction to this Report explores the general issues of knowledge generation and accumulation, and the conditions under which the results of scientific and engineering endeavours translate into business competitiveness at the micro level and growth and development at the macro one, including the cross-border distribution of the benefits of scientific and technological progress. Focusing on a specific aspect of technological development, this chapter looks at the role in that broad picture played by information and communication technologies (ICTs), which are generally considered to have been the prime mover in the powerful wave of innovation that transformed the global economy over the last quarter of the 20th century.[1]

The importance of innovation policies

Innovation, in a narrow sense, can be defined as the introduction by an enterprise of an entirely new product, service or productive process not previously used by anybody else. In a broader sense, the concept of innovation can be applied to the activities of enterprises which introduce for the first time products or processes that were not present in the particular market in which they operate, or in certain contexts, were merely new to the firm.[2] In that broader sense, innovation can also take place when new products or processes spread through imitation or when gradual improvements resulting in higher productivity are made. Innovation matters because, thanks to its positive effects on productivity, it is the major driver of long-term increases in output per worker and hence it determines the ability of the economy to support better living standards for all. Chart 4.1, taken from Hall and Mairesse (2006), illustrates the major relationships between innovation, productivity and firm performance.

Public policies in support of the innovative activities of enterprises are justified by the existence of a number of possible market failures that could result in private investment in innovation that would fall below the socially desirable levels. Such market failures can affect both financial markets and markets for goods and services. Intellectual property regimes and cross-border investment rules also play fundamental roles in the generation and diffusion of knowledge and innovation. The existence of a strong national scientific base is another crucial reason for policy interventions to support innovation.

R&D is often highly capital-intensive. At the same time, the uncertainty of innovation and the existence of large asymmetries in the information that is available to innovators and to the providers of capital make it more costly for innovative enterprises, particularly the smaller ones, to access capital to fund R&D activities and/or the practical implementation of their outcome. The fact that knowledge spillovers make it impossible for the financial returns of innovations to be fully appropriated by innovative firms adds to the risk that the operation of financial markets may result in suboptimal levels of R&D investment.

Failures that affect the operation of markets for goods and services may also hold back innovation. Competitive pressure is a strong motive for innovative behaviour among enterprises; removing factors that unnecessarily restrict competition may therefore provide a way to accelerate innovation. The reduction of the impediments that limit market size and hence the opportunities for innovators to recoup the sunk

Chart 4.1

Innovation and productivity

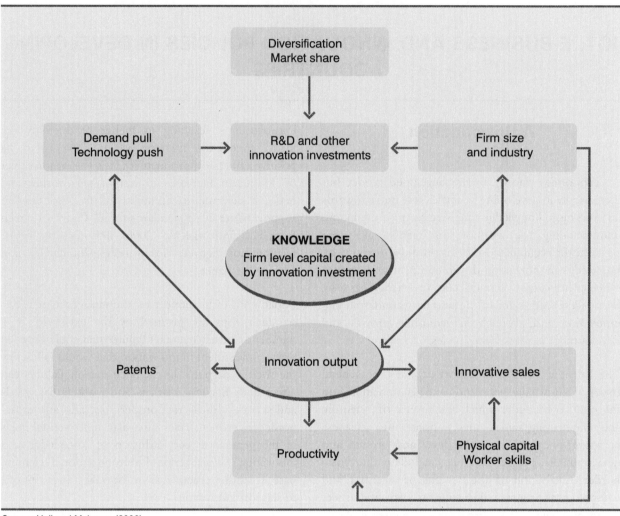

Source: Hall and Mairesse (2006)

cost of R&D activities is another measure to consider in innovation-supporting policies.

Patents, the main instrument of intellectual property regimes, offer innovators a time-bound monopoly in the commercial application of the knowledge they produce. This enables them to recover the investment that R&D activities represent and may thus provide an incentive for innovation. However, innovation can happen, and has happened, in the absence of strong patent protection.   A strong patent regime makes it more costly for other potential innovators to build on existing knowledge in order to generate their own innovation. The balance between the benefits and costs of patents from the point of view of the public interest (i.e. to support innovation and to facilitate knowledge spillovers) is not a straightforward one.

Knowledge spillovers may derive from the activities of publicly funded academic research centres and also from private-sector, commercially motivated research. Transnational corporations tend to be more active and obtain better results in their innovation efforts than domestic firms. [3] But it does not necessarily follow from this that policies to attract foreign direct investment (FDI) are the only, or even the best, instruments that developing countries have to facilitate the dissemination of innovation spillovers in their economies, because (a) transnational corporations are also better than domestic firms in preventing knowledge spillovers, [4] and (b) domestic enterprises in developing countries often lack the capacity to absorb potential spillovers without outside support. In order to diffuse knowledge and innovation across the economy, FDI policies are therefore often complemented with interventions to facilitate the interaction between local firms and innovative international firms. From the point of view of the reinforcement of the absorptive capacity of local firms, it is important to facilitate the emergence of a solid scientific base at the national level. Even in a highly globalized world, knowledge transmission often takes place within relatively small networks and does not always flow easily across borders.[5]

The application of ICTs to productive, financial, administrative and commercial activities has been particularly fertile in terms of innovation. ICT use by enterprises has helped them to become more efficient through business-process innovations (for example, electronic data interchange), and it has also resulted in the emergence of entirely new products or services (for example, online banking).

The importance of the dynamic relationship between the use of ICT and innovation is increasingly recognized at the policy level. A sign of this is the abundance of initiatives at all governmental levels aimed at supporting ICT-driven innovation. Another one is the emerging trend for policies to promote ICT use by enterprises on the one hand and policies to foster innovation on the other hand to be entrusted to the same policymakers or placed under the same overall political responsibility. For example, in around half of the countries covered in a recent survey by the European Union the ministry or agency that is responsible for innovation policy is also involved in e-business promotion.[6] Most often, innovation policies include or address aspects of the use of ICT by businesses. However, even when innovation and policies related to the use of ICT by enterprises share the same institutional framework, it is not necessarily the case that they are envisioned as a single set of policy objectives with a coherent arsenal of policy instruments to achieve them. The borderlines are uncertain and ministries and agencies dealing with matters such as industry, small and medium-sized enterprises (SMEs), education, scientific research and others may be involved at various levels.

Earlier editions of UNCTAD's *Information Economy Report* and its predecessor, the *E-Commerce and Development Report* have covered extensively the general effects that ICTs have had as drivers of globalization. They have also discussed specific aspects of how enterprises in developed and developing countries need to adapt their operations in order to ensure sustainable competitiveness in an environment that is heavily influenced by ICT considerations. This chapter complements that coverage, looking (from the development and policy perspectives) at the linkages between ICT adoption in general and e-business in

particular, and the facilitation of innovation processes in enterprises with regard to both ICT- and non-ICT-related aspects of their operations. The thrust of the chapter will be a description of existing best practices with a view to facilitating the choice of better support instruments and policies by developing countries.

B. Impact on innovation of the use of ICTs by enterprises

Economic globalization has significantly increased the competitive pressure on enterprises in many sectors. This comes as a result of, among other factors, the emergence of new, lower-cost producers, fast-changing demand patterns, increased market fragmentation and shortened product life cycles. In such an environment, innovation (either in terms of business processes or of final products and services) becomes crucial for the long-term competitiveness and survival of enterprises. It also enables them to climb the value ladder, a particularly important consideration for the enterprises of many developing countries. At the same time, the enterprises of those countries, particularly SMEs, face serious difficulties in benefiting from ICT-led innovation. For example, since R&D involve high fixed costs, they are intrinsically a high-risk activity and are subject to economies of scope that favour larger firms. Other general features of SMEs such as greater vulnerability to the essentially unpredictable market responses to innovative activity, or the greater difficulties they face in accessing financial and human capital, place them in a disadvantaged position with regard to engaging in innovative activities. When it comes to ICT-based innovation, policymakers need to take into account the general difficulties encountered by enterprises in developing countries, particularly in connection with ICT access and use by enterprises.[7] This section will consider the ways in which ICTs affect innovation, present examples of how ICTs may contribute to innovation in its different aspects, and explore a number of implications for developing countries.

How are ICTs connected with innovation?

Chart 4.2 presents a classic model of the interactions between markets, science and technology, and innovation in the context of the operation of markets. ICTs (irrespective of whether they are regarded as commodified tools or as drivers) can be introduced at any point in that model in order to bring about

Chart 4.2

Innovation system diagram

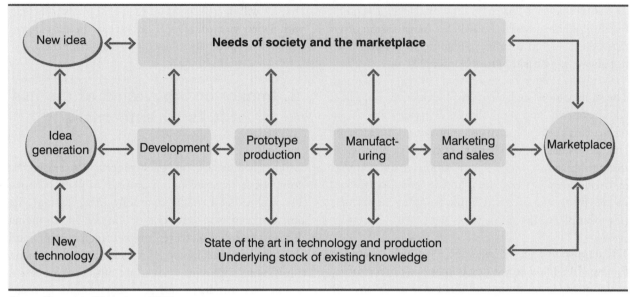

Source: Mowery and Rosenberg (1978).

innovation – product or process. They can be applied in, for example, marketing or communication, supply chain management, knowledge management, recombining existing technologies and product development.

Often ICT adoption by enterprises is regarded as a self-contained event that takes place in the course of an enterprise's development within a given technological, commercial and business environment. From that viewpoint, ICT use in businesses' operations would not be considered fundamentally different from, say, the purchase of more up-to-date machinery. In reality, however, if the full potential of ICT use in enterprises is to be realized, it must be understood and implemented as an integral part of a much broader process of transformation of business structures and processes, a transformation that necessarily involves the emergence of new or significantly improved products or services, and/or the implementation of new productive or managerial methods, namely innovations.

This chapter started with a reference to the value-creation role of knowledge as source of innovation and the central function that the latter performs in the development of a market economy, a view articulated in a path-breaking manner by Schumpeter (1943). However, the knowledge that is available to a society at a given point in time and the knowledge that is actually involved in innovation have not historically been identical magnitudes. Before the explosive development of ICTs, information and knowledge were much more

costly to acquire and, in particular, to disseminate. The consequence was that they were much less available to enterprises and society at large, and therefore the pace of innovation was slower: only a few people would have the necessary knowledge, and decision-making processes would be strongly hierarchical (which may have been the most socially efficient solution in an environment of high cost of information). The link between innovation and information remained at the top socioeconomic layers and reached the mass of enterprises only after significant delays. With the fall in the cost of access to information the situation changes radically. The possibility of innovating is open to much wider strata of economic actors. Indeed, for many firms innovation becomes not a possibility, but a necessity. When everyone can innovate everyone must innovate, or risk being undercut by competitors. Abundant information creates a pro-innovation force by itself.

While in an environment of scarce, costly information innovation tended to result more or less exclusively from advances in technology embodied in material capital (better machines), the abundance of information makes innovation in intangibles (for example, brands or business models) increasingly important, hence the importance of innovation in business processes and structures. This is particularly visible in the flattening of organizational hierarchies. As enterprises become more aware of the importance of innovation for their competitiveness, and the need to be proactive innovators, interactions among different categories of employees and managers evolve. Information has to be shared much more widely than in the past and ICT-enabled business processes tend to generate information flows that ignore formal hierarchies.

It is easy, almost trivial, to explain the many ways in which adopting ICTs can improve business performance, particularly by allowing companies to operate differently and at lower costs, and also to bring new products to the market. However, determining with precision what is going to be the net financial impact of particular ICT implementations, choosing the links in the value chain where interventions could yield better results and deciding on the sequencing and pace of adoption are not necessarily equally simple tasks. A decision to put in place e-business methods at a particular point in the diagram in chart 4.2 may require matching investments in skill acquisition, business process re-engineering or communications. All of those complements to ICTs represent additional expense, but without them e-business benefits may not materialize and the endeavour may worsen, rather than improve, the company's competitive stance.

Research tends to support the concept that e-business cannot be considered in isolation from other aspects of business change.[8] A common finding is that the positive effects of ICT use can include increased revenue and market share, greater adaptability to market conditions, shorter product development cycles, improved quality of after sales and, as a result, greater profitability. At the same time, research indicates that companies that complement their e-business projects with investments to enhance their workers' skills and to facilitate business process change are the ones that reap most of the benefits of e-business. In the absence of those conditions, the impact of ICT on the performance of enterprises tends to be significantly less.

Why ICT still matters for innovation

It has been argued that ICTs are now so widely prevalent in developed economies that their adoption can no longer be considered to be able to make a noticeable difference in innovation-based competitive strategies. According to this view, ICTs, like other "infrastructure" technologies (such as electricity), would have become essential for business operations and, at the same time, irrelevant in terms of strategic advantage.[9] UNCTAD data in chapter 1 of this Report show that approximately two thirds of enterprises in developed countries have an interactive website, although the average figure for

all the countries (including developing countries) for which information is available is around 30 per cent. Other UNCTAD data in the same chapter show how for the sampled countries only an average of 13.9 per cent sell online and 22.5 per cent place orders via the Internet. Therefore, it seems that any possible signs of a "commoditization" of information technologies are most likely not applicable to developing countries. Furthermore, even in the context of an advanced economy the effects of ICTs on innovation and hence on competitiveness depend on the extent to which the implementation of those technologies has led to actual changes in business processes. Large variations seem to persist in the modalities by which enterprises incorporate ICTs into their competitive strategies, including innovation-related activities. Another reason for caution is that while the theoretical connections between innovation and ICTs are not difficult to establish, there is some difficulty in gathering hard statistical information about how those two phenomena interact in the real life of enterprises. The statistical measurement of e-business is still in its early stages of development, the notions involved are not always homogeneously understood and technological progress keeps moving the goalposts.

Changing the innovation paradigm

In spite of the above considerations, it can be argued that there are fundamental differences between ICT and other "infrastructure" technologies that make ICT a force for change in production/exchange processes and an accelerator of innovation. In fact, what seems to be happening is that the contribution that ICT makes to the competitiveness of enterprises is not decreasing; on the contrary, the gap between the potential of ICT and the value it delivers to organizations is actually growing (Brown and Hagel, 2003). For example, in a recent survey of innovative enterprises in Europe, around half of all innovations introduced in recent years were ICT-enabled, and they seemed to be at least not inferior to other innovations in terms of impact on enterprise profitability. While ICT-enabled innovations were more visible with regard to process innovation compared with product and service innovation, the latter seemed to have a greater impact on profits (Koellinger, 2005).

In this context it can be useful to refer to the concept of techno-economic paradigm, according to which a widespread set of related innovations, developed in response to a particular cluster of technical problems and using the same scientific principles and organizational methods, provides the basis for scientific and productive activity at a particular point in time (Freeman and Perez, 1988). ICTs have profoundly changed the techno-economic paradigm within which innovation takes place today in developed and more advanced developing countries. Prior to those changes innovation revolved around concepts of mass production, economies of scale and corporate-dominated R&D. In the last three decades of the 20th century this has been replaced to a large extent by an emphasis on economies of scope, exploiting the benefits of interconnected, flexible production facilities and greater flexibility and decentralization of R&D and development. Flexibility, interconnectedness and collaboration rely on ICTs, which also play a fundamental role in facilitating research diversification and collaborative, interdisciplinary approaches.

Since the early 1990s, multinational corporations from developed countries and increasingly some advanced developing countries have increased their cross-border investment in the field of R&D, including in R&D facilities located in developing countries. For example, table 4.1, which is taken from a report published by Booz Allen Hamilton (2005), lists the top 10 global companies by R&D expenditure in 2004 and shows how practically all of them have established research facilities in developing countries. Those investments have acquired a greater significance than that of their

previous role as mere agents of transfer of technology to the local market, becoming the locus of technology development for local, regional and increasingly global markets. The fast-moving integration of those research activities has given rise to the emergence of global R&D networks (Serapio and Hayashi, 2004).

ICTs also enable faster cross-border knowledge dissemination, particularly within transnational corporations, but also by facilitating networking and partnering among smaller players. By investing in ICTs enterprises improve their capacity to combine disparate technologies in new applications. This is important not only from the point of view of ensuring that firms achieve an adequate spread of internal technological undertakings but also from the point of view of the need to engage in R&D partnerships. In that regard, the major benefit of adopting ICTs may not necessarily derive from the technology per se but from its potential to facilitate technological recombination and change.

The relative decline of economies of scale as the massively dominant factor of competitiveness provides another justification for the role of ICT in the emergence of a new innovation paradigm. Product specialization, which through mass production generates competitive advantage, can be complemented in terms of competitive strategies by product or market diversification, which places a premium on the notions

Table 4.1

Geographical distribution of R&D facilities of the top ten R&D companies in 2004

Company	Home country	Locations of R&D facilities	Newest locations
Microsoft	United States	United States (3), China, India, United Kingdom	India
Pfizer	United States	Unites States (5), Canada, China (planned), France, Japan	China
Ford	United States	United States, Germany	Germany
DaimlerChrysler	Germany	Germany (4), United States (4), China, India, Japan	China, Japan
Toyota	Japan	United States (2), Europe (2), Australia, Thailand	Thailand
General Motors	United States	11 centres, including in Australia, Brazil, China, Germany (planned), Republic of Korea, Mexico, Sweden (planned), United States	Germany, Sweden
Siemens	Germany	150 centres in 30 countries (15 per cent in Germany)	China, India, Russian Federation
Matsushita Electric	Japan	10 sites in the United States and centres in Canada, China, Europe and Malaysia	China
IBM	United States	United States (4), China, India, Israel, Japan, Switzerland	India
Johnson & Johnson	United States	Australia, Belgium, Brazil, Canada, China, France, Spain, Switzerland, United Kingdom, United States	United States

Source: Booz Allen Hamilton (2005).

of flexibility, customization and adaptability for which ICTs are particularly fitted.

Much as the capital goods sector provided the means by which innovation moved across sectors and industries in earlier innovation waves based on the steam engine and electrification, ICTs are playing a crucial transmission role in the new innovation paradigm. But, instead of being a mere instrument for the transmission of knowledge about how to apply technology to various fields of activity, ICTs facilitate the connection of different fields of innovation potential (inside the firm or among members of value networks). This results in broader scope for innovation.

The dominance of ICTs in the new innovation paradigm also influences the geographical distribution of innovative activities and centres of excellence. Companies that fully understand and implement ICT-enabled strategies can become capable of participating in geographically complex international networks made possible by, among other factors, the new combinations of activities, whose respective centres of excellence may have been sited in distant locations. ICT-intensive innovators can be effective in managing those geographically dispersed networks, and in recombining formerly distinct and distant learning processes. This observation is mostly applicable to the manufactures industries in general and more particularly the ICT equipment sector, in which global learning networks are a common consequence of participation in global networks for the sourcing of technology – a case in point being that of original equipment manufacturers in Taiwan Province of China and in mainland China. In this way, ICTs can be seen as the string that bundles together three trends in innovation in the globalizing economy that were sketched out in the preceding paragraphs: the emergence of R&D networks that bind specialized poles of innovation, the diversification of innovation activity at the level of the firm, and the emergence of technology-focused inter-firm alliances

Some implications for developing countries

What does all this mean for developing countries? Enterprises in those countries face particularly difficult conditions in their innovation activities. Major constraints include the following: in most cases they cannot rely on a strong local scientific research community; they have to be able to make a profit while operating in narrow, low-income markets; and they lack the funds to support in-house R&D activities, with

little or no venture capital ready to finance innovations. Clearly, ICTs do not provide immediate responses to those problems. However, some of the features of the ICT-led innovation paradigm that were discussed above could help create a more innovation-friendly environment in developing countries.

In the first place, the reduction in the cost of accessing and processing information and knowledge can help reduce the cost of generating the local scientific base of innovation. ICTs' significant enhancement of scientific-content accessibility should not be underestimated. The means for those involved in R&D in locations away from the leading scientific centres to remain abreast of the latest developments in any particular field of study are today vastly superior and cheaper than they were just two decades ago. In addition to providing easier access to the outputs of scientific research, ICTs are generating, stocking and enabling the manipulation of enormous databases that, by themselves, represent an opportunity for accelerated progress in natural and social sciences. At the same time, one should bear in mind that the generation and the accumulation of knowledge constitute complex social process whose success is not determined merely by the material availability of information. The tacit component of knowledge remains much less amenable to transfer through ICTs than its codified one. History, institutions, and even the physical and geographical conditions in which enterprises in developing countries operate, have a considerable influence on their capacity to move from acquiring information to learning.

Over a shorter time-span, the ICT-facilitated trend towards decentralization and diversification of research and innovation can benefit those developing countries where there is already some innovative activity by facilitating their integration into global learning networks and technology-based alliances. As technological change accelerates, competitive pressures force companies to augment their knowledge and capabilities. An increasingly important way to do that is to use "open innovation", a term that describes how even companies that are leaders in their field must complement their in-house R&D efforts with technologies developed by others, and open up their own technological knowledge to outsiders (Chesbrough, 2003). Traditional, inward-oriented technological capacity-building efforts tend to be complemented or even replaced with more outward-looking strategies that rely on the output of networks of universities, joint ventures, start-ups, suppliers and even competitors. This often results in technology-based alliances, which are being more and more used as instruments of learning.[10] In addition to providing a fundamental means of cooperation in R&D within alliances, ICTs themselves are among the technologies that are more frequently the subject of technology-based alliances.

Technology-based alliances are increasingly important means of facilitating cross-border technology and know-how transfer. In 2000 estimates of R&D activity that were based on the number of patents indicated that only about 10 per cent of the technology development work of transnational corporations was done away from their home country, and only about 1 per cent was taking place in the developing countries (Archibugi and Iammarino, 2000).   More recently, UNCTAD (2005) indicated that the share of developed countries in R&D expenditure had fallen from 97 per cent in 1991 to 91 per cent in 2002, while that of developing Asia had risen from 2 per cent to 6 per cent. According to Narula and Sadowski (2002), over 93 per cent of the technology partnerships they studied were between developed countries. The partnerships undertaken by firms from developing countries were signed almost exclusively with firms from developed countries. The countries most actively engaged in those agreements are the East Asian newly industrializing countries, together with several economies in Central and Eastern Europe. The participation of African firms was negligible.[11]

On the other hand, UNCTAD (2005) found that R&D was internationalizing rapidly and that between 1993 and 2002 the R&D expenditure of foreign affiliates worldwide increased from an estimated $30 billion to $67 billion (or from 10 per cent to 16 per cent of global business R&D). While the rise was relatively modest in developed host countries, it was quite significant in developing countries: the share of foreign affiliates in business R&D in the developing world increased from 2 per cent to 18 per cent between 1996 and 2002.

UNCTAD (2005) also indicated that the reported increase in the internationalization of R&D was affecting developing countries in a very uneven way, with developing Asia as the most dynamic recipient by far. The case of the R&D facilities established in India by a number of leading foreign companies suggests some factors that, over and above the lower salaries of scientists and engineers, may account for the decisions to engage in ICT innovation activities in developing countries. Such factors are often considered to include the existence of good-quality scientific and technical universities, which in turn results in the availability of a sufficient number of science and engineering graduates, and the pre-existence of a tissue of commercial and

academic entities that are relevant to the relevant area of work.

The interplay of the factors that have made possible the emergence of ICT R&D poles in some developing countries is a complex one. Archibugi and Pietrobelli (2002) suggest that the arrival of a leading foreign technology firm may "generate externalities and induce the public sector to give prominence to associated Faculties and other public research centres". They quote as an example the R&D facility established by Texas Instruments in Bangalore as long ago as 1985, which specialized in design circuits. While it is not possible to ascertain whether the specialized ICT hub that currently exists in Bangalore would have developed irrespective of that decision, it seems clear that the large number of ICT firms that have been created and developed there would not have emerged without active public support policies for the ICT sector, particularly those that relate to the generation of a large pool of qualified engineers.

ICTs also facilitate and influence the location decisions in relation to R&D activities of enterprises from developing countries, which may sometimes choose to locate some facilities in developed countries. This may be done in order to facilitate the absorption of best practice and its transfer to the home country production sites. For example, according to Serapio and Dalton (1999), firms from the Republic of Korea owned more R&D facilities in the United States than companies from several developed countries with economies of comparable size. Those facilities were mainly operating in the field of ICTs. More recently, UNCTAD (2005) reported that the foreign R&D activities of developing country TNCs were growing rapidly. That trend is driven by the need to access advanced technologies and to adapt products to major export markets. Some of those TNCs are targeting the knowledge base of developed countries, while others are setting up R&D units in other developing economies.

Open innovation approaches

Open source software represents an innovation-oriented form of ICT-focused collaboration that is of increasing interest to enterprises in developing countries. In this case, ICTs provide not only the origin of the conceptual framework for a new mode of production and dissemination of knowledge and innovation, but also the main tools through which that framework can be put into operation.

Open source software, which consists of collaborative projects involving people located around the world and who, had it not been for the Internet, would never have had an opportunity to pool their knowledge, is one of the major examples of Internet-driven social and economic change. Free and open source software (FOSS) provides the most visible model of that trend. A discussion of the economic rationale for the open source software model and the advantages it offers from the point of view of developing countries is to be found in UNCTAD (2003).

With regard to innovation, FOSS gives developing countries the opportunity to benefit in a number of ways. In the first place, thanks to FOSS, software developers are able to enhance their skills and to improve more easily on ICT (software)-based business models. Participation in FOSS projects facilitates the creation and consolidation of local pools of expertise in ICTs, and given the low barriers to access to that software, enables the emergence of new ICT-related business activities. This may result in the launching of new (innovative) products or services, geared to the specific needs of local demand, that may not have been possible with proprietary software.

FOSS also reduces the cost of access to ICTs, although it should be stressed that cost considerations are not the primary advantage of the FOSS model for developing countries.[12] Since ICTs, as discussed above, are powerful enablers and accelerators of innovation, FOSS can play a crucial role in improving innovation in developing countries by supporting the accumulation and dissemination of knowledge. As FOSS uses open standards, it provides all economic actors with access to this key technology on equal terms. This works against the emergence of monopolies, favours technological diversity and promotes the emergence of new innovative business models.

The approach followed in the implementation of open source software projects is now being applied in fields outside software development. In the case of open source software, individuals cooperate on line with their peers in the production of software that is openly available to anyone who is interested in carrying the work further. The same model of on line collaboration can be applied to any business model in which the end product consists of information.

The open approach to innovation is based on the notion that as knowledge becomes more widely available and increasingly varied in sources and content, companies cannot limit themselves to their own research in order to be innovative and stay competitive. They have to be able to integrate and exploit external knowledge and skills (Chesbrough, 2003). ICTs are used to integrate the originally highly dispersed information and capabilities in a trial process from which innovation emerges. This is as valid for the individual enterprise as for cooperative networks of enterprises organized online. This open cooperation process of production can take place on a volunteer basis, with the end product being made available to everyone at no cost (such as Wikipedia). Alternatively, some form of compensation may be demanded by contributors and the end product may be used in a commercial operation.

Open innovation is related to the idea that competent users should be integrated into the innovation process (von Hippel, 2005). This refers to the trend for users (whether enterprises or consumers) to develop themselves the product or service they need and then freely release their innovation, as opposed to the traditional, manufacturer-centred approach whereby a manufacturer identifies the users' needs, develops the products and profits from selling them. User innovation is common in many fields, from scientific instruments to computer applications (for example, e-mail) or sports equipment (von Hippel, 2005). Users tend to innovate collaboratively in communities, a process rendered much more effective by the Internet. They also tend to grant free access to their innovation because often it does not make sense to protect it. Their major motivation for innovating is the improved usefulness of the product, which can be enhanced even further if other users have access to it so that they can improve on it. In many cases, their reputation as a skilled user-developer also matters strongly to them. The innovativeness of the results of the process can be so great that some suppliers may even decide to concentrate exclusively on manufacturing, giving up all product development and limiting themselves to downloading user-developed designs from user community websites. As an example, von Hippel (2005) mentions cases in fields as different as kitesurfing equipment, spine-surgery devices and toys. It is clear that the social benefits of user-led innovation would be significantly reduced without ICTs that provide users with strong computing capabilities needed for product design at a reasonable cost, as well as fast, cheap and far-reaching means for collaboration in users' communities.

The growing reliance by online media on "user-generated content" and the emerging phenomenon of "crowdsourcing", whereby the Internet is used to involve interested individuals in the definition and development of new products or in the identification of a solution for a particular problem, are similar manifestations of the capacity of ICTs to democratize innovation processes.

Other business models have been developed for using the Internet to harness the resources of large pools of individuals or firms in order to address innovation problems in an open context. This includes NineSigma.com, which claims to have access to a global network of 1.5 million experts. According to the company, 60 per cent of the experts come from companies, 30 per cent are academic researchers and the remaining 10 per cent are with research laboratories, either publicly or industry-funded. Since those experts are free to forward the request for a solution to an innovation problem to other experts they know, as much as 40 per cent of solutions provided come from experts that were not originally contacted by NineSigma. The company reports that solutions have been provided by sources in a number of developing and transition countries, including Bulgaria, China, India, the Russian Federation, the Syrian Arab Republic and Yemen.[14]

Another example is Yourencore.com, which matches demand from companies looking for expertise regarding a particular scientific or engineering problem with supply of expertise from retired scientists and engineers who can thus earn extra income while working according to their own flexible schedule and staying intellectually active in their preferred professional field.

Based in India, Ideawicket.com has recently launched an "Open Innovation Portal", which provides a platform for individual innovators and corporations to exchange their innovation resources and requirements. Innovators are also able to give visibility to their creations and eventually make them available to interested enterprises.

A similar concept with a developing country involvement is the OpenBusiness project launched jointly by Creative Commons UK, Creative Commons South Africa and the Centre for Technology and Society at the FGV Law School in Rio de Janeiro, Brazil.[15] They describe OpenBusiness as a "platform to share and develop innovative Open Business ideas-entrepreneurial ideas which are built around openness, free services and free access".

Box 4.2 gives two concrete examples of how this kind of tool can be used to increase the efficiency of innovation activities even in companies that were well known as highly successful innovators in the pre-Internet paradigm.

Open approaches to innovation can take different forms. In some cases they may consist in a form of outsourcing of part or all of a company's R&D. Contracts and market conditions are the fundamental factors governing such projects. ICTs are helpful, but not necessarily crucial. In other cases, enterprises may go for a closer cooperative arrangement where reciprocity, long-term strategic considerations and

---

Box 4.1

The experience of InnoCentive

"Crowdsourcing" may be considered to be conceptually close to the business model of InnoCentive, an e-business venture that was launched in 2001 by the US-based multinational pharmaceutical company Eli Lilly. InnoCentive uses the Internet to help firms find scientific and technical expertise that they can use to meet innovation challenges. The "seeker" companies (which include very large transnational corporations) post on InnoCentive's website information about the problem they are trying to solve. Any participant in the network can attempt to provide a solution – for which they can be paid anything between $10,000 and $100,000. The occupations of the people who are providing these solutions cover a very broad spectrum, from people without formal expertise to scientists or engineers whose formal occupations do not fully satisfy their intellectual curiosity, or whose career paths have taken them far from their preferred domain of engagement, or others who are interested in the challenge of putting their specialization (say in physics) to use in a field outside its conventional application (biotechnology). Companies active in a completely different field can also provide innovative solutions to the "seeker" problem: a maker of tooth whitener can benefit from the experience of a laundry product maker, and polymers used in the semiconductor industry can help a maker of wrinkle-free shirts. InnoCentive is active in a number of developing and transition countries, including China, India and the Russian Federation. China in particular, where InnoCentive has established partnerships with 26 academic institutions, now represents over 30 per cent of the company's source of "solver" competition. The advantages of the participation of Chinese scientists in schemes such as these are not merely financial, since it allows them to gain exposure to, and an understanding of, the problems and methodologies of international companies in a wide range of industries.[13]

partnership are the keys. ICT-enabled cooperation becomes an important element in that case. Finally, taking the approach one step further they can go for what one could describe as "open access" innovation, along the conceptual lines of the FOSS model. In that case, innovation is the work of networks involving enterprises, users and eventually other stakeholders. Transparency, reputation and trust are the keystone concepts and ICTs are indispensable for the implementation of such projects.

The emergence of an ICT-based paradigm of innovation makes it advisable to gear government intervention to the promotion of cross-firm and cross-border knowledge flows, it being assumed that firms follow the model of a continually interactive search for better methods and improved products, and hence a Schumpeterian search for higher profits through experimental innovation. This could be more socially efficient, particularly from the standpoint of developing countries than the more traditional approach of protecting the exclusive right of the private owners of

the resources invested in the generation of knowledge to protect that knowledge. The next section looks at how Governments can shape their e-business and ICT policies so that their interaction results in faster, more development-relevant innovation.

C. Maximizing synergies between e-business and innovation policies

While e-business is approaching maturity status as an operational technique and there is no controversy about the role of ICTs in recent business innovation waves, the recognition at the policy level of the need to simultaneously consider e-business and innovation issues is recent and incomplete, and to a large extent led by developed countries. As developing countries adapt their national innovation systems to benefit from the dynamic interplay between ICT use by enterprises and innovation-led competitiveness policies they need to be aware of available experience in this regard and

---

Box 4.2

Use of open innovation by Procter and Gamble and Nokia

In 2002, marketing executives at Procter and Gamble, a US-based consumer goods transnational corporation, came up with the idea that a good way to make their potato crisps more enticing to customers would be to have text or images printed on them. When this turned out to be rather challenging technically, the company did not use its extensive internal engineering resources to solve the problem. Instead, it exploited the potential of global networking. A document detailing the technical challenge to be met was circulated throughout the company's global networks of individuals and institutions, the purpose being to find anyone who had already thought about the same problem. Eventually, they found a small bakery in Bologna, Italy, run by a university professor who also manufactured baking equipment. He had invented an ink-jet method for printing edible images on cakes and cookies, and this was rapidly adapted to print on potato crisps. Less than two years later the product was launched and it was delivering high revenue growth to the division concerned.

Procter and Gamble has a long history of in-house generated innovative output. For more than a century the company had followed a policy of heavy investment in internal R&D work. However, in 2000 it decided to radically change its innovation model, from its traditional continuous internal innovation approach to a "connect & develop" model. The reason for this was that the company needed to accelerate its innovation rate, which had stalled at a success rate of 35 per cent. This was considered to be a factor in the fall of its stock price. Since the most successful innovation projects seemed to result from linking ideas across internal businesses, it seemed plausible that an increase in the number of ideas coming from outside businesses would be equally productive. The target was to increase the rate of external innovation to 50 per cent. This was to be done by connecting and collaborating with suppliers, scientists and entrepreneurs and systematically exploring their business environment in search of innovation opportunities to which the company could apply its expertise and capacity. Through this approach, R&D productivity is said to have increased by about 50 per cent in the five years that followed and some 100 new products have been launched, for which some of the development was not done internally. In the words of Larry Huston, Procter & Gamble's vice president of innovation and knowledge, "It has changed how we define the organization. We have 9,000 people on our R&D staff and up to 1.5 million researchers working through our external networks. The line between the two is hard to draw."[16]

Another example of an innovation network is the "Nokia Developer Operations" peer production system for software development operated by the Finnish telecom equipment manufacturer. Nokia has created an electronic platform on the Internet that allows about 2.5 million users to obtain any kind of technical information about Nokia products that they may need in order to develop new solutions. This turns users into partners of the company. This network of users is actually integrating Nokia into a broader open source community, which makes it easier for the company to spot users' changing needs and tastes and to innovate in response. Nokia partners with large and small corporations and asks them to join the network.[17]

adapt the lessons from it to their specific needs and concerns. This section will present the main lines of convergence of e-business and innovation policies, and identify best-practice examples that may be relevant and transferable to developing countries.

Emerging approaches to innovation policy: integrating e-business

Innovation policy aims at increasing the amount and enhancing the effectiveness of innovative activity by enterprises. As discussed above, broadly defined innovative activity not only includes the creation, adaptation or adoption of new or improved products or services, but also refers to many other value-creation processes. Therefore, innovation policy needs to consider not only strictly technological innovation but also organizational (new working methods) and presentational (design and marketing) innovation. Given the strong links between ICT use by enterprises, competitiveness and innovation, it is clear that enterprises need to align policies in those fields.

In practice, the integration of policies to promote ICT use by enterprises within general innovation policies still has some way to go in most countries. Two broad categories of approach can be identified in this regard. A relatively small number of countries, mostly in Northern Europe, have adopted an approach to innovation policy that sees it as a cross-cutting issue that must be included in the policies of individual agencies, together with ICT and e-business policies. This often leads to a reduction in the number of independent initiatives undertaken by individual agencies and ministries. A second, more common approach still sees innovation as a dual phenomenon. On the one hand, it is perceived as being the result of R&D processes, support for which is in the hands of agencies dealing with research and education. On the other hand, it is also considered to be a tool for the enhancement of competitiveness and SME modernization and this aspect of innovation is handled by ministries in the sphere of economics, industry and trade. In this approach, programmes to support innovation may include e-business as one issue among others to be dealt with. The instruments used, however, may not be too different from the ones chosen by policymakers who focus primarily on e-business as an innovation driver. This is particularly true in the case of SME-targeting programmes, since for this group of enterprises adapting to the changes in the competitive environment that have been brought about by e-business is probably the priority innovation strategy.

While the links between the policy framework of innovation and that of e-business policymaking are becoming stronger in a growing number of countries, the range of policy interventions and actors remains wide, particularly since different levels of Governments and quasi-governmental agencies see innovation, competitiveness, e-business and/or entrepreneurship support as crucial to their development objectives. From the substantive point of view, the differentiation between ICT/e-business policies on the one hand and innovation policies on the other remains visible with regard to several important dimensions of the problems in question. For instance, standard innovation indicators are not frequently considered relevant to efforts to benchmark the development of the "information society". On the other hand, it is common to see e-business promotion plans that lack an explicit reference to innovation as an objective, even though different forms of innovation represent the core goal of those plans. In practice, this results in e-business measures that are implemented at the operational level by the same ministries or agencies as innovation measures, ones that address the same target population and share policy instruments, but lack a clear common conceptual framework and formal feedback mechanisms, and waste opportunities for synergies.

The importance of integration and coordination of policies from different ministries and at different levels must be stressed. Since e-business affects many sectors and processes, an integrated approach to the issues is more likely to produce practical results than a piecemeal one. A way to seek coordination can be to bind together all instruments for ICT use promotion and innovation support in one single public sector ministry. However, that does not always produce optimal results in terms of integration. A way to improve that is to bring together in policy formulation all different stakeholders. Up to three levels of policymaking can be distinguished in this field. One includes decision makers in national ministries and agencies. A second level is formed by entities such as internal agency working groups, inter-agency commissions and the like. The third level involves different ICT user groups. While the clear distinction between the policy and the operational levels should be maintained, the involvement of all those stakeholders is necessary, and there should be channels to ensure a reasonable balance between top-down and bottom-up policymaking. It is also important to support networking across sectors and across innovation themes.

Adapting best practice in e-business and innovation promotion

There is a major opportunity for developing countries to learn from best practice in the field of public support related to e-business and innovation. They should explore some of the existing mechanisms of governance and weigh up the pros and cons of pursuing national, regional and local agendas and also exploring alternative ways to distribute public, private or public/private responsibilities for implementation.

The first issue to consider is the integration of the institutional frameworks of innovation and e-business policymaking. In that regard, many of the developed countries, particularly in Europe, have entrusted overall policymaking for innovation and e-business to the same organizations. However, this does not imply that policies are necessarily coordinated. Interestingly, when one considers the case of the European Union, it is noticeable that among the newer member States the dominant trend is to give institutional pre-eminence (in the form dedicated ministries or commissions) to the issues of ICTs, the Information Society or e-business, while in the older member States – the EU 15 – those matters are more likely to be placed in the hands of the institutions generally in charge of innovation, scientific R&D and industry.

Norway and Iceland are examples of countries in which two different ministries are involved in those matters (in Norway the ministries of trade and industry and of education and research; in Iceland the ministries of education, science and culture and of industry and commerce). In other countries innovation and e-business policies are addressed in parallel. This is the case of the Netherlands, where the Ministry of Economic Affairs has responsibility for both policy types, but e-business issues are delegated to an external agency, while other ministries are also involved. This does not, however, ensure policy coordination and the linkage of innovation and e-business policies. As a response to the challenge of integrated governance an Interdepartmental Committee for Science, Innovation and Informatics (CWTI) has been to created to prepare the overall policy strategies on science, research, technology and innovation policy.

An example that has been regarded as representing best practice, addressing both the innovation and the ICT policy areas is the Italian Action Plan for ICT Innovation in Enterprises, which was launched in 2003 by the Ministry for Innovation and Technologies and the Ministry for Productive Activities. In addition to questions of ICT and innovation, the Plan dealt with other elements that play a role in improving the development potential of enterprises, such as education and training, R&D and entrepreneurship. It also included a consultative e-business committee composed of representatives of academia, business associations, financial institutions and trade unions.

Malaysia, a leading performer in terms of innovation and e-business adoption among developing countries, provides an example of efforts to integrate ICT and innovation policymaking. ICT policy-making was transferred from the former Ministry of Energy, Communication and Multimedia to the Ministry of Science, Technology and Innovation (MOSTI). MOSTI's mission statement reads as follows: "harnessing Science, Technology and Innovation (STI) and human capital to value-add the agricultural and industrial sectors and to develop the new economy, particularly through information and communications technology (ICT), and biotechnology".[18] Thus, ICT policy is conceived as an integral part of STI policies. Among other aspects of ICT policymaking, MOSTI is responsible for the formulation and implementation of national policy on ICT and the encouragement of R&D and commercialization in ICT.

The case of Mexico illustrates a different approach to the relationship between innovation and ICT policymaking. In response to strong competition from other developing countries, Mexico has undertaken a number of initiatives to upgrade the innovative potential of its economy. Many of those initiatives to expand its science and technology base are undertaken under the leadership of the National Council for Science and Technology (CONACYT), an agency that reports directly to the President of the Republic. Other government agencies are also implementing programmes to enhance the productivity and competitiveness of the private sector, enhancement being considered one of the strategic objectives of the national development plan. Several of those programmes relate to ICT and e-business issues. There are several other ministries and agencies, in addition to CONACYT, that have responsibilities connected to ICT policies. They include the Ministry of the Economy, the Ministry of Communications and Transport, the Ministry of Education, the development bank and the 32 State governments.

The role of the State governments in the Mexican case is an example of a situation that is found across developing and developed countries. Regional and local governments play an important role in implementing innovation and e-business policies, although in most countries the largest role in policy definition and coordination remains at the national level. Regional and local levels of policy delivery often seem to be highly relevant, as in most countries most enterprises are of local or regional size and only relatively few operate at the national level. Programmes commonly rely on local or regional networks of stakeholders or communities of practice for the delivery of support measures. While many factors influence the ability of firms to innovate, some of the most direct ones are often not easily handled at the national level. Clusters, for example, are a powerful instrument in innovation policies which is normally managed at the regional level. This does not mean that there is no role for policy at higher levels of aggregation: the operation of clusters can be improved through networking, the identification of best practice and common learning. National frameworks provide useful support for those aspects of innovation and e-business policies.

Major support instruments for e-business innovation

Innovation takes place primarily within firms. The priority concern of Governments therefore is to establish and implement sound innovation policies that enable firms to maximize innovation. As mentioned in the introduction to this chapter, creating an environment that facilitates innovation often requires the addressing of a number of market failures that may result in suboptimal levels of innovative activity by the private sector. Those failures may relate to the operation of markets (financial or goods and services) or limit the presence of knowledge spillovers. Instruments commonly used to address such problems include the direct funding and subsidizing of R&D activities, the provision of support for venture capital funds, the facilitation of transfer of technology and partnership networking mechanisms, incubators and clusters, demand-side management, the provision of data, analysis and studies, and the establishment of centres for demonstration and testing. E-business policy interventions are very frequently present in most if not all of those categories of instruments. For example, in R&D funding, public funding is frequently granted for research in areas such as software, expert systems or Internet technologies. E-business is also frequently chosen as a field in which multi-stakeholder partnerships involving enterprises, business associations, educational institutions and government agencies aim at facilitating transfer of technology and innovation.

Funding of R&D activities tends to be the innovation policy instrument most frequently used. Technology transfer, partnerships and networking are other policy tools that are frequently used. Frequently, programmes supporting e-business and innovation cover more than one policy area in scope. Projects and funding for the support of feasibility studies are also well represented, possibly as a consequence of the ongoing experimental nature of e-business and e-commerce development. Box 4.3 presents some examples of specific programmes that link e-business and innovation.

In some cases policy interventions do not establish a clear-cut distinction between e-business goals and broader societal objectives such as the improvement of e-skills, e-government initiatives or ICT infrastructure and access issues. In that context, e-business is sometimes perceived not so much as a policy area in itself but as rather an instrument to achieve goals in respect of a number of Information Society issues.

From the point of view of the relationship with innovation policies, the traditional approach still tends to see e-business more as part of the ICT policy field than as a component of the innovation policy set. One reason for this is that ICT has been a major force for change in many policy areas beyond innovation (health, employment, security, education and so forth). In practice, however, the application of policies that may a priori be categorized as coming within the scope of either innovation or e-business policy often results in the development of responses to economic problems that involve a subject crossing from one policy field to the other. It is increasingly difficult to establish a clear-cut distinction between those two cross-sectoral groups of policies. As empirical evidence for the impact of ICT and e-business on economic performance mounts, the logic of reinforcing innovation policy by incorporating e-business and ICT support policies becomes stronger.

While it can be expected that e-business and innovation policies will tend to become more closely integrated in the medium and long term, it also clear that the connection between e-business policies and general ICT policies will remain stronger in many countries for some time. A reason for this is that the implementation of ICT and e-business investment by a company or government agency may not translate into an immediately noticeable improvement in their capacity to innovate. In fact, the potential to transform business operations may not even be the explicit policy objective of ICT adoption, although in the end their application may generate very significant innovations. Cost-

---

Box 4.3

Examples of programmes linking e-business and innovation

In Chile the Innova programme, which is coordinated by the Ministry of the Economy and involves several other ministries and agencies, has the general objective of promoting technological innovation and includes significant ICT and e-business components. Together with instruments addressing issues of infrastructure and access to ICT, the programme also provides incentives and technical support to help enterprises undertake the process of innovation and organizational change that is needed to successfully absorb information technology and develop innovative business models.[19]

"Webworks" is an Enterprise Ireland regional development scheme. It aims to expand existing clusters of technology-based companies in order to achieve a critical mass in informatics, e-business, digital media or health sciences. This model gives preference to start-ups with R&D capabilities and a large potential for export sales. As part of cluster-building, it provide flexible, highly wired office accommodation and management support.[20]

In Singapore, the Infocomm@SME programme of the Infocomm Development Authority explicitly aims at accelerating both the adoption of ICTs and their innovative use by SMEs. Its action plan combines initiatives to facilitate ICT access and use by enterprises with other instruments aimed at using ICT to catalyse sectoral transformation in specific growth sectors, as well as with programmes targeting SMEs that can use ICT and e-business to improve their innovative performance and become role models for other enterprises.[21]

The United Kingdom's ICT Carrier programme is an example of an initiative aimed at transferring ICT into UK engineering industries and driving its uptake. This is a programme where the tool is ICTs but the objective is to "promote enterprise, innovation and increase productivity". Particular emphasis is placed on the integration of e-business and e-manufacturing into engineering industries and their immediate supply chains.[22]

---

cutting, improving the relationship with customers or many other straightforward business needs may be the explicit reasons for e-business adoption. Sometimes the objective can be of a more strategic nature, such as driving and facilitating profound changes in the organization.

Getting the SMEs on board

Given the huge weight of SMEs in developing economies, particularly from the point of view of employment, the question of how to address their particular needs in the field of ICT and innovation should be carefully addressed.

The first point to consider is the way in which innovation and e-business affect the competitive and innovative behaviour of SMEs. In that respect it is important that policymakers keep in mind the need to adapt their instruments as ICT-driven innovation spreads in the economy and creates network business models that may affect the relevance of policies and interventions. A related business environment factor to consider in the design of innovation support programmes targeting SMEs is that as e-business practices generalize, SMEs find that innovation becomes more easily visible to, and eventually imitable by, competitors.

Another consideration that is specific to SMEs and may require special efforts from innovation support policies concerns the need to make clear to the SMEs the relevant benefits that a particular intervention delivers to them and the specific manner in which their competitive potential will be enhanced as a result of it. The amounts involved may be significant.

Helping SMEs identify how innovating through e-business can be actually done in their particular environment can be a time-consuming and expensive exercise that needs to be conducted in an individualized manner. Policies should aim at helping SMEs integrate ICT and e-business considerations as a fundamental element of their enterprise development plans. This means that programmes need to focus on what ICT and e-business can do to solve real problems that SME managers can easily identify. An adequate outreach strategy in that regard needs to make extensive use of the language that enterprises understand best: the financial performance benchmarks that they are used to. When SMEs can make a clear connection between their performance benchmarks against those of their competitors and their relative position in terms of ICT, e-business use and innovation the vital importance of the questions involved comes across to managers immediately.

An important lesson that countries which are considering putting in place support programmes in this field need to keep in mind is that for initiatives to succeed they need to remain in place for a reasonable period of time. The value of any set of ICT innovation support measures can be judged on a rational basis only once some impact measurement has been undertaken, and this takes time. However, it is not uncommon that programmes in this field are terminated before their effects on enterprises can be assessed. This makes it difficult to replicate and scale up successful initiatives, and to accumulate and disseminate best practice. At the same time, it is also important that policies adapt and change in response to practical experience. Striking the right balance between the needs for policy stability and flexibility and evolution calls for mechanisms that allow feedback from end-users to reach policymakers, and frequent and meaningful interaction between all stakeholders.

D. Conclusions and policy recommendations

The main purpose of this chapter was to present the case for a much closer relationship between policies to promote innovation and policies to facilitate the adoption and use of ICT by enterprises. This is based on the premise that in today's innovation paradigm ICTs play a central role. In a growing number of sectors the capacity to implement e-business practices is a fundamental determinant of competitiveness, which is the ultimate purpose of innovation policies. Bringing innovation policies closer to ICT and e-business policies will require the development of a framework in which the relationships between the concepts of ICT, e-business and innovation become clearer to the stakeholders.

As discussed in section B, an innovation policy framework that fully takes into consideration the changes generated by ICT must give prominence to open approaches to innovation, which present significant advantages for developing countries. In order to support open innovation, which relies on cooperation, policymakers need to take into account the trade-offs that exist between the competitive incentives to innovate and the ability to cooperate with other enterprises. Partnerships and alliances, between the private and the public sector and between enterprises, become particularly useful in that regard and ICT provides platforms that make them easier to implement.

Another area in which policy can support the development of open innovation is the intellectual property regime, a topic that is far too complex to be dealt with in the context of this chapter. However, it should be noted that while in a closed innovation framework intellectual property is used to capture value by excluding competitors from knowledge, in an open innovation approach intellectual property instruments can be used to create a "knowledge commons" in which collaboration can generate value. Policymakers should pay attention to the balance between the use of intellectual property to protect market positions and the need to facilitate cooperation.

An open innovation approach requires some adaptation of innovation support policies. The emphasis should move from supporting individual firms to providing more assistance to innovation networks. Many funding or incentive instruments to support innovation currently orient firms towards building closed, in-house R&D capacity and do not encourage them to look for networking opportunities. In many countries, enterprises that rely on cooperation with external networks in order to innovate are not eligible for innovation support measures, which are geared to firms with internal R&D facilities only. Supporting networks of innovation also involves strengthening networking skills, particularly among SMEs; this includes aspects such as human capital management, intellectual property management and trust-building.

Public research institutions and academic centres have a key role to play in advancing open innovation. The academic incentive system may benefit from reconsidering how to stimulate the emergence of collaborative strategies, to foster knowledge-sharing and to engage with industry.

User-driven innovation is a form of ICT-enabled open innovation that is increasingly important. Policy can support it in a number of ways. One is to promote the use of open standards, particularly through public procurement policies. Another one is to undertake efforts to assess the extent to which user innovation is actually implemented in order to adapt overall innovation policy. Policymakers should also consider the possible negative effects on user-driven innovation that may result from imposing restrictions on the ability of users to modify products that they own.

Finally, general environment conditions that are frequently mentioned as necessary for the development of an information society are also important for the development of open innovation. These include an educated population, IT infrastructure of good quality, efficient capital markets and the creation of a high-trust business environment.

The overall conclusion of this chapter is that as the effects of e-business adoption spread across the economy and affect most sectors and aspects of business process an integrated approach to e-business and innovation policies is needed. Experience shows that the effectiveness of innovation policy and e-business-related policies improves when they are conceived and implemented in a closely coordinated manner. But the changes that ICT have introduced in the functioning of modern economic systems are such that this coordination between both policy areas is likely to gradually lead to a significant level of integration among them.

In practice, the integration and coordination of policies in those two fields amount to the creation of an institutional framework of cooperation among the different ministries and different levels of government that are often involved in them. Coordination can be achieved by integrating all instruments for e-business and innovation support into a single ministry or agency, although this does not always result in closer coordination. It can also be achieved by creating institutional mechanisms that allow the full involvement of all stakeholders, with the right balance between bottom-up and top-down policymaking.

Enough experience of policy approaches and instruments is already available to suggest that for developing countries ICT-enabled innovation policies provide an opportunity to explore open innovation approaches that may be more suited to their concerns than traditional ones. Sharing of experiences and learning to learn from others will be crucial in that regard. This is a process that UNCTAD may be well placed to facilitate at the global level as part of its work in the field of science, technology and innovation, in cooperation with other relevant organizations at the regional level.

References and bibliography

Archibugi D and Iammarino S (2000). Innovation and globalisation: Evidence and implications. In Chesnais F, Ietto-Gilles G and Simonetti R (eds.), European Integration and Global Technology Strategies. Routledge. London.

Archibugi, D and Pietrobelli, C (2002). The globalisation of technology and its implications for developing countries: Windows of opportunity or further burden? *Technological Forecasting and Social Science*, vol. 70, no. 9, November, pp. 861–884.

Booz Allen Hamilton (2005). *The Booz Allen Hamilton Global Innovation 1000: Money Isn't Everything* (by Barry Jaruzelsky, Kevin Dehoff and Rakesh Bordia). Available at http://www.boozallen.com/media/file/151786.pdf.

Brown J and Hagel J (2003). Letter to the Editor. *Harvard Business Review* July 2003.

Cantwell J A (1999). Innovation as the principal source of growth in the global economy. In: Archibugi D, Howells J and Michie J (eds.), *Innovation Policy in a Global Economy*. Cambridge: Cambridge University Press.

Carr N (2003). IT doesn't matter. *Harvard Business Review*, May.

Cassiman B and Veugelers R (2004). Foreign subsidiaries as channel of international technology diffusion: some direct firm level evidence from Belgium. *European Economic Review*, 48:2, pp. 455–76.

Chesbrough H (2003). *Open Innovation: The New Imperative for Creating and Profiting from Technology*. Harvard Business School Press.

Criscuolo C and Martin R (2004). Multinationals and US productivity leadership: evidence from Great Britain. CEP Discussion Paper No. 672, http://cep.lse.ac.uk/pubs/download/dp0672.pdf.

Criscuolo C, Haskel J and Slaughter M (2005). Global engagement and the innovation activities of firms. National Bureau of Economic Research Working Paper No. 11479.

Freeman C and Perez C (1988). Structural crises of adjustment, business cycles and investment behaviour. In: Dosi G et al. (eds.), *Technical Change and Economic Theory*, Pinter, London.

Hall B and Mairesse J (2006). Empirical studies of innovation in the knowledge driven economy. *Economics of Innovation and New Technology*, pp. 15, 289–299.

Huston L and Sakkab N (2006). Connect and develop: inside Procter and Gamble's new model for innovation. *Harvard Business Review*. March

Koellinger P (2005). Why IT matters: an empirical study of e-business usage, innovation and firm performance. German Institute for Economic Research Working Paper 495. Berlin.

Mowery DC and Rosenberg N (1978). The influence of market demand upon innovation: a critical review of some recent empirical studies. *Research Policy*, April.

Narula R and Sadowski M (2002). Technological catch-up and strategic technology partnering in developing countries. *International Journal of Technology Management*, vol. 23, no. 6.

National Science Foundation (2000). Science and Engineering Indicators, Washington, DC: National Science Board.

Schumpeter J A (1943), *Capitalism, Socialism and Democracy*. Allen and Unwin, London.

Serapio MG and Dalton DH (1999). Globalization of industrial R&D: an examination of foreign direct investment in R&D in the United States. *Research Policy*, 28, 303–316.

Serapio MG and Hayashi T (eds.) (2004). *Internationalization of Research and Development and the Emergence of Global Research and Development Networks*. Elsevier Research in International Business, vol 8.

UNCTAD (2001). *E-commerce and Development Report 2001*.

UNCTAD (2003). *E-commerce and Development Report 2003*.

UNCTAD (2004). *E-commerce and Development Report 2004*.

UNCTAD (2005). *World Investment Report 2005*.

UNCTAD (2006). *Information Economy Report 2006*.

UNCTAD (2007). *The Least Developed Countries Report 2007*.

Von Hippel E (2005). Democratizing innovation. Available at http://web.mit.edu/evhippel/www/democ1.htm.

Wired.com (2006). The rise of crowdsourcing. Issue 14.06, June   http://www.wired.com/wired/archive/14.06/crowds.html.

Notes

1. There is a rich and growing literature on the effects of ICT on productivity and overall economic performance. Contributing to it from the viewpoint of trade and development, UNCTAD (2001) includes a simulation of the effects of various hypotheses about e-business adoption on the performance of developing economies, UNCTAD (2003) includes a summary review of the most relevant literature on the question of ICT and productivity and growth, while UNCTAD (2006) outlines the major features of national ICT policies for the development of the information economy, as well as their poverty-reduction dimensions. UNCTAD (2005) focuses on transnational corporations and the internationalization of R&D, including its implications for innovation and development.

2. See UNCTAD (2007).

3. See, for example, Criscuolo and Martin (2004) and Criscuolo, Haskel and Slaughter (2005).

4. See, for example, Cassiman and Veugelers (2004).

5. See UNCTAD (2005) for an extensive discussion of the role of FDI in research, development and innovation in a developing country context.

6. See http://www.trendchart.org.

7. See chapter 2 in UNCTAD (2004) for a survey of the major SME-specific issues of e-business in developing countries.

8. See for example UNCTAD (2004) or the results of the OECD Electronic Commerce Business Impacts Project available at http://www.oecd.org/document/21/0,3343,en_2649_34449_2539157_1_1_1_1,00.html.

9. Carr (2003).

10. See National Science Foundation (2000) for data on the growth of strategic technological alliances.

11. It should be borne in mind that some modalities of innovation that may be more prevalent among developing country enterprises are more difficult to capture with patent statistics. For example, this can be the case of "creative imitation", whereby an existing product is introduced into a new market, it being adapted to the needs and environment of a lower-income market.

12. Cost considerations as a factor in favour of FOSS are all the less important in view of the fact that in many countries most software used was actually obtained at zero cost of acquisition, because of piracy. See for example the 2006 global piracy study conducted by the Software Business Alliance and IDC: w3.bsa.org/globalstudy//upload/2007-Global-Piracy-Study-EN.pdf.

13. See www.chinaeconomicreview.com/cer/2007_03/Problem_solvers.html.

14. See www.businessweek.com/innovate/content/jun2007/id20070611_139079_page_2.htm.

15. See www.openbusiness.cc

16. See Wired.com (2006).

17. See http://www.forum.nokia.com.

18. See www.mosti.gov.my.

19. See http://www.innovacion.cl/chileinnova/sec_chile.php?id_seccion=9.

20. See http://www.webworkscork.com.

21. See http://www.ida.gov.sg/Programmes/20060926120315.aspx?getPagetype=34.

22. See http://www.ictcarrier.co.uk/

Chapter 5

E-BANKING AND E-PAYMENTS: IMPLICATIONS FOR DEVELOPING AND TRANSITION ECONOMIES

A. Introduction

In its *E-Commerce and Development Report* (the predecessor of the *Report*) in the years 2000 and 2001 UNCTAD made one of the first attempts to define electronic banking (e-banking) and electronic payments (e-payments) and to assess the future of this new, even for developed countries, stage in the development in finance (UNCTAD, 2000, 2001). By looking at the state of the art in e-banking and e-payments at the time, it tried to understand the prospects for their dissemination worldwide, and especially their implications for international financial flows and financial systems in developing and transition economies. The main focus was on the analysis of the Internet or online banking and payments arrangements created by banks and related entities. Since the beginning of this decade e-banking and e-payments have grown rapidly and become widespread financial services not only in developed economies, but also in many developing and transition economies. The customer base for those services is experiencing double-digit growth, particularly in many dynamically growing developing countries.

Financial service providers, and especially banks, had been using electronic messages for quite some time before the introduction of the Internet. Those messages were transmitted through proprietary software systems, also known as intranets, through modems linked to clients' personal computers or phone lines permitting them to consult accounts and make changes in payment orders in an offline regime. The Internet revolution started the process of replacing those traditional methods of electronic communications by Internet Protocol (IP) based systems. The literature is increasingly focusing attention on this newer and more widespread development of e-banking and e-payments, while continuing to discuss also the role of payment cards, automated teller machines (ATMs), telephone banking and mobile banking (m-banking) or m-payments. The latter can use IP and other communication protocols and are relatively more important in the context of developing countries.

The use of modern information and communication technologies (ICTs), and especially the Internet, by financial services providers greatly increased their communications capacities and speed, decreased the transaction costs of financial operations and permitted the networking of a host of players in various financial schemes. As a result, it brought about major business process innovations. For example, ICTs accommodated the explosion of large-value international payments traffic especially during last two decades, in particular thanks to the introduction of new online payments protocols and real-time gross settlement (RTGS) systems.

E-banking and e-payments, both corporate and retail, proved to be less costly for the commercial banks and at the same time more convenient for businesses, Governments and households. The use of e-finance relies on bank deposits and diminishes the role of cash money (notes and coins). However, it has created another set of security challenges such as the need for protection against emerging cybercrime, which has introduced further innovations allowing more secure methods of e-banking and e-payments.

The intensive use of ICTs also facilitated the transformation of traditional bank-related debt such as various loans and mortgages into securities circulating in capital markets. That increased trading and securitization activities of banks coincided with their relatively reduced role as deposit-taking institutions, characterized by long-term relations with their clientele.

Banks and payment card providers remain at the core of e-banking and e-payments. However, relatively new players such as non-bank money transfer operators, mobile phone operators and e-payment technology vendors are also coming into the picture, trying to carve out niches or special value-added operations from the main players or to conclude various cooperative arrangements with them.

The financial flows between developed and developing countries are also increasingly taking place in the framework of major online inter-bank transfer systems. While those systems are facilitating the transmission of the main private and public finance flows such as bank credits, foreign direct investment (FDI), portfolio investments and official development assistance (ODA), ICTs are no less important for retail and small-volume financial transfers to households and small businesses in those countries.

The most important small-scale private financial transfers are migrant remittances, which are increasingly relying on online money transfer systems, with a consequent saving of resources for both originators and end-users of those funds. Since remittances have considerably exceeded ODA in their overall volume and are currently the largest non-debt-creating financial flows, the prospects for the moving of this major source of development finance towards electronic channels constitute an important element in analysing the implications of e-finance for developing countries.

Making e-banking and e-payments more affordable to banks and their clients in developing countries (and to less affluent parts of the population in developed countries) is still a major issue. Furthermore, giving to small and medium-sized enterprises (SMEs), micro-enterprises and individuals (part of whom are "unbanked", i.e. have no bank accounts) better access to simple forms of e-banking and e-payments or m-payments is also becoming an increasingly major challenge, which has started to be addressed very recently.

While acquiring knowledge of modern e-banking and e-payments techniques is still the main hurdle for developing economies, the lack of deeply rooted proprietary systems of finance might become a positive factor and enable countries that are not really attached to traditional systems to leap frog to e-finance. Exploiting that potential will require that the financial sector of developing and transition economies have the capacity to move rapidly towards modern ICT-based systems.

This chapter starts by looking at the main manifestations of financial innovation such as online banking and payments, m-payments and e-money, as well as the impact of ICTs in facilitating securitization of credit instruments and the need to address the security challenges of e-finance operations. It reviews the state of play in Internet banking and e-payments and examines their implications for developing countries. In particular, it discusses the role of ICTs in facilitating

money transfers resulting from migrant remittances, e-banking and e-payments for SMEs and microfinance. Lastly, it tries to assess the future of e-banking and e-payments, particularly in the context of developing countries and countries with economies in transition. In that connection a number of policy and institution-building measures necessary for supporting the development of those services are proposed. Thus, the chapter contributes to the Report, which is putting emphasis on the role of innovations in economic development, by reviewing ICT-generated innovations in banking and payments, as a case study of financial innovations.[1]

B. ICTs and innovations in banking and payments

The active use of ICTs during the last two decades further released the forces of globalization, liberalization and technological change. ICTs were particularly driving major innovations in business processes, and especially in financial intermediation.[2] One of the most innovative industries in terms of using ICTs, the financial sector, increased its share of value added particularly in developed economies. For example, in 2006 its share in the United States GDP exceeded 7 per cent, that is it more than doubled during the last 50 years.[3]

Another indicator of the so-called financial deepening of the economy is the ratio of financial assets (bank deposits, government debt securities, private debt securities, equity) to the economy. The volume of financial assets currently exceeds gross domestic product (GDP) by a factor of 3 to 4 in the main developed market economies. Financial deepening is also becoming more and more evident in China and other emerging Asian economies (Indonesia, Malaysia, Philippines, Republic of Korea and Thailand), where it represents double the local GDP. The same process is still lagging behind in Latin America (except Chile), Eastern Europe and especially Africa (McKinsey, 2007).

Meanwhile, the use of the Internet as an increasingly important channel of distribution for financial services, and, in particular, e-banking and e-payments, was the most important innovation making it possible to move liquidity and various financial instruments online at much lower transaction costs and higher speed, with a consequent change in the nature of monetary transmission and financial intermediation. The possibility of intensive use of ICTs made it possible also to improve transparency in financial operations. As a result information asymmetry in financial markets was reduced and they were made more complete. In addition to the migration of traditional banking and payments to the Internet, the use of ICTs facilitated the large-scale introduction of securitized forms of traditionally bank-based debt instruments, thus allowing capital to be allocated at lower cost and with a better risk profile. Active participation in that process by insurance companies, pension funds, private equity funds and hedge funds was also supported by the use of ICTs. Those major changes provided not only new opportunities, but also created challenges, the greatest of which was the problem of ensuring the security of online financial operations. To respond to security threats, financial service providers had also to introduce innovative solutions to protect their clients against inventive cybercriminals.

The manifestations of financial innovation are having an increasingly direct impact on developing countries, which are not overburdened by older technologies and can more easily start their e-banking and e-payments operations in a new environment and with new technologies.

1.   From brick to click: emerging e-banking, e-payments and e-money

In the heyday of the Internet revolution expectations regarding the development of e-banking and e-payments were highly optimistic, and not without reason. It was a trial-and-error process, with some models being increasingly marginalized or even destroyed during this major and mainly successful process of creation. Thus, models that did not survive or keep pace with the dynamics of overall trends included the majority of new purely Internet-based banks and e-cash ventures. At the same time, responding to the challenges of Internet-only banks, nearly all major traditional banks introduced their e-banking operations starting, with moving online the inter-bank payments traffic and supporting online the operations of their clients. With time, major banks expanded their e-banking activities, in many cases at the expense of traditional banking ones (known as "brick and mortar"), and created a twin model also known as "click and mortar" or "brick and click", with the "click" part profoundly innovating and still continuing to transform the banks' overall business models.

The introduction of new technologies especially in the initial stages, involves major investments and sunk costs

that can be recovered with increasing economy of scale. Research comparing Internet banking experience with the experience of old and newly chartered traditional banks in the United States, during the period 1998-2003 showed that Internet-only banks had a lower return on assets (ROA), a lower return on equity (ROE) and lower net interest margins, and had to offset an inferior accounting performance, higher loan losses and higher borrowed fund costs with higher capital ratios, even in comparison with newly chartered traditional banks. The authors assumed that Internet-only banks can become successful competitors if they are able to improve economies of scale, gain adequate market share and combine new products with traditional banking products (Cyree, Delcoure and Dickens, 2005).

The Internet as a financial services delivery channel is acquiring its market share mainly through innovations in traditional banks' organizational structures. The integration of new Internet technologies also required training and retraining of employees such as account officers, integrating new "front office" remote services into the "back office" services menu, moving towards holistic and standardized IT systems, and making the necessary changes in the process of communications and management (Novametrie, 2004).

E-banking delivery channels for both major and retail clients are based more or less on the same principles of online operations with clients' accounts, including account consultation, transfers, the use of digitized giros, bills, cheques and other forms of payment, and placing orders for stocks, bonds, derivatives and other securities. At the same time regrouping these and other orders and channelling them between banks generate a huge payments traffic that is handled by major inter-bank payments systems.

The automation of cashless payments and the dramatic reduction in their transmission costs were ICTs' main contribution in the payments systems. Moving inter-bank payments traffic to the Internet was also a formidable innovation challenge, one which the international banking community has been addressing, particularly during the current decade. That in turn brought about an explosion in the value of payments traffic as a result of financial innovations. Thus, during the last 25 years the value of payments traffic in G10 countries increased 13-fold, and its volume threefold. A more rapid increase in larger value payments, which in many major economies accounted for three quarters of GDP, explains the predominance of those payments over retail payments, with the latter representing only a very small percentage (BIS, 2007). A major innovation here was the move from deferred net settlements (DNS) systems, whereby the settlements were made by the end of the day, to real-time gross settlement (RTGS) systems. The possibility of settling in the real-time regime was a major step forward in managing systemic risks related to the time needed to clear and settle the inter-bank payments.

In retail e-banking and e-payments there were two main trends in the introduction of innovative solutions: the modification of traditional payments methods such as credit transfers, direct debits, card-based payments and their adaptation to various IP-based applications, and the development of such relatively new instruments as e-money [4] and other pre-funded payments, cumulative collection or integrated payments, payment portals, and (becoming increasingly important) mobile payments. One of the results of the expansion of e-banking and e-payments was substantially lower costs related to initiation and handling of paper orders. The introduction of various e-payments methods decreased in G10 countries the share of payments by paper-based cheques and credit transfer orders from 60 per cent in 1989 to 20 per cent in 2005 (BIS, 2007).

Given the relatively greater importance of mobile telephony in developing and transition economies, it is important to stress the role of m-banking and m-payments as a major phenomenon in e-payments-related innovations. The possibility of using cell phones and other electronic devices as prepaid payment cards was a major step forward in mainstreaming poor and "unbanked" population into e-payments. In the case of m-payments the central role is mainly played by mobile network operators (MNO), while banks are backstopping them by keeping the consolidated accounts and the float accumulated by those mobile operators.

One of the key functions of e-payments is to support B2B, B2C, P2P and other e-transactions. At the same time e-payments can also service offline business transactions that accept deferred payments.

The introduction of e-banking and e-payments has decreased the tendency of economic agents to keep cash, namely notes and coins issued by Central Banks (CBs), and increased the importance of commercial bank money. Indeed, the value of deposits kept by non-banks in commercial banks in G10 countries increased from 30 to 50 per cent of GDP, while the share of currency remained stable at 7 per cent (BIS, 2007). At the same time, many attempts were made, especially at the height of the Internet revolution, to replace cash with various types of e-money.

Understanding innovation related to e-banking, e-payments and e-money will be incomplete without looking at how they impact on monetary circulation and transmission. Thanks to the Internet, e-banking and e-payments are increasing the speed and diminishing the costs of financial transactions such as payments, as compared with the similar operations intermediated by other mediums such as cash or traditional bank transfer. Because of the greater speed of transfers, e-banking and e-payments are also increasing the velocity of money and hence decreasing the demand for money in circulation at a given level of economic activities.

While e-banking and e-payments are becoming more important in the operations of financial intermediaries, they also have to cope with the inertia of habits and the preferences of retail and corporate customers while servicing their needs. Those preferences are determined by many economic factors, including the propensity of many households and SMEs to engage in cash transactions in order to preserve the anonymity of their transactions, especially in the context of the informal economy. In such a situation e-money can serve as a medium for a particular part of a transaction requiring an electronic transfer of funds, which then might be again transformed into cash. Also, money might be transferred through a sequence of electronic transfers, part of which are not Internet-based and hence are more expensive. In other words, since they are the most efficient way of effecting payments and other financial operations, the IP-based e-banking and e-payment systems can either replace more traditional and more expensive ones, or coexist with them according to the preferences of monetary circulation participants, including those in the "unbanked" population.

There were also attempts to replace currency (notes and coins in circulation) issued and backed by CBs with electronic carriers of anonymous private e-monies as a means of payment embodied in easily portable and mobile electronic devices. In that connection, it is important to understand that private financial intermediation does not really bring about the creation of e-monies that coexist or even replace public e-money backed by a CB and used by banks in their online transactions. In fact, the Bank for International Settlements (BIS) uses the narrow concept of e-money, describing it as a "stored value or prepaid product in which a record of the funds or value available to the consumer for multipurpose use is stored on an electronic device in the consumer's possession" (BIS, 2004, p. 2). In other words, it is a transfer of value originated either from fiat money (cash) or from bank accounts into an electronic device embedded in a card, or a purse, or software linked to a PC. Having been transferred it becomes e-money making possible further payments. As those devices are produced by private entities, regulatory issues of ensuring their usability at face value may arise. Also, most of e-money systems are not anonymous – in other words the convenience of having e-money comes at a price (Arnone and Bandeira, 2004). As a result, the volumes of e-money are still very small and this type of innovation is still at a very early stage compared with, for example, Internet banking or m-payments.

The proliferation of parallel trusted systems made it possible to consolidate payments, clear mutual debts of, for example, auction participants and settle the remaining debts at the end of regular periods. In some private exchange systems, private e-monies are usable between participants, but unlike currencies are not legal tender. Such mutual debt-clearing systems should normally reduce the demand for money in circulation. The non-bank electronic billing and clearing systems are acting as consolidators of mutual claims between participants of a given network and are then netting out the residual debts in the system by using cash and bank accounts, in other words generally accepted means of payment.

So far, the expectation that e-money would spell the end of notes and coins has proved to be premature. While some elements of those processes are still at an early stage, e-payments, especially in the retail sector, have continued to be centred on payment cards, which have been increasingly becoming smart cards adapted for the Internet. Meanwhile, the expectation that payment cards would be all the more important by now has also proved to be exaggerated.

To assess the role of e-banking and e-payments in monetary circulation it is important to understand which part of payments and other transfers, intermediated mainly by commercial banks, is passing through the Internet and thus enjoying the low transaction costs and speed of that huge network. In other words such elements of M1 as coins and notes, could become, sight deposits and e-money, that is be easily transformed through online banking and payment channels into their purely electronic forms and then revert to more traditional forms based on the preferences of the beneficiaries of those funds.

The modernization of payments systems that rely to a considerable extent on the availability of low-cost, real-time online communications and networking requires better regulation of payments systems in various countries and its coordination, and closer links between regulators and financial service providers in order to rapidly overcome problems due to possible disruptions and crises in various parts of the system. If those requirements are met, monetary transmission and payments flows will become smoother and more stable.

2.   ICTs and securitization of debt instruments

The development of standardized and secure online messaging, and the substantial reduction of transaction costs and information asymmetry related to financial operations, gave a boost to financial innovation in various areas. In particular, that innovation is taking place in the transfer of risks from banks to capital markets through the development of new techniques for securitization of loans. That allows commercial banks to remove a part of the risks from their balance sheets and hence reduce the volume of capital that they have to maintain in order to meet regulatory requirements. The result is a thriving financial derivatives market and an overall increase in the securitization of traditional bank loans and other debt-related instruments.

While ICT-based innovation in finance was one of the drivers of financial globalization and the increase in primarily private international financial flows, it further facilitated the process of liberalization and deregulation in financial services by blurring the frontiers between commercial and investment banking. More commercial parts of global banks started to resort to securitization of their loan portfolio, while investment banks and hedge funds were instrumental in marketing so-called credit default swaps and collateralized debt obligations and other debt-based securities, supporting the development of debt-related derivative instruments. The need to move a part of assets out of balance sheets to meet the stringent capital and asset quality requirements of Basel II has only accelerated that process.[5]

Modern ICTs and, in particular, the Internet facilitated the securitization of traditional bank-related debt instruments and the dispersion of credit risks among non-bank financial institutions such as insurers, private equity and hedge funds. They thus permitted the development and pricing of a host of complex instruments and 24-hour trading of huge volumes of securities.[6] The digitization of financial operations and possibilities of creating hybrid products outside the scope of regulated banks triggered the creation of such instruments as collateralized debt obligations, credit derivatives, and structured investment products. In many cases these were managed by specialized companies and funds, special purpose vehicles and monoline insurance companies. While risk dispersion from banks due to financial innovation might be considered a positive trend per se, financial regulators are now concerned as to whether in the event of financial crisis the risk will flow back to the banking system and whether the latter will have enough capital and liquidity to absorb the stress related to such an eventuality.[7] In that regard overcoming information asymmetry at the level of global financial markets is important for the purpose of adequately assessing the distribution of risks throughout the international financial markets.

It is also thanks to modern ICTs that banks' trading activities related to various debt and other financial instruments became relatively more important than longer-term deposit-taking and credit-providing activities. While analysing the importance of the information revolution for financial innovations, especially in capital markets, is beyond the scope of this chapter and will, it is hoped, be tackled in forthcoming research, it was important to stress here the role of ICTs in facilitating securitized debt instruments, which are increasingly competing with commercial bank-based lending and borrowing operations.

3.   E-finance and security challenges of cyberspace

A major challenge for e-banking and e-payments that requires further innovative approaches stems from the need to tame and contain the negative effects of rapidly growing cybercrime. Various actors involved in developing new security systems include technology vendors, security system designers, banks and non-bank financial service providers.

While according to an optimistic scenario about the use of e-banking in the United States the number of users will rise from 56 per cent of households in 2006 to 76 per cent in 2011[8], other sources are noting a tendency for some groups of customers to reduce or even cease their e-banking and e-payment activities because of security concerns. According to another source, 73 per

cent of users in the 18-50 age group are concerned about identity theft.[9]

The issue of security on the Internet has dominated more of the literature since the definition of e-banking moved from ATMs and telephones to the Internet. Security is all-important with online banking and it might determine its success or failure. In fact, some consumers stop paying bills and conducting other operations online as the (perceived) risk increases.

The security risk can cause banks to lose the gains of Internet banking if problems are not properly addressed. One of the risks is identity theft, which can occur in two ways: "phishing", namely sending authentic-looking-emails, which trick consumers into giving password details by replying to the e-mail; and "spoofing", which uses a fake website that makes customers believe they are using a real site. There are also key logging, software tracking and remembering passwords and numerous other viruses and trojans (Clayton, 2006).

Another worry from the security point of view is the "trust gap", an unproven but suspected divide between those who use the Internet for other tasks and hence trust online banking and those that do not (Fox and Beier, 2006). This means that, to encourage Internet banking, one may first need to encourage other activities on the Internet.

Once the banks start to outsource IT-related operations to other firms, there is the increased risk that the latter may not be sufficiently regulated and that the transfer of information to them and between them and the banks may increase the security risk. Consequently there is a need to develop approaches to make sure that a combination of enough technical expertise, security support and oversight is in place when one is engaging in outsourcing and offshoring in e-finance (Mu, 2003).

The security problem should, however, not be overestimated. Many innovative schemes have been developed to solve the problems (Singh and Malhotra, 2004, Rombel, 2003), with more, including biometrics, in the pipeline (Arumuga, 2006). Furthermore, some authors claim that the slower growth of Internet banking in the United States is not linked to security concerns but to the macroeconomy and market saturation (Rombel, 2005).

While ICTs help to improve credit risk management and hence the situation regarding information asymmetry, the technology itself can represent a risk and according to the BIS falls into the category of banks' operational risks. From the regulator's perspective, provisions to deal with the eventuality of service disruption for technological, including security, reasons are addressed in the operational risks cluster and are a prerogative of the Basel Committee on Banking Supervision.

Major providers of payments systems such as the Society for Worldwide Interbank Financial Telecommunication (SWIFT), payment card providers such as Visa and Mastercard or peer-to-peer payments service providers such as Paypal are also taking major steps to introduce better-performing security systems so as to protect their clients. For example, Visa and Mastercard have improved security on the web with systems such as Verified by Visa and MasterCard SecureCode (OECD, 2006), while Paypal has made its security procedures more stringent.[10]

Banks, other firms, institutions and Governments should adopt a risk management approach to information security (UNCTAD, 2005b). Its particular feature is that it involves an economic assessment of the information assets at risk, which is conducted before possible solutions are looked at. It should clearly outline security risks and compare these with the investment needed to guard against them, all the time considering the value of the underlying information assets. Policy processes at the national or international level may consider possible action to improve incentives for investing in better information security. More practically, in order to use a risk management approach, it is fundamental to define risks, to evolve ways of keeping risk perceptions current, and to measure or develop methodologies to quantify risks. It is immediately apparent that the task in question may be more difficult with regard to information security in financial services. Part of the problem lies with the ever-expanding scope of use of information technology in managing financial operations. Governments and financial industry supervisors and regulators may choose to mainstream certain aspects of financial information security risk measurement and methodologies, including the provision of quantitative data, in order to assist in policy development and implementation, as well as international cooperation on this crucial issue.

C. Recent trends in e-banking and e-payments

As was shown in the previous section, the traditional commercial bank services for corporate and household customers are undergoing dramatic changes that are moving rapidly towards e-banking. Given the relatively greater importance of commercial banking in the financial sector of developing countries, understanding the current trends in e-banking and e-payments, and deriving lessons from them, could be of the utmost importance for the further modernization of banking in those countries. Consequently there is a need to analyse the current trends and best practices in e-banking and e-payments, bearing in mind their applicability in developing and transition economies.

1. The state of play in Internet banking

E-banking combines informational and transactional facilities. While the risks related to informational websites are limited to incorrect information or negative perceptions of the host bank, the risks with transactional facilities are more considerable. Those facilities should be functional and user-friendly, and customers should be sure that their security and privacy will not be infringed. In a wider perspective, e-banking risks include the following categories: transaction or operations risk, credit risk, funding and investment-related risks, legal and compliance risks, and strategic risk (Ainin, Lim and Wee, 2005). Thus, the use of e-banking poses additional operational and hence reputation risk for existing banks, especially when they become vulnerable to fragile technology, customer confusion and hackers (FFEIC, 2003). Banks that start e-banking may also face initially high costs and technical problems, but those that wait, may lose customers to those who capture the market first. As the experience of many pioneers in e-banking shows, properly tackling the above-mentioned risks from the outset and managing them in the course of the further development of e-banking is important for maximizing the benefits of e-banking in the longer run.

The literature on e-banking mainly discusses the patterns of emerging virtual banks and of established banks diversifying from a purely branch-based approach to a twin online and more streamlined branch system that makes it possible to reduce high staff costs and overheads in a branch (Akinci, Aksoy and Atilgan, 2004). The banks' strategies have also been reflecting consumers' preferences for a mix of delivery channels: the increasingly popular "click and mortar" approach. This model is becoming increasingly the norm in Europe and the United States, since consumers, although increasingly attracted by e-banking, continue to value personal interaction and are concerned about security risks on the Internet (Novametrie, 2004).

It combines face-to-face interaction and other advantages of traditional banking with the 24/7 availability and low, non-distance defined costs associated with Internet banking. Adelaar, Bouwman, and Steinfeld (2004) looked at a Dutch cooperative banking network and found that, since costs are much lower on the Internet, the banks had begun offering the most expensive services exclusively on the web. For other services customers can complete forms online and then finish the transaction in a branch. There is, however a risk that such "click and mortar" models will not be sustainable or will contribute to increasing the digital divide, since while educated and affluent customers will benefit from online services, the poor will lose out as fewer branches will be at their disposal. As a result in the early stages of e-banking there were calls for State protection of traditional banking to protect the poor (Hawkins and Mihaljek, 2001).

It is now common knowledge that transaction costs related to Internet banking operations are much lower than those incurred at a branch or on the telephone. Many sources, including UNCTAD (2001) quoted the famous example showing that a payment operation that costs a branch one dollar could cost one cent if effected via the Internet. With these savings and relatively low set-up costs it is not hard to see the attraction of e-banking (Adelaar, Bouwman, and Steinfeld, 2004); (Dandapani, 2004). Simpson (2002) used data from 1999 for a sample of American banks and banks in emerging markets to assess the different cost profiles and risks in e-banking and in traditional banking. He found that banks in the United States, which have reached an advanced stage in the use of e-banking have lower overhead costs and are considered to have a lower banking risk attached to them when compared with banks in emerging markets, which do not use e-banking. Hence, banking risk and overhead costs could be reduced through greater use of e-banking.

The profitability of Internet-only banks has, however, been questioned, and statistical tests were used to compare it with that of established banks. In one model profit efficiency for Internet-only banks was higher, whereas in another it was lower, suggests that, at least in the short run, Internet-only banks are not necessarily viable and are more likely to be sustainable as one of the channels in a multi-channel banking services delivery model (Cyree, Delcoure and Dickens, 2005). In the case of Italy, studies have shown that there is a strong positive link between profitability and Internet banking activities (Hasan, 2002). Internet banks may also have a revenue structure different from that of traditional banks, relying less on interest income

and core deposits than other banks (Furst, Lang, and Nolle, 2002).

Internet banking cannot be forced on consumers, as many of them have the alternative of using their "brick and mortar" banks or telephones, and incentives are therefore needed. The factors influencing consumers to take up online banking vary between countries and depend on national culture, demographics and the structure of the economy (Brown et al., 2004). Li and Worthington (2004) examined data from 15 developed and 12 developing economies and found that the most important factor influencing the use of Internet banking is ownership of a personal computer: the more people in an economy who own a personal computer, the more Internet banking customers there will be. A lesser, but still significant correlation also exists between Internet banking use and the number of Internet hosts.

In developed economies the demand is for quality Internet banking provision, and this is reflected in its supply. For example, in the United States, Wells Fargo has developed a system called "web collaboration" which allows bank staff to see the same screen as the customer sees as they talk on the telephone. Wells Fargo has also introduced the "desktop deposit system" to allow online deposits without the need for extra software, and the "my spending report", which allows consumers to view simultaneously several accounts on one screen shot (Rombel, 2003). In Switzerland, UBS, which is one of the largest global banks in terms of assets under management, has developed a sophisticated Internet banking menu for all types of customers, which allows them to make a variety of online transfers and payments, consult investment offerings and buy various assets. In parallel, in addition to ATMs, a system of so-called Multimat machines was installed in branches with screens enabling customers to undertake operations similar to Internet banking operations. As a result, the bank's branches are now staffed more with personnel who are becoming financial advisers who help clients to make more complex financial decisions such as obtaining credit or investing in various financial instruments proposed by the bank. Withdrawing funds or making payments is becoming an automated process. However, the e-banking systems face new problems – for example, clients sometimes have to wait to access to machines of which there are only a few, or customers, who are less knowledgeable about technology have to wait even longer to receive assistance.

In the area of risk management the use of technology means that the market risk incurred by banks (the risk that asset prices will change) can now be better identified and as a result of which a better risk-return trade-off is created. Customers also benefit from the banking system's greater resilience to risks. Banks are also able to evaluate and manage credit risk through the use of various instruments for debt securitization and the development of the derivatives markets (Bernanke, 2006).

2.   Wholesale e-payments

According to the Boston Consulting Group's regular global payments report, the volume and value of both domestic and cross-border payments in the Americas, Europe and the Asia-Pacific region will continue to experience dynamic growth. The number of domestic payments is expected to grow from 213 billion operations in 2003 to 414 in 2013, with an increase in value from $1,731 trillion to $3,146 trillion, while cross-border payments will display a similar growth pattern rising from nearly 2.5 billion transactions with a value of $318 trillion to 6.8 billion with a value of $489 trillion (BCG, 2006, p. 33). Such volumes of payments traffic need to be underpinned by modern inter-bank electronic transmission technologies and techniques, based mainly on IP.

To ensure the fluidity of inter-bank payments, clearing and settlements systems of various levels are put in place, comprising mainly local and regional automated clearing systems that are backed up with larger RTGS systems mainly controlled by CBs. While banks domestically or regionally are interconnected through automated clearing houses (ACHs), at the global level such major inter-bank payment networks as SWIFT, Fedwire (US-centred Federal Reserve system), TARGET (Trans-European Automated Real-time Gross-settlement Express Transfer system) and some other RTGS systems represent the core of cross-border flows of payments. Messaging and transfers between financial institutions intermediated by those and other payment systems have been profoundly transformed by Internet-based technologies. In that respect, it is relevant to provide further information about the ability of CBs and commercial banks to keep up with the rapid technological changes and innovation triggered by the Internet revolution.

SWIFT, the largest global messaging system between banks (including those from developing and transition economies), started in 2002 its move from a "store and forward protocol" network to a "session initiation protocol" (SIP) network in the framework of SWIFTNet – the advanced IP based messaging solution of SWIFT. This was completed in 2004, and SWIFT is currently working with the Clearing House Payments Company (a leading system for settling and clearing globally United States dollar payments[11]) to make the two networks interoperable, as the clearing house also upgrades to SIP. The new SIP network allows complete tracking of a transaction throughout the process. It also acts as an "actual private network", ensuring that messages are sent quickly and securely, with the component "FileAct" allowing bulk messages and cheque imaging. SWIFT supplied banks with specific pieces of software to enable them to transfer to the new system and in 2005 began fining banks, that had not yet migrated.[12] Developing countries were slower to switch over because of SWIFT's technical requirements (Hadidi, 2003).

SWIFTNet is used by CBs including those of the United Kingdom, France and the European Central Bank. The Euro Banking Association (EBA) and several ACHs also use it, and it will be at the core of the Single Euro Payments Area (SEPA) when the latter becomes active in 2008. In addition to servicing financial service providers SWIFTNet is also used by major corporations to obtain information for their multiple bank accounts.

In January 2007 SWIFT took another step forward with its business model and launched a new access model called SCORE (Standardized Corporate Environment), which enabled, for the first time, corporations to interact in a standardized way with participating SWIFT financial institutions. Previously corporations could connect to SWIFT only through the MA-CUG (Member Administered Closed User Group) and Treasury Counterparty (TRCO) models, which continue to be available for corporations not eligible for SCORE. This new model is expected to cut down administration costs and simplify the process of standardized messaging as it essentially bypasses the banks when the corporation wants to send a message.

SWIFT's longer-term strategy (SWIFT 2010) is to increase the number of messages it carries and also the type of messages it can carry. The new SIP system allows several different types of messages to be transmitted. SWIFT has evolved to encompass securities with "Wall Street related" messages, which account for almost half of its network traffic. This longer-term strategy is focused on securities and derivatives, once SWIFT feels it can add value through standardization and hence challenge Euroclear's domination.[13]

3.   Retail e-payments

Regarding the retail e-payments, the literature reported *inter alia* the "non event" of the move to new methods e-money (Bounie and Gazé, 2007; OECD, 2006). In developed countries most e-money models have failed while traditional payment methods, including credit transfers, cheques, direct debits and payment cards, have evolved into their electronic forms, this evolution entailing their active migration to the Internet. Nevertheless, credit cards, giro and in some countries cheques are the main methods used in retail payments. New integrated or cumulative payment services such as Paypal and new payment portals for consumers and small businesses have also been introduced (OECD, 2006). Moreover, mobile telephony is increasingly becoming a preferred platform for m-banking and m-payments, which may use payments messaging systems that are in many cases compatible with the Internet.

Online credit transfers and giro, bills and cheque payments are the main Internet-banking services provided to retail customers. Another service that banks provide on a recurrent basis is direct debits, namely payments by banks to various service providers according to the customer's instructions. Standing orders are similar to direct debit operations and are initiated by the payer rather than the bank. Electronic bill presentment and payments (EBPP) services offered by financial and other institutions are also emerging as a major method of making and tracking online payments.

The global payments card market continues to be dominated by a few players, such as Visa, MasterCard, American Express and Diners. The legal decision handed down as the result of the challenge of American Express concerning MasterCards and Visa's practices, especially by setting of high interchange fees, was supposed to lead to increased competition in the market.[14] However, so far, those fees continue to show rise. According to the Nilson Report, a consultancy specializing in consumer payment systems research, the fees that the United States merchants are paying to card companies and their issuing banks have been constantly increasing during the present decade. Thus, the weighted average fee rose from 1.52 per cent in 2000 to 1.88 per cent in 2006, while the volume of fees collected for the same years jumped from $24 billion to $56 billion. As a result, the merchants filed

a class action against major card companies. In spite of the increased competition from various debit card networks, automated clearing house networks and systems such as Paypal, it is still the issuing banks that have the bargaining power when deciding which card to issue. As a result, the major part of interchange fees received by payment card companies goes to the issuing banks.[15]

The Initial Public Offering (IPO) of Mastercard is not only providing the means to pay the legal charges related to the above process, but is also a part of a change in MasterCard's business strategy to a more open form of business and a new corporate governance structure intended to meet competition in the credit card market. After the IPO it changed its official name to "MasterCard Worldwide" and now has a three-tiered business model as a franchiser, processor and adviser.[16] American Express and Discover take a more segmented approach, looking at the upper and lower ends of the market respectively (Simpson, 2005). These three players are all public companies (Discover is a part of Morgan Stanley), while Visa like SWIFT is still a cooperative of banks servicing over 20,000 financial institutions. At the end of 2006, Visa, the largest global payments card company, announced restructuring whereby Visa Canada, Visa USA and Visa International would be merged to create a global public corporation called Visa Inc., the eventual goal being to launch IPOs. Meantime Visa Europe will probably continue to remain a membership organization. In consequence, Visa expected to accelerate product development and innovation.[17]

Although they are convenient means of payment, credit cards continue to be expensive not only for merchants but also for those cardholders that use them as a short-term consumer credit instrument. The interest on credit card overdraft is at least 15 per cent per annum including in countries where interest rates, mortgage rates included, are much lower. In spite of the unsecured nature of card-related debt, such a large difference is not justified, especially in view of the new possibilities arising from current small borrowers' credit-scoring techniques. In fact, one of the reasons for mortgage-related debt increase in the United States, and one of the underlying causes of its current crisis, is the massive refinancing by US households of their consumer debts (i.e. mainly credit-card-related debts) through incremental borrowing on the mortgage side using the collateral of the house.

To counter this lock-in effect of credit card market oligopoly, competing debit card systems, frequently initiated by major retailers, are emerging and new card issuers are entering the market. However, they are far from being the leaders in acquiring a critical mass of clients, and hence convincing the banks to issue the cards on their behalf. Moreover, the major credit card companies and major issuers are also entering into ventures of presenting major retailers, and in such cases they agree not to charge annual fees and considerably reduce the interest rates charged in the case of overdraft.

Despite the problems outlined above, the convenience of electronic payments proved to be a stronger argument for consumers, and in the United States for example, in 2003 the number of electronic transactions (card and direct debit transactions) totalled 44.5 billion, and overtook for the first time the number of cheques paid, which stood at 36.7 billion.[18]

Lastly, mobile hansdsets are increasingly becoming a means of payments for their owners – either through their use as a credit card, with the customer being billed by a mobile operator at the end of the month, or as a prepaid card. Prepaid phones, by definition, use the latter form of m-payment. The active involvement of mobile network operators in such payments has created some legal uncertainty as far as their role in the financial sector is concerned, and regulators are trying to address this question without impeding innovation in that sector. In fact, competition from non-banks in payments services is exactly what

European Commission is going to encourage in its new payment-services directive, which will come into force in 2009.[20]

D. E-banking and e-payments for development

1. E-banking and e-payments in developing and transition economies

Since banks in developing countries and in countries with economies in transition did not have an abundance of traditional proprietary technologies and hence less attached to particular banking techniques, they were also in many cases more open to introducing the most recent Internet-based technologies. The fact that they were less burdened with tradition in some cases made it possible introduce new technologies more rapidly, and also adapt to them more rapidly. However, that requires more vigorous training and retraining of employees. This is why it is important to understand how the financial institutions in those countries were managing to strike a balance between the opportunity to be more innovative and the challenge to service a population with a lower per capita income and large share of the "unbanked". In particular, it is interesting to identify how using ICTs and providing e-banking and e-payment services could improve the access to finance of SMEs and microenterprises, which

---

Box 5.1

Alternative payments systems[19]

The largest alternative Internet payment provider with a strong position in the Internet auction market is PayPal (which belongs to e-Bay). Customers hold accounts with PayPal and send by e-mail payment or withdrawal-related instructions. They fund those accounts either directly from their bank account through ACH transfer or by using cards or cheque deposits (BIS, 2004). PayPal has doubled the number of its accounts in just two years and has now over 143 million accounts in 190 countries (of which 35 million are in Europe). PayPal is no longer just in the business-to-consumer market. Beginning in mid-2004, it began directly offering its services to small businesses as a conduit for business-to-business transactions that do not involve an e-Bay purchase. In fact, nearly 40 per cent of its traffic is not related to E-Bay. However, most e-commerce sites do not suggest PayPal as a payment agent. Fund transfers via web telephony such as Skype and mobile phones are also fields that the company is exploring. PayPal is neutral in its banking arrangements; it is not aligned with a specific bank or consortia of other payment service providers.

Other potentially emerging players in this field include Google and Yahoo. For example, Google's Checkout uses its powerful advertising engine to encourage businesses advertising in Google to also use the services of Checkout. Microsoft is also planning to launch an online micro-payments service.

To create trust and build confidence in third-party service providers, especially aggregators, CBs need to monitor them further and provide clear guidelines and regulatory framework for such services.

represent the core of those economies. This underlines the importance of further research into the use of e-banking and e-payments in SME finance, as well as in microfinance.

There are a number of features that the customer values in a financial service. These include acceptability, accessibility, affordability and ease of use. As far as businesses are concerned they look for a number of other features, ones that will help ensure that they provide a cost-effective service. They include functionality, segmentation (the ability to sell different products to different groups of customers), competitive fees and charges, increasing efficiency, controlled development costs, distribution among the population and partnerships with other companies. Financial literacy is important for increasing e-banking among the poor, and there are a number international programmes, for example set up by the World Bank, to increase this literacy (Hadidi, 2003).

Much of the literature on e-banking focuses on case studies and comparisons in developed countries, where it is well established and in some cases is not far form the point close to saturation. However, e-banking is also increasingly taking root in many developing and transition economies, and in some of them its penetration exceeds OECD median indicators.

For example, Internet banking has been surprisingly successful in Estonia. The first bank to introduce Internet banking was Hansabank in 1993. There are several reasons for this success. First, since the country had previously had a command economy, the traditional commercial banking did not exist, and hence consumers did not have time to grow attached to the branch model used for many years in other countries. As a result, consumers did not value or expect much from the bank branch experience. Second, banks were still developing and did not have full country coverage. They lowered their costs and improved their image. The success of Internet banks has been largely due to their wide appeal and the Internet technology available for reaching sparsely populated areas; that availability has been due in turn to government policies to deregulate, increase competition in the telecom markets and increase Internet use (Kerem, 2003). In fact Estonia was ranked quite high – ahead of some OECD countries – in the Networked Readiness Index of the World Economic Forum and the E-Readiness Ranking of the Economist Intelligence Unit.[21]

Internet banking has grown rapidly in Brazil and has been the fastest-growing banking medium (Diniz,

Porto and Adachi, 2005). In Turkey e-commerce is also most developed in the banking sector and driven by reduced costs, transaction costs on the Internet being only 5 per cent of those at a branch. The first bank to use the Internet in Turkey was Isbank in 1997 (Akinci, Aksoy and Atilgan, 2004).

The Nigerian Inter-Bank Settlement System (NIBSS) has recently developed the Electronic Fund Transfer Service, which focuses on giro and ACH functionalities. NIBSS is collectively owned by Nigerian banks and is responsible for transactions including final settlement with the Central Bank. Launching ACH and adopting the electronic cheque presentment with image exchange system have shortened the clearing cycle still further, as well as reducing fraud and cutting costs.[22] Despite great ambitions for Internet banking in Nigeria, these were thwarted by technical problems. This was particularly the case with the national telecommunication network, when banks have been led to seek alternatives to it, but in an uncoordinated and hence costly manner. Nevertheless, consumer perception of the effect of IT on the banking industry stayed positive (Idowu, Alu and Adagunodo, 2002).

Thanks to the policies of the Central Bank and the well-developed business sectors, South Africa is particularly well suited to electronic banking, which includes ATMs and other features. Those features make them attractive to poorer customers (Cracknell, 2004). However, there is also the psychological barrier, especially among the poorest (Hadidi, 2003). Singh and Malhotra (2004) conducted a survey to determine why some South African customers are reluctant to bank online. They found that the more affluent and younger are the groups that are most frequently banking online. For those groups that did not bank online, the main reasons for not doing so were fears about security and lack of knowledge. At the end of 2003 online bank accounts in South Africa topped the million mark for the first time, amid claims that the media had grossly exaggerated security fears but been mainly ignored by the public.[23]

The Reserve Bank of India (which is the Central Bank) has been heavily involved in the development of e-banking and has drawn up guidelines for banks to follow when providing Internet banking services. The development of Internet banking in India began with the foreign and private banks, with the public banks lagging behind (Singh and Malhotra, 2004).

In Romania foreign banks also led the way in electronic banking, beginning in 1996, and local banks followed later; however, usage rates have remained low (Octavian and Daniela, 2006). Banks need the approval of the Ministry for Information Technology and Communication before they can offer Internet banking services. There are currently 28 banks offering such services. Most of the approved banks use VeriSign to guarantee their sites and also employ methods such as Digipass Pack to identify users and secure transactions.[24]

Siriluck and Speece (2005) suggest that Asian cultures attach greater importance to interpersonal contacts and relations than those in the West, and as a result Internet banking faces an extra hurdle. Malaysia is especially interesting for Internet banking because of the high level of technical education and the quality of the technology available (Goi, 2005). In a survey of over 500 people in that country the use of online banking was seen to be concentrated around younger and higher-income groups. The quality of online banking websites was high, although not all banks provided all services and options via their websites. Most Internet customers had a positive perception of Internet banking; however, there were also security concerns (Ainin, Lim and Wee, 2005). Foreign banks were allowed to provide banking over the Internet as from January 2001 (a year after domestic banks). But this delay did not protect domestic banks from foreign competition, even though foreign banks offered less technologically sophisticated services (Yeap and Cheah, 2005). Regulation took the form of the Central Bank's minimum guidelines on the provision of Internet banking issued in 2000, which require banks to meet with customers face to face before an Internet account can be opened or credit given (Goi, 2005). The Central Bank of Malaysia has developed research collaboration between Asian CBs and the BIS on technology issues, and has also established minimum guidelines and standards for risk management as well as monitoring of compliance and risks.

Electronic cards are becoming one of the channels for e-payments in countries with a low level of banked population. Figure 1 shows transaction flows related to regional payment cards project of the West African Central Bank (WACB) created within the framework of the Economic and Monetary Union of West Africa. The common currency in the Union helps banks in the region to avoid foreign exchange risk in their mutual transactions and has enabled WACB to develop a card-based payment system called "Monétique UEMOA" (in addition to other two regional RTGS and ACH systems). The system has an inter-bank card payments processing centre (CTMI in French; see chart 5.1)

that allows 63 commercial banks in the region to issue payment cards for corporate and individual users.

The phenomenal growth of mobile telephony in developing countries has also brought about a dynamic expansion of so-called m-banking and m-payments.[25] They are becoming very popular in small purchases through m-commerce as well as in P2P transfers such as remittances, or in B2B payments, especially in microfinance (see following sections). Since most mobile phone owners in developing countries are "unbanked", they participate in e-payments by buying prepaid cards at various points of sale and then make their m-payments. Given the relatively high level of mobiles' penetration in developing countries, they can become the main channel for delivering online payments, especially in countries and regions with few branches and ATM networks. Chart 5.2 shows the relationship between mobile phones, ATMs and bank branches in Thailand and South Africa is similar to that of the United Kingdom, although the penetration is still at a relatively lower level. At the same time, in the case of China, Kazakhstan and Egypt and to a greater extent other African countries, the penetration of mobile handsets is far ahead of that of bank branches and ATMs.

A mobile payment scheme called M-PESA is being successfully run in Kenya by Vodafone, local Safaricom with the support of the United Kingdom Department for International Development. Around 150,000 M-PESA customers with the help of around 500 agents, can make payments or send money using their mobile phones. The customers' phones are charged with prepaid funds in Safaricom shops, petrol stations and other points of sale as soon as the owners of the mobile phones, whose identities are verified via their mobile numbers, pay the cash. M-PESA has developed its own financial system, including clearing and settlement, and keeps the float in a single account in the Commercial Bank of Nairobi. As long as it does not pay interest or invest this money, the Central Bank of Kenya tolerates such financial activities.[26]

The following sections will focus on three important aspects (from the development perspectives) of financing destined to less endowed groups of population and enterprises and namely remittances as well as e-finance for SMEs and microenteprises. The introduction of digital delivery channels in those areas is providing new opportunities to combat poverty by improving access to finance through the better use of new more accessible for those groups tools of ICTs.

Chart 5.1

Architecture of the CTMI-UEMOA

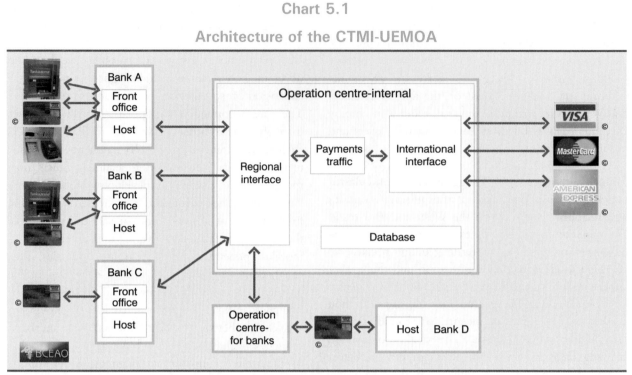

Source: A Altine, Opportunities and challenges of e-finance in the UEMOA subregion, West African Central Bank, 2006
(see www.uneca.org/e-trade/).

Chart 5.2

Penetration of bank branches and mobile phones*

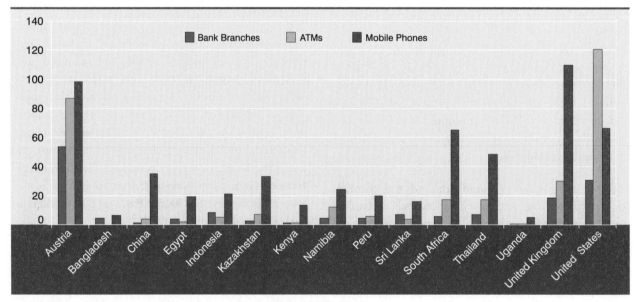

Sources: Beck, Demiurg-Kunt and Martinez Peris (2005); ITU (2007)
*Bank branches and ATMs per 100,000 people; mobile phones per 100 people.

2. ICTs and remittances

Remittances, representing mainly small-scale transfers by migrants, expatriates and to a lesser extent charities, are becoming an increasingly important source of external financing for many developing and transition economies and as such merit careful examination. As a distinct item of balance-of-payments (BOP) financing their volumes in the case of many recipient countries are comparable to other sources of foreign exchange such as exports, foreign private capital inflows and official aid. However, unlike others, those flows, also known as unrequited or unilateral transfers (i.e. not incurring debt-related obligations), are a result neither of earnings by country residents due to exports of goods and services, nor of their borrowing or investment-attracting activities.

According to the World Bank, remittances sent by migrants to developing countries reached $206 billion in 2006 (World Bank, 2007, p. 54). However, it is estimated that the real level of current remittances could be 50 per cent higher owing to the major role of various informal remittances networks (Coyle, 2007). In other words, the overall volume of all remittances might come close to that of FDI. According to UNCTAD, FDI flows to developing economies totalled $379 billion (UNCTAD, 2007).

Compared with other types of financial flows, remittances are the least cyclical source of BOP financing. They play a stabilizing and even countervailing role when countries, in adverse economic circumstances, suffer from reductions in flows of external finance coming from other sources and especially from short-term portfolio capital. That positive effect is due to the

underlying motivation behind those transfers, which is primarily to support families back home. Remittances allow the consumption level of the poor to be sustained, giving them better access to basic facilities, including health and education facilities, in recipient countries. Remittances also help to launch and sustain small family businesses.

Historically the main part of those funds was transferred through informal systems such as *hawala*. *Hawala* is a form of informal networking whereby *hawala* brokers intermediate transfers without actually moving funds. They work on the basis of trust with minimal documentation, and each tries to have enough funds to meet the demand of correspondent brokers (Fugfugosh, 2006). As in the case of credits, informal

Chart 5.3

Financial flows to developing countries ($ billion)

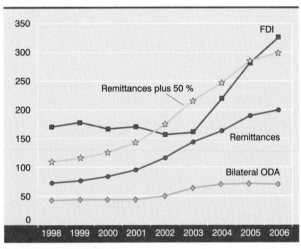

Source: World Bank (2007)

Box 5.2

Defining remittances in the balance of payments

In 2008 the new IMF BOP Manual is expected to take into consideration recommendations by the United Nations Technical Subgroup on the Movement of Natural Persons and replace the term «worker remittances» by three separate terms:

- Personal transfers (PTs), which are transfers in cash or in kind received or made by resident households from or to non-resident households. Sources of income, relationship and purpose of transfer are not taken into account.

- Personal remittances (PRs), which are the sum of PTs, net compensation of employees and capital transfers – that is, they involve all household-to-household transfers and take the perspective of the recipient country.

- Total remittances (TRs), which are the sum of PRs and social benefits paid to households from other institutional sectors. They also take the perspective of the recipient country.

Source: Migrant remittances to regions neighbouring the EU. ECB Monthly Bulletin, February 2007, pp. 91–92.

intermediaries were charging the remitters passing through their channels high fees. At the same time, the fees collected by major formal money transfer service providers were also initially quite high, considerably exceeding, for example, the level of interchange fees paid by merchants for credit-card processing. However, with the advent of the Internet and further dissemination of electronic means of payments, various money transfer service providers started to carve large chunks of payments out of the cash-based systems. Competition, and in particular the use of the Internet, made it possible to considerably reduce fees charged by such major operators as Western Union and Moneygram. The competitors from regions within the main corridors of remittances were quite creative in using various delivery channels, including Internet-based networks or m-payments, and hence considerably cut their costs of operations. They were also less strict than major Western operators in their perceptions of risks, given the extremely short periods of time needed to complete the e-remittance operations.

As the example for 2004 in table 5.1 shows, regional systems such as Anelik and Unistream were charging much less for transfers from the Russian Federation to Armenia than was the case for Western Union. That competition has paid off and while for the same corridor Anelik is still charging around 3 per cent for the transfer of, for example, $200, Western Union cut its fee to around 5 per cent, which is much less than in 2004.

In countries with large migrant populations banks are trying to attract migrants as bank clients and are using remittances as a tool to attract more deposits from them. For example Chase has introduced a scheme for

the free transfer of money to Mexico, enabling cheque-account customers to make a limited free transfer of funds (up to $1,500 per month) to its partner bank Banorte, which has 1,000 branches and 2,800 ATMs in Mexico. The recipients of those funds can either keep them in accounts in Banorte or collect cash without paying a fee. According to the Central Bank of Mexico, remittances from the United States totalled $23 billion in 2006 and are expected to reach $24,5 billion in 2007.[27]

A relatively new phenomenon is the use of mobile phones as a means of transferring money between countries or within developing countries from urban to rural areas. Sometimes those areas may even lack bank branches, and the only way to cash the transfer may be to use the specialized points of sale that also sell prepaid cards.

3. E-banking and e-payments for SMEs

To access finance, SMEs have to compete with households and large corporate businesses in order to gain a share of the financing disbursed by banks. Traditionally, banks were biased against SMEs given the relatively higher costs of acquiring and processing transactions for SMEs in the light of a wide geographical spread and the greater number of low-volume transactions. That resulted in high unit transaction costs related to SMEs' financing. Modern e-banking with online credit risk databases and much lower transaction costs related to data processing and data mining is providing new opportunities for improving SMEs access to, in particular, short-term working capital and trade finance through e-banking and other e-trade finance services. It is becoming possible for SMEs to manage their own accounts and payments traffic online, while banks can not only provide finance, but also help them manage their cash flow and ensure a quicker turnaround of working capital.

For financial institutions, servicing this sector requires the finding of solutions that take into account the specific risk-return perspective of the sector. For the most part, the challenges to SME funding result from lack of information, such as unclear financials, and the fact that information about SME borrowers is being limited. There is also a lack of comparative data on SMEs. Banks in most developing countries do not have access to transaction data and credit history (unlike in many developed countries) through credit bureaux and other credit management service providers

Table 5.1

The cost of transferring $50 and $200 to Armenia in 2004

	Cost as %	
	$50	$200
Anelik	1.5–3.0	1.5–3.0
HSBC	70.9	18.5
Unibank (Unistream)	1.0	1.0
Western Union	26.0	11.0

Source: Roberts & Banaian (2005).

such as credit insurers and factors. Developing such services primarily through the appropriate use of ICTs – offering financing and credit management through open networks such as the Internet – could overcome impediments related to SME financing.

E-banking benefits are especially apparent to banks when doing business with SMEs. SME business was labour-intensive and offered relatively low profits for banks, whereas SMEs were finding the service provided by banks too "one size fits all", inefficient and supply-driven. E-banking and other e-finance techniques might address those problems through automated, 24-hour procedures and lower costs. SMEs can then use their new efficiency and access to credit at lower rates to fuel growth in developing countries (Mu, 2003). However, as Wendel (2006) and others emphasize this debate is not much different from the debate related to retail banking. The role of the traditional branch, especially when dealing with SMEs suggest that the "click and mortar" approach will continue.

In addition, the reduction in costs associated with the possibility of having better-organized information flows, greater transparency and hence a better loan-tracking process can be considered to be positive externalities of ICT adoption in SME financing. The new e-financial tools that are designed for SMEs and their needs are supposed to help in gaining those advantages. Even though the development of those systems is not very time-consuming, its finalization requires quite a long time, and the systems cannot start to be remunerative until their integration process has been completely finished.

Investments in this field cannot be remunerative unless they are associated with a change in business process (Firpo, 2006) and improvements in skills at different levels. As a consequence, ICT use can be considered a new approach for both human capital and the organization of SMEs, and the means to achieve better financial results. Local requirements are certainly one of the most important issues for the success of this technology in developing countries. Delivery channels and product mix are two important aspects for the success of each company. However, the high costs of the comprehensive adoption of ICTs often imply the need for public financial support or various forms of public–private partnerships with development and export-import banks and other supporting institutions.

In order to create a public record of past payments SMEs in developing countries need to move out of informal economy. By fully formalizing themselves SMEs might create for themselves a public reputation as of good-quality borrowers. Under such circumstances financial institutions can create electronic credit histories and Internet-based credit risk databases in helping creditors to provide capital with fewer and less stringent requirements. A responsible behaviour by firms can lead banks to trust SMEs more, requiring fewer guarantees for loans and creating conducive borrowing environment. As a consequence, such a change in SMEs way of operating can have important effects for the whole banking sector. Indeed, there is a need for a larger formal economy, characterized by a business-friendly regulatory and institutional environment, including a more competitive banking sector that will provide credits to SMEs with less onerous collateral requirements and with lower interest rates. A larger formal market, including the majority of SMEs, would also attract other financial service providers that will contribute to credit risk management and financing; this will lead to more competition in the financial sector.

The lending infrastructure plays a very important role in the efficiency of ICT use. Indeed, the introduction of internationally acceptable accounting and payment performance information of a good standard are fundamental for credible financial statements and are a precondition for the feasibility of loan contracts. Also, the legal and institutional infrastructure strongly influences the adoption of ICT necessary for backing up e-finance for SMEs. Commercial codes, property rights and bankruptcy laws affect the degree of confidence in financial contracts. Moreover, even tax and regulatory environments can alter SME credit avilability in a negative way. In some cases, regulatory restrictions on the entry of foreign banks into the domestic market and other restrictions on foreign banks' operations may negatively affect SMEs' access to credit. As a consequence, the availability of credit for SMEs depends on the quality of regulatory and institutional environment in general and lending insitutions in particular.

The structure of financial institutions impacts on SME lending and can lead to different treatment for SMEs operating in similar markets. Large institutions can run economies of scale and offer competitive interest rates, even though they often analyse firms only from the financial ratios perspective. However, large institutions certainly have more agency problems than smaller ones characterized by fewer internal hierarchical separations. As a consequence, the latter can manage soft information more easily than larger institutions. In developing countries it seems consistent for foreign-owned institutions to have more problems than domestic ones in managing soft information. Nevertheless, a combination of better technology, easier capital market access, greater capacity to diversify portfolios and better working skills can make foreign-owned banks more efficient. Also State-owned banks in developing countries, in order to succeed in financing SMEs, need to pay more attention to modernizing their delivery channels through better use of ICTs. The use of various modern ICT-intensive delivery channels and more competition might attract new lenders in the SME finance market and result in better access for SMEs to e-banking and other e-finance facilities.

4.   Microfinance

Microfinance, by definition, constitutes credit and other financial services of a low monetary value given to microenterprises and households primarily to encourage their productive activities as a means overcoming the poverty trap. In addition, the aim of microfinance institutions (MFIs) is to reach out to the poor, who have been left outside the traditional mainstream financial services. This implies that a very high volume of small transactions is a defining characteristic of MFI operations.

As in the case of SMEs, the main reasons why commercial banks do not regard the population that microfinance targets as worthwhile customers is the fact that serving this population involves a high volume of low-valued transactions. Traditionally high transaction costs and other market failures were a barrier to supplying small-scale credits, even though the interest rates charged could be relatively high. However, the cost of formal microfinance credits are still normally much lower than the usurious terms charged by informal creditors, especially in rural areas (see chart 5.4). In fact, the main target populations for microfinance are peasants, and tapping into that market is a strategic goal for microfinance (Ferro, Luzzi and Weber, 2006).

One of the reasons for the cost of microfinance that is acceptable to borrowers is the fact that many MFIs receive funds from NGOs or donor agencies. However, that is starting to change. While many are still donor- or public-investor-dependent, others are becoming more market-oriented, profitable and (it is to be hoped) sustainable. The proliferation of various types of MFIs is bringing about a rapid increase in their numbers in developing countries. Indeed, as chart 5.5 shows, such early adopters of MFI as Bangladesh and other countries in South and South-East Asia are leading in terms of penetration.

One of the key factors distinguishing top-level MFIs from their peers is the adoption of technology. In fact, some MFIs are becoming quite sophisticated in their use of ICTs (Ivatury, 2004). The obvious benefit that the MFI industry derives from ICTs is the reduction in costs terms of the MFIs' back-office operations

---

Box 5.3

Using ICTs in some key SME financing channels

E-trade finance based on financial statements and payments record. Some specialized financial service providers provide short-term working capital and trade finance on the basis of combining dynamic analysis of a given SME borrower's cash flow with its financial statements and ratios. As SME receivables are the primary source of repayment, obtaining online information on the borrower's flows of receivables and payables makes it possible to ensure that there are no abnormal fluctuations threatening debt repayment. It also makes it possible to send early warning signals to the borrower in such cases. Moving all that information online and using models to trace abnormalities in cash flow, permit banks to automate the process and to continue financing the successful SMEs.

E-credit information and SME credit scoring. To obtain further financing it is in the interest of SMEs to provide information enabling credit histories to be compiled and thus showing their capacity to repay loans on time. E-credit information also includes information-gathering from independent sources. As a result SMEs receive credit scoring or rating based on credit history databases and on that basis obtain better access to credit.

E-credit insurance and e-factoring. These are the channels adapted to SMEs' needs, protecting suppliers' receivables and making it possible to provide supplier credit to buyers. To provide credit protection through credit insurance or to discount accounts receivables, service providers need to have large credit risk databases composed mainly of SMEs, and a knowledge of buyer's country political risk. All those information and transaction tools are currently available online.

and of delivering their services, which should mean lower fees for MFI's customers. However, there are additional benefits derived from the introduction of ICTs, such as easier tracking of loans and repayments, standardized processes and better flows of information in organizations, and more transparency. Also, MFIs may be able to offer additional features and services that would not have been possible without the adoption of ICTs.

With regard to clients, ICTs can be used to expand the range of products and services offered by MFIs to their clients. The main technologies that have been identified are ATMs, point-of-sale (POS) devices and mobile phones.

There are different types of ATMs: they range from those that dispense only cash to those that offer a full range of services, such as accepting deposits and transferring money to different accounts. ATM networks are generally quite expensive to own and operate, although costs are falling thanks to improvements in technology, which may no longer be as prohibitively expensive as before (Baur, 2006). An ATM network allows customers to conduct transactions at their own convenience, and provides banks with a cheaper way of handling a high volume of transactions.

A POS network involves devices installed in stores and other appropriate outlets which enable customers to make their payments and transactions using their cards. These cards can be of the traditional magnetic-strip variety, or the more advanced, and more expensive, "smart cards" with an embedded chip, which can contain detailed transaction records and personal data to enhance security and facilitate transactions. Cards can also store value on a prepaid basis and be used as a means of payment (Cracknell, 2004). POS devices tend to be cheaper than ATMs, and the outreach of a network of POS devices installed in small stores and shops can be greater than the outreach that can be achieved through an ATM network. As in the case of ATMs, such a network allows the MFIs to process a large volume of transactions more cost-effectively than over-the-counter operations. However, a POS network tends to be limited regarding the type of transactions that can be handled, and there are operational, and even regulatory, complexities that arise from the fact that retail outlets are in effect acting as agents for the MFIs.

Lastly, mobile banking is an especially attractive channel in countries where a fixed-line network is not well established. In much of the developing world, mobile

Chart 5.4

Average annual interest rates

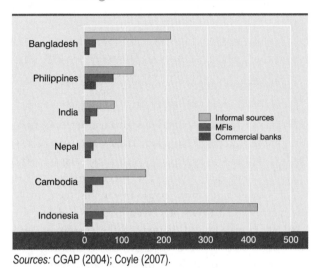

Sources: CGAP (2004); Coyle (2007).

Chart 5.5

MFI penetration rates

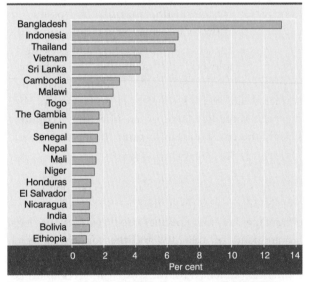

Sources: Daley-Harris et al. (2005); Claessens and Feijen (2006).

telephony has overtaken landlines in terms of access, volume of calls, and even affordability and reliability. People that own mobile phones but who have never had a bank account are now a widespread phenomenon. Some MFIs have tapped into this opportunity so as to offer microcredit loans and other financial services via menu-driven systems or via SMS text messaging. The advantages of these channels include simplicity and user-friendliness, and the fact that the infrastructure is in many cases already in place (Cracknell, 2004).

Apart from lowering transaction costs, the primary additional benefits of having these new channels of delivery are the increase in the volume of transactions by increasing their accessibility and expanding outreach, especially to more remote places, and, by extension, the increase in the convenience for customers.

As in e-banking in general, a number of factors have been identified that determine the value attached by customers to moving away from cash to electronic solutions in microfinance. The additional features offered by the new technology to customers include acceptability, accessibility, affordability and ease of use (Cracknell, 2004). Stemming from these features, the key drivers in determining the success or failure of client adoption of a new technology within the context of microfinance are its perceived increase in value; the level of education and training of the customers, and thus their comfort level in using this technology; its usability; the cultural fit of the technology within the community; and whether or not clients have sufficient trust in the new technology (Ivatury, 2004).

To reach greater levels of efficiency in organizations, ICTs should be integrated into the management information system (MIS) to make it possible to automate the flow of information and thus ensure the smooth running of agreed procedures and networking between employees. Although this is technically incorrect, the MIS of a bank or financial institution is treated like its back-office operations. For an MFI, there are three key areas of an MIS: the accounting system, the credit and savings-monitoring system, and the client impact data-gathering system.[28]

Existing software packages for the accounting system are abundant and easily found. They differ widely depending on the specific needs as well as legal requirements of the MFI and the country in which it operates. The decision on a credit and savings monitoring system, however, is often more complicated. There are no real standards or guidelines. This is because a well-thought-out system should fit the needs and operations of the MFI using it. MFIs differ from one to another and hence there is no off-the-shelf software that can address the requirements of every MFI. They are at the same time less complex in terms of operations, than commercial banks (Ahmad, 2006). The software should facilitate transfer of data to and reconciliation with the accounting system. Integrated systems do precisely that, but they are usually quite expensive, especially if a high level of customization is required. The third system, for client impact, is usually designed informally, if at all (CGAP, 1998, 2001).

In order to maximize the benefit of cost efficiency and for the technologies to be sustainable, there needs to be a sufficiently large volume of participants. This calls for a shared infrastructure among MFIs. However, experience has shown that sometimes MFIs resist cooperation for fear that their competitive advantage will be lost (Firpo, 2006).

There are benefits in addition to increased efficiency and lower operating costs that MFIs can reap from technology – for example more informed decision-making, increased flexibility, and better transparency and reporting, the latter being especially important for MFIs that are dependent on donor funds (CGAP, 2006).

Generally, MIS can be categorized into three types: manual systems that are still in being used by very small organizations; semi-automated systems, which do not completely satisfy organizations' information requirements; and fully automated systems. Most MFIs are in a semi-automated mode. No matter what the level of automation is, for an MIS to be effective, it must be cost-effective and fulfil business requirements, while being flexible regarding future changes, reliable, simple to use, scalable to accommodate the growth of the business, and integrated amongst the branches of the business to produce a single consolidated picture (Ahmad, 2006).

Because many MFIs have branch offices, as part of their mandate to reach out to geographically remote areas, the issue of networking needs to be addressed. The World Bank suggests that each branch should at the very least have its own MIS and database. If resources allow, the MFI should consider networking all branches together, which would allow real-time access to up-to-date information, which in turn would greatly improve information efficiency. However, setting up and maintaining a network can be quite costly, adding another level of complexity to its day-to-day operation. Moreover, many of the branches in the rural areas of the developing world are constrained by the level of reliable communication infrastructure available, and establishing a network may not be feasible at all. Furthermore, in locations with harsh environments, the need to support and maintain hardware equipment in the branches may even outweigh any benefits provided by the technology (Gagnon 2006).

The recurring lesson from the experience in the field has been that ICTs are not a fix-all solution. A careful analysis of an MFIs' operations and its exact ICT needs should be conducted before an investment decision on the implementation of a new system, be it on the client delivery side or in the back-office. So far, even among leading MFIs, using technology productively is still an exception rather than the rule (Ivatury, 2004). Therefore, while it is recognized that most MFIs do eventually need an MIS system, it may not always be abundantly clear to decision makers that such an investment is needed. A balance needs to be sought between solutions that are appropriate and those that are state-of-the-art but are of limited practical use because of constraints in the local context. Such constraints could be anything from an inadequate telecommunications infrastructure to the prevalence of illiteracy. Technologies need to be not only innovative, but also appropriate and designed to scale (Firpo, 2006).

Since in general the cost associated with building the infrastructure for the new technology is too high for MFIs to bear on their own, external funding sources are often needed (Firpo, 2006). The management of the MFI and the donor organization or outside investor need to ask questions about the proper fit of any proposed adoption of technology with the strategy and structure of the institution, as well as carefully weighing the monetary cost and the ongoing resource requirements needed to maintain the system. As part of that process, Ivatury makes four recommendations about how to overcome the common challenges that an MFI faces in tackling technology projects. First, manual processes should be improved and streamlined; second, the decision makers need to ask the right questions; third, independent expert advice should also be sought; fourth, there should be plans to train staff on an ongoing basis.

E.   Regulatory issues related to e-banking and e-payments

E-banking and e-payments are increasingly being scrutinized worldwide by regulators that are trying to ensure financial stability without impeding technological innovation, especially in the banking industry. The main CBs are coordinating their efforts in this field with support of BIS through the Basel Committee on Banking Supervision (the latter deals with IT-related operational risks), the Committee on Payment and Settlement Systems and other forums. Consensus achieved here is not binding for developing and transition economies' CBs, but the latter normally consider above frameworks as those generating best practices and normally follow their recommendations.[29] The CB's have a dual role as facilitator and regulator

of e-banking. The United States Federal Reserve is an important example, as it operates an ACH and Fedwire. It has introduced FedLine for the web and other electronic products (Blakeley and Matsuura, 2003). Regulation is especially important now since e-banking may be subject to security breaches that can undermine confidence and thus the stability of the banking sector. In terms of regulations the Basel Committee has laid down guidelines, which the CBs have elaborated on (Bernanke, 2006). For example, the Federal Financial Institutions Examination Council in the United States has developed them in more detail and the Singapore Monetary Authority has developed its policy statement on Internet banking. But more legislation is needed, including on such essential elements as electronic signatures (making them legally binding), money laundering and others (FFIEC, 2003).

There are several tools that regulators can use. These include existing regulations and laws adapted to the requirements of above-mentioned Committees, and making sure that banks have adequate facilities for training staff in new technologies. The legal definition and recognition of new methods of conducting transactions and the international harmonization of regulations are important for preventing a "race to the bottom" and also properly defining the roles of host and home country jurisdictions as regards international banks and cross-border customers (Nsouli and Schaechter, 2002; Mu, 2003). It is important that new risks and challenges be integrated into the evaluation criteria for banks. Ironically, regulation is most needed where it is hardest to implement, as Ezeoha (2006) explains in the case of Nigeria, where there is a need to protect an emerging banking industry and also the country's international reputation by addressing "Internet fraud" and deeper problems in the politico-economic system.

However, even once laws are in place careful implementation is needed. In South Africa banks have been accused of unjustifiably requesting online customers in cases of identity theft to sign away their rights to money that might be eventually returned. Also, the standards of banks' security training and advice on their websites are still relatively low (Clayton, 2006). Another regulation-related issue is the need for a comprehensive system. For example, in India the data protection law emphasizes that hackers should be punished. However, firms are under no obligation to prevent them from being able to hack into the systems in the first place. Thus the law does not really serve the consumers' interest (Gupta, 2006). Here also legislation prevents anyone other than a security guard and the customer from being present at an ATM, and thus new customers cannot be taught to use ATMs. In spite of major advances in financial sector liberalization, Indian banks were not allowed to compete with rural regional banks, and hence expand a network of ATMs to optimize their value (Cracknell, 2004).

In South Africa security is still a major problem, especially since organized crime groups have developed software to steal passwords and identities (Meyer, 2006). The law is uncertain, both in its terminology and whether or not it applies to online banking. South African banks have experimented with several measures. Nedbank has an SMS authorization system whereby a payment needs to be secured by the customer's providing a unique reference number sent to his or her mobile phone before it can be completed. This is mobile-phone-based and hence avoids viruses and data miners on the customer's computer. Standard Bank, the second largest bank by market share (Singh and Malhotra, 2004), has also developed a similar system, whereby passwords are supplied to phones and not via a computer (Rombel, 2005).

In relation to self-regulation, the World Bank has made a number of policy recommendations for e-banking. These include establishing a comprehensive security control process, centralizing back office operations; developing an automated credit authorization system by developing an appropriate credit scoring system; comprehensive oversight of outsourcing partners; and proper inclusion of e-banking risks in the overall risk assessment. Authorities should provide oversight of security controls and of outsourcing and partnership arrangements, provide the infrastructure and human capacity for technology adoption, and create appropriate regulatory frameworks. By doing that they strengthen the payments and settlement systems, as well as improve transaction-reporting services. Also, they should use new technology to better disseminate information, especially relating to credit, and to establish set up registers for collateral. Regulation by the public sector is needed, they argue, in order to organize security procedures and standards for electronic signatures (Mu, 2003).

As regards at the "bigger picture", the role of CBs has also changed with technology (Quaden, 2002). They need to adapt monetary-policy-setting rules related to the use of new technologies. Since e-banking makes transactions cheaper and easier, users of money do not have to hold cash balances for as long, and incur the losses of doing so. There is also increased access to interest-bearing assets via the Internet, especially to households and SMEs, and thus there is more incentive not to hold cash balances. This means that money demand is more sensitive to interest rate changes (Fullenkamp and Nsouli, 2004). The role of the CBs is to ensure, together with monetary stability, the stability of the banking sector. This means they must adapt to the changes in banking technologies and techniques. That includes developing secure and technologically advanced payments and settlement procedures, especially cross-border ones. Broadening the scope of CB research beyond the effect of their policy on the real economy is one of the important steps in that respect. The continued implementation of the Basel accords should assist in this transition (Bernanke, 2006). Finally, there is a need to recognize the bypassing of banks by non-bank electronic payments providers (Bounie and Gazé, 2007), and thus a need to develop a regulatory oversight over those institutions as well.

F. Conclusions

E-banking is increasingly becoming Internet banking or online banking. Other electronic means such as ATMs continue to play a major role, especially in the context of developing countries. Moreover, various forms of basic card-based operations and m-payments are becoming important elements of e-payments in many of those countries.

Domestic and regional banks in developing countries have tried to catch up with international banks and have developed websites that perform various tasks ranging from provision of information and product marketing to e-payments and complex online trades. More facts are confirming the early findings that Internet banking (if only as part of "click and mortar" business models) can increase efficiency and hence profitability, and provide services not provided before. Moreover, while online banking can be initially biased towards more sophisticated and affluent customers, its dramatically lower unit transaction costs can permit much better access to finance and e-finance for SMEs and micro-enterpises as well as households in general. Thus there are possibilities for expanding the customer base with further efficiency gains for both banks and customers.

Customers can be encouraged to take up e-banking if they are supported in beginning to use the Internet for other activities. Programmes to increase computer literacy and assist banks in developing secure, user-friendly websites should be put in place. They should target in particular those segments of the population that have been slow to accept Internet banking. This can be done as a cooperative effort by banks, national authorities and international organizations.

In the case of ATMs and m-payments the issue is less of encouragement, and more of providing infrastructure and service, which are already attractive.

The security risks do not disappear as Internet banking becomes established. This leaves room for the introduction of better security systems by financial service providers and their technology vendors, better regulation in finance and telecommunications, and improved coordination between them. Information and educational programmes empowering consumers with better ways of securely using their means of electronic access to finance should also be put in place. Regulation needs to be more than the passing of laws and issuing of guidelines, and should look at implementation and tackling the root causes of problems.

The major inter-bank messaging networks, and in particular SWIFT, have evolved into online services and will continue to develop as such. The challenge for them is in the trade-off between further expansion and the need to maintain a high level of security. As developing country CBs and banks are developing their own ACHs and RTGS systems and are increasingly participating in regional and global RTGS systems, it is important that their efforts be coordinated, such coordination helping them to increasingly integrate into the global payments system.

The e-payments issue also includes the future of payment cards that are increasingly adopting smart-card technologies. The market is an interesting one. While it is still dominated by a few players, it is also accepting new players, who compete vigorously and have pushed for the altering of business models in recent years. It remains to be seen where the payment card markets go from here and how they adapt to emerging challenges.

The emergence of mobile devices as a means of e-payments and the development of m-banking might also have a promising future, especially for the

"unbanked" population, which is still in a majority in many developing countries.

Once reliable and operational infrastructure is in place, e-banking can increase efficiency and improve service, or simply provide a service where there was none before. The challenge here is the correct level and style of regulation and encouragement for e-banking by CBs.

While e-banking penetration might reach in the coming decade saturation point in some developed markets it still has to be introduced in the majority of developing countries and has a long way to go before those countries mature in terms of using e-banking. While many countries face the dilemma of the relatively low cost of labour and the need for a high level of initial investment in ICTs, the importance of having a modern financial sector increasingly pushes for the second solution at least at the level of major local banks. The same is also true for remittances, SME finance and microfinance, provided that ICTs with the appropriate level of sophistication are put in place in each case.

As for microfinance, basic and low-cost ICT solutions could be the right answer. Also, private – public partnerships to serve the financial needs of the poor is one of the keys to addressing the issues of adequate investment in MFIs and their technological infrastructure, including hardware and software applications and trained personnel.

More generally, CBs and ministries of finance should carefully consider integration and intertwining of strategies to further develop the use of ICTs in financial services. The creation of a supportive environment by policymakers in developing and transition economies to facilitate the development of what is the most information-intensive sector of the economy is essential for encouraging financial service providers and their competitors such as mobile network operators to start and further develop the e-banking and e-payment services.

References and bibliography

Adelaar T, Bouwman H and Steinfeld C (2004). Enhancing customer value through click-and-mortar e-commerce: implications for geographical market reach and customer type. *Telematics and Informatics* 21, pp. 167–182.

Ahmad A (2006). Management information systems (MIS) for microfinance. Foundation for Development Cooperation, Brisbane, Australia.

Ainin S, Lim CH and Wee A (2005). Prospects and challenges of e-banking in Malaysia. *Electronic Journal on Information Systems in Developing Countries,* vol. 22, no. 1, pp. 1–11.

Akinci S, Aksoy S and Atilgan E (2004). Adoption of Internet banking among sophisticated consumer segments in an advanced developing country. *International Journal of Bank Marketing,* vol.22, no. 2/3, pp. 212–232.

Arnone M and Bandiers L (2004). Monetary policy, monetary areas, and financial development with electronic money. *IMF Working Paper* WP/04/122.

Arumuga S (2006). Effective method of security measures in virtual banking. *Journal of Internet Banking and Commerce* 11, 1.

Basel Committee on Banking Supervision (2001). Risk management principles for electronic banking.

Baur T (2006). Set to leapfrog? *i4d,* vol. IV, no. 5 pp.13–15.

BCG (2006). Navigating to in: global payments 2006. By Viner N, Creyghton A, Carl Rutstein C and Storz N. May.

Beck T, Demirguc-Kunt A and Martinez Peris MS (2005). Reaching out: access to and use of banking services across countries. World Bank, Research Department, September.

Bernanke, BS (2006). Modern risk management and banking supervision. *BIS Review,* 52.

BIS (2004). *Survey of Developments in Electronic Money and Internet and Mobile Payments.* Committee on Payment and Settlement Systems, Bank for International Settlements, March.

BIS (2007). Statistics on payments and settlements. *BIS Quarterly Review,* Part 4, pp. 41–53, June.

Blakeley CJ, and Matsuura JH (2003). Electronic banking: how e-government can foster e-commerce. *Government Technology International,* August.

Bounie D and Gazé P (2007). Payment and the Internet: issues and research perspectives in economics of banking. In: Brousseau E and Curien N ed. *Internet and Digital Economics,* Cambridge University Press: 569–588.

Brown I, Hoppe R, Mugera P, Newman P and Stander A (2004). The impact of national environment on the adoption of Internet banking: comparing Singapore and South Africa. *Journal of Global Information Management,*12, 2, pp.1–26.

CGAP (Consultative Group to Assist the Poor) (1998). *Management Information Systems for Microfinance Institutions: A Handbook.* Technical Tool Series No.1, CGAP/World Bank.

CGAP (2001). *Using Microfin 3: A Handbook for Operational Planning and Financial Modelling.*Technical Tool Series No. 2.

CGAP (2004). Interest rate ceilings and microfinance: the story so far. Occasional Paper 9.

CGAP (2005). Funding microfinance technology. Donor Brief No. 23.

CGAP (2006). Using technology to build inclusive financial systems. Focus Note No. 32.

Claessens S and Feijen E (2006). Financial sector development and the Millennium Development Goals. World Bank Working Paper No. 89.

Clayton C (2006). Banks "fail" to protect online clients. *Personal Finance,* 8 July.

Coyle D (2007). Overview: the transformational potential of m-transactions. Vodafone Policy Paper Series, No. 6, July.

Cracknell D (2004). Electronic banking for the poor: panacea, potential and pitfalls. *Small Enterprise Development,* vol. 15, no. 4, pp. 8–24.

Cyree K, Delcoure N and Dickens R (2005). Is Internet-only banking the goose that lays golden eggs? An examination of performance and prospects for the future. University of Missisipi and University of South Alabama.

Daley-Harris S et al. (2005). State of Microcredit Summit Campaign Report 2005.

Daley-Harris S et al. (2006). State of Microcredit Summit Campaign Report 2006.

Dandapani K (2004). Success and failure in web-based financial services. *Communications of the ACM,* vol. 47, no. 5, pp. 31–33.

Diniz E, Morena Porto R and Adachi T (2005). Internet banking in Brazil: evaluation of functionality, reliability and usability. *Electronic Journal of Information Systems Evaluation,* vol. 8, no. 1, pp. 41–50.

Dospinescu O and Rusu D (2006). The adoption of electronic banking services in developing countries: the Romanian case. Future of Banking After the Year 2000 in the World and in Czech Republic, Social Science Research Network, New York.

Ezeoha AE (2006). Regulating Internet banking in Nigeria: some success prescriptions – part 2. *Journal of Internet Banking and Commerce,* vol.11, no.1.

Federal Financial Institutions Examination Council (FFIEC) (2003). *E-Banking IT Examination Handbook.*

Ferro LG and Weber S (2006). Measuring the performance of microfinance institutions. Social Science Research Network, New York.

Firpo J (2006). Banking the unbanked: technology's role in delivering accessible financial services to the poor. Microdevelopment Finance Team. Development Gateway Foundation, Washington.

Fox S and Beier J (2006) Online banking 2006: surfing to the bank. Pew Internet & American Life Project, 14 June.

Fugfugosh M (2006). Informal remittance flows and their implications for global gecurity. GCSP Policy Brief No. 11, Geneva, September.

Fullenkamp Cl and Nsouli SM (2004). Six puzzles in electronic money and banking. IMF Working Paper 04/19.

Furst K, Lang WW and Nolle DE (2002). Internet banking: developments and prospects. *Journal of Financial Services Research,* vol. 22, no. 1–2, pp. 95–117.

Gagnon G (2006). Innovations in information technologies. The Microcredit Summit Campaign.

Goi CL (2005). E-banking in Malaysia: opportunity and challenges. *Journal of Internet Banking and Commerce,* vol. 10, no. 3.

Gupta A (2006). Data protection in consumer e-banking. *Journal of Internet Banking and Commerce,* vol. 11, no.1.

Hadidi R (2003). The status of e-finance in developing countries. *The Electronic Journal on Information systems in Developing Countries,* vol. 11, no. 5, pp.1–5.

Hasan I (2002). Do Internet activities add value? The Italian bank experience. Federal Reserve Bank of Atlanta, Berkley Research Center, New York University.

Hawkins J and Dubravko M (2001). The banking industry in the emerging market economies: competition, consolidation and systemic stability – an overview. *BIS Papers* No. 4.

Idowu PA, Alu AO and Adagunodo ER (2002). The effect of information technology on the growth of the banking industry in Nigeria. *Electronic Journal on Information Systems in Developing Countries,* vol. 10, no. 2, pp.1–8.

ITU (2007). World Telecommunication/ICT Indicators Database.

Ivatury G (2004). Harnessing technology to transform financial services for the poor. *Small Enterprise Development,* vol. 15, no. 4, pp. 25–30.

Kalan G and Dilek A (2005). Assessment of remittance fee pricing. Background Paper for the World Bank's Global Economic Prospects, 2006.

Kerem K (2003). Internet Banking in Estonia. PRAXIS Working Paper 7.

Li S and Worthington AC (2004). The relationship between the adoption of Internet banking and electronic connectivity: an international comparison. Queensland University of Technology Discussion Papers in Economics, Finance and International Competitiveness. Discussion Paper No. 176.

McKinsey (2007). Fulfilling the potential of Latin America's financial systems. By Andrande L, Farell D and Lund S. *McKinsey Quarterly,* May.

Meyer J (2006). Cash, the web and the mousetrap. *Sunday Tribune.* 16 July, South Africa.

MIT Sloan School of Management (2004). Global e-readiness – for what? readiness for e-banking". MIT Sloan Working Paper 4487.

Mu Y (2003). E-banking: status, trends, challenges and policy issues. *Financial Economist.* World Bank.

Novametrie (2004). A new wave of Internet banking? Survey. White Paper.

Nsouli SM and Schaechter A (2002). Challenges of the "e-banking revolution". *Finance and Development,* vol 39, no 3, pp. 48–51.

OECD (2006). Online payment systems for e-commerce. Working Party on the Information Economy. DSTI/ICCP/IE(2004)18/REV1.

Orozco M (2004). The Remittance Marketplace: Prices, Policy and Financial Institutions. Washington, DC: Pew Hispanic Center.

Orozco M (2006). International flows of remittances: cost, competition and financial access in Latin America and the Caribbean – toward an industry scorecard. Report presented at the meeting on "Remittances and transnational families" sponsored by the Multilateral Fund of the Inter-American Development Bank and the Annie E Casey Foundation.

Quaden G (2002). Editorial: central banking in an evolving environment. *Journal of Financial Regulation and Compliance,* vol. 10, no. 1, pp. 7–13.

Ratha, D and Riedberg, J (2005). On reducing remittance costs. Background Paper for the World Bank's Global Economic Prospects 2006.

Roberts, BW and Banaian K (2005). Remittances in Armenia: Size, Impacts, and Measures to Enhance their Contribution to Development. Yerevan: Armenian International Policy Research Group.

Rombel A (2005). Next step for Internet banking. *Banking ServiceZone.Org.*

Simpson B (2005). Cards recharged. *Credit Card Management,* vol. 18, no. 1, pp. 32–39.

Simpson J (2002). The impact of the Internet in banking: observations and evidence from developed and emerging markets. *Telematics and Informatics,* vol. 19, pp. 315–330.

Singh B and Malhotra P (2004). Adoption of Internet banking: an empirical investigation of Indian banking sector. *Journal of Internet Banking and Commerce,* vol. 9, no 2.

Siriluck R and Speece M (2005). The impact of web-based service on switching cost: evidence from Thai Internet banking. ACM International Conference Proceeding Series, vol. 113.

UNCTAD (2000). E-commerce and financial services. Section IV, Chapter 2, *Building Confidence: Electronic Commerce and Development,* United Nations, 2000.

UNCTAD (2001). Managing credit risks online: new challenges for financial service providers. Chapter 7, *E-Commerce and Development Report 2001*, United Nations, New York and Geneva.

UNCTAD (2002). E-finance for development: global trends, national experiences and SMEs. Chapter 6, *E-Commerce and Development Report 2002*, United Nations, New York and Geneva.

UNCTAD (2005a). E-credit information, trade finance and e-finance: overcoming information asymmetries. Chapter 3, *Information Economy Report 2005,* United Nations, New York and Geneva.

UNCTAD (2005b). Protecting the information society: addressing the phenomenon of cybercrime. Chapter 6, *Information Economy Report 2005,* United Nations, New York and Geneva.

UNCTAD (2007). *World Investment Report 2007*, United Nations, New York and Geneva.

World Bank (2007). Financial flows to developing countries: recent trends and prospects. Chapter 2, *Global Development Finance Report 2007.*

Yeap BH, and Cheah KG (2005). Do foreign banks lead in Internet banking services? *Journal of Internet Banking and Commerce,* vol 10, no. 2.

Notes

1. See the Introduction to this Report.

2. The current Report is also covering the issue of ICT induced innovations in business processes in Chapter 4, "ICT, E-Business and innovation policies in developing countries".

3. Kevin M. Warsh, Financial intermediation and complete markets, *BIS Review* 59/2007.

4. E-money may be defined as a monetary value stored in an electronic device in the possession of its owner.

5. Basel II is the framework, revised by the Basel Committee on Banking Supervision, governing the capital adequacy of banks (for more details see: http://www.bis.org/publ/bcbsca.htm).

6. M Wolf. Unfettered finance is fast reshaping the global economy. *Financial Times*, 18 June 2007.

7. P Tucker. A perspective on recent monetary and financial system developments.*BIS Review,* 44/2007.

8. C Graeber. US online banking: five-year forecast, *Forrester,* 2 April 2007.

9. Has your computer been hijacked? *E-Marketer,* 7 June 2007.

10. See www.visa.com, www.mastercard.com, www.paypal.com.

11. See www.theclearinghouse.org.

12. Payments industry sees strength of SwiftNet tool. *The Banker,* June 2005; The race belongs to the SWIFT. *Bank Systems & Technology*, CMP, 27 June 2003.

13. See www.swift.com.

14. M Vickers. Plastic under attack. *Fortune*, 17 May 17 2006.

15. Cashiered. *The Economist*, 7 April 2007, p. 70; see also www.nilsonreport.com.

16. See www.mastercard.com.

17. See www.visa.com.

18. See http://www.federareserve.gov/BoardDocs/Press/other/2004/20041206/default.htm.

19. Online Payments: A Battle at the Checkout, *The Economist* May 5, 2007, p.p. 75-76; www.paypal.com; http://finance.yahoo.com; http://checkout.google.com.

20. http://ec.europa.eu/internal_market/payments/framework/index_en.htm.

21. See www.weforum.org ; http://globaltechforum.eiu.com.

22. See www. nibs-plc.com.

23. See www.theworx.biz/bank04.htm.

24. See http://www.ceris.ro/e-banking; www.verisign.com; www.digipasspack.com.

25. For more details on mobile penetration, see chapter 6: "Mobile telephony in Africa: a cross-country comparison" of this Report.

26. Dial M for money. *The Economist*, 30 June 2007, pp. 74 – 75.

27. Mexico finance: Chase offers free money transfer. *EIU Industry Briefing*, 25 June 2007.

28. See www.microfinancegateway.org.

29. See: www.bis.org.

Chapter 6

MOBILE TELEPHONY IN AFRICA: CROSS-COUNTRY COMPARISON

A. Introduction

Mobile telephony is the most important mode of telecommunications in developing countries. While Internet access has become a reality for many businesses and public institutions, and for individuals with higher levels of education and income, for the vast majority of the low-income population, mobile telephony is likely to be the sole tool connecting them to the information society in the short to medium term.

In 2002, UNCTAD's *E-Commerce and Development Report*[1] considered the growth of wireless communications and their role in increasing ICT use by business and consumers – frequently referred to as "m-commerce". A number of policy issues were discussed, including the liberalization of telecommunications markets, licensing new mobile operators, creating independent regulatory bodies that would establish a fair and competitive market for mobile services while supporting compatible standards, and facilitating interconnection among mobile services providers.

The discussion concluded that while e-commerce activities involving firms were mainly undertaken using fixed-line networks, individuals would find mobile services increasingly attractive as an entry point for using ICTs, either to order and purchase products and services – the so-called B2C context – or to improve their livelihoods through better communication in their communities and households. It was reported that there were still a number of obstacles to the provision of mobile services that needed to be overcome in developed nations, and in developing ones in particular. Most of those obstacles were related to practical problems in making electronic payments on mobile handsets and networks, and were linked to concerns about the security and privacy of transmitting personal and financial data.

Most importantly, the technical ability of mobile handsets to provide a rich Internet content has yet to match that of a personal computer using fixed-line broad-band networks access.[2] This has led mobile providers to explore a different spectrum of services that do not necessarily compete with those offered by a personal computer with a fixed-line Internet connection. Some of those services may not directly result in immediate transactions but would enable or improve the conduct of activities such as intra-firm communication among management and employees, the implementation of sales and marketing programmes, and the provision of after-sales customer services.

In 2006, the UNCTAD *Information Economy Report* observed that mobile communications were growing at a remarkable rate in developing countries, and that mobile telephony continued to be the only ICT sector where developing countries were quickly catching up or even in some ways overtaking developed countries.

Mobile connectivity sidesteps some important obstacles to other types of connectivity, but most notably to the deployment of fixed-line infrastructure, which can be hampered by, among other things, cost and the remoteness of certain areas. In Africa, mobile phones have proved so successful that in many cases they have replaced fixed lines.

This chapter starts by looking at the origins and economic nature of mobile telephony. It describes several examples from developing countries where mobile telephony was used to enhance entrepreneurial and market efficiencies and therefore improve value creation and, ultimately, the economic welfare of their communities. The chapter then examines the development of mobile telephony markets in several sub-Saharan African countries and in doing so considers the main indicators for mobile telephony penetration rates, subscribers and the trend in mobile prices. It concludes by describing policies that Governments may consider in order to enhance the positive development implications of improved mobile telephony network coverage and a growing subscriber base.

B. Mobile telephony: stylized facts

1. Origins and effects

Mobile telephony was invented by AT&T in 1947. Initially known as "radiophones", the technology developed from exchange-based radio links systems to cellular networks during the 1980s.[3] Cellular networks were developed to allow users to move from one cell – and its geographical coverage limit – to another cell without a break in the call, as a result of which true mobile telephony was made possible. The first-generation commercial mobile networks were the analogue Nordic Mobile Telephone system and the Nippon Telegraph and Telephone system, which were introduced in 1979 and became fully operational in the early 1980s.

These "1G" networks were replaced by digital "2G" networks which allowed greater call capacity. 2G digital networks and handsets also reduced power consumption and thus enabled their miniaturization and portability. The invention of the modern mobile handset is often attributed to Martin Cooper of Motorola.[4] Today, we are witnessing the deployment of 3G networks and handsets that are intended to enable the reception and transmission of broadband data that, in turn, enable mobile Internet connectivity and the development of content-intensive services.

Mobile telephone technology has enabled relatively low network build-out costs and these have resulted in the rapid growth of mobile telephony to the point where it is growing significantly faster than fixed telephony, both globally and in developing countries.

Table 6.1 contains data for 2001 and 2005 that confirm those observations. Table 6.2 provides a comparison by continent of mobile teledensity, described as the number of subscribers per 100 inhabitants, for the years 2002 and 2006.

The year 2001 may be seen as an interesting baseline year as already in 2002 mobile subscribers overtook fixed-line subscribers worldwide (ITU, 2003). In 2005, the worldwide number of mobile phone subscribers passed the two billion mark, with Asia accounting for more than 40 per cent of them. Private research estimates that by the end of 2006 the number of global mobile phone subscribers was approximately 2.6 billion.[5]

In developed countries growth in the mobile phone industry will come from the increased offer and use of innovative services, from SMS and affordable roaming to Internet access and music downloads. For example, it is expected that more than one third of all Europeans will have Internet-enabled phones by 2010 (Kelley and McCarthy, 2006), while a recent survey indicates that 29 per cent of Internet users in France, Germany, Italy, Spain and the United Kingdom regularly access the web from their mobile phones, compared with only 19 per cent in the United States.[6]

Mobile connectivity is very much responsible for the current surge in ICT utilization and, as such, introduces science, technology and knowledge (STK) inputs into value-creation processes in such sectors as industry, services, health and cultural and government services. The actual implementation and effect of mobile technologies within a specific process or value chain will differ. Businesses may see measurable pro-

Table 6.1

Growth in the number of mobile telephone subscribers worldwide, 2001-2005

	Mobile subscriptions	Mobile subscriptions	Absolute change	% change	Source of growth
	2001	2005	2001 – 2005	2001 – 2005	2001 – 2005
Developing countries	388 674 941	1 167 050 600	778 375 659	200.26	65.28
Africa	26 091 686	134 296 038	108 204 352	414.71	9.07
Asia and Oceania	278 511 819	793 375 236	514 863 416	184.86	43.18
Latin America and the Caribbean	84 071 436	239 379 326	155 307 890	184.73	13.03
Transition economies	22 325 131	185 068 576	162 743 445	728.97	13.65
Developed countries	553 610 317	804 830 507	251 220 190	45.38	21.07
World total	964 612 390	2 156 951 688	1 192 339 293	123.61	100.00

Source: UNCTAD summary based on the ITU World Telecommunication/ICT Indicators database.

Table 6.2

Mobile subscribers per 100 population worldwide in 2002 and 2006

	Subscribers per 100 pop.		% change
	2002	2006	2002 – 2006
Developing economies	11	29	179
Africa	4	21	356
Asia	11	28	162
Latin America and the Caribbean	19	53	183
Transition economies *	12	69	479
Developed economies	64	91	42
World	19	41	116

Source: UNCTAD summary based on the ITU World Telecommunication/ICT Indicators database.

* South-East Europe and CIS

ductivity improvements and managerial efficiencies. Governments may move closer to established goals of universal access, and communities may improve their welfare through better self-organization and improved empowerment, the latter being achieved by diversifying and improving the quality of their sources of information.

2. Accessibility of mobile telephony

Schemes that make mobile telephony more accessible and affordable have contributed to the growth of mobile services in developing countries. From a business model perspective and technology perspective, the advantages of mobile telephones are twofold. The first is that they enable prepayment and thus do away with the need to establish post-paid accounts and the financial infrastructure that enables such accounts, thus avoiding problems with non-payment. This advantage is doubly important in regions where large populations would not necessarily qualify as creditworthy because of their relative poverty. The second advantage is that the technology enables a fairly rapid development of the supporting wireless network and thus overcomes the drawback of inadequate fixed-line infrastructures. As a result, near-zero waiting times for new mobile subscriptions are common – at least once the network is in place – and represent a dramatic improvement in the wait and cues for fixed-line accounts. In many countries prepaid services are used to provide mobile public payphones, which improve connectivity and accessibility in rural areas.

For example, in 2004 almost 88 per cent of mobile subscribers in Africa used prepaid services that were tailored to low-income markets (ITU, 2006). However, prepaid fees are higher than post-paid ones, and this underscores the fact that low-income users need access at low prices. Other important factors that have contributed to the popularity of prepaid services are the ability to control costs, and, when savings and earnings are low, using the handset and network to only receive calls. Of equal importance are the reduced requirements for sign-up, as there is no need to submit often non-existent financial or physical data on, for example, bank accounts or postal addresses. Furthermore, prepaid mobile services do not typically require credit checks and do not propose long-term service contracts, as a result of which they are available to people who do not have steady income streams. Finally, prepaid mobile services reduce the amount and expense of various administrative activities for mobile operators as they do not need to engage in billing and money collection, and do not need to risk-manage their exposure to non-payment. While the absolute numbers of prepaid subscriptions may increase, their relative share in mobile accounts will decrease as a result of the interaction between a maturing mobile market and the developing economy.

While the question of whether fixed-line and mobile services are complementary or competitive substitutes is inconclusive in developed country markets (Madden and Coble-Neal, 2004), in many developing countries mobile telephones are generally a substitute for fixed-line telephony. This is a result of not only the low fixed-line penetration rates but also the substandard quality of service and long delays for obtaining fixed-line subscriptions. Competition in mobile telephone markets contributes to lower access costs and introduces innovation for the benefit of users. For example, the growth of mobile telephony in Asia is, in part, due to highly competitive markets, and this has led to lower prices for calls and mobile handsets. In fact, enhanced competition positively affects mobile teledensity in developing countries in general (World Bank, 2006).

3. The economic context of mobile telephony

The underlying proposition is that, of all ICTs, mobile telephony has the most immediate potential to stimulate growth in the developing countries, and especially in Africa, in particular in sectors where entrepreneurship and access to market information are important factors used to create value and economic benefit.

Mobile telephony has real economic consequences, particularly for micro-entrepreneurs. The relevance of mobile phones for small businesses in developing countries was noted in UNCTAD's *Information Economy Report 2005*, which suggested that the importance accorded to those economic benefits is reflected in the larger share of income that developing country users

spend on telecommunications as compared with users in developed countries. Like any other ICT investment, mobile telephony can lead to economic growth in several ways. Investment in network infrastructure and related services creates direct and indirect employment opportunities. The use of mobile telephony in the conduct of business reduces the costs and increases the speed of transactions. Those effects will be more pronounced for economic activities that have a greater need for information or where added information enables increasing returns to scale.

There is a broad consensus in current literature that the introduction and the growth of mobile telephony have important economic effects for developing countries, however much estimates may vary and methodologies differ. Hausman (2002) suggests that mobile telecommunications services create very large gains in consumer welfare. Aochamub, Motinga and Stork (2002) consider that there is an important bidirectional relationship between growth in the telecommunication sector and economic development. Thompson and Garbasz (2007) found that countries with less developed ICT use, and therefore emerging mobile services markets, may experience stronger feedback of growth in mobile ICT use into general economic growth. Keck and Djiofack-Zebaze (2006) analysed the relationship between mobile use and

Chart 6.1

Mobile teledensity and GDP per capita

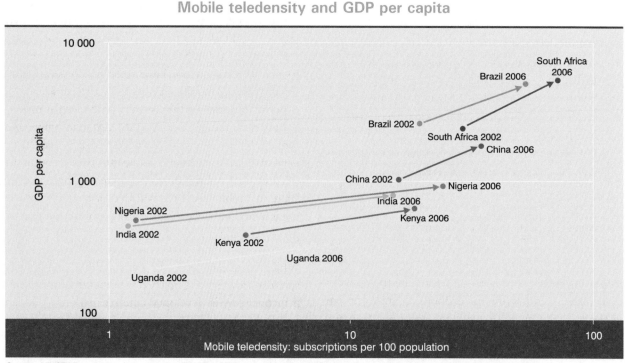

Source: UNCTAD summary based on the ITU World Telecommunication/ICT Indicators database and UNCTAD Globstat database.

growth in African member countries of the WTO, and suggest that increasing access to mobile networks by 1 per cent may translate into a 0.5 per cent increase in real GDP per capita. Chart 6.1 presents a comparison between the growth of mobile teledensity and GDP per capita in several developing countries.

While overall mobile price trends are downward, Africa still suffers from higher prices compared with the world average. Enhanced competitiveness, as measured by the number of operators, may work to adjust prices towards global prices and thus improve penetration rates. Regulation aimed at securing affordable rates and improvements in quality of service should also reduce regulatory uncertainty for mobile operators and promote investment. The fact that mobile markets have eclipsed the fixed-line sector in developing countries may be better explained by new investment and the increasingly competitive nature of mobile telephony markets. Overall economic growth and development may be playing a role in the adoption of mobile telephony. However, most of the recent literature explores how technology affects growth and development, and not the other way around.

The general UNCTAD policy recommendation that ICT adoption be part of an integrated e-strategy and overall national development strategy is equally tabled in current literature. Unless policymakers accept and implement that recommendation, it is unlikely that investments in mobile services will bring economic and social development to Africa and, in particular, to its rural areas. Transport and financial services, education and health are among the many necessary and complementary elements for successful development policies with important ICT and e-strategy components.

4. Use of mobiles

In developing countries the potential impact of mobile telephones is quantitatively different from that in the developed world. For the developing world mobiles have become an essential entry point into the information society. Mobile telephony is the critical tool that creates business opportunities, enables efficient sharing of information and intelligence, and empowers households and communities to stay connected. That said, it is still mainly the booming urban communities that are best served by mobile providers in developing countries.

Mobile telephones can also be used to facilitate or even generate business since they provide an opportunity to check current market prices for agricultural and other commodities relevant to rural economies, and currency rates, as well as to confirm payments, enquire about weather patterns, and more generally keep in touch with customers and stay informed about transport logistics.[7] For example, in Uganda, improved market and price information using short message services (SMS) on mobile telephones has diminished the role of middlemen and has resulted in an increase in farmers' and fishermen's income as their bargaining power.[8] A selection of the various modes of use of mobile telephones is presented in more detail in box 6.1.

Both Sinha (2005) and Meso, Musa and Mbarika (2005) explain that lower absolute penetration rates of mobile subscriptions in developing countries must not be taken literally, and consequently their impact on society should not be underestimated. In developed countries the operational model for mobile telephony is one of individual ownership of one or more subscriptions. However, in developing countries, because of their portability and ability to function using prepaid subscriptions, mobiles can be shared in terms of access and payment, with the positive side effect of reinforcing community linkages.

The Grameen Phone company is perhaps the best-known example of community use models. Owned by the micro-credit pioneer Grameen Bank and the Norwegian telecom Telenor, Grameen Phone set out in 1997 to bring telephones to rural Bangladesh. Today, it is the largest mobile telephone provider in the country. The Village Phone Program offers access to telecommunication services in remote, rural areas to people who normally cannot afford to own a telephone, while providing more than 200,000 women Village Phone operators in rural areas with an income-earning opportunity.

In the United Republic of Tanzania, the mobile operator Vodacom has a similar initiative, called Simu Ya Watu (People's Phone). Its aim is to create public call stations using the mobile network to provide public fixed telephones. The service is aimed specifically at rural and densely populated, but under serviced, areas of the country. The business model requires the involvement of local entrepreneurs who purchase a single or multiple call line unit, and is said to be designed for shops, restaurants and other business establishments with fixed premises that would like to offer public call services to their customers. The benefits for customers would be lower and more flexible call charges and the ability to make phone calls without owning a mobile telephone.[9]

Box 6.1

Examples of mobile telephone use in a development context

Every day a new example of innovative and productive mobile telephony use in developing countries surfaces in the media. What is interesting about many of them is their focus on the economic realities of everyday life. What follows is a diverse selection of various activities that is by no means exhaustive but only illustrative of the possibilities that ICTs create for economic development. Their common linkage is how mobile technology affects better information and financial flows that reduce market inefficiencies and improve the earnings of users and their communities.

Mobiles improve earnings of fishermen in Kerala, India*

An abundant catch of fish for Kerala fishermen usually meant that local beach market prices would be low. Fishermen then faced the dilemma of accepting a smaller return on their time and effort, or investing additional labour and fuel – and taking the relevant risks – to seek opportunities in a more distant market where the local catch might not have been so good and therefore the prices were possibly higher.

According to Jensen (2007), on average 5 to 8 per cent of the total catch ended up being thrown away. At the same time, owing to information inefficiencies, some fish markets were undersupplied and prices of fish varied substantially along the Kerala coastline. Since mobile telephones were introduced in Kerala in 1997, anecdotes have surfaced about how fishermen used them for price discovery and inventory management: how much fish to catch and to which of the beach markets to take the catch. However, Jensen's research adds a formal microeconomics analysis to existing anecdotal evidence.

Having access to information about catch volumes and beach market prices at various locations resulted in fishermen taking on the additional risk of venturing beyond their home markets, the final result being a reduction in catch wastage and in price volatility. Thanks to improved market efficiency, fishermen's profits ros4e by an average of 8 per cent while fish prices fell by 4 per cent thus benefiting consumers. Improved earnings more than compensated for the cost of the mobile telephone handsets and subscriptions. The enhanced flow of information helped local markets work more efficiently and, consequently, improved welfare.

Mobile telephones for weather and market information for farmers in Rajasthan, India**

The timing of the annual onset of the often unpredictable monsoon rains is crucial for the farming communities of Rajasthan as it dictates when they should sow their crops and when to take their produce to market. The accuracy of the forecast is critical for the Indian economy, and the timing and intensity of the monsoon can have significant effects on Indian GDP. While farmers in the past relied on faith and ceremony to perceive the nature of oncoming monsoons, mobile technology is today taking the lead in the form of a pilot scheme launched in June 2006 by Reuters India. A local-language text message service offering weather and market price information will be offered to 200 farmers in the state of Maharashtra. Following the trial, a full roll-out will be offered, targeting farmer landowners with three acres or more and an average annual income of $2,000.

According to Professor Venkata Reddy, Associate Professor of Marketing at the University of Agricultural Sciences in Bangalore, market inefficiencies due to a lack of information result in a waste – totalling up to $12 billion – of India's fruit and vegetable production. Farmers are not aware of market conditions and prices in neighbouring markets and therefore do not venture to invest to diversify their distribution network. In order to be successful, the Reuters service will need to overcome the relatively poor coverage of mobile networks and the dearth of handsets in rural areas: less than 2 per cent of the 100 million mobile phones in India are owned by people living in rural areas.

Local agricultural content in Uganda***

In Uganda the CELAC (Collecting and Exchanging Local Agricultural Content) project is helping farmers organize their production and distribution using information provided through mobile telephone technology. Under this scheme, farmers regularly receive and send vital information from their network and other networks affiliated to them. For example, WOUGNET (Women of Uganda Network) assists by translating information into local languages for farmers based in northern Uganda. In the past farmers in rural Uganda relied heavily on agricultural extension workers for knowledge support regarding their livestock or crops. Unfortunately, the experience was often not timely and frequently lacking. Community radios that would have helped to bridge the vacuum also suffered from coverage problems and did not provide possibilities for feeding back information into the network.

To remedy that problem, the CELAC project collects and exchanges local agricultural information using diverse technologies, media and networking opportunities. According to the project organizer, the Busoga Rural Open Source and Development Initiative, CELAC aims to benefit in particular women farmers. It uses various information and communication technology (ICT) methods to foster knowledge-sharing in an effort to enhance poverty reduction and food security, especially among women farmers. Regarding mobile technologies, the project uses SMS to send text messages on local agro-related information to the project's database of phone numbers every Monday. This service is designed for farmers, community development workers, agricultural extension workers and any other interested persons. CELAC has its own website that is used to share information on the crop and animal farming practices that have proved successful among Uganda's farmers. CELAC uses open development mediums, in particular free and open-source software for knowledge-sharing and information management and encourages the use of empowering participatory development frameworks methods enabled by ICTs.

Box 6.1 (continued)

Boosting profits using mobile telephones in South Africa****

About a hundred rural African farmers from the Makuleke region in South Africa are experimenting with mobile technologies in order to access information about market conditions in Johannesburg. This reduces the need to travel and move inventory, with frequent and substantial waste and pilferage during the nine-hour "share-taxi" voyage, and with little certainty as to sales and income. Using a virtual trading facility installed on mobile phones provided by the project sponsors, Vodacom, farmers can sell their produce directly from their small farms.

By checking prices in the Johannesburg markets by mobile telephone, farmers can avoid paying excessive commissions to intermediaries. At the same time they are able to negotiate from an improved position, fully aware of market and price conditions, and with a consequent reduction of the information divide between them and larger industrial farmers. With mobile telephone use growing rapidly, the exaggerated information asymmetries caused by rural isolation and poverty will, if not disappear, at least be substantially reduced.

Senegal market information on mobile telephones*****

In May 2005, the Senegalese company Manobi established a commercial trading platform for farmers and fishermen. The platform allows users to access information on the Internet and through Internet-enabled phones as well as by requesting price data and trading using simpler SMS text messages. The SMS version of the platform, Xammarsé, was developed in partnership with Sonatel (principal telecommunications provider of Senegal), the Ministry of Commerce and the National Agricultural Saving Bank.

More than 3,400 producers, middlemen, traders and hotel keepers receive by SMS daily data on the prices of products of interest in selected local markets. About 5 per cent of Xammarsé users typically will make additional SMS requests to receive price updates. Better information allows farmers to improve their negotiating position with middlemen and increase their earnings. To gather data, Manobi employees conduct daily surveys in Dakar's markets using pocket computers and mobile connections. Mobile technology enables them to store up to 15,000 data entries per day and thereby keeps the Xammarsé database permanently updated.

Mobile telephone services and microfinance in Kenya******

Microfinance is the provision of financial services to poor people. Micro-credit, micro-savings and micro-insurance are essential support services to enable poor people to trade and take part in the mainstream economy. Realizing the potential of mobile technology for extending financial services beyond urban areas, Vodafone/Safaricom in 2003 initiated a pilot mobile-telephone-based project in Kenya. While the initial objective was to create efficiencies to reduce the cost of loan disbursement and recovery, the technology was found by users to be convenient for person-to-person transfers. Since early 2007 the project has been commercialized and is currently subscribed by more than 175,000 users.

To implement this scheme, Vodafone/Safaricom partnered with the Commercial Bank of Africa, Citibank, DFID-FDCF and the Faulu micro-finance company to design and test M-PESA micro-payment platform. M-PESA allows customers to use their mobile telephones like a bank account and debit card. Customers credit their accounts with their prepaid time vendor and can, in addition to spending their credit on calls and messages, transfer funds to another subscriber, or make small or micro-payments for goods and services without the need for cash.

* http://www.economist.com/finance/displaystory.cfm?story_id=9149142; Jensen (2007); http://www.ncaer.org/downloads/lectures/popup-pages/PressReleases/popuppages/PressReleases/7thNBER/RJensen.pdf.

** http://business.timesonline.co.uk/tol/business/markets/india/article691684.ece; http://www.nextbillion.net/blogs/2006/08/01/in-india-weathermen-are-the-new-priests.

*** http://www.celac.or.ug/; http://blogs.bellanet.org/index.php?/archives/198-Farming-on-mobile-phones-in-rural-Uganda.html; http://www.comminit.com/experiences/pds2006/experiences-3873.html.

**** http://digital-lifestyles.info/2005/07/05/african-farmers-boost-profits-with-mobile-phones/ ; http://www.nextbillion.net/newsroom/regional/subsaharanafrica.

***** http://www.ictparliament.org/CDTunisi/ict_compendium/paesi/senegal/SEN05.pdf; http://www.manobi.net/news.php?M=5&SM=2&idnews=2006_01_02_manobi&lang=en; http://www.manobi.net/.

****** http://www.financialdeepening.org/default.asp?id=694&ver=1; http://www.vodafone.com/start/responsibility/our_social___economic/socio-economic_impact/micro-finance.html; http://www.iht.com/articles/2007/07/08/business/micro09.php; http://www.nextbillion.net/news-room/2007/07/09/in-poorer-nations-cellphones-help-open-up-microfinancing.

Besides voice and SMS communication, beeping – ringing a number according to a pre-agreed code (e.g. one beep or ring would mean "I have arrived") and hanging up before the call is connected – is often used when incomes do not permit the luxury of completed calls.[10] As beeping is not charged it allows a certain redistribution of wealth to lower-income users as they free-ride the established mobile infrastructure. It is the least costly form digital communication and will be found anywhere where consumers are highly price-sensitive. However, its content is not as meagre as it initially seems. Each time a call is beeped and missed the mobile handset logs the calling number (i.e. identity) and the time of the communication.

Building on discussions of various modes of use and ownership models, Donner (2005) points out that much of the current research tends to relate more to urban users. Empirically, this makes sense, given that the urban milieu has been the cradle of mobile growth and information would be easier to obtain. However, by focusing on urban users in the developing world, researchers may underestimate the true development impact of mobile use. Such discussion may lead to questioning the utility of mobile telephony for the rural environments by arguing that most rural communication is local. Other issues raised are that public mobile and shared mobile telephones can be unreliable and frustrating for users. Also, non-voice services may be a red herring for rural populations, given lower literacy levels, and policy should refocus on voice-based communication rather than SMS and Internet-based e-mail. Because many economically disadvantaged users may not be mobile – for example, subsistence farmers – the quality of mobility in mobile telephony may be less relevant than usually assumed.

The examples in box 6.1 provide evidence that dispels to a certain extent such short-sighted notions and advances our understanding about just how interdependent, and therefore interconnected, rural communities may be. They also remind us of the role and value of information for productive efficiency even in the oldest and most iconic of all economic ventures, namely fishing. Fishermen, farmers, SMEs and large-scale businesses all require information for decision-making. In order to provide products and services to customers, knowledge-based systems based on appropriate technologies can be used even in the most turbulent environments and sectors. The ability of farmers and fishermen alike to forecast changes in weather and in other natural occurrences and prepare timely responses will determine their success. This ability is tied solely to their capacity to retrieve

useful information, and therefore the enabling role of ICTs and mobile telephony in particular should not be underestimated. Indeed, mobile technologies today provide the building blocks for enhancing the productive capacity of all economic sectors in developing countries.

5. Mobile telephony and GATS

Commitments in telecommunications services were first made during the Uruguay Round (1986–1994). In extended negotiations during 1994–1997, members of the General Agreement on Trade in Services (GATS) successfully negotiated on basic telecommunications services. Since then, new commitments have been made either by new members, upon accession, or in a unilateral fashion by an existing member. These commitments are still guided by the WTO Reference Paper on Regulatory Principles for Basic Telecommunications that was produced by the WTO Negotiating Group on Basic Telecommunications and contains definitions and principles for national regulatory frameworks for basic telecommunications service.[11] The Reference Paper is a negotiated text containing a summary of regulatory principles and definitions which a number of WTO member countries have incorporated into their GATS commitments for telecommunications.

The broader significance of the Reference Paper is that many countries regard it as a model for regulatory frameworks for telecommunications services that could also be regarded as a best-practice example for other service sectors. More specifically, the Reference Paper establishes a common set of regulatory principles relevant for opening telecommunication markets to competition. These include eliminating restrictions on interconnection, establishing and maintaining the independence of regulatory authorities, and public availability of licensing criteria. It has been suggested (Satola, 1997; International Chambre of Commerce, 2005) that Reference Paper implementation would be an important element for attracting investment in the telecommunications sector of developing countries, and thus Governments establishing national telecommunication legal and regulatory framework should use it as their lead reference.

C. Country analysis

1. Mobile telephony in Africa

The growth in the number of mobile subscriptions in Africa has indeed been swift as table 6.1 indicates. According to ITU statistics, between 2000 and 2006 the number of mobile subscriptions increased more than 12 times, from 15,633,872 to 189,497,105 and in 2006 represented approximately 62 per cent of total mobile and fixed-telephone subscriptions. By 2006, all African countries had active mobile services operators, compared with only one out of five in 1993. In spite of such progress, the number of telephones per 100 inhabitants is still low, since approximately 80 per cent of the African population lives in rural areas to which mobile networks and services do not extend. Consequently, the extension of mobile services into rural Africa may be an important component for continued growth and acceleration of ICT use. However, the rapid growth of mobile telephony in Africa emerges from initially very low penetration rates.

Even if mobile penetration in African urban conglomerations may be near the averages found in other developing countries, improving rural penetration is a more daunting task because of inherent difficulties caused by the lack of distribution channels, education and poverty (Anderson, 2006). Still, when we consider teledensity intensification in Africa, it is easily the fastest growing region in the world to the extent that between 2000 and 2005 the expansion of new mobile subscriptions equalled the number of fixed lines that were built in the past 100 years (ICT4D, 2005). On the demand side, commonly cited contributory factors are the decrease in relative prices of handsets, calls and connection charges and the use of more affordable second-hand handsets and equipment, inexpensive handsets specifically designed for low-income countries and the lower level of ICT skills required in order to operate a handset, compared with using the Internet. On the supply side a quicker return on investment, easier resolution of geographical obstacles and the simplicity of the prepayment model are often noted (Engvall and Hesselmark, 2005; Waverman, Meschi and Fuss, 2005).

The prices of mobile services will remain relatively high for local incomes in developing countries, as certain cost components (e.g. network hardware) will be sourced at global prices, and this will have a negative effect on access and affordability. In response,

solutions such as mobile phone sharing or new services such as mobile payphones[12] are being introduced to increase access. However, the most important impact on increased access and affordability will come with greater investment in the mobile sector. Box 6.2 describes an interesting example of investment and entrepreneurial activity in the African mobile sector.

To examine the development of mobile markets in Africa several datasets will be considered. Mobile penetration rates, the number of mobile subscribers and the trend in mobile prices will be examined in four countries of the sub-Saharan region, namely Kenya, Nigeria, South Africa and Uganda. Those countries were selected because they provide four examples of possible approaches to mobile market development and policy, and the outcomes point to varied mobile telephony adoption indicators and growth rates, thus illustrating the general discussion of underlying trends and relationships with country reviews in some detail. The supportive data, if not indicated otherwise, are based on indicators for mobile telephony from the ITU World Telecommunication ICT Indicators database.

2. Nigeria

Overview of the telecommunications sector

As recently as 2001, Nigeria was one of Africa's most underserved telecommunications markets. Today it is one of the fastest-growing mobile markets in the world, with a 125 per cent average annual rate of growth in the number of subscribers for mobile and fixed lines. In the past three years, there was a dramatic decrease in waiting lists for telephone lines and a reduction in local, national and international telephone tariffs, which are currently the lowest in Africa. Furthermore, mobile operators have started targeting residential and business customers, changing the landscape of the telecom market and contributing to the development of Nigerian economy.

Before the liberalization of the Nigerian telecommunications sector, telephone lines were scarce and expensive, and waiting times were long. In 1992, the Nigerian Communications Commission (NCC) was established to improve consumer service, begin with the deregulation of the telecommunications industry and introduce greater competition in the sector. In 1999, the federal Government decided to pursue a more aggressive mobile telephony development policy, as a result of which, if one is to judge by the number of mobile

---

Box 6.2

Investment in African mobile telephony: the case of Celtel*

Celtel was founded by a Sudanese-born British Telecom engineer, Mohamed Ibrahim, in 1998 and is today owned by Mobile Telecommunications Company (MTC), a Kuwait-based telecom firm operating in 21 countries and providing telephony services to over 20 million subscribers.** Early on, Celtel correctly assumed that people felt the need to communicate with each other equally in Africa as anywhere else, but that Africans felt frustrated because of underdeveloped fixed-line telephony. Mobile telephony was able to satisfy this pent-up demand, and Celtel moved to invest in the first mobile networks in Uganda and Zambia in the late 1990s.

Realizing that roaming interconnectivity was crucial for many African subscribers whose family, social and business communities transgress political borders, MTC/Celtel launched in 2006 its «One Network» borderless network for East Africa. One Network enabled its subscribers in Kenya, Uganda and the United Republic of Tanzania to use their subscriptions in any of those countries to make calls at local rates without roaming surcharges, as well as to receive incoming calls free of charge and recharge their prepaid accounts with any of the local top-up cards. In June 2007 One Network was being extended to six more African countries, the plan being to ultimately provide a seamless network in all 15 African countries of operation.

MTC and Celtel, before and after the acquisition, are examples of successful private sector investment and operation in the mobile telephony sector in developing countries. Celtel has to date invested more than $750 million in Africa and has influenced the continuously improving perception of Africa as a continent of investment opportunities. According to its founder Mohamed Ibrahim, as recently as 1998 nobody was keen to invest in sub-Saharan Africa apart from South Africa. The negative image of this region, which is conveyed by the daily difficulties facing its impoverished people, led to a greatly exaggerated perception of investment risk.

* See http://www.celtel.com/en/our-company/leadership/mo-ibrahim/index.html ; http://www.celtel.com/en/our-company/index.html and http://www.moibrahimfoundation.org/downloads/news/TheEconomist_240507.pdf.

** MTC is a public company listed on the Kuwait Stock Exchange with market capitalization exceeding $25 billion. Through its Celtel acquisition, MTC now operates in 15 countries in Africa: Burkina Faso, Chad, Congo, the Democratic Republic of the Congo, Gabon, Kenya, Madagascar, Malawi, Niger, Nigeria, Sierra Leone, Sudan, Uganda, the United Republic of Tanzania and Zambia.

---

operators and licence issues, Nigeria may now have the most liberalized telecoms market in Africa.[13]

The first mobile licences were issued in 2001 to Econet Wireless Nigeria and MTN Nigeria Communication. A year later a third company – Mtel – entered the market, followed by V-mobile and Glo Mobile in 2003. All operators were fully established by 2003, after two years of testing and market research. As chart 6.2 indicates, in 2002 MTN took the lead in the mobile sector and was soon followed by Econet.

While the mobile sector experienced unprecedented growth, the fixed-line market serviced by the monopoly provider Nigerian Telecommunications Ltd (NITEL) remained stagnant, owing to lack of infrastructure investment as well as administrative and bureaucratic inefficiency. After prolonged difficulties and delays in the privatization process, which had started in 1999, NITEL was finally sold to Transcorp in 2006, together with its mobile subsidiary, Mtel.[14] Late in 2002, the NCC granted Globacom a license for a second national operator in order to improve competitiveness by providing an alternative network to the, at the time,

Chart 6.2

Mobile telephone market share from 2001 to 2004

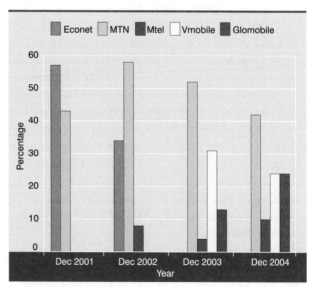

Source: NCC (2005).

government-owned NITEL. Globacom paid the $200 million required for the license, which gave it the right to operate a national carrier service, digital and mobile services, long-distance connections and fixed wireless access services. NITEL remains the dominant fixed-telephone operator with 556,590 connected lines (76.8 per cent of the total).

Between 2001 and 2004, Nigerian Communications Commission has issued 523 new telecoms licenses of various types, including many that commission companies to invest in developing parts of the physical network, interconnections and exchanges. However, not all of these licences have become fully operational and many types of licenses are held by a single company.

Today in Nigeria there are approximately 30 fixed and mobile service companies, including four GSM operators, as well as at least 80 Internet Service Providers (ISPs) and VSAT companies. Table 6.3 shows the growth in the number of operators in Nigerian telecommunications sub-sectors.

Mobile market development

The introduction of several operators in the mobile industry has fostered development and enhanced operators' competitiveness, which has had a positive effect on the entire telecom sector. The main outcome has been a major increase in the number of mobile subscribers. Towards the end of 2004, there were 10.2 million active mobile telephone subscriptions compared with 3.1 million in 2003, representing an increase of 291 per cent over the previous year. The increase of fixed-telephone lines connected during the same period was less spectacular, growing by 17.8 per

Chart 6.3

Fixed and mobile telephone lines and subscriptions in Nigeria, 2001-2004

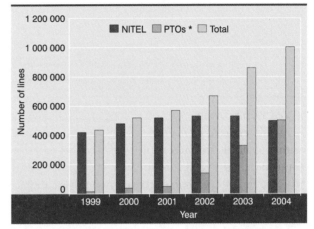

Source: NCC (2005); ITU World Telecommunication/ICT Indicators database.
* PTOs = public telecommunications operators.

Chart 6.4

Teledensity and subscriber growth in Nigeria, 2001–2004

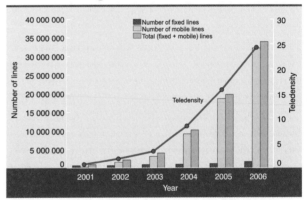

Source: NCC (2005); ITU World Telecommunication/ICT Indicators database.

Table 6.3

Number of operators and service providers in Nigeria

Service category \ Year	1999	2000	2001	2002	2003	2004
National carriers	1	1	1	2	2	2
Mobile (GSM) telephony	1	1	3	3	4	4
Fixed telephony	9	16	16	17	20 **	24 ***
VSAT networks	N/A	N/A	N/A	N/A	51	52 *
Internet services	18	30	30	35	35	36
Total	29	48	50	57	112	118

Note: * confirmed, ** including 3 fixed wireless access (FWA) operators; *** including 6 fixed wireless access (FWA) operators.
Source: NCC (2005).

cent from 872,500 subscriptions in 2003 to one million in December 2004. As a result, teledensity increased significantly from 1.89 per 100 inhabitants in 2002 to 3.36 in 2003. Chart 6.4 shows the growth in the total number of connected phone lines (fixed and mobile) and teledensity in Nigeria.

The growth of the mobile sector has resulted in a decrease in mobile tariffs and prices, wider coverage, better quality of service and more innovative products, as well as improving employment opportunities. When Nigeria liberalized its market in 1999, telephones were neither available nor accessible to the majority of the population, partly owing to the high tariffs charged by NITEL. In 2001, with the entry of two private mobile operators, MTN and Celtel, in competition with the incumbent Mtel, mobile prices quickly dropped to below the prevailing mobile and fixed-line range, with a consequent fall in fixed-line tariffs by almost 90 per cent before the end of 2004. Charts 6.5 and 6.6 describe these important price trends.

The entry of the fourth mobile operator, Glo Mobile, in 2002 completed the change from a controlled to a competitive telecommunications market in Nigeria. Glo Mobile introduced a "per second billing" option and other innovative packages aimed at reaching low-income segments of the population, thus contributing to further lowering mobile tariffs. From an initial $0.38 per minute for basic prepaid service, mobile call costs have dropped to between $0.19 and $0.26 per minute. However, in 2003 the incumbent operator rebalanced tariffs, with a consequent 50 per cent increase in price for local calls, counterbalanced by a reduction in

international call rates (which dropped from $0.72 to $0.33 per minute). Lately, there have been continued price reductions among all operators, matched with ever more complex service packages.

The benefits that subscribers are gaining from lower tariffs and greater choices in Nigeria are the direct outcomes of the liberalization process and the resulting competition among private operators. In particular, reductions have been the result of increased competition triggered by the launch of NITEL's new tariff structure in May 2004 and by Globacom and Celtel marketing strategies.

Chart 6.6

Connection and airtime charges for mobile telephones in Nigeria

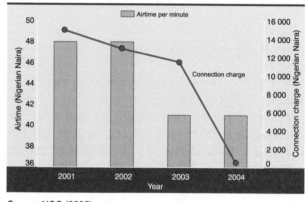

Source: NCC (2005).

Chart 6.5

Fixed and mobile telephone connection charges in Nigeria

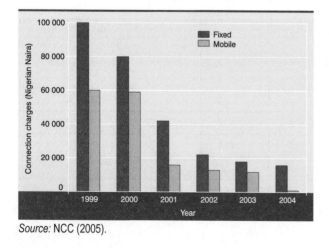

Source: NCC (2005).

Beyond the immediate providers who are the first echelon of contact with consumers, the development of the mobile sector and the fall in network infrastructure prices are to an important degree enabling the benefits of price reduction to be passed on to consumers. However, the main problems facing the Nigerian telecom sector remain the provision of universal access and the scarcity of investment in mobile telephony for rural areas. This is quite important since 54 per cent of Nigerians live in rural areas. While competition and market liberalization may benefit consumers in urban areas, they have little effect on rural access. Therefore, Government needs to consider adopting a more active policy stance through incentivizing rural investment or, indeed, providing public investment funding for infrastructure development.

3. Kenya

Overview of the telecommunications sector

Between 1995 and 2004 Kenya experienced remarkable growth in the proportion of mobile subscribers among its population, and this has been credited with giving an important boost to activities in the small business sector. According to the Government's 2005 Economic Survey, mobile phones had a positive impact in the development of the small business sector as they created approximately 437,900 new jobs. This is in line with research conducted by Waverman, Meschi and Fuss (2005) suggesting that a developing country which added 10 more mobile phones per 100 population between 1996 and 2003 would have enjoyed per capita GDP growth that was 0.59 per cent higher than otherwise. This increase is mainly determined by the productivity and efficiency gains stemming from the use of mobile phone technology by local entrepreneurs (Eagle, 2005). From an ICT adoption perspective, Kenya is not only one of the fastest-growing mobile markets in the region, but also has one of the largest Internet communities in Africa. It has great potential for further ICT adoption and ICT-based developments, as mobile and fixed-line penetration are only at around 10 per cent and 1 per cent respectively.

When mobile telephones were first introduced in the Kenyan market in 1992, they were so expensive that only the wealthiest of the population could afford them. Considering that 64.8 per cent of the Kenyan population lives in rural areas and only 40 per cent manages to achieve incomes above the poverty line, the wealthy represented a small market. This resulted in a marginal mobile subscriber growth of less than 20,000 for the period from 1993 to 1999.[15] In 1999, with the establishment of the Communications Commission of Kenya (CCK), based on the Kenya Communications Act enacted in 1998, the level of competition in the mobile market increased. The newly privatized Safaricom Limited and a new market entrant, Celtel Kenya (formerly known as KenCell Communications), were licensed by the CCK to provide mobile services. The CCK, as a matter of policy, expects competitive market forces to determine prices. However, mobile operators are required to present their prices to it before they can apply them.[16]

In terms of licence agreements, the two operators have covered the majority of the areas required, adding new districts according to business growth. For example, Celtel significantly exceeded its licence targets from its very first year of activity, while Safaricom remains the operator with the highest number of subscribers. Chart 6.7 describes and compares the subscriber growth of Celtel and Safaricom.

Mobile teledensity has shown continuously superior growth over the years when compared with fixed-line teledensity, which has actually decreased. As a result, total teledensity (i.e. mobile and fixed) increased from 8.8 lines per 100 population in April 2003 to 14.7 lines per 100 population in May 2005 as a result of the extraordinary increase in the number of mobile subscribers. Despite this growth it is important to note that the great majority of fixed line subscribers – 94 per cent – still live in urban districts, while only 6 per cent of fixed lines reach out to rural areas. The latter also explains the decrease in the absolute number of fixed-line subscribers, this decrease suggesting that mobile telephony is a substitute for fixed-line telephony. Chart 6.8 illustrates the rural-urban divide in terms of mobile access in Kenya.

Chart 6.7

Mobile networks in Kenya

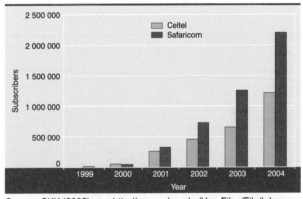

Source: CKK (2005); see http://www.cck.go.ke/UserFiles/File/telecom.pdf.

Chart 6.8

Urban and rural fixed connections in Kenya

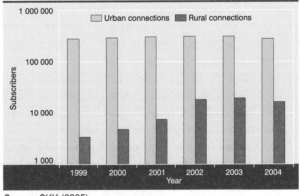

Source: CKK (2005).

Mobile market development

Following the licensing of Safaricom Limited and Celtel Kenya, the number of mobile subscribers grew significantly, did as the geographical areas covered by mobile networks. Furthermore, connection tariffs dropped from $142 in 1999 to $33 in 2005. Call prices also dropped from $0.40 per minute in 1999 to $0.20–0.32 per minute in 2005. Since Kenya still has a single incumbent fixed-line provider, the most recent tender for a second national operator having failed in March 2007, it is not surprising that long-distance and international tariffs have decreased by only 25–40 per cent, while local call prices have risen by 20 per cent. Therefore, it is no surprise that already in 2001, three years after the introduction of competition, the number of mobile telephony subscriptions had started overtaking the number of fixed telephony subscriptions, and towards the end of 2005 the number of mobile subscribers was ten times larger than the number of fixed-telephone lines. Chart 6.9 maps this development during the last seven years.

Chart 6.9

Comparison of growth of fixed and mobile networks in Kenya

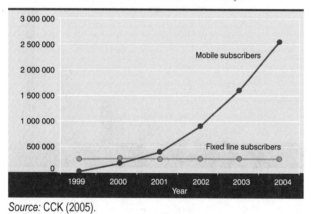

Source: CCK (2005).

In June 2004, Telekom Kenya's monopoly was unbundled and three new kinds of licences (i.e., Network Facilities Providers, Applications service Providers and Content Service Providers) were introduced by the government, somewhat opening up the market. Currently, the licensing of a third mobile operator and the liberalization of VoIP telephony are under review. The third licence was originally awarded in 2003 but revoked shortly afterwards by the Ministry of Information and Communication, because of issues related to shareholding requirements concerning local investor holdings. The licence has been reinstated in early 2007 and Econet Wireless is now planning to deploy its network early in 2008.[17]

As a result of the strong growth of the mobile sector, fixed line traffic and revenue levels have decreased in 2004 and 2005 with local call charges falling by 15.6 per cent, compared to 31 per cent for long-distance calls and about 33 per cent for international calls. Prices for fixed-to-mobile calls fell by 7.4 per cent over the same period. While changes in mobile prices in recent yeas have been characterized as insignificant, a 28% reduction of international mobile call charges by Safaricom Ltd has eased prices for consumers calling abroad. Overall, Kenya has experiences an increase in teledensity to 18.5 mobile subscribers per 100 population in 2006, compared to 7.8 in 2004 and 13.5 in 2005.[18]

4. South Africa

Overview of the telecommunications sector

The mobile sector in South Africa has played a significant role in boosting overall telecommunications access. Mobile telephony is increasingly more affordable and accessible thanks to continuous innovation in tariff structures and the introduction of prepaid services.

Today, 43 per cent of the South African population has access to mobile services. In 1993, two GSM public land mobile network licences were issued to Mobile Telephone Networks Ltd (MTN) and Vodacom Ltd, which then started providing mobile telephony on a national level. The actual provision of mobile services was introduced in mid-1994 and by the end of the 1990s the number of subscribers had grown to over two million.

The liberalization of the telecommunications sector took almost a decade. In 1996, the new Telecommunications Act allowed further mobile network competition and a year later Telekom, the incumbent fixed-line operator, was partially privatized. However, because of an exclusivity agreement that was established for fixed-line services for Telekom in order to increase its attractiveness for investors, no real progress was made in improving competitiveness before its expiry in May 2002. Moreover, the fact that only two licences were issued for mobile telephony leads on to wonder whether if an additional one or two licences could have more significantly improved affordability, given the potential of the South African market and its overall higher level of development.

The Telecommunications Act of 1996 also established

the South African Telecommunications Regulatory Authority (SATRA), which became responsible for regulating the telecommunications market in the public interest. It was succeeded by the Independent Communications Authority of South Africa (ICASA) in July 2000 pursuant to the provisions of the Independent Communications Authority of South Africa Act No. 13 of 2000, which merged the functions of SATRA with those of the Independent Broadcasting Authority. The two bodies were merged in expectation of the convergence of telecom and broadcast technologies.[19]

In 1997, the introduction of a prepaid option for mobile calls boosted the number of subscribers to 1.9 million, or 4 per cent of the population, a figure that exceeded initial expectations. Given such positive results, in November 1998 licence clauses for the third and fourth mobile providers were released. The market continued to grow steadily and by the end of 1999 there were roughly 3.2 million subscribers, or 7 per cent of the population. Following the numerous regulatory challenges in 2000, the South African mobile industry continued to grow as new products and services, such as the "happy hour tariff", were introduced (Vodacom, 2005).

Chart 6.10

South African telecommunications providers' earnings growth*

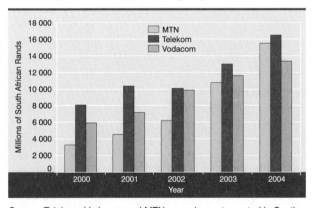

Source: Telekom, Vodacom and MTN annual reports quoted in South Africa Foundation (2005).

* Earnings before interest, tax, depreciation and amortization.

Mobile market development

In 1993 two mobile operators, Vodacom and MTN, were licensed to operate mobile telephony services, and they operated as a duopoly until 2001. In November 2001, a third mobile licence was granted to Cell C, allowing full competition in the market (Gillwald and Kane, 2003).

By 2001, the South African mobile market had over 30 per cent of the total telephony market in terms of total revenue and more than three times the number of subscribers than the fixed network, its growth having thus exceeded all expectations.

At the present time there are four mobile operators in South Africa, with Virgin Mobile having acquired a licence in 2006. Vodacom and MTN together have a market share of 85 per cent. On the other hand, Cell C, which started in 2001, and the most recent operator, Virgin Mobile, which started operations in 2006, have a combined market share of 15 per cent, a fact that suggests that they are having difficulties in attracting both new and existing subscribers. One possible reason is that operators may find it challenging to compete in terms of price price since mobile tariffs need to be filed with and approved by ICASA.

While the regulatory wisdom is that this may protect the operators during the duopoly period from below-cost pricing as a result of price wars and consequently financial and service delivery problems, it is questionable whether such a practice is ultimately positive from a consumer welfare perspective. As a result, the costs of telecommunications in South Africa are usually more than they could be when there is more liberal price-based competition. Specifically, mobile call tariffs are still high compared with fixed-line rates, and the fast growth of mobile telephony can be attributed solely to service quality and the convenience of prepaid mobile services. Furthermore, all four mobile network operators depend on the sole national operator Telekom for interconnection and are consequently constrained by its pricing structure (South Africa Foundation, 2005).

Chart 6.11

Fixed and mobile telephone subscriber growth in South Africa

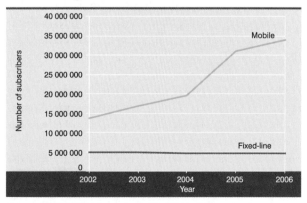

Source: ITU World Telecommunication/ICT Indicators database.

While there has been a continuous decline in total fixed teledensity, it has been compensated for by continued growth in the number of mobile subscribers. As already noted, the latter has been possible thanks to the rapid diffusion of prepaid services and a general increase in competitiveness with the additional licences. According to market research conducted by Bmi-TechKnowledge (2005), 75 per cent of mobile subscribers use prepaid services and more than 90 per cent of new connections are prepaid. Furthermore, Cell C estimates that 98 per cent of its subscribers are prepaid users (Gillwald and Kane, 2003).

Chart 6.12

Comparison of cost of fixed and mobile networks in South Africa, 2001–2006

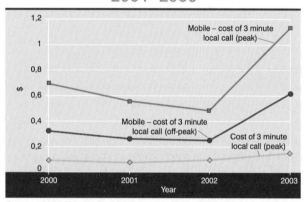

Source: 2002 Telkom Annual Report, in Gillwald and Kane (2003); ITU World Telecommunication/ICT Indicators database.

Although universal access has increased noticeably, there are still important differences in access and services between rural and urban households. The introduction of prepaid service has helped lessen those differences, since the number of prepaid subscribers on both national networks is now greater than the number of contract subscribers. Moreover, new strategies have been adopted in order to increase accessibility. For example, Vodacom has established phone shops in refurbished shipping containers located in poorer communities. They not only offer prepaid calls that cost less than a third of the commercial rate, but also generate job opportunities for the local community (Panos Report, 2004).

High mobile call prices are reflected in high levels of use of public access phones, even among those living in areas that have mobile phone network coverage (Gillwald, Esselaar, Burton and Staveou, 2004) and underscore questions raised about the value of the "mobility" in mobile telephony for low-income regions or populations (Donner, 2005). As illustrated in chart 6.12, fixed-line connections are still more affordable.

Chart 6.13

Fixed and mobile telephone subscriber growth in Uganda

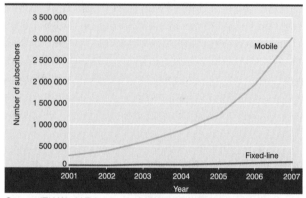

Source: ITU World Telecommunication/ICT Indicators database.

At the moment, the South African telecommunications market is undergoing radical changes, since Telekom has been floated on the Johannesburg Stock Exchange and a second national operator, Neotel, is expected to become fully operational in mid-2007. While some argue that this will result in the entry of another fixed-line operator and in the consequent creation of a duopoly, which could to a certain extent improve the openness and competitiveness of the South African telecommunications market, others cite the fact that Neotel plans initially to use mobile technology for the last-mile infrastructure since ICASA was unable to remove Telekom's control of the local exchange from its last mile connections and make the last-mile infrastructure commercially available to Neotel. This has led to suggestions that it is in fact yet another mobile operator instead of a much-needed competitor.

5. Uganda

Overview of the telecommunications sector

According to the ITU (2002) Uganda can be considered a model for other developing countries that aim to develop and seek development impact from their mobile telephony sector. Although more than 82 per cent of the population lives on $1 a day or less and telephony may seem to be a luxury, in the past 10 years the number of mobile telephone subscriptions has grown significantly (Gamurorwa, 2004), as the indicated in chart 6.13.

In 1995 the Ugandan Government issued a second network licence to Clovergem Celtel Ltd to foster competition and thus open the market to multiple mobile operators. In 1997 the Government decided to liberalize the telecommunications sector and therefore created an independent regulatory agency, Uganda Communications Commission (UCC). 1998 is considered to have been a turning point in the Uganda mobile industry since it is the year in which the second network licence was sold to South African mobile operator Mobile Telephone Networks (MTN). During the first year of operation, MTN had 36,500 subscribers (i.e. two thirds of the number of main lines served by the incumbent operator) and by 1999 the number of MTN subscribers surpassed the total number of main lines. In only six years the mobile subscriptions rose

from 17,098 in 1992 to 98,900 in 1998.[20] In 2000, the Ugandan fixed line operator UTL was privatized, with the sale of 51 per cent to a consortium made up of Telecel (South Africa), Orascom (Egypt) and Detecon (Germany, a division of Deutsche Telecom) (Shirley, 2001; UCC, 2005).

Recent research (Williams, 2005) suggests that there may be a positive link between mobile penetration and FDI. Uganda can be considered a practical example of this supposition. The rapid growth in the mobile market was triggered by a "Big Push Strategy" introduced by

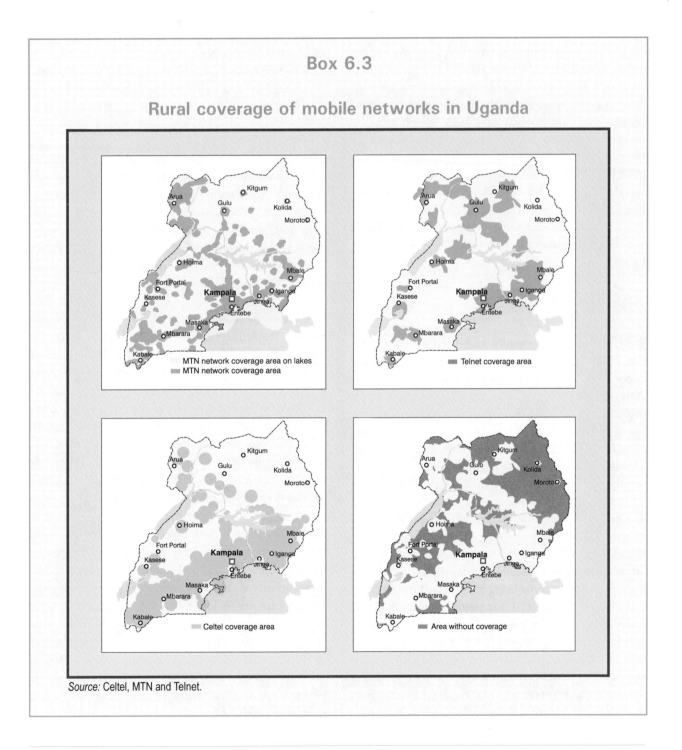

Box 6.3

Rural coverage of mobile networks in Uganda

Source: Celtel, MTN and Telnet.

the Government in 1998, and to be implemented during the period 2000–2005 in order to foster investment and build ICT infrastructure.[21] It should be noted that Uganda does not impose any limitations on foreign investors' ownership stake (Gamurorwa, 2004).[22]

In the past six years the mobile market has grown significantly: in November 2004 there were nearly 1,040,127 mobile subscribers compared with about 71,056 fixed-line customers. According to UCC, this may be because the past performance of fixed-line telephone networks was associated with great inefficiency and poor customer service. Furthermore, the introduction of prepaid plans and the development of innovative products, such as packages tailored to the low-income population, have had a positive impact on the spread of mobile telephony. Overall, the UCC estimates that there has been significant progress in terms of the freedom to make calls, and improved connectivity and mobility. Mobiles telephones are an extremely important means of communication in Uganda, since fixed-telephone lines are still lacking in most people's homes. In particular, mobiles have been very beneficial to local business, even in rural areas. Box 6.1 in part B.4 of this chapter gives an example of the use of mobile technology in improving knowledge and efficiency in rural and agricultural activities in Uganda.

Mobile market development

With new market entrants and licences issued for the Ugandan mobile market, mobile prices and connection charges dropped considerably (Shirley, 2001). Competition among the two mobile service providers, in addition to the privatized incumbent, has led to increasingly affordable prices and better quality of service. While national GDP has had an average growth rate of 5 per cent since 2000, the mobile telephony sector has experienced a growth of about 25 per cent per year since 1998, when MTN started to operate in the country. Despite the rapid growth in mobile telephony, the majority of the population still does not have access to basic telephony as the infrastructures are still inadequate and tariffs are still too high, per capita income being rather low ($200 in 2004) and over 39 per cent of the population living below the poverty line. Even if there has been a downward trend in prices, mobile penetration remains low (3 per cent), a fact which indicates that affordability is still a key issue. Box 6.3 provides a graphic representation of, and a commentary on, rural coverage by the main providers.

In 2001, local fixed telephony tariffs increased slightly, since MTN used GSM to offer fixed-telephony services. On the other hand, international tariffs have on the whole been reduced. In December 2003, international fixed and mobile tariff rates in Uganda were almost half of those in Kenya, as chart 6.18 illustrates.

D. Cross-country comparison

The rapid spread of mobile telephony in low income countries, namely Uganda, Nigeria and Kenya, has led to a much higher percentage growth in the number of mobile subscribers than in wealthier countries such as South Africa, where average income per capita is four to five times higher. Between 2002 and 2004, the percentage of mobile subscribers grew by 494.25 per cent in Nigeria, 172.93 per cent in Uganda, 107.37 per cent in Kenya and 42.73 per cent in South Africa.

The cost of mobile telephone calls has changed in recent years and in particular from 2002 onwards. Although Nigeria and Uganda show the same trend, as tariffs for both peak and off-peak calls have decreased since 2002, for Nigeria the fall has been more dramatic. In South Africa both tariffs have increased, while Kenya shows a decrease in costs for off-peak calls and a much more significant increase in the cost of peak ones.

Rates for the adoption of mobile telephones and mobile teledensity differ vastly from country to country. Chart 6.14 describes the effect of the entry of a second mobile operator in the market on the number of subcribers. Table 6.4 provides data on the changes in mobile teledensity during the period 2002 – 2006. While South Africa has a teledensity of more than 10 telephone lines per 100 inhabitants, Nigeria and Kenya have a density of between 5 and 10, and Uganda less than 5. Moreover, divergences within the same country are equally as large, a fact which suggests that current policies are effective in promoting ICT access in urban areas, while there is still a policy gap as regards rural areas. For example, Nigeria has a national teledensity of approximately 5.2, with some urban areas having a density of as 25, while in most of the rural ones teledensities are as low as 0.1. The overall telephone coverage is less than 45 per cent of Nigeria's total surface area. In South Africa the situation is rather different. While national teledensity is quite high at 46, rural areas have values of 5. On the whole, mobile penetration in rural areas of the countries under examined is still low.

Chart 6.14

Growth in number of mobile subscribers

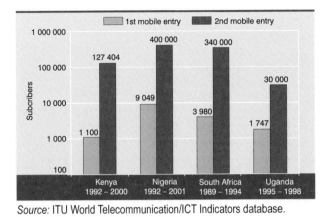

Source: ITU World Telecommunication/ICT Indicators database.

Chart 6.15

Call costs in 2003: mobile versus fixed lines*

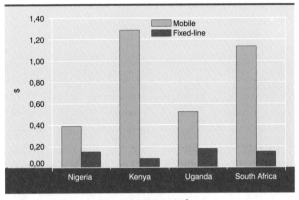

Note: * Three minutes, local, peak-time, in $.
Source: ITU World Telecommunication/ICT Indicators database.

Table 6.4

Change in mobile teledensity, 2002-2006

	2002	2003	2004	2005	2006
Kenya	3.8	5.0	7.8	13.5	18.5
Nigeria	1.3	2.6	7.2	14.1	24.1
South Africa	29.4	35.9	41.3	65.4	71.4
Uganda	1.5	2.9	4.2	5.3	6.7
Africa	4.5	6.1	9.0	14.8	20.5

Source: ITU World Telecommunication/ICT Indicators database.

While some research suggests that high costs (Panos Report, 2004) are the main factor in low mobile telephone penetration in rural areas, supply-side issues, namely a simple lack of commercial incentive to provide geographical extension and coverage, may also be valid, as well as issues related to appropriate modes of use relevant to rural populations, highlighted in part B.4. Whatever the case, connecting rural areas continues to be a major issue because, at current teledensity and penetration rates, approximately 70 to 80 per cent of the African population still lacks basic telecommunications infrastructure. This reinforces their economic marginalization and sustains the unattractiveness of rural development as far as market-based, private-investment-led solutions are concerned. Moreover, this may be a manifestation of market failure which needs to be addressed through government intervention by means of policies that improve incentives for investment, and ultimately by investing public funds to develop a supply capacity.

On the other hand, pro-competitive policies, including market liberalization, competition and the establishment of independent regulators, have improved significantly telecommunications services in urban areas.

E. Conclusions

This chapter has described how mobile telephones contribute to digital empowerment. Firstly, mobile telephony enables increasing digital inclusiveness through a platform that is growing in sophistication. Short message services have introduced simple wireless text and data transfers. Mobile handsets are growing in sophistication and functionality, such as digital photography and multimedia messaging, and increasingly provide useful functions previously available only with more costly personal digital assistants or specialized mobile appliances. Secondly, mobile handsets provide a starting point for digital literacy.[23] For many individuals and communities, once the initial hurdle of ICT acceptance has been overcome, the adoption of higher-level technologies may be less daunting.

It has been demonstrated that a major driving force behind the rapid growth of mobile telephony use in Africa has been the improved supply of services, which has been achieved by issuing licences to a larger number of entrant providers. In the countries examined in this chapter there are different degrees of market competition. Moreover, the difference in growth, adoption and competitiveness levels can be examined by looking at the dynamics regarding the entry of the first mobile operators and subsequent ones, the resulting developments in pricing, and comparing these data with the growth of mobile subscriptions. Even countries with low per capita income have achieved a higher level of growth in the number of mobile subscribers than wealthier countries thanks to the implementation of policies supporting mobile services development and competition in mobile markets. In contrast, South Africa's telecom market, where competition policy was implemented more slowly, has not shown a dramatic decrease in mobile prices expressed in prices of handsets, connectivity charges and call tariffs.

A number of issues raised in chapter 6, dealing with web services of the UNCTAD *Information Economy Report 2006* equally apply to mobile telephony. The issues of interconnection and the network effects of increasing outreach and access to mobile telephony necessarily entail a market environment increasingly marked by cooperation within, and competition among,

networks and, in the case of the present chapter, mobile telecommunications firms. This implies that the successful mobile providers and, more importantly, the growing mobile networks will be those that have the capacity to generate and sustain trust among business partners and consumers.

Regulatory authorities have a crucial role to play in this regard and Governments of developing countries should support the establishment and development of regulatory environments that provide mobile firms, investors and consumers with the confidence and trust that will facilitate ICT-enabled development and its positive implications for overall economic development. The regulatory environment is not static. Rather it is a set of parallel processes that aim to balance the interests of users and consumers, investors, firms and their employees. While consumer welfare is the bottom-line parameter, Governments can achieve their policy ambitions only by giving adequate attention to all processes, as the public interest embraces the totality of the regulatory and economic outcome.

Although privatization of State monopoly operators, market liberalization, competition and the establishment of independent regulators have significantly improved telecommunications services in urban areas, the accessibility of mobile services in rural areas is still low. From a technology perspective, mobile telephones are a workable solution for the development of the information society in rural districts since they can reach wider areas at lower costs. However, economic incentives seem to be still insufficient, and there appears to be no clear consensus as to which policies, if any, can improve this situation.

References and bibliography

Aochamub A, Motinga D and Stork C (2002). Economic development potential though IP telephony for Namibia. Discussion paper No. 2002/84, United Nations University, http://www.wider.unu.edu/publications/dps/dps2002/dp2002-84.pdf.

Allen JR (2003). Rural access to ICTs in Southern Africa. Telecommunications Policy Research Conference, ID 263, http://tprc.org/papers/2003/269/RuralAccessICTs2.pdf

Anderson J (2006). A structured approach for bringing mobile telecommunications to the world's poor. The Electronic Journal on Information Systems in Developing Countries, 27, 2: 1–9, http://www.ejisdc.org/ojs2/index.php/ejisdc/article/viewFile/282/178.

Bmi-Techknowledge (2005). Communication Technologies Handbook: Nigeria, (p. 328–353), http://www.bmi-t.co.za/bmi/Content/Handbooks/handbooks%20pics/2005/Nigeria05.pdf.

Communications Commission of Kenya (CCK) (2005). *Annual Report 2004/05*, http://www.cck.go.ke/annual_reports/cck_annua%20_report_05-06.pdf.

Donner J (2005). Research approaches to mobile use in the developing world: a review of the literature. International Conference on Mobile Communication and Asian Modernities, City University of Hong Kong, 7–8 June, 2005, http://www.jonathandonner.com/donner-mobrev.pdf.

Eagle N (2005). Pervasive computing in the midst of pervasive poverty: Entrepreneurship and education though mobile phones in Kenya. MIT, Nokia Kenya proposal.

Econ One Research (2002). A case study in the private provision of rural infrastructure, http://www.itu.int/ITU-D/treg/Events/Seminars/2005/Thailand/Reference%20Material/Reference%20Material%20-%20Uganda%20Experience.pdf.

Engvall A and Hesselmark O (2005). Internet for everyone in African GSM networks. Scanbi-Invest, Stockholm, http://www.scanbi-invest.com/ebc/GPRS_report2.pdf.

European Union Committee (2007). Mobile phone charges in the EU: curbing the excesses, European Union Committee of the House of Lord of the United Kingdom, HL Paper 79-I, http://www.publications.parliament.uk/pa/ld200607/ldselect/ldeucom/79/79i.pdf.

Gamurorwa A (2004). Uganda: A success story? In: Panos Report 2004. *Completing the Revolution. The Challenge of Rural Telephony in Africa.*

Gebreab A (2002). Getting connected: competition and diffusion in African mobile telecommunications markets. World Bank Working Papers, http://econ.worldbank.org/files/15963_wps2863.pdf.

Gillwald A, Esselaar S, Burton P and Staveou A (2004). Towards an e-index for South Africa: measuring household and individual access and usage of ICT. LINK Centre, Johannesburg, South Africa.

Gillwald A and Kane S (2003). South African telecommunications sector performance review, http://link.wits.ac.za/papers/tspr2003.pdf.

Goodman J (2005). Linking mobile phone ownership and use to social capital in rural South Africa and Tanzania. In: Vodafone Policy Paper Series, vol. 3, Africa: the impact of mobile phones.

Hausman J (2002). Mobile telephone. In: *Handbook of Telecommunications Economics*. Elsevier Science B.V., http://econ-www.mit.edu/faculty/download_pdf.php?id=464.

International Chamber of Commerce (2005). ICC policy statement on WTO telecom. Commission on E-Business, IT and Telecoms Services Negotiations, Document No. 373/463, http://www.iccwbo.org/uploadedFiles/ICC/policy/e-business/Statements/373-463_WTO.pdf.

ITU (2001). Regulatory implications of broadband workshop. Case study: broadband – the case of South Africa, http://www.itu.int/osg/spu/ni/broadband/workshop/southafricafinal.pdf

ITU (2002). *World telecommunication report*, http://www.itu.int/ITU-D/ict/publications/wtdr_02/.

ITU (2003). Mobile overtakes fixed: Implications for policy and regulation, http://www.itu.int/osg/spu/ni/mobileovertakes/Resources/Mobileovertakes_Paper.pdf.

ITU (2006). *World Telecommunication/ICT Development Report 2006. Measuring ICT for social and economic development*, http://www.itu.int/ITU-D/ict/publications/wtdr_06/.

Jensen R (2007). The digital provide: information technology, market performance and welfare in the South Indian fisheries sector, *Quarterly Journal of Economics*, vol. CXXII, Issue 3, http://www.mitpressjournals.org/doi/pdf/10.1162/qjec.122.3.879.

Keck A and Djiofack-Zebaze C (2006). Telecommunications services in Africa: the impact of multilateral commitments and unilateral reform on sector performance and economic growth. WTO Staff Working Paper ERSD-2006-10, http://www.wto.org/english/res_e/reser_e/ersd200610_e.pdf.

Kelley CM and McCarthy C (2006). The Chinese and Australians Soak Up Broadband, Forrester Research. http://www.forrester.com/Research/Document/Excerpt/0,7211,39378,00.html.

Love D (2005). An overview of the South African telecommunications industry: from pre-1994 policy-making to gloomy 2005 realities. Research Report, University of the Witwatersrand, South Africa.

Madden G and Coble-Neal G (2004). Economic determinants of global mobile telephony growth. *Information Economics and Policy*, 16, pp. 519–534.

Meso P, Musa P and Mbarika V (2005). Towards a model of consumer use of mobile information and communication technology in LDCs: the case of sub-Saharan Africa. *Info System Journal*, 15, pp. 119–146.

Monti M (2004). Competition for consumers' benefit. European Commissioner for Competition Policy, European Competition Day, Amsterdam, 22 October 2004, http://ec.europa.eu/comm/competition/speeches/text/sp2004_016_en.pdf.

Muthaiyah S (2004). Key success factors of 3rd generation mobile network services for m-commerce in Malaysia. *American Journal of Applied Sciences*, 1(4), pp. 261–265.

Nigerian Communications Commission (NCC) (2005). Trends in telecommunications markets in Nigeria, 2003-2004, http://www.ncc.gov.ng.

Oestmann S (2003). Mobile Operators: Their contribution to universal service and public access. Intelecon Research & Consultancy Ltd, http://www.inteleconresearch.com/pdf/mobile%20&%20us%20-%20for%20rru.pdf.

OFTA (2003). Report on the effectiveness of competition in Hong Kong's telecommunications market. Office of the Telecommunications Authority of Hong Kong SAR China, http://www.ofta.gov.hk/en/report-paper-guide/report/rp20030620.pdf.

OECD (1996). Competition in telecommunications. OCDE/GD(96)114. http://www.oecd.org/dataoecd/34/50/1920287.pdf.

Onwumechili C (2005). Reaching critical mass in Nigeria's telephone industry. *Africa Media Review*, vol. 13, no. 1, pp. 23–40, http://www.codesria.org/Links/Publications/media_review1_05/chuka.pdf.

Panos Report (2004). Completing the Revolution. The Challenge of Rural Telephony in Africa, http://www.researchictafrica.net/images/upload/The%20Challenge%20of%20Rural%20Telephony%20in%20Africa.%20Panos%20Report.pdf.

Röller L-K and Waverman L (2001). Telecommunications infrastructure and economic development: a simultaneous approach, *American Economic Review*, vol. 91, no. 4., http://www.jstor.org/view/00028282/di020246/02p01767/0.

Samuel J, Shah N and Hadingham W (2005). Mobile communications in South Africa, Tanzania and Egypt: Results from community and business surveys. In: Vodafone Policy Paper Series, vol. 2, Africa: The impact of mobile phones.

Satola D (1997). Legal and regulatory implications of implementing WTO telecommunication commitments in developing markets. World Bank Legal Department, Legal Reform and Private Sector, http://siteresources.worldbank.org/INTLAWJUSTICE/Resources/WTOTelecomDevelopmentMarkets.pdf.

Scott N, Batchelor S, Ridley J and Jorgensen B (2004). The impact of mobile phones in Africa. Prepared for the Commission for Africa, CNTR 026, http://www.commissionforafrica.org/english/report/background/scott_et_al_background.pdf.

Shirley MM (2001). *The Effects of Regulation and Competition on Telecommunications in Africa.* International Society for New Institutional Economics, Berkeley, California.

Sinha C (2005). Effect of mobile telephony on empowering rural communities in developing countries. International Research Foundation for Development, Conference on Digital Divide, Global Development and the Information Society, http://www.irfd.org/events/wf2005/papers/sinha_chaitali.pdf.

South Africa Foundation (2005a). Telecommunications prices in South Africa: An international peer group comparison. Occasional Paper, 1/2005.

South Africa Foundation (2005b). Reforming telecommunications in South Africa: twelve proposals for lowering costs and improving access. Occasional Paper, 2/2005.

Thompson HG Jr and Garbacz C (2007). Mobile, fixed line and Internet service effects on global productive efficiency. Information economics and policy, http://dx.doi.org/10.1016/j.infoecopol.2007.03.002.

Uganda Communications Commission (UCC) (2005). *Recommendations on Proposed Review of the Telecommunications Sector Policy: Policy Review Report,* http://www.ucc.go.ug.

UNCTAD (2006). *Information Economy Report.* UNCTAD/SDTE/ECB/2006/1, http://www.unctad.org/en/docs/sdteecb20061_en.pdf.

Varoudakisa A and Rossotto CM (2004). Regulatory reform and performance in telecommunications: unrealized potential in the MENA countries. *Telecommunications Policy*, 28, pp. 59–78. http://www.sciencedirect.com/science/journal/03085961.

Vodacom (2005). Group interim results for the six months ended September 30, 2005, http://www.vodacom.co.za.

Waverman L, Meschi M and Fuss M (2005). The impact of telecoms on economic growth in developing countries. In: Vodafone Policy Paper Series, vol. 2, Africa: The impact of mobile phones, http://www.itu.int/osg/spu/dtis/documents/Papers/vodafonepapers.pdf.

Weigel G and Waldburger D (2005). ICT4D – Connecting people for a better world, section II.1: Innovating for equitable access. Swiss Agency of Development and Cooperation and the Global Knowledge Partnership, http://www.globalknowledge.org/ict4d/index.cfm?menuid=45

Williams M (2005). Mobile networks and foreign direct investment in developing countries. In: Vodafone Policy Paper Series, vol. 2, Africa: The impact of mobile phones.

World Bank (2006). Information and communications for development: global trends and policies. World Bank, Washington, DC.

Notes

1. That report was the thematic predecessor of the current *Information Economy Report.*

2. This is mostly related to the physical limitations of small screens, keyboards and scrolling devices, as well as the important differences in how such hardware is designed for various handsets.

3. See Historical Timeline at About Bell Labs/Bell Labs History at http://www.alcatel-lucent.com.

4. See interview with Martin Cooper at http://news.bbc.co.uk/1/hi/uk/2963619.stm; for more information on early standards development see http://news.bbc.co.uk/1/hi/uk/2963619.stm and http://www.rfidc. com/docs/introductiontomobility_standards.htm.

5. See http://www.pcworld.com/article/id,127820/article.html.

6. See http://www.comscore.com/press/release.asp?press=1041 or http://www.emarketer.com/Article. aspx?id=1004233.

7. See Waverman, Meschi and Fuss (2005) and Sinha (2005).

8. See http://news.bbc.co.uk/2/hi/africa/3321167.stm or http://www.interaction.org/ict/success_text_ Kenya.html.

9. See http://www.vodacom.co.tz/docs/docredir.asp?docid=3281 and http://news.bbc.co.uk/2/hi/ business/4145435.stm.

10. See http://www.taipeitimes.com/News/biz/archives/2007/02/25/2003350026 , http://research.nokia. com/people/jan_chipchase/JanChipchase_SharedPhoneUse_vFinal_External.pdf , or http://www. panos.org.uk/newsfeatures/featureprintable.asp?id=1206 .

11. For the full text of the Reference Paper see http://www.wto.org/english/tratop_e/serv_e/telecom_e/ tel23_e.htm.

12. A mobile payphone is a privately owned handset and subscription that has a definite physical location, for example attached to a general store or a Internet/phone café, which non-subscribers can use for a fee. In this sense, it is not truly mobile; rather, it uses mobile connectivity to establish a payphone service where fixed line infrastructure is lacking or of poor quality.

13. For more details on the pre-1992 period see Onwumechili (2005).

14. Transnational Corporation of Nigeria PLC (Transcorp) is a domestic corporate conglomerate that has investments in oil, electric power, tourism and communication companies. See http://www. transcorpnigeria.com/.

15. See http://www.cck.go.ke/market_information-telecommunications/.

16. See http://www.cck.go.ke/telecommunications-tariff_regulation/.

17. See http://www.cellular-news.com/story/25056.php .

18. See CCK (2005); additional data are UNCTAD calculations based on ITU World Telecommunication ICT Indicators database.

19. See http://www.icasa.org.za/Content.aspx?Page=17.

20. See Econ One Rsearch (2002, p. 23).

21. See http://www.mtti.go.ug/docs/MTCS%20for%20the%20private%20sector.pdf.

22. See https://www.chogm2007.ug/index.php?option=content&task=view&id=76.

23. It should be noted that basic digital literacy in the sense of being able to use basic mobile telephone functions does not mean and should not be confused with computer literacy and the truly significant ICT human capacity building needed develop an information society. This is an ambition that outstrips the purpose and potential of mobile telephony. However, it is for many people a first step.

Chapter 7

PROMOTING LIVELIHOODS THROUGH TELECENTRES

A. Introduction

The multi-purpose applications of ICT and the rapid fall in its costs have allowed innovative uses of that technology in many poverty reduction programmes. Governments worldwide[1] are developing and implementing ICT policies to support economic and social development. However, so far, the impact of ICT strategies and programmes in providing economic opportunities for people living in poverty has been limited. Attempts to provide wider access to ICT, including the establishment of public ICT access spaces (telecentres) and programmes to develop ICT skills, have not been sufficient.

This chapter reviews the role of a key policy instrument (telecentres) in promoting livelihood opportunities for people living in poverty, in line with earlier UNCTAD analytical work on ICT and poverty reduction. In the 2006 edition of this Report, UNCTAD proposed a pro-poor ICT framework[2] to evaluate to what extent a policy or programme supports people living in poverty. The framework highlights 12 dimensions (the 12 Cs) that ICT programmes and policies supporting poverty reduction should consider (connectivity, content, community, commerce, capacity, culture, cooperation, capital, context, continuity, control and coherence). This chapter considers the "commerce" component of that framework – that is, the need for ICTs to be relevant to the economic lives of their users.

To examine how ICTs can support means of living for people living in poverty, this chapter will use the concept of livelihoods. "A livelihood comprises the capabilities, assets (including both material and social resources) and activities required for a means of living" (DFID, 1999). In other words, what is important for a person is not only the wage income that he or she may receive but also all the other assets (e.g. subsistence food, savings, physical assets, access to government support) and capabilities (e.g. education, access to markets, access to information, ability to communicate) that he or she may need to have in order to earn a living.

Sustainable livelihoods approaches[3] put people living in poverty at the centre of development programmes and emphasize that programmes supporting poverty reduction should be based on the specific context, and people's vulnerabilities, capabilities and assets, and should address the influence that institutions (whether Governments, civil society or private companies), organizations and social relations have on the ability of men and women to pursue a strategy to escape from poverty. Thus, using telecentres for promoting livelihoods[4] requires an understanding of how they can support people's capabilities and assets, how they can address their vulnerabilities, and how they can work in and influence the specific institutional and social context shaping those livelihoods.

Section B examines telecentres as a specific approach for using ICTs in order to support livelihoods. The first part provides an introduction to the concept of telecentres and of how telecentres can support livelihoods. The second part discusses the findings of the survey conducted by UNCTAD (2007) among telecentre networks. The subsections that follow present, respectively, examples of best practices regarding how telecentres are promoting livelihoods, and a discussion of the challenges ahead. Section C provides a checklist of elements to take into account when using telecentres for supporting livelihoods. This checklist expands the commerce component of UNCTAD's 12 Cs pro-poor ICT framework.[5] Some general policy recommendations addressed to Governments and telecentre networks for enhancing the role of telecentres in promoting livelihoods are also made.

B. The case of telecentres

1. How telecentres can promote livelihoods

Telecentres are public facilities where people can access the Internet, computers and other information and communication technologies to gather information, communicate with others and develop digital skills (telecentre.org). A telecentre may be, for example, a public library providing Internet access and basic digital literacy training courses (e.g. Biblioredes in Chile) or a community development centre providing, among other services, Internet access to meet the information needs of the community (e.g. the Pallitathya telecentre in Bangladesh featured in box 7.1). Cybercafes are "privately owned, primarily urban establishment providing limited services, such as e-mailing and browsing" (Fillip and Foote, 2007, p. 19). Telecentres may come in varied forms and names (e.g. village computing, information kiosks, community multimedia centres), but they have a common goal: to serve the community and support local development. This developmental characteristic distinguishes telecentres from cybercafes. Telecentres are often a key policy or programme to bridge the digital divide, and some Governments include a telecentre programme as part of their national ICT policy.[6]

The first telecentre programmes in developing countries started in the late 1990s. Financial, political and sociocultural sustainability concerns pose a challenge to the existence of telecentres and some of them have had to be discontinued or reduced in scale. Nevertheless, telecentres continue to play a critical role in supporting an inclusive society. Given the limitations of the private sector as regards serving less profitable areas, telecentres are increasingly recognized as a public good and a development instrument worthy of public support (Fillip and Foote 2007). Telecentres, since they support underserved communities, will often necessarily require financial subsidies. Governments'

Box 7.1

How telecentres are supporting livelihoods: some examples

Pallitathya – network of rural information centres (Bangladesh)

Mr. Nurul Islam Khan produces rice, together with beans, bitters and bottle gourds, on his land. He supports a six-member family with an average monthly income of BDT 8,333 ($120). One day he found that his cultivated vegetables were attacked by harmful insects. He sought information from the Pallitathya Kendara (the local rural information centre) and used audiovisual CDs to obtain the desired advice. Mr. Khan successfully applied the prescribed insecticides and saved his crop and, above all, his livelihood. He was able to prevent a total loss of BDT 8,500 ($123 USD) without paying any fee for the service offered by the Pallitathya Kendra. He thinks that the information services provided can save farmers from potential loss, and he suggests that farmers like him would benefit immensely if the infomediaries of the centre regularly visit the farmers in the area to provide them with the necessary information.

Ms. Najmunnahar looks after her family, which has an average monthly income of BDT 2,000 ($29). She heard about the different services offered at the local Pallitathya Kendra from her neighbours and from the centre's infomediaries during their field mobilization. She visited the centre to collect printouts of different embroidery designs and paid BDT 5 (7 US cents). Ms. Najmunnahar used the designs to make beautiful rural blankets known as *Kantha*. Her designs brought her praise, and they have given her a certain standing in the neighbourhood. She is very pleased with the service received and will be glad if the centre can provide training for making clothes and accessories.

Ajb'atz' Enlace Quiché Association (Guatemala)

Ajb'atz' Enlace Quiché is an association developing the capacities of Mayan communities through ICTs. It focuses on the promotion of the Mayan culture and linguistics and the provision of educational services. In the association's community telecentre, adult students on the «Up to date with ICT» training course find the new computer skills very useful in several aspects of their lives: "as a teacher is very useful to teach children and also to prepare material for the lectures", "now I can help my children with their homework and ICT skills".

Partnerships for e-Prosperity for the Poor (Indonesia)

Telecentre Muneng, in East Java province, has inspired the establishment of Gabungan Pemuda Terampil dan Kreatif (GPTK), a cooperative of cricket breeders. After finding through the Internet access facilities offered by the telecentre a new method for breeding crickets, four breeders established GPTK in December 2005 with an initial investment of 3 million rupiah ($330) and a working capital of 300,000 rupiah ($33). The new method of breeding proved to be more productive than the conventional one. The founders also learned how to find buyers through the Internet. Substantial requests and orders for crickets from several big cities in Java had surprised them and made them more enthusiastic and convinced of the potential of cricket breeding.

Source: Pallitathya (2006).
         Cowell and Marinos (2006).
         Pe-PP (2007).

financial support for telecentres does not need to be permanent. In the longer term, telecentres may be able or expected to be self-sustainable, and the private sector may start offering some of the services in an affordable manner. Moreover, today, telecentres often benefit from an enhanced environment in which to operate and from a greater understanding of best practices for dealing with sustainability.[7] Although there is no one sustainable telecentre model, sustainability can be achieved through the provision of value-added services, by working in partnership with others in the public and private spheres, and by timely scaling-up to build on existing experiences, capacities and resources (Fillip and Foote, 2007).

How can telecentres support the livelihoods of people living in poverty? Telecentres may, for example, support the development of technical and business skills, provide access to key information, facilitate access to government services and financial resources, and/or provide micro-entrepreneurship support. Box 7.1 provides examples of a telecentre in Guatemala that supports teachers in developing ICT skills and telecentres in Bangladesh that help farmers reduce vulnerabilities and women enhance their income-generating activities. Telecentres are also centres for dissemination and consumption of science, technology and knowledge. As box 7.1 shows, telecentres facilitate access to knowledge and support the dissemination of productive methods.

The examples in box 7.1 are an illustration of how telecentres promote economic opportunities and sustain livelihoods. However, as the results of an earlier UNCTAD survey (2006b) among Chilean telecentres show (see box 7.2), the promotion of economic activities through telecentres may not be widespread or may pay limited attention to the needs of some user groups. In the first place, while telecentres in Chile have been successful in promoting e-government services, only a small number of them are designed to support economic activities, and capacity-building programmes in trade and business are not mainstreamed. Secondly, there is limited understanding of how women and men benefit from telecentres, and thus women may have fewer opportunities for using ICT to support their livelihoods.

A study by Parkinson and Ramírez (2006) has assessed to what extent the Aguablanca telecentre in Cali, Colombia, is supporting livelihoods. It shows that the telecentre supported only some of the livelihood strategies employed by the community. Residents used the telecentre mainly to develop social assets

(social relationships) and to increase financial assets in the long term by investing in education to improve formal employment prospects, even when current opportunities in the community lie primarily in informal employment. Other livelihood strategies were not being served through the telecentre, for example (a) increasing financial assets in the short term through the conversion of human or physical capital (i.e. formal employment or informal employment), (b) reducing reliance on financial assets by developing other assets (i.e. house ownership), and (c) reducing the necessary recourse to financial expenditure. The unemployed "rarely used the telecentre for searching a job and most considered it inappropriate for that use". The self-employed "rarely used the telecentre, or had any ambition to use it, in support of their business needs".

Telecentre managers and promoters can conduct similar analysis to better understand how specific telecentres can support local livelihoods. Such analysis, although context-specific, can help inform broader strategic decisions regarding the role of telecentre networks and support policymakers in their analysis and promotion of policies that can support the livelihoods of people living in poverty. Chart 7.21 defines the elements to be considered in assessing the livelihood dimension of telecentres, from the point of view of commerce, within the broader 12 Cs pro-poor ICT framework developed in the *Information Economy Report 2006*.

The following section provides a broader analysis of how telecentres, as institutions, are supporting the livelihoods of people living in poverty. The analysis is based on key elements of the sustainable livelihoods approach, which includes the broader definition of poverty, an acknowledgement of people's different assets and the strategies women and men can follow, and a holistic approach (where in addition to building capacities, development efforts take the context, structures and processes into account). However, UNCTAD's analysis focuses on how telecentres support economic activities, and does not explore other areas important for sustaining livelihoods, such as physical well-being, which are not UNCTAD's areas of expertise.

2. Telecentres' impact on supporting economic opportunities for sustainable livelihoods

To examine, at the broader level, to what extent current telecentres are supporting livelihoods, UNCTAD, in

Box 7.2

A gender perspective on supporting livelihoods through ICT: the case of Chilean telecentres

In 2006, UNCTAD conducted a study on Chilean telecentres and their contribution to poverty, in particular among women. Chile was selected as a case study because the Government has put in place a broad ICT strategy for development that includes support for Chilean telecentre networks. Chile>s development in the last decade, including in the area of access to and use of ICTs, has been notable. However, wide gender and economic disparities, including in ICT access and use, persist (Cecchini, 2005). Recent data (Cecchini, 2005; PNUD, 2006) show substantial inequalities in ICT access owing to income and geographical differences. For instance, in 2003, among the richest decile, three quarters had mobile telephone access and almost half had Internet access, while among the poorest decile, only one quarter had mobile telephone access and 1.4 per cent had Internet access (PNUD, 2006). In 2000, only 0.8 per cent of rural households had access to the Internet compared with 9.4 per cent of urban households (Cecchini, 2005, p. 29), and at the regional level, in 2003, while Santiago's Metropolitan region had a 9 per cent Internet penetration rate, the Maule region Internet penetration rate was 1.7 per cent (PNUD, 2006, p. 39).

The research used UNCTAD's 12 Cs pro-poor ICT framework as the point of departure for a survey and follow-up interviews among stakeholders from different Chilean telecentre networks. The 15 in-depth responses, albeit from a limited number of participants,[8] provide a flavour of the understanding of different stakeholders (i.e. national and regional telecentre coordinators, individual telecentre managers and users) about the capacity of telecentres to support livelihoods.

The Chilean telecentre network is a resourceful initiative: (a) it has strong and continued political support; (b) it strenuously engages with different actors; (c) it has implemented a large-scale ICT literacy campaign; and (d) it has had successful experiences in sustainability and community involvement

However, its relevance for reducing poverty among women is limited by several factors: (a) some stakeholders understate, where others take for granted, the importance of telecentres for reducing poverty; (b) there is limited understanding of how poor men and women use and benefit from telecentres; (c) gender is not mainstreamed – for instance, no specific training, content, evaluations or resources have been developed/earmarked for (poor) women – and the involvement and ICT capacity of women's institutions and organisations are limited; and (d) telecentres have yet to provide specific skills and livelihood opportunities for poor women.

Regarding the commerce dimension, Chilean telecentres participate in the broad promotion of a number of e-government services related to economic activities, and there are examples of successful adhoc collaboration with partners promoting the use of telecentres in that regard (for example, to complete tax returns online). There are also telecentres, even though limited in number, that support economic activity, such as those from SERCOTEC (Chile's Technical Support Service) and from FOSIS (Chile's Social and Solidarity Investment Fund).

However, support for economic activities, including for poor women, is limited. The training provided is not sufficient to support livelihoods: "in 18 hours it is difficult to transmit the practical use of ICTs" (questionnaire respondent). Collaboration is adhoc and depends on the ability/ willingness of the partners. As one respondent put it: "To promote livelihoods, telecentres can only support other entities in promoting and providing access to their online services, and the issue is the ICT capacity of these other entities. For instance, telecentres do not provide employment opportunities but could support local public employment offices if they used ICTs". As noted in an evaluation study carried out by the Chilean Undersecretariat of Telecommunications (SUBTEL, 2005, p. 74), the main failing of Chilean telecentres is their inability to have a noticeable economic impact.

The major challenges for Chilean telecentres with regard to supporting women's economic opportunities, as highlighted by respondents, are the following:

- There is limited awareness among women living in poverty of ICT opportunities.

- Few women have a regular job.

- Local governments are not particularly interested in supporting poor women entrepreneurs.

- Managers are not trained in this area.

- Capacity-building programmes in trade and businesses are not mainstreamed.

- There is a lack of resources.

collaboration with telecentre.org,[9] carried out a survey (in 2007) among telecentre networks[10] on the impact of telecentres on promoting economic opportunities.

The questionnaire targets telecentre networks rather than individual telecentres for a number of reasons. One reason, in addition to practical reasons, is that telecentre networks play an important role in creating dynamics within the telecentre movement, in pooling resources to, for example, develop content and training materials, in sharing best practices, in creating partnerships and, in summary, in being able to bring about changes.

The questionnaire (in annex 2) was sent to 84 coordinators and leaders of telecentres[11], and over a quarter of them (22) responded. The respondents represented 22 different networks in 21 countries across Africa (7 responses), America (6 responses), Asia (7 responses) and Europe (2 responses), which serve over 7 million users a year. The large majority of the respondents were from telecentre networks in developing countries. There were four responses from two developed countries (Canada and Spain), and these are also valuable since those networks also aim at supporting vulnerable communities. The complete list of respondents is provided in annex 3.

The following subsections present and review the survey's findings.[12] The survey should not be regarded as a statistical representative account but as an overview of how telecentres are supporting economic activities in different circumstances. The results are contrasted with the existing literature on telecentres and complemented with information provided by the telecentres themselves. As the responses to the questionnaire come from diverse telecentre networks with different objectives, sizes, and experiences, and while the analysis takes into account the spectrum of networks, some of the results may not accurately reflect individual experiences.

2.1 Profile of telecentre networks that responded to the questionnaire

In terms of size, most networks (73 per cent) have fewer than 100 telecentres and a third have fewer than 25 telecentres but three large networks each have between 300 and 900 telecentres. Most of the networks are young: over half of them are less than five years old (established in 2003 or after), and over 80 per cent of them were established after the year 2000. In 2006 six networks were established; furthermore, there are indications that at least two other networks are being established in Rwanda and the Dominican Republic,

Chart 7.1

Main sources of finance

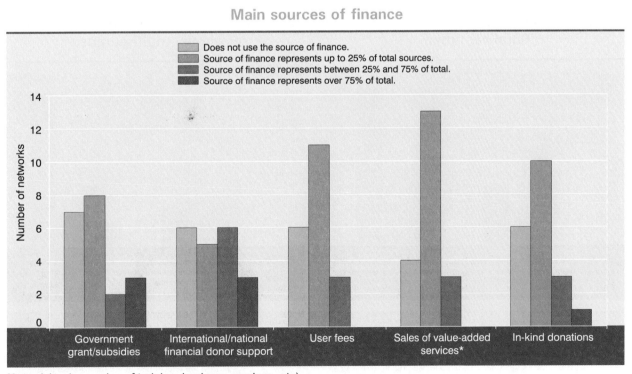

Note: * (such as sales of training, business services, etc)

and that discussions are taking place about setting up one in Indonesia. All of this suggests that there is an ongoing expansion in the establishment of telecentre networks.

In terms of budget size, nearly half of the networks reported that the annual cost of running the telecentre network was below $50,000 while for five networks the annual cost was between $250,000 and $5 million, and for four networks it was above $5 million. The networks generally diversify their sources of finance. Chart 7.1 shows that only one third of the networks are predominantly funded (over 75 per cent of total funds) by one source of finance, and when that is the case, they largely depend on government grants (three networks) or donor funds (three networks). The chart also shows that over two thirds of the networks charge user fees or charge for the sale of services to finance their activities. However, for most of those networks, user fees or the sale of services represent less than a quarter of all financial sources. Additionally, survey findings show that more than 60 per cent of the networks use four or more different sources of finance. There is only one exception: one network is entirely funded by government grants/subsidies. On average,[13] donors and Governments are the largest providers of funds (providing, respectively, 30 per cent and 24 per cent of the funds of a network), followed by user fees,

sales and in-kind contributions, which provide around 15 per cent each.

The telecentres served through these networks are mainly rural and multi-purpose (that is, telecentres, in addition to access to telephones, computers, the Internet and/or radios, offer access to other value-added services such as training and business support services). Charts 7.2 and 7.3 provide an overview of the composition of telecentre networks.

In nearly a third of the networks there is a balanced representation of female staff – that is to say, women represent from 40 to 60 per cent of the network staff (see chart 7.4). However, one third of the networks still have a limited female ratio (women represent less that 30 per cent of the staff).

2.2   Users

To understand how telecentres support livelihoods, the first question is as follows: Whose livelihoods do telecentres support?

The 17 networks that reported data support over 7 million users per year. Half of the networks have fewer than 15,000 users per year, while the other half have over 100,000 users per year. The survey asked the telecentre network leaders to identify their two

Chart 7.2

**Composition of telecentre networks
Rural vs. urban**

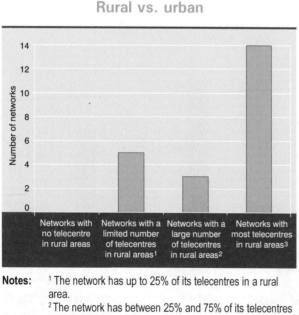

Notes: [1] The network has up to 25% of its telecentres in a rural area.
[2] The network has between 25% and 75% of its telecentres in a rural area.
[3] The network has over 75% of its telecentres in a rural area.

Chart 7.3

**Composition of telecentre networks
Basic vs. multi-purpose**

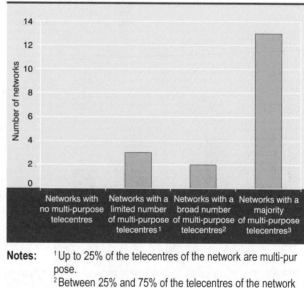

Notes: [1] Up to 25% of the telecentres of the network are multi-purpose.
[2] Between 25% and 75% of the telecentres of the network are multi-purpose.
[3] Over 75% of the telecentres of the network are multi-purpose.

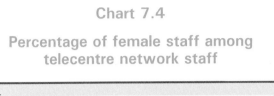

Chart 7.4

Percentage of female staff among telecentre network staff

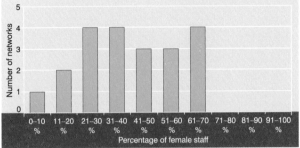

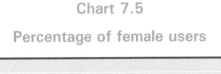

Chart 7.5

Percentage of female users

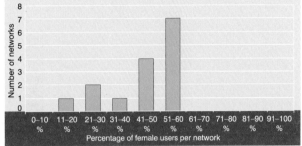

or three main user groups. Responses to the open question varied widely: some networks highlighted students, others the unemployed or new immigrants (in developed countries), but most responses did not identify the two or three main group of users.

Regarding female users, on average,[14] women represent around 40 per cent of users, and the majority of telecentre networks have a balanced ratio (40 to 60 per cent) of male and female users. However, there are great differences in female user ratios across networks (ranging from 15 per cent to 60 per cent) and three networks have a female user ratio of below 30 per cent (see chart 7.5). Other studies (e.g. Kumar and Best, 2006) also highlight significant variations in the gender balance among telecentre users to the detriment of female users.

It should be noted that the number and the gender of direct users do not necessarily reflect the total number and the gender of beneficiaries. For instance, a study in the Solomon Islands indicates that women often send their husbands to the radio station to transmit a message that they (the women) want to send (Chand et al., 2005, p. 39).

Who uses and who benefits from telecentres largely depends on the network's aims and design. For example, in the telecentre of the Ajb'atz' Enlace Quiché association in Guatemala (Cowell and Marinos, 2006), an association that focuses on supporting education and training by and for Mayan people, three quarters of users come from an indigenous group (a reflection of the wider community) and the large majority of users are students, and among adults, nearly half of them are teachers. User demographic analyses reveal that all-purpose telecentres (telecentres that not specifically target one group of users) are not necessarily used by the whole community. For example, a recent survey

(Pun et al., 2006) shows that users from Nepal's Wireless Networking are mainly young (83 per cent were under the age of 30), male (72 per cent) and educated (all users are literate, although the national literacy rate is only 53 per cent).

2.3   Services provided

A second step to understanding how telecentres support livelihoods is to examine the services offered by them. What range of services do telecentres offer? Do they provide specific services to support the economic activities of the community?

General services

A review of the types of technologies available across the telecentres reveals the kind, and delivery format, of services that can be provided. For instance, telecentres offering broadband Internet access can provide a wider number of services (e.g. downloading online training materials, VoIP) and perform a broader range of activities (e.g. buying and selling online) than telecentres offering limited access to the Internet.

Over 80 per cent of the networks provide access to computers in more than 75 per cent of their telecentres, that indicates that providing access to a computer is an essential service (see chart 7.6). All the networks that responded to the questionnaire provide access, at least in some of their telecentres, to the Internet via dial-up[15] or broadband, or both. However, broadband access to the Internet is still not available in a quarter of the networks. Few networks offer access to radio broadcasting, and when they do, it is to a limited extent – only one network offers it in more than 25 per cent of its telecentres. Access to telephone, fax and photocopier services varied widely across networks.

Chart 7.6

General services provided by the telecentre networks

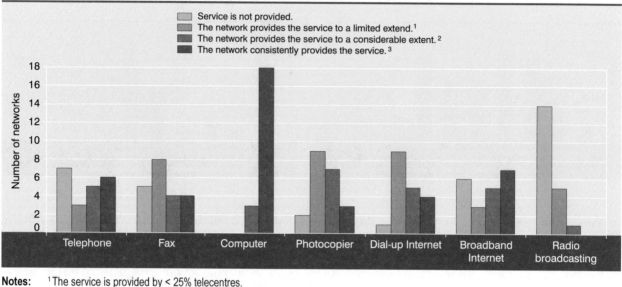

Notes: [1] The service is provided by < 25% telecentres.
[2] The service is provided by 25% – 75% of the telecentres.
[3] The service is provided by >75% of the telecentres.

The range of services provided by telecentres evolves with technological developments and as different technologies become more affordable. For example, the telecentre network in the United Republic of Tanzania reports that telephone services are no longer offered because of the emergence of mobile telephones and an increasing use of VoIP.

Training services

Training services can help develop competencies that telecentre users need in order to conduct economic activities – that is, to enhance their human capital, which they may be able to exchange in the short or longer term for financial assets. Training, to be relevant, needs to build the competencies required for the economic activities that trainees undertake or will undertake and, additionally, trainees must be able to apply those skills. The questionnaire enquired about the different types of training services the telecentres provided and to what extent they were provided across the network.

Chart 7.7 shows that telecentres mainly focus on providing basic ICT skills:[16] all networks provide basic ICT skills and three quarters of them consistently provide such training through their telecentres. More advanced ICT skills training in the use of advanced or sector-specific ICT tools and in the use of advanced functions of generic ICT tools is provided in nearly all of the networks, but a large proportion of them

provide it only to a limited extent (i.e. in less than 25 per cent of their telecentres).

The development of basic literacy skills (the ability to read and write) is supported by numerous telecentres. Half of the networks provide training in those skills either consistently (i.e. in more than 75 per cent of the telecentres) or in a considerable number (25 to 75 per cent) of telecentres. This is one indication that networks are reaching people living in poverty.

Training to develop skills important for developing economic activities, including e-business skills,[17] general business skills (e.g. marketing and management) and occupation-specific skills (e.g. farming, crafts and tourism), is provided in over half of the networks to a limited extent (less than 25 per cent of their telecentres). This means that there is scope for expanding the provision of that type of training by replicating within networks services already being provided by telecentres in the same network. Only one or two networks consistently provide (i.e. in more than 75 per cent of their telecentres) training in e-business or broader business skills. Some networks do not provide training services for e-business skills (three networks), general business skills (six networks) and occupation-specific skills (three networks).

What type of business-related training are telecentres providing? The telecentre network of Asturias (Spain) organizes training courses for businesses and

Chart 7.7

Training services provided by the telecentre networks

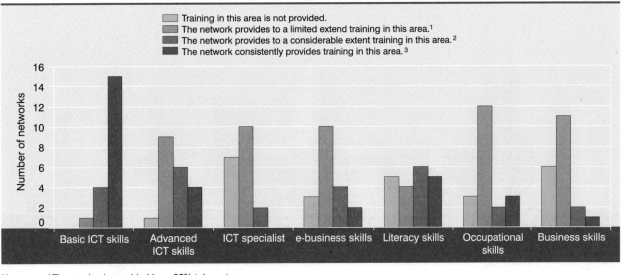

Notes: [1] The service is provided by < 25% telecentres.
             [2] The service is provided by 25% – 75% of the telecentres.
             [3] The service is provided by >75% of the telecentres.

entrepreneurs to become familiar with searching for employment opportunities and recruiting personnel, keeping financial records, conducting business communications, searching online for products, services and competitors, managing security (how to make security copies and protect from cybercrime), obtaining digital certificates, digitalizing documents and organizing business trips. After those training courses, the telecentres noted a significant increase in the number of wage employees and self-employed users (Fundación CTIC, 2006). In Bangladesh, the Community Information Centres organize a training programme on "how to start up a new business", which explores information available on the web and how the telecentre can help users request a trade licence or a bank loan (Shahid Uddin Akbar, SEBA Coordinator, GPCIC, UNCTAD questionnaire, 2007).

Business-related services

To further examine the extent to which telecentres support economic activities, the questionnaire enquired about the business support services being offered by telecentres.

The service most consistently provided, and by far, is "searching for information". All networks provide this service and 13 networks do so in the majority of their telecentres (chart 7.8). The other two business-supporting services more widely available are "searching for/advertising jobs" and providing

access to "government services" (a third of the networks provide those services in the majority of their telecentres). Other services often supported are searching for "professional/sector-specific information", "typing", "business communications" and supporting "employment opportunities". Some networks provide support for "designing and creating websites", "developing content", "buying and selling", and developing "business opportunities".

In Indonesia, to empower and mobilize poor communities for economic activities, and on the basis of the findings of ethnographic research, each telecentre has an infomobilizer – that is, a person that supports the development of the community by, among other things, using and promoting the use of relevant information. The infomobilizer helps the community/ village identify its needs and its opportunities to improve livelihoods (for example, acquiring new agricultural skills, expanding the marketing of village products and linking to experts for information). Moreover, websites to promote village products and tourism are developed.

Among responding telecentres, there is limited support in the areas of "advertising", "accountancy", "banking", "microfinance", supporting the carrying out of "payments", "export/import and trade facilitation", "data management and storage", "taxation", and "innovation/research and development". Each of those services is provided in less than half of the networks.

Chart 7.8 a, b

Business-related services provided by the telecentre networks

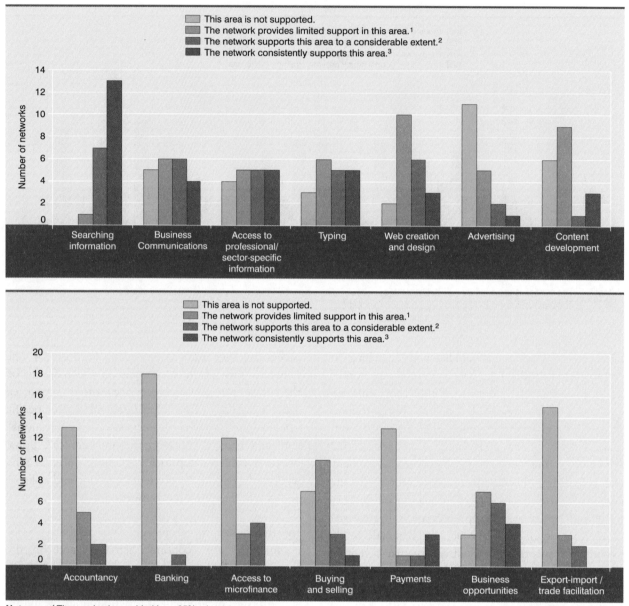

Notes: [1] The service is provided by < 25% telecentres.
              [2] The service is provided by 25% – 75% of the telecentres.
              [3] The service is provided by >75% of the telecentres.

Successful experiences of other networks in supporting some of these services (such as taxation in the case of Chile (box 7.3)) suggest that there is substantial scope for expansion. However, expanding the offer of business-related services requires the existence of certain conditions. For instance, supporting tax filing or other government services may need to be led by the Government. Also, some of the above areas (e.g. trade facilitation and banking) require specialized skills or additional infrastructure not always available.

Business-related services can be supported in a number of ways:

(a) Through specific training courses;

(b) As part of broader training courses – for example, a general course on ICT may show how to search the Internet for general and sector-specific information, conduct payments online and access different e-government services;

(c) Adhoc support provided by staff;

(d) Provision of the service by the telecentre itself.

Survey responses (chart 7.9 a, b, and c) show that to support business-related services telecentres mostly provide the service itself or provide adhoc support.

Chart 7.8 c

Business-related services provided by the telecentre networks

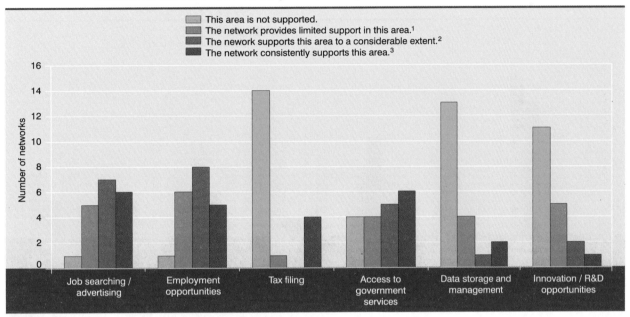

Notes: [1] The service is provided by < 25% telecentres.
   [2] The service is provided by 25% – 75% of the telecentres.
   [3] The service is provided by >75% of the telecentres.

Box 7.3

Making life easier: government services provided through the telecentres

Chile: telecentres support citizens in completing their tax declarations.

The Government of Chile, as part of a campaign to ensure that all tax declarations are completed online, has developed a cooperation agreement with telecentre networks across the country. Telecentres provide free access, or access at a reduced fee, access to the Internet and the technical tools needed for completing the declaration. The tax office provides training to telecentre staff so that they can support clients completing their tax declaration online. In 2007, for the fifth year running, the tax office programme allowed and helped citizens to complete their tax declaration throughout 578 telecentres operated by INJUV, Biblioredes, Sercotec, Confedech and Conupia, as well as 73 cybercafes.

Jhalawar, India: Janmitra improves citizens' access to government services in rural areas.

Janmitra is a network of 30 kiosks providing one-stop access points for various government services, such as land records and application forms. Users can, for a small fee, ask for land records and the patwari (local revenue official) visits the kiosk twice a week to sign the records, which are then delivered to the user by the kiosk owner either by hand or by courier. All other land-related records are manually provided by the patwaris.

An early needs assessment helped identify how the ICT project could support livelihoods in the rural areas, and experience confirms that such projects do not necessarily need to be connected to the Internet. Despite numerous challenges, the integration of various e-services and offline activities can generate enough revenue at information kiosks. For the Janmitra project, the franchisee model (based on enterprises charging users for their service in enabling the latter to access government services) has worked better than if the telecentres had been manned by government-employed staff. However, e-governance services are possible only when government departments streamline and e-enable them to function. A crucial lesson learned from the Janmitra project is that a close partnership with government is essential.

The sustainability of the project will now depend upon its level of institutionalization.

Source: Gobierno de Chile (2007).
         Harris and Rajora (2006).

Chart 7.9 a, b and c

Approaches used by telecentres to support business-related services

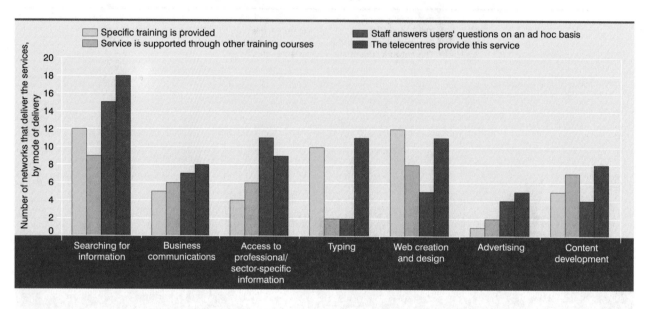

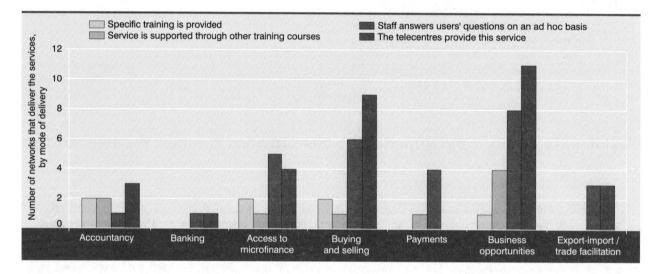

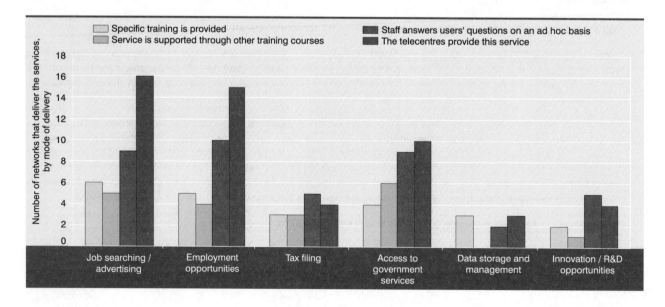

Those services are supported to a lesser extent with specific training or as part of other training courses. Specific training is mostly offered for "searching information", "creating websites", and "typing", and secondarily for supporting "job searching" and "employment opportunities", for conducting "business communications" and accessing "professional or occupational specific information". No telecentre network offers training in "banking", or "export/import and trade facilitation", and only one network offers training (as part of a broader training course) in conducting payments online.

Overall, there is limited explicit effort to build capacities in several business-related areas and to mainstream business-related training, the reasons probably being the greater complexity (telecentres have started by providing more basic services) and the specific knowledge requirements. Telecentre staff can provide adhoc support in a convenient manner, but it is unrealistic to expect that they can provide in-depth support in all the different areas. Rather, telecentres "in the future will not be a one-stop shop that requires a 'superman' office person, but a place where computing is a means for delivery of services from many back ends that support individual retail outlets" (John Sherry, quoted in Bell, 2006 p. 18). Some basic e-business training could be developed by the telecentre networks or integrated into more general ICT training courses (e.g. how to buy and sell online, online payments, etc.). Several telecentres are starting to include a wider range of specialized training services for micro-entrepreneurs. In Guatemala, Enlace Quiché (box 7.1), in collaboration with Fundación Omar Dengo, is developing a training course for small and medium-sized enterprises (SMEs) in Central America to support productivity, facilitate their administrative and productive processes and support the development of management capacities, including leadership, entrepreneurship and technical skills. Training of greater complexity, particularly when it deals with specialized topics (e.g. exporting and importing goods), may be better developed and provided in collaboration with specialized organizations such as trade-supporting organisations, business-promoting agencies, banks, and so forth. Later on, this chapter examines the type of organizations that telecentres are working with, providing an idea of which organizations could be further involved to support business-related services.

Often, as respondents concede, telecentres would like to offer business-related services, but they are limited by their "lack of resources and partners to support such projects" (Kenneth Chelimo, National

Chart 7.10

Extent to which supporting economic activities is an objective of the telecentre networks

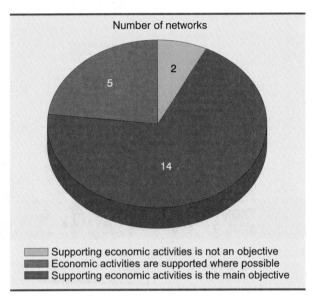

Number of networks

2
5
14

■ Supporting economic activities is not an objective
■ Economic activities are supported where possible
■ Supporting economic activities is the main objective

Coordinator of Kenya's Network of Telecentres), and "the lack of information available in digital formats and the low levels of local business development and connectivity" (questionnaire respondent). There an expectation ("this is an area we have just started working in", Julia Pieruzzi, Uruguay National Director, CDI) that as networks acquire experience and as Governments develop online services, a wider range of business support services will be provided across the telecentres.

2.4   Economic focus: objectives and sectors

While most of the telecentre networks that have responded to the questionnaire support economic activities where possible (see chart 7.10), less than a quarter of the networks reported that supporting economic activities is their main objective. Two telecentre networks (Red Conecta and Fundación CTIC in Spain) have a clear mandate that does not include supporting economic opportunities beyond supporting employment access through the development of ICT skills and job advertising. Several telecentres stress the importance of their social and educational goals, and the relevance of services in those areas for developing economic opportunities. Thus, as supporting economic activities is not the main goal for many organizations, resources are often not focused in these areas.

Chart 7.11

Economic sectors supported or serviced by telecentre networks, by level of support

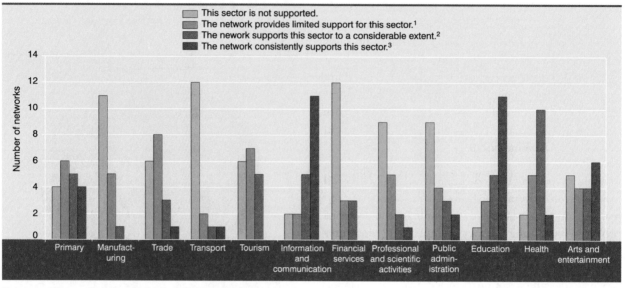

Notes: [1] The service is provided by < 25% telecentres.
[2] The service is provided by 25% – 75% of the telecentres.
[3] The service is provided by >75% of the telecentres.

Telecentres have the potential to support different economic areas. For example, in the aboriginal northern communities of Canada, the Community Access Programme of Nunavut helps artists in the booming local crafts industry to create websites,[18] to sell their products and to trade through eBay. Users in the tourism sector use the centres to create websites for advertising purposes. As the Secretary-Treasurer of N-CAP explained (UNCTAD questionnaire, 2007), local business people who cannot afford a computer and visiting scientists needing connection to the Internet use the facilities to gather information and manage e-mails. The telecentres also support visiting scientific researchers in various ways and with some specific projects. Many telecentres offer basic literacy training and access to distance education, as well as access to both government and non-government health information. Several of the telecentres specialize in film production and editing.

The survey asked each telecentre network whether their telecentres provide support or offer services related to twelve broad economic sectors. The two economic sectors in which telecentre networks provide most support or offer most related-services are, obviously, the information and communication sector, and the educational sector. Those sectors are consistently supported or serviced by over half of the telecentres (chart 7.11). Another two widely supported sectors are arts and entertainment and the primary

sector (agriculture, fishing etc.), which are consistently supported or serviced in, respectively, a third and a fifth of the telecentres. The health sector is supported in nearly all of the networks; however, only two networks support this sector in the majority of their telecentres. Rather surprising is the fact that only two networks consistently support the public administration sector and half of the networks still do not support it. The tourism sector is supported by over two thirds of the networks, but no network consistently supports that sector. Trade and tourism are supported to a limited extent in 40 per cent of the networks, a fact which indicates that there is potential for sharing practices among telecentres within the same network. The manufacturing sector and professional and scientific activities are rarely supported, and services or support regarding the financial services sector are scarcely provided. In general, the responses from the survey indicate that telecentres support some economic sectors, but there is potential to intensify the support and offer related services for a number of sectors, such as public administration, trade and tourism, and to explore opportunities for telecentres to provide services important for conducting business, such as financial services.

To understand to what extent people living in poverty benefit from the services offered by telecentres, the survey enquired how often, and how, services are targeted to specific vulnerable groups (women, the

Chart 7.12

Social groups targeted by telecentres

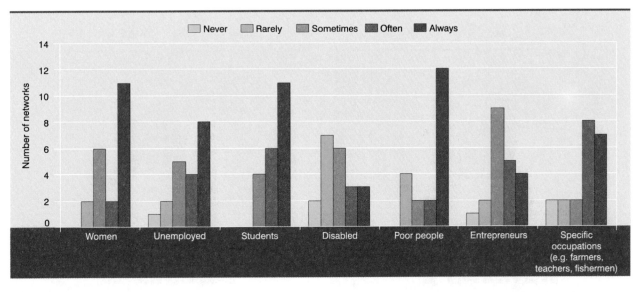

unemployed, the disabled, people living below the national poverty line), to businessmen/women, and towards specific occupations. Since several reports often suggest that a key user group is students (e.g. Kumar and Best, 2006), the UNCTAD survey also enquired about that group. Results confirm that students are the group most often targeted, as well as women and people living in poverty (chart 7.12). Over half of the networks always target their services to people living below the national poverty line, women and students. And over one third of the networks always target their services to the unemployed and to specific occupations. The groups that are less often targeted are disabled people and entrepreneurs.

An overwhelming majority (over 90 per cent) of the telecentre networks indicated that each of the following methods of targeting services is employed in their network:

-   The training content is adapted to the target group.

-   Courses take into account the needs of specific groups when designing the training (i.e. timing of the courses, location, etc).

-   Some courses are specifically designed for the target audience.

However, only some networks (less than a third) offer financial support to target their services. Other methods of targeting services indicated by respondents, and not

mentioned above, include targeted marketing and the fact that the courses are free for the users.

The only obvious conclusion that can be drawn from those responses is that telecentres use a variety of methods to target their services, but not financial support. Telecentre networks would need to conduct more specific and local analyses to be able to assess how effective the different methods are in targeting different community groups.

2.5   The environment: important factors and challenges

The context in which telecentres work shapes their role and ability to support livelihoods. For example, the regulatory environment and the quality of the general infrastructure condition the type of technologies that are available in a community and their affordability. The level of development of the private sector, the public sector and civil society conditions the range and type of services that are available and can be offered through the telecentres.

The survey asked network leaders to identify which factors are important for a telecentre with regard to supporting livelihoods and to what extent they are in place. Charts 7.13 a, b, c and d show how different environmental factors were rated. The brown shape shows how important a factor is believed to be, while the orange shape indicates the extent to which it is present. The closer the brown/orange shape is to the outside of the net, the more important the factor is or

Chart 7.13 a, b, c, d

Environment conditions: their importance and the extent to which they are being met

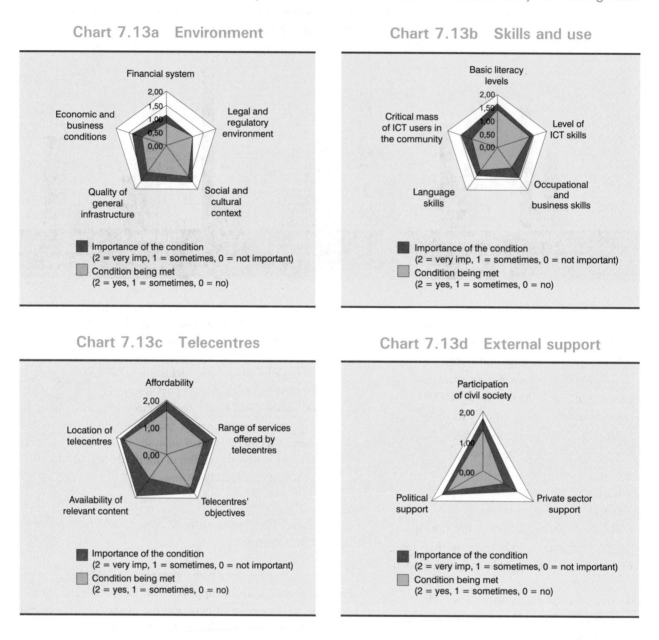

Chart 7.13a   Environment

Financial system

2,00
1,50
1,00
0,50
0,00

Economic and business conditions

Legal and regulatory environment

Quality of general infrastructure

Social and cultural context

■ Importance of the condition
(2 = very imp, 1 = sometimes, 0 = not important)
□ Condition being met
(2 = yes, 1 = sometimes, 0 = no)

Chart 7.13b   Skills and use

Basic literacy levels

2,00
1,50
1,00
0,50
0,00

Critical mass of ICT users in the community

Level of ICT skills

Language skills

Occupational and business skills

■ Importance of the condition
(2 = very imp, 1 = sometimes, 0 = not important)
□ Condition being met
(2 = yes, 1 = sometimes, 0 = no)

Chart 7.13c   Telecentres

Affordability

2,00

1,00

0,00

Location of telecentres

Range of services offered by telecentres

Availability of relevant content

Telecentres' objectives

■ Importance of the condition
(2 = very imp, 1 = sometimes, 0 = not important)
□ Condition being met
(2 = yes, 1 = sometimes, 0 = no)

Chart 7.13d   External support

Participation of civil society

2,00

1,00

0,00

Political support

Private sector support

■ Importance of the condition
(2 = very imp, 1 = sometimes, 0 = not important)
□ Condition being met
(2 = yes, 1 = sometimes, 0 = no)

the more it is present. Thus, the brown area surface between the two shapes highlights the relative extent to which a particular condition is being met. In other words, the larger the surface, the greater the impact of improvements in the environment factor will be. The charts present average opinions and they may thus mask substantial differences across different networks.

Regarding the broader environment (chart 7.13a), responses highlight two issues as important and not being met: first, the quality of the general infrastructure and, second, the economic and business conditions. Interestingly, telecentre leaders do not consider the legal and regulatory environment as important as other

general broad conditions. This may be explained by the fact that telecentre leaders may be inclined to consider more important those issues that are closer to their daily obligations and their area of influence.

Chart 7.13b, which looks at the skills available and at ICT use in the community, points out that basic literacy skills (the ability to read and write), ICT skills, and language skills (other than with regard to the mother tongue) are factors that, although important, are currently present, while the areas that would require more attention are the development of occupational and business skills and the development of a critical mass of ICT users in the community. The development of a critical mass of ICT users, which is non-existent particularly in rural areas, is an obstacle that telecentres themselves try to address by providing training in ICT skills.

Telecentre features (such as the objectives of the telecentres, their location, the range of services offered and their affordability) and the availability of relevant content are considered highly important by telecentre leaders (chart 7.13c) – which is logical given their closeness to them. More importantly, respondents highlight the lack of relevant content as a condition not being met to support livelihoods, followed then by the range of services offered by telecentres and affordability. The fact that telecentre leaders consider that there is a greater gap in the provision of relevant content than in the affordability of telecentres reinforces the observation that "the content and services sector supporting village computing is still in its infancy" (Bell, 2006, p. 24) and the importance of having locally developed content, which is of good quality, and relevant to the livelihoods of the users. "Content and services that are particularly relevant and that have an increased potential to generate revenue are those that are integrated into business and trading activities" (Bell, 2006, p. 24).

Chart 13.d explores the type of support important for telecentres. The participation of civil society (ownership of telecentre programmes) is rated as the most important, followed by political support. However, the greater gap is in the support received from the private sector. Chart 13.d confirm the importance of local ownership (Amariles et al., 2006) and political support[19] for creating an enabling environment (e.g. Gerster and Zimmerman, 2003) for telecentres, but also highlight the fact that there is scope for engaging further with the private sector.

Most telecentre networks receive some form of government support. However, public financial support is not available for telecentres in poorer countries (e.g. Bangladesh, Congo, Kenya, Nepal, Mali, United Republic of Tanzania). The ministries that most often provide support are the Ministry of Telecommunications or ICT, followed by the Ministry of Education, the Ministry of Social Affairs/National Development and the Ministry of Agriculture.

Among respondents, Chile – a State with a strong government vision to promote and adopt ICT (EIU, 2007, p. 3–4) – is the country where most ministries are involved in supporting telecentre networks. For instance, only respondents from Chile acknowledge

receiving support from the Ministry of Finance/Economy/Trade.

The Partnerships for e-Prosperity for the Poor (Pe-PP) in Indonesia is a programme funded by the United Nations Development Programme (UNDP) and implemented by the National Development Planning Agency. The national Government provides strategic support, involving many ministries, such as the Ministry of Communication and Information, the Ministry of Research and Technology, the Ministry of Education and the Ministry of Agriculture. It also collaborates by sharing content, training programmes, and networking telecentres and by funding buildings, operational costs (that is, Internet connection) and human resources. One of the objectives of the Pe-PP is to mainstream ICT for poverty reduction in the national poverty reduction programme

2.6   Use and impact

To understand the impact of telecentres, the survey asked telecentre leaders about the purposes for which the different user groups use the telecentre, and about their perception of the positive and negative impact that telecentres may be having in the community.

In terms of use, the responses were very clear. The three more common purposes of using a telecentre are "personal communications", "searching for information" and "receiving training". Other purposes, including "buying and selling goods and services", "business communications" and solving "administrative matters", are rarely perceived as one of the two key purposes (see chart 7.14). This is consistent with other findings. A study of Chilean telecentres (SUBTEL 2005) shows that according to users, telecentres' main impact on their quality of life was in the area of information and knowledge, as well as communication. Another study, this time of telecentres in five African countries[20] (Etta and Parvyn-Wamaliu 2003), noted that they were mainly used for communication and entertainment, rather than for economic purposes. More specific analysis will be necessary in order to understand why in some cases telecentres mainly support informational and entertainment activities (Is it because of a lack of services available and a lack of support for conducting economic activities? Is it because there are other local organizations better placed to support such activities?), and to explore the scope and mechanisms for promoting a more varied and deeper use of telecentres.

Chart 7.14

Key purposes of using a telecentre

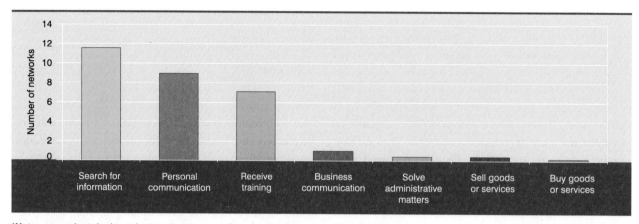

*Note: respondents had to select two key purposes for using the telecentres. The chart shows the average number of networks that highlighted the purpose as one of the 2 key purposes for using the telecentre

Chart 7.15

Key purposes of using a telecentre, by user group

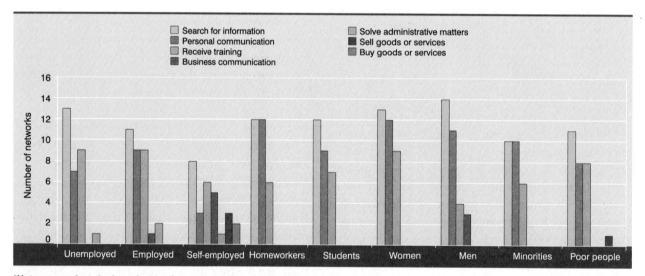

*Note: respondents had to select two key purposes for using the telecentre. The chart shows the number of networks that highlighted the purpose as one of the two key purposes for using the telecentre.

By user group (see chart 7.15), the perception is that self-employed users make more diverse use of the telecentre, and that for that group "business communications" are more important than "personal communications". Other groups perceived as having, to a limited extent, other key purposes besides "personal communication", "searching for information" and "training" are the unemployed ("solving administrative matters"), the employed ("solving administrative matters" and "conducting business communications"), men ("conducting business communications"), and people living in poverty ("selling goods and services" and "solving administrative matters"). People engaged in family duties, students, women and minorities are perceived as having only three key purposes (the above three common ones) for using the telecentre. The perceived differences between use by men and use by women are that (a) men may have as a key purpose to conduct business communications, but not women; and (b) women are more interested in receiving training than men.

These results are based on the perceptions of telecentre leaders, and may differ from the actual situation. Being aware of those perceptions is valuable because perceptions guide the decisions that telecentre

Chart 7.16

To what extent the telecentre networks help users to...

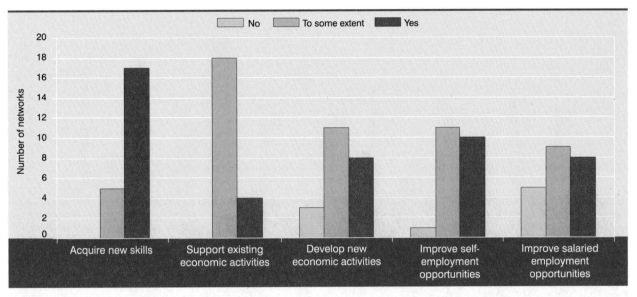

leaders make. However, all telecentre managers should corroborate perceptions with actual information, and telecentre networks should monitor on a regular basis the purposes for which different groups use the telecentre.

To understand what type of impact telecentres have in supporting the economic activities of their users, the UNCTAD questionnaire (2007) asked telecentre leaders to what extent telecentre networks help users to:

- Acquire new skills;
- Support existing economic activities;
- Develop new economic activities;
- Improve self-employment opportunities;
- Improve salaried employment opportunities.

According to telecentre leaders, the greatest impact comes from the acquisition of new skills, followed by improvement in self-employment opportunities (chart 7.16). Over three quarters of the respondents are confident that telecentres are helping users acquire new skills and nearly half of them indicated that telecentres are improving self-employment opportunities. All respondents stated that telecentres are supporting existing economic opportunities, while three of them indicated that telecentres are not supporting the development of new economic opportunities. There are substantial differences in the perception of the extent to which telecentres improve salaried employment opportunities: according to

nine respondents, telecentres are improving salaried employment opportunities, while for five respondents they do not. That difference can be partially explained by the context in which the telecentre operates: telecentres operating in more developed economies can better support salaried (often formal) employment opportunities, while telecentres operating in informal economies will have fewer opportunities to provide access to salaried employment opportunities.

Participants were also asked about changes observed in the lives of telecentre users, including changes in income levels and distribution, quality of life, access to public goods and services, coverage of basic needs (i.e. housing, health, nutrition), consumption, social relations and confidence levels. Most respondents highlighted improvements in confidence and changes in behaviour regarding the use of ICT. Half of the respondents cited improvements in income levels and employment, as well as in skills development. Seven respondents cited improvements in the coverage of information needs, and four mentioned improved access to public goods and services. For example, the Grameenphone Community Information Centres provide income opportunities for the entrepreneur managing the telecentre, who is able to recover his investment in one year, and for the end-user, who gets a fairer deal in economic transactions where the middleman is eliminated (AMM Yahya, Director of the GPCIC, UNCTAD questionnaire, 2007). With increased income, people can afford basic necessities such as food, shelter, health facilities and education. Also, information on public goods and services is

Chart 7.17

Organizations telecentres work with

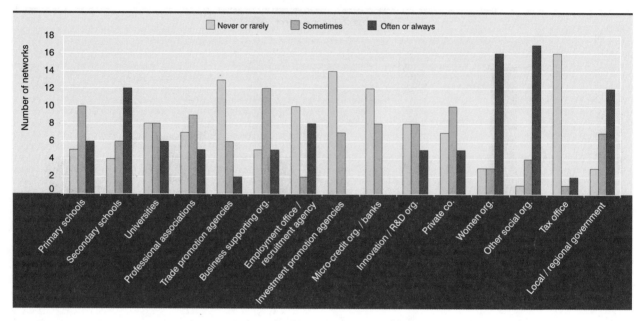

now available, and access to digitized forms used by the Government has been enhanced. There is also the possibility of accessing information on housing opportunities (real estate, house-building loans) as well as health information and services (AMM Yahya, Director of the GPCIC, UNCTAD questionnaire, 2007).

Most respondents did not indicate any negative impact of telecentres in the community. Those that indicate such an impact highlighted access to pornographic sites and cybercrime and "creating another area of necessary spending" (Joseph Sekiku, Interim Chairman, Tanzania Telecentre Network). A more precise analysis could identify other less visible negative impacts that may have important implications for broader community development, including further marginalization of non-users where "the learning obtained by those closest to the Telecentre in itself becomes another expression of power and control that interferes with the participation of those in the community that need it most" (Mardle, 2003).

2.7 Telecentre management: partnerships and assessments

As telecentre networks are too often short of financial and human resources, and their capabilities rest on the ability to work in partnership with other organizations and to leverage support, an examination of the

organizations that telecentres work with provides a good indication of the areas in which telecentres can support services. The findings of the survey show that over two thirds of the networks work regularly (that is, often or always) with social and women's organizations (chart 7.17). Over half of the networks regularly work with the regional or local government and with secondary schools. Over a third of the networks work regularly with the employment office/a recruitment agency, but only two networks always do so. Only around a quarter of the networks work regularly with primary schools, universities, professional associations, business-supporting organizations, private companies and innovation organizations. And only a very limited number (or none at all) work regularly with trade promotion agencies, microcredit organizations, the tax office or investment promotion agencies.

In general, telecentre networks work with social and educational institutions and, to a lesser extent, with organizations that promote economic activities (such as professional associations or business-supporting organizations). Therefore, there is scope for working with the latter organizations (for example, chambers of commerce or Trade Points[21]) in order to, for instance, share/provide training programmes and business-related services.

However, where the local and regional development of economic organizations is very limited – for instance, a respondent indicated that "some of the

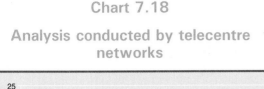

Chart 7.18

Analysis conducted by telecentre networks

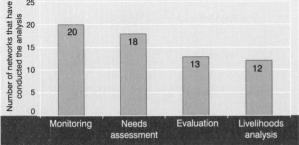

and what is the contribution to livelihoods, and it has a monitoring system.

Several networks conduct studies on an informal basis (for example, they are conducted by students or at the local level). Only five networks reported having the studies disaggregated by gender, income/poverty levels, education levels and/or occupation, and one respondent highlighted the difficulties in having data disaggregated by income levels. Those results raise questions about the ability of telecentre networks to target their activities without disaggregated data. Thus, as suggested earlier, it is important that monitoring activities and evaluations disaggregate data by gender, education, occupation and income/poverty levels. Telecentre networks can prepare some guidelines on how to collect disaggregated data, particularly by income/poverty levels, using some of the tools used by other developmental programmes and the telecentre's collective experience.

Telecentre networks were asked to describe the results of their studies. The responses were too limited, but some of the highlights were as follows:

- Networks located in developed countries, where the formal economy is more developed, noted that telecentres have provided employment-related skills (ICT skills, job searching).

- Community needs are great and the capacities of telecentres limited; thus, it is important to work with others.

- Two networks (in two developing countries) reported that beneficiaries are younger, educated males, while in more advanced economies they where reported to be unemployed aboriginals, and/or low-income rural dwellers and people with disabilities.

- There are difficulties in capturing the specific contribution of telecentres in terms of improvement of livelihoods, including limited resources to capture that impact at the household level.

above organizations are hardly present at the regional even national level" – it will be unrealistic to expect telecentre networks that have no specific objectives with regard to supporting economic opportunities and that are not embedded in an economic activity to successfully provide services in those areas.

Evaluations, assessments and monitoring reports help understand what are the needs of a community, how well a telecentre is performing and who is using the service. They are important in designing and upgrading the activities of the telecentres. Moreover, they provide an indication of telecentres' objectives and management style and, in particular, of how they are supporting economic activities. The UNCTAD survey asked telecentre networks whether they had undertaken any study in any of the following areas:

- Needs assessment (an assessment of the needs of the community);

- Monitoring (assessing who uses the telecentres and for what);

- Evaluation (a study of the effectiveness, sustainability and impact of the telecentres);

- Livelihoods analysis (understanding the livelihoods of a community).

The large majority of networks have conducted an assessment of the needs of the community and have monitored who uses the telecentres, and for what (chart 7.18). A smaller number of telecentre networks, although still the majority, have undertaken an evaluation or a livelihoods analysis. For instance, the Grameenphone Community Information Centre in Bangladesh has conducted several studies to understand the economic sector, business patterns, information gaps, what are the sustainability factors, which services are in demand,

Without more information on the outcomes of the evaluations carried out, it is not possible to assess the contributions that such evaluations make to understanding how telecentres support livelihoods. Nevertheless, the responses show that telecentres have different capacities with regard to undertaking monitoring and evaluation activities that affect the ability to target services and thus to have an impact on the livelihoods of men and women.

Chart 7.19

Key areas in which network leaders would like to receive more support

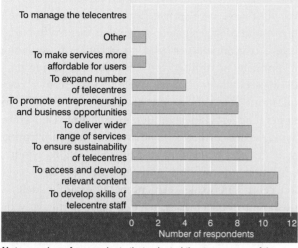

Note: number of respondents that selected the area as one of the three areas in which they would like to receive support.

Chart 7.20

Which institutions should be further involved to improve livelihoods?

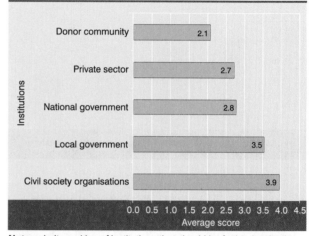

Note: priority ranking of institutions that should be further involved to improve livelihoods (5 highest priority - 1 lowest priority).

2.8 The future

The survey asked telecentre network leaders to identify three areas in which they would like to receive more support (chart 7.19). The two areas selected most often were: support for developing the skills of telecentre staff, and support for accessing and developing relevant content. The latter was consistent with earlier responses that identify the availability of content as an important condition not being met. Half of the leaders would also like to receive support for ensuring the sustainability of telecentres and delivering a wider range of services. Eight respondents would also like to receive support for promoting entrepreneurship and business opportunities. Telecentres leaders are not particularly interested in receiving support for making services more affordable for users, a finding that is not consistent with the earlier finding that affordability is an important condition not fully met.

These responses tend to highlight the interest of telecentres in providing a more in-depth and quality service, rather than in expanding the telecentre network. The sections that follow provide some suggestions about how to support access to relevant content, deliver a wider range of services, and promote entrepreneurship and business opportunities.

Telecentre network leaders believe that the institutions that should be further involved in improving livelihoods are primarily civil society organizations and local governments (see chart 7.20). Respondents gave a lower priority to the involvement of the private sector

and the national Government. This result is somewhat inconsistent with earlier responses stating that the private sector is an important condition not being met. A possible explanation may be the lower levels of confidence regarding the likelihood of private sector involvement. In any case, a more detailed analysis would be necessary in order to ascertain what specific support may be required from different institutional actors for a given context. For example, in Chile, telecentres are strongly supported by the national Government, and stakeholders would like to see greater involvement of local governments and civil society organizations.

In summary, the survey's findings show that telecentre networks come in different sizes and formats but are mostly rural and multi-purpose, providing access to a range of services. Computer access is widely offered, while access to the Internet, particularly broadband, is more limited. Most telecentres provide training in basic ICT skills, few provide advanced ICT skills training, and a number of networks consistently support the development of basic literacy skills. Training to develop general business, specific e-business and occupational skills is provided consistently in a small number of networks and to a limited extent in others. The business-related services most often supported, generally through telecentres providing the service themselves or offering adhoc support, are searching for information, searching for employment and access to government services. Specific training in business-related services is mostly offered for searching for information, creating websites and typing. The three key purposes of using a telecentre are searching for information, personal communications and receiving training.

The majority of networks support economic activities where possible, but this is not their main objective. The main economic sectors serviced by telecentres are the ICT and the educational sectors. With regard to supporting livelihoods, the single most important factor that needs attention is the availability of relevant content. Other key factors requiring support are the quality of the general infrastructure and of economic and business conditions, as well as the development of a critical mass of ICT users, of a wider range of services and of occupational and business skills. There is also scope for greater involvement of the private sector, in particular business-supporting organizations, and civil society, in that order.

3. Best practices and opportunities

On the basis of the above findings, as well as available literature on telecentres, this section provides an overview of best practices for supporting livelihoods, particularly economic activities, through telecentres. The approaches highlighted may not necessarily apply to all telecentre contexts but provide an indication of how livelihoods can be better supported.

Making life easier: facilitating livelihood strategies

Successful innovations and business propositions are those that make life easier for users and customers. Telecentres can make life easier and provide value added for their users by supporting their economic activities and livelihoods strategies. To do so, they must focus on the provision of services rather than on the provision of connectivity.

Value can be created by providing facilitated access to information and to more or enhanced government services, and by allowing transactions (Fillip and Foote, 2007). By providing access to government services such as land records in India (Harris and Rajora, 2007), offering customized services (such as specific agricultural information in the Pallitathya Kendra in Bangladesh) or saving time spent in acquiring market information in rural Nepal (ENRD, Nepal UNCTAD questionnaire, 2007), telecentres make life easier. The services that provide value added depend on the context. Some may already be offered satisfactorily by

other institutions or their provision may be determined by other institutions (such as, e-government services). "ICTs alone cannot improve the service delivery to rural poor. Significant re-engineering of backend processes and introduction of services that directly contribute to the poverty alleviation are needed to make such initiatives sustainable" (IIM, 2003).

Telecentres unable to provide value-added services become irrelevant. For example, an evaluation of the Gyandoot telecentre network in India showed that the rural poor did not perceive the telecentre as a platform for seeking government services, because there were alternative and preferred ways for them to obtain those services (Conroy, 2006).

Responses from UNCTAD questionnaire (2007) show that there is scope for providing a wider range of value-added services. For instance, the availability of training to develop skills important for undertaking economic activities (such as e-business skills) is still limited (see section 2.3). To continue providing value, telecentres have to plan on providing a continuum of services, from basic ICT skills to more specialized training, and support customers in using those skills and trying out improvements in their daily activities. In summary, telecentres can offer additional value by providing e-business skills training, supporting the use of ICT for specific or sectoral activities, and facilitating access to markets, finance and knowledge relevant to the livelihoods of the community.

Embedding ICT in economic activities

E-Choupal is one of India's most successful programmes using ICT to support the economic activities of people living in rural areas. Through "network orchestration"[22] it caters for underserved rural markets and helps farmers halve transaction costs. E-Choupal[23] is a commodity services programme that supports farmers through over 5,000 information kiosks providing real-time information on commodity prices, customized agricultural knowledge, a supply chain for farm inputs and a direct marketing channel for farm produce. Because the network is strongly embedded in a specific economic activity, it enables its participants to derive economic opportunities.

A similar project increasingly embedding ICTs in economic activities relevant to the poor – although not strictly a network of telecentres – is the Dairy Information Service Kiosks (DISK). DISK is a pilot programme that supports milk cooperatives in making better use of existing information on the quality, quantity and price of the milk created by automated machines of dairy cooperatives, as well as in improving access to information on dairying and milch cattle, as well as other services.

Both networks are based on market approaches supported by a larger organization (the private ITC and the Anand District Milk Cooperative). In a vertical integration model, the fact that there is a telecentre is secondary. The focus is rather on the exploitation of ICTs to support a specific economic activity, whether trading commodities or supplying milk. E-Choupal has developed a one-stop shop offering a wide range of services for farmers. The downside of telecentres embedded in one particular economic activity is that those not part of the activity will be excluded, and unless the services offered are expanded to support other areas, the benefits are limited to the specific economic activity.[24]

None of the telecentres that responded to the UNCTAD questionnaire (2007) are strongly embedded in one specific economic activity. Except when an existing economic organization such as an occupational association (i.e. a fishermen's association) or a private business, launches it, telecentres follow a more diversified strategy for supporting economic opportunities. In this case, providing value-added services (see earlier section) and developing niches of economic opportunities (see next section) are the two options to strongly support economic opportunities.

Developing niches of economic opportunity

One way of creating economic opportunities is to support clusters of economic activity. By developing support and knowledge in one area, a telecentre can, by virtue of concentrating resources and developing specialized know-how, provide additional economic opportunities.

Telecentres can play a role in developing clusters of economic opportunity by providing access to knowledge and opportunities for the development of expertise and relationships. For example, in Clyde River, a small impoverished community of 820 people in Nunavut (Canada), nearly two thirds of the adults have not finished high school and are not working in salaried positions nor are they self-employed. As Darlene Thompson, Secretary-Treasurer of N-CAP, explains, "following an increased interest in film production and edition, the telecentre looked for additional funding to purchase filming equipment and provide sector-specific training. As a result, there is a core group of young people trained in the industry, and film companies are taking an interest in using this community for making films given the availability of trained personnel." In response to growing demand from community members and visitors, and to an increasing number of job opportunities, the network has also developed a programme to support scientific research work, which offers services for visiting researchers (such as access to the Internet, rental of wireless modems, printers) and training in basic research methods, including logistical coordination and global positioning systems, for community members.

Other responses from the survey (UNCTAD, 2007) provide examples of economic sectors in which telecentres are working, but there is not much evidence about telecentres specializing in key economic sectors other than ICT and education. As indicated earlier (section 2.4), there is scope for expanding support to sectors such as trade and tourism. Developing niches of economic opportunities for the local community can help provide a reason for using the local telecentres and can thus help reinforce their sustainability. If specific know-how and valuable services were developed, users would be ready to pay more for them.

Providing specific support for those that need it most

If telecentres are to support people living in poverty, market approaches alone will not suffice. Specific efforts are needed to support those in weaker positions. For instance, "the poor are under-represented in accessing the MSSRF knowledge centres in India, as information about crop production and market prices is of interest mainly to landowners and not to the poor, who are predominantly landless" (Conroy, 2006, p. 25). Specific support to those that need it most can be in the form of infomediaries, specific programmes targeted to groups in a disadvantaged position and diversification of services to support the economic activities of the poorest.

Community infomediaries are the linkage between the knowledge and content available through ICTs and individuals. They are familiar with the Internet and ICT and help translate specific needs into information needs and solutions. In Indonesia, each telecentre of the Partnerships for e-Prosperity for the Poor has a manager, an IT administrator and a community development specialist (infomobilizer) that helps integrate "access to information and communication technology with community empowerment activities" (Pe-PP, 2007, p. 2). In Bangladesh, D.Net works with infomediaries recruited locally and trained in basic ICT issues, documenting processes, mobilization and marketing the telecentre. It helps villagers ask the help desk questions about their livelihood, or they may themselves look for answers in the content database (Hasan, 2006). Infomediaries are particularly important in communities with low literacy levels, strong preferences for face-to-face communications[25] and greater needs for accompanied support.

Specific programmes targeted at disadvantaged community groups are necessary in order to support equality of benefits and prevent exclusion patterns from being exacerbated. For example, women are often at a greater disadvantage than men in benefiting from ICT, and thus specific efforts are needed. In addition to mainstreaming gender concerns in telecentres – by, inter alia, ensuring balanced participation in telecentre use and management – there is scope for providing specific training and services for women and for working with institutional mechanisms that are gender-sensitive, such as women's self-help groups (Roman and Colle, 2002).

Supporting access to relevant information and knowledge.

Information and knowledge provide the foundation for economic development. Providing access to information and supporting the development of information and knowledge remain a key pillar for supporting livelihoods.

To facilitate access to relevant information some telecentre networks have focused on developing content for users. For instance, the Swaminathan Foundation's village computing project in India devotes specific efforts to creating, repackaging and disseminating content to its telecentres (Bell, 2006). The Manage cyber extension initiative in India (Conroy, 2006) has developed CD-based learning packages for, inter alia, making pickles, and women are now able to sell pickles in the local market. Other networks, such as the Grameenphone Community Information Centre (GPCIC), work with third-party content providers (AMM Yahya, Director of the GPCIC, UNCTAD questionnaire, 2007).

Telecentre networks can also facilitate access to relevant content by supporting users in developing their own

content. For example, Biblioredes in Chile supports users in developing their own websites, and provides free web hosting services in the network portal www.biblioredes.cl. "The web sites created by users, see for example www.biblioredes.cl/atr.cl, allow users to participate in networks, communicate with people with similar interests and share local content of high relevance for their communities" (Enzo Abbagliati, National Coordinator, Biblioredes, UNCTAD questionnaire, 2007)

Users often require support and/or have specific needs for information and knowledge not readily available. To help meet those needs, several initiatives have put in place help desks. For example, in Chile, SERCOTEC, a public institution supporting SMEs with telecentres integrated into its support offices developed in 2002 an online advisory service[26] enabling users to post online specific questions to advisers. The latter are committed to answering those questions within 48 hours. The questions and their answers, as well as users' evaluation of the latter, are posted online (i.e. relevant questions are stored in the frequent asked questions section) in order to promote a wider dissemination of knowledge and greater accountability. Collaboration agreements with other organizations (such as universities, government institutions, banks and business-related associations) have brought in experts able to advise a wider range of specialized subjects, including taxation, employment legislation and financial services. The online help desk is embedded in the institution and its development costs (for the last version of the portal) were around $60,000 and the annual operation costs are around $35,000 (excluding staff salaries). Outside advisers work on a pro-bono basis.

Using experience to advocate for an enabling environment for economic activities

Telecentres often operate in less than ideal environments with inadequate regulations for conducting e-business or, more generally, inappropriate telecommunications regulatory environments and trade regimes. Telecentre networks can use their experience and advocate for policy changes. Networks in Nepal and India are advocating for changes in government policies to promote a more conducive environment enabling users to benefit from ICT and participate in the information economy.

In India, to operate in a particular state, e-Choupal needs to ensure that the Agricultural Produce Marketing Committee Act is reformed before it can purchase grains directly from the farmers. The Act requires that certain grains be purchased from the official middleman (mandi). ITC, the parent company, has successfully convinced policymakers in different states to amend or allow specific exemptions to the Act since farmers can be better served by e-Choupal (Harris and Rajora, 2007; Conroy, 2006).

More broadly, in Nepal, ENRD has lobbied the Government to deregulate the import and use of the industrial, scientific and medical bands, so as to make VoIP free, at least for calls from computer to computer and from computers to Nepal Telecom landline telephones. It has also argued for decreasing ISPs' license fees, so that small business entrepreneurs can start ISP companies also in rural areas, and for subsidizing in each district of Nepal an organization willing to establish a community internet service provider (CISP) and use wireless technology to connect remote villages (Pun et al., 2006).

Developing a holistic but precise understanding

To support economic opportunities, a holistic but precise understanding of community livelihoods and economic activities is essential. Most telecentres have developed some type of monitoring or assessment of their activities. Networks that are able to fully understand the environment and community needs, as well as the impact of their activities, can confidently provide relevant services. Understanding how ICT services support local economic activities is a first step. An evaluation of the Solomon Islands' People First network (PFnet), a rural connectivity project, showed that there is scope for promoting PFnet services for business activities that enable users to earn a livelihood, by creating awareness and training people in new ways of accessing information and opportunities" (Chand et al., 2005).

Understanding who benefits and how from a telecentre programme is also indispensable for supporting the livelihoods of men and women living in poverty. For instance, field evidence from Uganda reveals that "the use of ICT facilities was and is male dominated with women often focusing on uses that integrate ICT in their already existing purposes more than for entrepreneurial activities. The latter uses are known to generate higher returns than the former" (Mandana and Amuriat, 2006).

Survey responses (UNCTAD, 2007) and the material provided indicate that telecentres often still have to develop a more precise understanding of the impact of their activities on different groups. For instance, six out of 23 respondents did not provide the percentage of female users. Consequently, there is room for concern about how a telecentre can support the livelihoods of men and women without having basic data on who is using the telecentre.

Financial and human resources constraints are often a deterrent with regard to carrying out monitoring activities, assessments and analytical studies. However, several networks (e.g. ENRD in Nepal) work with postgraduate students to carry out monitoring and evaluation studies. Similarly, there are universities that are carrying out studies on, and supporting, the work of telecentres. In Chile, the Instituto de Educación Informatica in the Universidad de la Frontera[27] has supported the design, development and evaluation of a regional telecentre network (Red de telecentros de la Araucanía).

4. Major barriers and challenges

What barriers prevent telecentres from effectively supporting the livelihoods of people living in poverty? This section examines the challenges faced by telecentres in promoting economic opportunities, elaborates on the conditions enabling telecentres to support livelihoods and highlights the role of national and local governments in the elimination of those barriers.

The capacity of telecentre networks remains a main challenge

Capacity issues, including financial sustainability problems, already overstretched managers or limited human resources capacities, are a constraint on the expansion of telecentre activities and the provision of business-related services. Although not addressed in this chapter, the shortcomings of telecentres, as well as a wide range of best practices and opportunities, have been widely debated and illustrated in many studies.[28] As Proenza (2003) puts it, "Installing a telecentre is easy, the hard part is to keep it running".

Many telecentres are still in the process of being established and require time to build experience, diversify activities and accumulate knowledge on the developmental impact of their activities before moving into more complex activities. In that regard, telecentres can benefit from working in partnership with other organizations and networking with other telecentres. For example, in order to be sustainable, the Community Learning and Information Centres (CLIC) of Gao and Mopti in Mali share their broadband connection with other NGOs in the village (UNCTAD questionnaire, 2007). In more concrete terms, telecentres could share resources, such as training material and know-how, in the specific area of supporting economic activities. Older telecentre networks, on the contrary, can benefit from experience and their continuing existence. For example, the Government of Chile has included in the terms of its latest subsidies offer for the establishment of telecentres a new requirement, namely that awardees spend one year building up and supporting the transfer of the telecentre to a local organization (SUBTEL 2007). This new element has been incorporated on the basis of evaluation findings that indicated that local ownership was very important for the sustainability of the telecentre and that local organizations had limited capacities. For the telecottage network in Hungary, drawing support from the private sector became easier once the programme was present in 5 to 10 per cent of communities, and much easier when the Government decided to support and use the network (UNDP, 2006).

Governments can play a key role in building capacities, even when not directly sponsoring a telecentre programme – first, by facilitating the establishment and functioning of telecentres; and secondly, by supporting the development of ICT skills directly relevant for economic activities. To support the spreading of the benefits of ICT for economic activities, Governments could mainstream e-business skills training programmes (as some Governments, for example the Government of Chile, have done for digital literacy) for telecentre managers and for telecentre users.

Limited availability of content and services, and limited capacity of the Government to develop e-government services in the short term

Survey responses (UNCTAD questionnaire, 2007) indicate that the availability of content relevant for supporting livelihoods is one of the crucial factors still not satisfied with regard to supporting livelihoods (see section 2.5) "The content and services sector supporting village computing is still in its infancy" (Bell, 2006, p. 24).

The private sector has yet to become a significant provider of content for rural economic activities. Telecentres with insufficient economies of scale are unable to develop, and maintain, relevant content. Furthermore, although Governments are responsible for developing e-government services to provide wider access to Government services, facilitate economic activities, and increase accountability and transparency, they often do not have the means to do so.

Developing content is costly. For example, Chile's SERCOTEC online help desk costs $35,000 a year, while the total annual operation costs of half of the networks that responded to the questionnaire is below $50,000. That means that maintaining an online help desk is the equivalent of at least 70 per cent of annual operation costs, an expense which many cannot afford. The cost of developing and accessing content gives rise to two problems, namely:

(a) Some Governments charge, and allow others to re-charge, for accessing government e-services, such as downloading official forms. While this may make sense in terms of financial sustainability, it hinders the spread of government services to those that need them most.

(b) Telecentre managers, as in the case of CLIC in Mali, may be "under pressure to sell services that bring in revenue and have little incentive to focus on content dissemination" (Bell, 2006).

The broader economic and business structures and conditions are not present

Survey responses (UNCTAD questionnaire, 2007) highlight economic and business conditions, as well as the quality of the general infrastructure, as the key environment factors that are not in place (see section 2.5).

The expansion of ICT-enabled economic opportunities requires specific skills and infrastructure (i.e. online payment systems, transactions platform, access to microfinance) that are often not present. In the poorest countries, "the formal private sector hardly exists in many districts, and where it does it is also struggling" (respondent, UNCTAD questionnaire, 2007). Telecentres in more isolated areas with limited economic infrastructure and institutional presence may find it more difficult to find partners to develop economic activities.

In that regard, Governments play a major role in developing a general enabling environment conducive for e-business and for telecentres to support economic activities. Governments can encourage the development of general infrastructure by, inter alia, developing policies that support a competitive telecommunications market and establishing incentives or obligations to serve marginal areas. Governments influence broader economic and business conditions and, more specifically, the development of business-supporting services and the provision of information crucial for developing economic activities (such as the SERCOTEC help desk in Chile). They can assist sectors that are of interest to those living in rural areas and telecentre users, and, as major economic actors, they can support the development of SMEs by putting in place mechanisms that facilitate State purchases from SMEs.

While some telecentre programmes, for example e-Choupal, are effectively lobbying public institutions to make changes to the regulatory framework in order to develop a more enabling environment for e-business, smaller telecentre networks or those with limited presence at the regional/national level may have difficulties in promoting a more enabling environment.

Ensuring equality of benefits

ICT benefits derive from the appropriation of technology, rather than from the consumption of technology products. When e-business services have attached a fee and/or are embedded in a particular economic activity there is a risk that only some groups will benefit from the telecentres. For example, it emerged from a survey of pro-poor ICT projects in India that users did not believe that the benefits were evenly spread among the member communities (Harris and Rajora, 2006, p. 3). To ensure equality of benefits, ICT policies and programmes must support the development of e-business skills and capacities, increase access to e-government services and expand ICT programmes to different fields of activity.

While the findings of the report (Harris and Rajora, 2006). draw attention to the possibility of unequal benefits, the statement is based on the perceptions of a group of users, and not on empirical data. The availability of data, both quantitative and qualitative, is important and necessary to ensure equality of benefits. The collection of basic data on access to telecentres, their use and their impact, disaggregated by gender, age,

income and social background, can still be improved. For example, the survey (UNCTAD, 2007) indicates that there are gaps in the availability of basic data such as gender-disaggregated statistics on telecentre access. Only 16 out of 22 telecentres provided user data disaggregated by gender.

In their study of five telecentres in Tamil Nadu (southern India), Kumar and Best (2006) note that telecentre users are overrepresented by younger, educated, male, Hindu users and (in some of the telecentres) by the more socially and economically advantaged castes. Using the diffusion theory of technological innovation, they suggest that to support the adoption of the technology (that is, the use of telecentres) by a broader set of community members, telecentres must:

(a) Address their lack of perceived compatibility with the situation of local women (by ensuring that both content and institutional settings are adequate for women and other marginal groups);

(b) Redress the perceived complexity by providing user-friendly content and services;

(c) Strengthen the way in which innovation is communicated and shared within the communities, by working through local champions from the marginalized community groups and targeting marginal communities in their marketing efforts.

To include those with fewer resources and capacities, complementary efforts (including human and financial resources) are required. Reaching the poor involves additional resources for adapting content and activities (for example, adapting content to other formats more suitable for the illiterate), diversifying the areas of activity (so that activities relevant to the poorest are also provided), working with community infomediaries and reaching local champions, changing perceptions that ICT are for the wealthy and educated, and conducting assessments on the socio-economic impact of the telecentres on the different groups in the local community.

The next section provides a checklist of key factors influencing telecentres' ability to support livelihoods and provides recommendations for policymakers and telecentre managers that build on the findings of the questionnaire and the best practices and challenges presented.

C. Recommendations to support livelihoods through telecentres

Checklist of important factors for telecentres to support livelihoods

This chapter has explored how telecentres can support economic activities and the livelihoods of people living in poverty, and chart 7.21 provides a checklist of factors that policymakers and practitioners should

consider for telecentre networks to support livelihoods. The checklist develops the commerce dimension of UNCTAD's 12 Cs pro-poor ICT framework,[29] and suggests that to support livelihoods through telecentres action is needed at three levels:

- At the micro level, it will be necessary to conduct livelihood assessments to understand the livelihood strategies that vulnerable communities follow.

Chart 7.21

12 Cs pro-poor ICT framework
Exploring the C of Commerce

	Connectivity	Content	Community	Commerce	Capacity	Culture	Cooperation	Capital	Context	Continuity	Control
MACRO LEVEL				**Policies & programmes supporting economic activities** • Enabling environment: regulation and policies promoting an enabling environment for economic development in marginal urban and rural areas. Legislation enabling online transactions; policies promoting rural entrepreneurship • Direct support programmes: economic development programmes, e-business skills literacy programme • E-government services and content: development, customization, cost of access • Policies supporting equity of benefits, including female participation and control in economic activities							
MESO LEVEL				**The telecentre network as an institution supporting livelihoods** • Area of influence: rural/urban, economic sectors • Scope of influence: mandate, resources (staff, budget), support received • Business-related services & training provided (e-business, entrepreneurship, occupational) • Use: who uses the telecentre and for what purposes? • Impact and assessments (livelihoods assessment, monitoring, evaluation) • Relationship with other institutions supporting economic activities (private sector, civil society, donors, government departments)							
MICRO LEVEL				**Assessing local livelihoods** • Assets (human, financial, social, physical), capabilities and activities of communities' members • Vulnerabilities • Institutions (Government, market) • Organizations • Social relations (gender, power relations) **Exploring how a telecentre can support those specific livelihoods**							
Vision				◆ Is supporting economic opportunities a goal of the network?							
Assumptions				◆ E.g. Telecentres can effectively support livelihoods and economic opportunities ◆ E.g. Different community groups can benefit from the activities of the telecentre							
Conflicts				◆ **Telecentre model:** Which telecentre model best serves the livelihood strategies? What are the implications for other dimensions of the framework? ◆ **Scaling-up:** When? How? How many services can/should the network provide? ◆ **Different needs:** Whose economic activities should be supported? ◆ Government support: Open content or fee-based content and services?							

Source: UNCTAD.

- At the meso level, an assessment of the role that the telecentre network as an institution has in supporting such livelihoods is needed – namely, an evaluation of the telecentre's scope, mandate, resources and impact, as well as of the relationships with other organizations, to highlight the areas in which the telecentre network is best placed to support livelihoods, and the options available for cooperating and collaborating with other institutions.

- At the macro level, the different policies and programmes help telecentres support livelihoods should be considered, including:

  - E-business capacity-building programmes;

  - E-government content and services;

  - Policies promoting an enabling environment for economic development in marginal areas;

  - Telecentre programmes.

Additionally, policymakers and telecentre managers should consider the vision of the telecentre and to what extent the role of the telecentre is to provide economic opportunities. They should also be aware of the assumptions that the telecentre programme departs from (for instance, the telecentre network as public service should benefit different community groups). Finally, they should consider the key conflicts arising from the choices made such as the model of the telecentre, the approach to scaling up activities, the priority needs to be addressed, and the type of government support to be provided. This would, therefore, require consultations with all stakeholders to assess the role and limitations of telecentres in supporting livelihood opportunities for people living in poverty.

The 12 Cs pro-poor ICT framework does not prescribe any specific course of action; rather, it is a tool to help stakeholders examine and debate how telecentres (or other ICT policies and programmes) are supporting or can support livelihoods and economic opportunities.

Main findings on how telecentres are supporting livelihoods

Results from the survey conducted among telecentre networks and a review of the literature show that most telecentres' efforts have focused on providing access to

ICTs and developing basic ICT skills. The key purposes of using telecentres are searching for information, personal communications and training. While survey findings show that telecentres are primarily used for informational and educational purposes, telecentres are also being employed for economic purposes. There are some good examples of how telecentres are providing access to business-related services, most notably access to government services, employment-related information (in more developed economies), sector-specific information and business communications.

Those programmes embedding ICT in existing economic activities provide most economic opportunities as they make improvements in the supply chain, facilitate access to specialized knowledge and reduce transaction costs. As these programmes may benefit specific community groups (i.e. farmers with access to land, workers in the formal economy), the telecentre networks need to make additional efforts to reach other vulnerable community groups and those working in the informal sector.

There are opportunities to provide support for using ICT for economic activities by expanding training in occupational and e-business skills, providing a wider range of services such as access to finance or sectoral expertise, enhancing access to public services in general, and supporting services in a broader range of economic sectors relevant for different groups in the community. The availability of information and services in appropriate format is the single most important element impacting on telecentres' ability to support livelihoods.

Key recommendations to Governments and telecentre networks

How can international organizations, national Governments and civil society further support livelihoods through telecentres? Bearing in mind the results of the questionnaire and the best practices described earlier, what follows are a number of policy recommendations to improve ICTs' impact in providing economic opportunities. The recommendations are addressed to Governments and to telecentre networks separately. However, both groups of stakeholders, and other players such as local civil society organizations, donors or business associations, may have a role to play in some of the recommendations. For instance, recommendations addressed to telecentre networks may be taken up by those Governments, donors or civil society organizations directly sponsoring and supporting telecentres networks. Regarding recommendations addressed to Governments, actors from the telecentre movement have a valuable role to play in advocating and supporting government policies through the provision of inputs, including studies on local needs, for policy development and the implementation of such policies. Similarly, business associations can make valuable contributions to the development of e-business skills and services.

Recommendations primarily addressed to telecentre networks (meso and micro levels)

1. Provide value-added services that have a direct impact on the livelihoods of the local community, and develop telecentres' capacity to support economic activities by offering access to business-related services and by developing the capacities of telecentre staff in the field of e-business. Survey findings (UNCTAD, 2007) indicate that there is scope for expanding the range of business-related services offered (e.g. to include support for conducting payments, accessing microcredit or buying and selling) and the sectors supported (e.g. to expand support for the trade and tourism sectors). Where possible, develop niches of economic opportunities based on the potential of the local context. Develop staff's capacities in the niche area and partner with experts and organizations in that field.

2. Mainstream e-business skills programmes to develop entrepreneurship and business-related skills. Provide a wider choice of training, including training in e-business skills, occupational skills and general business skills, relevant to the local livelihoods, to promote the development of entrepreneurial skills and support specific sectors of interest to the local economy. Network leaders can support the adaptation and replication of training already provided in some telecentres to the rest of telecentres in the network.

3. Enhance the understanding of the local context and livelihood needs and strategies, as well as the potential that ICTs offer by carrying out, together with other local development players, livelihood assessments. Continue to support the use of ICTs for poverty reduction by commissioning, and sharing, independent evaluations of the impact of telecentres in local livelihoods on a regular basis. Use such studies to advocate for an enabling business and telecommunications regulatory environment and structures.

4. Ensure that the activities of telecentres also support the economic activities of those in weaker positions, including women, and employ community infomediaries to reach more vulnerable communities or groups. Develop and provide specific targeted training and services for those at a disadvantage. In that regard, partner with civil society leaders, including those from women's organizations, and with champions from vulnerable groups who could become involved in the work of the telecentre and make a valuable contribution to the assessment and design of services relevant for supporting livelihoods and creating awareness of the economic opportunities that telecentres can offer. Telecentre networks where female users represent less than 40 per cent of the total number of users should pay particular attention to ensure that women are able to use the services of the telecentre.

5. Engage with organizations supporting economic activities. Survey findings (UNCTAD, 2007) indicate that there is scope for engaging with a wider range of organizations, such as universities and organizations supporting economic activities (e.g. professional associations, business-supporting organisations, micro-credit institutions). Support the development of linkages with the private sector, including increased collaboration with micro-finance institutions, and promote the contribution of industrial sector actors. Similarly, work with other telecentres to share resources and expertise for supporting local livelihoods; in particular, as some telecentres develop business skills training programmes, consider sharing curricula and materials for business skills development.

Recommendations primarily addressed to Governments (macro, meso and micro levels)

The following recommendations address the role of the Government in setting out a conducive environment for the development of e-business among smaller and rural enterprises and in providing direct financial support for telecentre networks. Public financial, and non-financial, support can, in conjunction with the virtuous circle of offering additional value-added services, play a crucial role in ensuring the sustainability of telecentres and thus the continuation of services of public interest.

1. Develop and promote relevant e-government content and services that support economic activities and livelihoods. First, develop services to support

economic activities, in particular those of SMEs and micro-entrepreneurs. Increase the number of services related to enterprises (tax declarations, inscriptions, trading documents etc.) available through the Internet and/or the telephone, and strive to streamline back-end processes and reduce red tape. A comprehensive national e-government strategy can provide the overall framework for improving access to public services, and should include specific plans to support the facilitation of economic-related activities. Governments should support the areas of e-business, e-trade and e-finance, and not only e-education and e-health.

Secondly, develop sectoral information strategies (for example, as an element of e-government strategies) and, on the basis of best practices, put in place different mechanisms, such as help desks or information centres, to develop and provide relevant and customized information. The specific sectors to be supported as a matter of priority should reflect the needs and economic activity of telecentre users.

2. Support the development of e-business skills. As the findings of the questionnaire shows, digital literacy programmes are broadly available through telecentres. However, there is still limited support for developing e-business skills. Establish an e-business skills capacity development programme that includes specific support for trainers (i.e. telecentre staff). The curricula of the e-business skills capacity development programme should contain specific modules on e-business, such as conducting transactions online, accessing services available online (e.g. e-banking, trade facilitation services, taxation), web creation and design, advertising and content development, as well as accessing relevant information sources online. The programme may also include broader business skills modules (e.g. accountancy, trading), modules to develop entrepreneurship skills (e.g. project management) and modules to develop sector-specific skills relevant to the livelihoods of telecentre users.

3. Develop appropriate conditions for e-business in rural and marginal urban areas by putting in place appropriate business-supporting structures (e.g. rural business information centres and extension workers/infomediaries) and by facilitating the development of complementary services (e.g. access to finance). More generally, support the development of infrastructure to increase the benefits and reduce the costs of using ICTs to develop economic activities in rural and marginal

areas. Those two recommendations aim at addressing the two environmental factors, namely economic and business conditions, and general infrastructure, highlighted in the findings of the questionnaire (UNCTAD, 2007) as critical for telecentres to be able to support livelihoods.

4. Provide strategic financial support for telecentre networks to scale up their activities and develop value-added services, such as e-business training or customized support through community infomediaries, that support the economic activities of local communities. Provide seed money to enable telecentres to develop expertise in an economic activity, as in the example of the CAP telecentre in Nunavut, Canada, that can support local livelihoods.

Through the experience of telecentres the chapter has highlighted two processes whereby science, technology and knowledge are disseminated. The first process involves existing economic activities, for example the Indian examples of e-Choupal and the Dairy Information Service Kiosks, where ICT embedded in agricultural and economic activities has facilitated access to knowledge and increased benefits. The second process involves shared ICT access models, for example the telecentre Muneng in Indonesia, which enable access to knowledge and innovation. However, the dissemination of knowledge using ICT and participation in the development of knowledge are far from automatic. The Internet revolution can benefit poorer community only when it offers access to relevant content and knowledge, when it is affordable to use, when it is accompanied by relevant applications and when its dissemination is supported through skills development efforts.

In conclusion, telecentres are a valuable institution for supporting sustainable livelihoods. However, their ability to do so depends on their capacity to become an institution supporting local development, and not only access to ICT. Survey findings show that those telecentres that are capable of drawing strategic support from a wide selection of stakeholders, for example the public sector, civil society and business associations, and involving and catering for the local community, and that have particular expertise are able to support livelihoods.

Annex 7.1

ICTs and poverty reduction. Case study: Chilean telecentre network
Overview of respondents

Respondent	Job title	Organization	Works at the following level	Works with
1	Regional Operation Manager, Biblioredes	DIBAM	Regional (Region VII)	Biblioredes Region VII
2	Regional Operation Manager, Biblioredes	DIBAM	Regional (Region II)	Biblioredes Region II
3	Executive Director	Corporación Maule Activa	Regional (Region VII)	Maule Activa Network
4	Regional Technical Secretary	SUBTEL	Regional (Region IV)	Region IV telecentres
5	Regional Technical Secretary	SUBTEL	Regional/Local (Region X)	Region X telecentres
6	Laboratory Manager, Biblioredes	DIBAM	Local	Biblioredes
7	User, Biblioredes	DIBAM	Local	Biblioredes
8	Coordinator	La Araucanía	Regional (Region IX)	La Araucanía
9	Telecentre Manager, Biblioredes.	DIBAM	Local	Local library
10	Coordinator, National Programme for Telecentres	SUBTEL	National	National Telecentres Network
11	Information Systems Manager	INJUV	Regional	INJUV
12	Public Administrator ICT Area	Ministry of Education	National	National ICT Literacy Campaign National telecentres and specific networks Other stakeholders
13	Head of Programmes	INJUV	National	INJUV
14	Regional Adviser	SUBTEL	Regional/local (Region IX)	Region IX telecentres
15	Telecentre Manager	Corporación Maule Activa	Regional/local (Region VII)	

Annex 7.2

UNCTAD questionnaire

Promoting men's and women's livelihoods through ICTs:
the case of telecentres

Note: You are not obliged to answer all the questions. This questionnaire has been drafted thinking in a broad number of telecentres networks with different characteristics. If you feel that a question is not relevant to your context, you can leave it blank.

To answer the questions: follow the instructions in orange, mark your answers with an X and delete what is not appropriate.

Please return the completed questionnaire by Monday, 23 April 2007

Definitions

Livelihoods: The skills, resources (both material and non-material) and economic activities (self-employment and/or wage-employment) necessary to fulfil the needs of person or family.

ICT: Information and communication technology.

A.   ABOUT YOUR TELECENTRE NETWORK

A.1 Which year was your network established? [Select one]

O 1980	O 1981	O 1982	O 1983	O 1984
O 1985	O 1986	O 1987	O 1988	O 1989
O 1990	O 1991	O 1992	O 1993	O 1994
O 1995	O 1996	O 1997	O 1998	O 1999
O 2000	O 2001	O 2002	O 2003	O 2004
O 2005	O 2006	O 2007		

A.2 Composition of the network

Total number of telecentres participating in your network	[nº]
Number of telecentres located in **rural** areas	[nº]
Number of telecentres located in **urban** areas	[nº]
Number of **basic** telecentres (telecentres that only offer access to telephone, computer, internet and/or radio)	[nº]
Number of **multipurpose** telecentres (telecentres that only offer access to telephone, computer, internet and/or radio)	[nº]

A.3 Staff

Total number of staff across the network	[nº]
Estimated % of female users	[nº] per cent

A.4 Users

Estimated number of annual users across the network	[nº]
Estimated % of female users	[nº] per cent

A.5 Please briefly describe your network. Which are the main objectives and features of your network ?

[Write here]

A.6 Which is the annual cost of running your telecentre network
(including individual telecentres) ? Select one

O < 10,000 USD

O 10,000 USD - 50,000 USD

O 50,000 USD - 250,000 USD

O 250,000 USD - 1 million USD

O 1 million USD - 5 million USD

O > 5 million USD

A.7 Main sources of finance. Indicate estimated percentage (e.g. 30% users free, 70% government grant / subsidies)

Government grant / subsidies	[n°]	%
International / national financial donor support	[n°]	%
User fees	[n°]	%
Sales of value-added services (i.e. sales of training, business services, etc.)	[n°]	%
In-kind donations	[n°]	%

B.   SERVICES OFFERED BY TELECENTRES

B.1 General services: Are the following services offered across your network ? Select one answer for each service

	Service not provided	Service provided by < 25% of telecentres	Service provided by 25% - 75% of telecentres	Service provided by > 75% of telecentres
Telephone	O	O	O	O
Fax	O	O	O	O
Computer	O	O	O	O
Photocopier	O	O	O	O
Dial-up Internet (Total capacity in both directions < 256 Kps)	O	O	O	O
Broadband Internet (Total capacity in both directions = or > 256 Kps)	O	O	O	O
Radio broadcasting	O	O	O	O

Additional comments: Please comment or give examples of how these services support livilihoods

[Write here]

B.2 Training services: Does your network offer training services to develop the following skills?
Select one answer for each service

	Service not provided	Service provided by < 25% of telecentres	Service provided by 25% - 75% of telecentres	Service provided by > 75% of telecentres
Basic ICT user skills (to use generic tools (e.g. e-mail & web, word processors, spreadsheets, presentation tools))	O	O	O	O
Advanced ICT user skills (to use advanced / sector specific tools & advanced functions of the generic tools)	O	O	O	O
ICT specialist skills (to develop, operate & maintain ICT systems)	O	O	O	O
e-business skills (to exploit business opportunities provided by ICTs (e.g. to buy & sell online)	O	O	O	O
Basic literacy skills (read & write)	O	O	O	O
Occupation-specific skills (e.g. farming, crafts, tourism...)	O	O	O	O
Business skills (e.g. marketing, management)	O	O	O	O

Additional comments: Please comment or give examples of how these training services support livilihoods

[Write here]

B.3 Business support services: Do the telecentres provide training or other forms of support for any of the following services ? Select one option. **How are these services supported ?** Select all the answers that apply for each service

	Service supported by... Select one option				How is the service supported ? Select all the answers that apply for each service			
	Service not provided	Service provided by < 25% of telecentres	Service provided by 25% - 75% of telecentres	Service provided by > 75% of telecentres	With specific training	Service is supported in other training courses	Staff answer users' questions on an ad hoc basis	The telecentres provide this service
Searching for information	○	○	○	○	❑	❑	❑	❑
Business communications	○	○	○	○	❑	❑	❑	❑
Access to professional / sector-specific information	○	○	○	○	❑	❑	❑	❑
Typing	○	○	○	○	❑	❑	❑	❑
Web creation & design	○	○	○	○	❑	❑	❑	❑
Advertising	○	○	○	○	❑	❑	❑	❑
Content development	○	○	○	○	❑	❑	❑	❑
Accountancy	○	○	○	○	❑	❑	❑	❑
Banking	○	○	○	○	❑	❑	❑	❑
Microfinance (access to)	○	○	○	○	❑	❑	❑	❑
Buying & selling	○	○	○	○	❑	❑	❑	❑
Payments	○	○	○	○	❑	❑	❑	❑
Business opportunities	○	○	○	○	❑	❑	❑	❑
Export-import / trade facilitation services	○	○	○	○	❑	❑	❑	❑
Job searching / advertising	○	○	○	○	❑	❑	❑	❑
Employment opportunities	○	○	○	○	❑	❑	❑	❑
Tax filing	○	○	○	○	❑	❑	❑	❑
Access to government services	○	○	○	○	❑	❑	❑	❑
Data storage & management	○	○	○	○	❑	❑	❑	❑
Innovation / research & development opportunities	○	○	○	○	❑	❑	❑	❑

Additional comments: Do you offer business support services ? Please describe them

[Write here]

B.4 Do the telecentres provide support or offer services related to the following economic sectors ? Select one answer for each sector

	No	In < 25% of telecentres	In 25% - 75% of telecentres	In > 75% of telecentres
Primary (agriculture, fishing, etc.)	O	O	O	O
Manufacturing	O	O	O	O
Trade (wholesale & retail)	O	O	O	O
Transportation	O	O	O	O
Tourism	O	O	O	O
Information and communication	O	O	O	O
Financial services	O	O	O	O
Professional & scientific activities	O	O	O	O
Public administration	O	O	O	O
Education	O	O	O	O
Health	O	O	O	O
Art & entertainment	O	O	O	O

B.5 How often are services targered to the following groups ? Select one answer for each group

	1. Never	2. Rarely	3. Sometimes	4. Often	5. Always
Women	O	O	O	O	O
Unemployed	O	O	O	O	O
Students	O	O	O	O	O
Disabled	O	O	O	O	O
People living below the national poverty line	O	O	O	O	O
Businessman / businesswomen	O	O	O	O	O
Occupations (e.g. farmers, teachers, fishermen)	O	O	O	O	O

B.6 How do the telecentres target their services ? Select all the options that apply

○ Training content is adapted to the target group

○ Courses take into account the needs of specific groups when designing the training

○ Some courses are specifically designed for the target audience

○ Financial support

○ Other (please specify_____)

C. THE ENVIRONMENT

C.1 Which of the following conditions are important for a telecentre to support livelihoods? Are these conditions being met ? For each condition, indicate its importance and to which degree the condition is being met

	Condition important ?			Condition met ?		
	1. Not important	2. Sometimes	3. Very important	1. Not	2. Sometimes	3.Yes
Basic literacy levels	○	○	○	○	○	○
Level of ICT skills	○	○	○	○	○	○
Language skills (other than mother tongue)	○	○	○	○	○	○
Occupational & business skills	○	○	○	○	○	○
Legal & regulatory environment	○	○	○	○	○	○
Quality of general IT infrastructure	○	○	○	○	○	○
Economic & business conditions	○	○	○	○	○	○
Financial system	○	○	○	○	○	○
Social & cultural context	○	○	○	○	○	○
Location of telecentres	○	○	○	○	○	○
Range of services offered by telecentres	○	○	○	○	○	○
Affordability of telecentres' services	○	○	○	○	○	○
Critical mass of ICT users in the community	○	○	○	○	○	○

Availability of relevant content	O	O	O	O	O	O
Political support	O	O	O	O	O	O
Private sector support	O	O	O	O	O	O
Participation of civil society	O	O	O	O	O	O
Telecentre objectives	O	O	O	O	O	O

Any additional comments on the factors that are important for your network of telecentres ?

[Write here]

C.2 What type of government support does your network of telecentres currently receive ?
Describe the type of government support you receive and from which institutions.

Government institutions that provide support (i.e. Ministry of telecommunications, Ministry of Social Affairs, tax office etc.)	[Write here]
Type of support each of these institutions provides (strategic support, financial support, specific collaboration, training etc.)	[Write here]
Level of government support (local, regional, national)	[Write here]

D.   THE MANAGEMENT

D.1 How often does your network work the following organizations ? Choose one answer for each type of organization

	1. Never	2. Rarely	3. Sometimes	4. Often	5. Always
Primary schools	O	O	O	O	O
Secondary schools	O	O	O	O	O
Universities	O	O	O	O	O
Professional / industry associations	O	O	O	O	O
Trade promotion agency	O	O	O	O	O
Business supporting organizations	O	O	O	O	O
Employment office / recruitment agency	O	O	O	O	O

Investment promotion agency	O	O	O	O	O
Micro-credit organization / bank	O	O	O	O	O
innovation / research & development promotion organizations	O	O	O	O	O
Private companies	O	O	O	O	O
Women's organizations	O	O	O	O	O
Other social organizations	O	O	O	O	O
Tax office	O	O	O	O	O
Local / regional government	O	O	O	O	O

Additional comments: Describe your relationship with those organizations that are important for your telecentre network. Do you work with other organizations not listed here ?

[Write here]

D.2 Have you undertaken any study of...
Please select one option for each type of analysis

	Yes	No
... the needs of the community (needs assesment)	O	O
... who uses the telecentres and for what ? (monitoring)	O	O
... the effectiveness, sustainability and impact of the telecentres ? (evaluation)	O	O
... the livelihoods of a community ? (livelihoods analysis)	O	O

D.3 Please briefly describe the results of the analyses

[Write here]

If possible, please send a copy of these analyses to marta.perez.cuso@unctad.org

D.4 Are these analyses disaggregated by gender, income / poverty levels, educational levels and/or occupation (student, employed) ?

[Write here]

D.5 To what extent supporting economic activities is an objective of your network ?

Select one   ○ (a) Supporting economic activities is not an objective

○ (b) Economic activities are supported where possible

○ (c) Supporting economic activities is the main objective

please explain your answer

[Write here]

E.   THE USE AND IMPACT OF TELECENTRES

E.1 Who are your main groups of users ? (e.g. unemployed, employed, self-employed, students, retired, people engaged in family duties, women, men, etc.)

[Write here]

E.2 For which purposes do the following groups use the telecentres ? Select the 2 most important purposes for each group

	Personal communication	Business communication	Buy goods or services	Sell goods or services	Search for informantion	Solve administrative matters	Receive training
Unemployed	○	○	○	○	○	○	○
Employed (salaried)	○	○	○	○	○	○	○
Self-employed	○	○	○	○	○	○	○
People engaged in family duties	○	○	○	○	○	○	○
Students	○	○	○	○	○	○	○
Women	○	○	○	○	○	○	○
Men	○	○	○	○	○	○	○
Minorities	○	○	○	○	○	○	○
Users living below the national poverty line	○	○	○	○	○	○	○

E.3 Would you say that your network helps users to...
Please select one option

	No	To some extent	Fully
... acquire new skills ?	O	O	O
... support existing economic activities ?	O	O	O
... develop new economic opportunities ?	O	O	O
... improve self-employment opportunities ?	O	O	O
... improve salaried employment opportunities ?	O	O	O

E.4 What changes have you observed in the lives of telecentres users ? (e.g. changes in: income levels and distribution, quality of life, access to public goods and services, coverage of basic needs (housing, health, nutrition), consumption, social relations, confidence etc.)

[Write here]

E.5 Please describe a best practice example of how your network (or a telecentre in your network) is supporting livelihoods

[Write here]

E.6 Have you observed any negative impact of the telecentres on the local communities (both users and non-users) ?

[Write here]

E.7 Please name 3 areas in which you believe the network could strongly promote livelihoods in the next two years

[Write here]

E.8 In which of the following areas would you like to receive more support ? Choose 3 areas

O Support to promote entrepreneurship and business opportunities

O Support for ensuring the sustainability of telecentres

O Support to develop the skills of telecentre staff

O Support to expand the number of telecentres

O Support to deliver a wider range of services

O Advisory support on the management of telecentres

○ To make services more affordable for users

○ Support to access and develop relevant content

○ Other (please specify_____)

E.9 Please rank in order of priority the institutions that should be further involved to improve livelihoods through your network of telecentres (1 = highest priority, 5 = lowest priority). Please ensure each institution has a different rank

Civil society organizations / local community	[Rank]
Local governments	[Rank]
National Government	[Rank]
The private sector	[Rank]
The donor community	[Rank]

E.10 Any other comments or information you would like to share with us ?

[Write here]

F.   YOUR DETAILS

F.1 Please tell us about yourself

Name	[Write here]
Surname	[Write here]
Position	[Write here]
Telecentre network	[Write here]
Country / countries	[Write here]
Telephone	[Write here]
Telecentre network website	[Write here]

F.2 PRIVACY STATEMENT

UNCTAD and telecentre.org will keep confidential all information that might identify a respondent with his or her responses, unless the respondent waives confidentiality for specified uses.

Do you give us permission to individually identify you or your network ?

IMPORTANT: THIS IS A COMPULSORY QUESTION TO ANSWER

o Yes   o No

Thank you for your collaboration. Your insights are very valuable for this study, and I will shortly share and discuss with you the preliminary findings.

Annex 7.3

Telecentre networks that completed the questionnaire

Telecentre network	Country	Website
GPCIC (Grameenphone Community Information Center)	Bangladesh	www.gpcic.org
Pallitathya Kendra	Bangladesh	www.pallitathya.org
Learning Enrichment Foundation	Canada	www.lefca.org
N-CAP	Canada	nu.e-association.ca
Programa Biblioredes, DIBAM	Chile	www.biblioredes.cl
Red de Telecentros de La Araucanía	Chile	www.redcomunitaria.cl
Réseau des Télécentres Communautaires du Congo	Congo	
Principado de Asturias	Spain	www.asturiastelecentros.com
Red Conecta (Fundación Esplai)	Spain	www.redconecta.net
Ajb'atz' Enlace Quiché	Guatemala	www.enlacequiche.org
TARAhaat	India	www.tarahaat.com
Partnerships for e-Prosperity for the Poor (Pe-PP)	Indonesia	www.ict4pr.org
Kenya Network of Telecentres (KenTel)	Kenya	www.ken-tel.org
FETEMA	Mali	www.fetema.org; www.afriklinks.org
DidiBahini	Nepal	www.didibahini.org
E-Network Research and Development (ENRD)/Nepal Wireless Networking Project	Nepal	www.enrd.org
Philippine Community eCenter Network	Philippines	under construction
GDCO	Sudan	www.gedarefcity.org
Telecenter for Development	Sudan	
Tanzania Telecentre Network	United Rep. of Tanzania	
CDI - Comité para Democratización de la Informática	Uruguay, Brazil, Chile, Ecuador, Mexico, Argentina, Colombia	www.cdi.org.uy
Plus 1 network that preferred not to be individually identified		

References[30]

Amariles F, Paz O, Russell N and Johnson N (2006). The impacts of community telecentres in rural Colombia. Available at http://ci-journal.net/index.php/ciej/article/view/256/270 [17 September 2007].

Badshah A, Khan S, and Garrido M (2003). Connected for development: information kiosks and sustainability. ICT Task Force Series 4. Available at www.unicttaskforce.org/perl/documents.pl?id=1361 [17 September 2007].

Bell T (2006) Village computing: a state of the field. Reflections on the village computing consultation. Grameen Foundation, Washington DC, November 2006. Available at www.villagecomputing.org/docs/A_State_of_the_Field.pdf [17 September 2007].

Cecchini S (2005). 'Oportunidades digitales, equidad y pobreza in América Latina: ¿Qué podemos aprender de la evidencia empírica?' Serie de Estudios Estadísticos y Prospectivos 40. CEPAL, Santiago de Chile, diciembre 2005. Available at http://www.eclac.cl/ [17 September 2007].

Chand A, Leeming D, Stork E, Agassi A and Biliki R (2005). The impact of ICT on rural development in Solomon Islands: the Pfnet case. March 2005. Available www.usp.ac.fj/jica/ict_research/documents/pdf_files/pfnet_report.pdf [17 September 2007].

Chapman R, Slaymaker T and Young J (2001). Livelihoods approaches to information and communication in support of rural poverty elimination and food security. Overseas Development Institute, London. Available at www.livelihoods.org/hot_topics/docs/SPIS_Complete.pdf [17 September 2007].

Conroy C (2006). Telecentre initiatives in rural India: failed fad or the way forward? Working Paper 4 Natural Resources Institute, University of Greenwich, UK, December 2006. Available at www.nri.org/news/documents/wp4_telecentres_review_dec2006.pdf [17 September 2007].

Cowell A and Marinos S (2006). Mejorando los Cursos y Servicios del CETEBI Comunitario Ajb'atz' Enlace Quiché: Un diagnostico de las Necesidades y Opiniones de los Estudiantes. Quiché, Guatemala, November 2006.

Commonwealth Telecommunications Organization (CTO) (2005). The economic impact of telecommunications on rural livelihoods and poverty reduction: a study of rural communities in India (Gujarat), Mozambique and Tanzania. Report of DFID KaR Project 8347. Available at www.cto.int/Portals/0/docs/annual%20report.pdf [17 September 2007].

Department for International Development (DFID) (1999). Sustainable livelihoods guidance sheets. Available at www.livelihoods.org/info/guidance_sheets_pdfs/section1.pdf [17 September 2007].

Ducombe R (2006). Analysing ICT applications for poverty reduction via micro-enterprise using the livelihoods framework. Development Informatics Working Paper Series Paper No. 27, Manchester, UK. Available at http://www.sed.manchester.ac.uk/idpm/research/publications/wp/di/documents/DIWkPpr27.pdf [17 September 2007].

E-Choupal (2006). ITC e-Choupal presentation at UNCTAD's Expert Meeting on Enabling Small Commodity Producers in Developing Countries to Reach Global Markets, Geneva, 11–13 December 2006.

EIU (2007). The 2007 e-readiness rankings: raising the bar. A white paper from the Economist Intelligence Unit, London. Available at http://graphics.eiu.com/files/ad_pdfs/2007Ereadiness_Ranking_WP.pdf [17 September 2007].

Etta F and Parvyn-Wamahiu S (eds.) (2003). *Information and Communication Technologies for Development in Africa, Vol 2, The Experience with Community Telecentres*. Available at www.idrc.ca.

Fillip B and Foote D (2007). Making the connection: scaling telecentres for development. Information Technology Applications Centre, Academy for Education Development, Washington DC, March 2007. Available at http://connection.aed.org/main.htm [17 September 2007].

Fundación CTIC (2006). Las Redes de Telecentros en España. Una Historia Por Contar. Asturias. Available at www.fundacionctic.org [17 September 2007].

Gerster R and Zimmerman S (2003). *Information and Communication Technologies and Poverty Reduction in Sub-Saharan Africa. A Learning Study (Synthesis). October 2003*. Available at http://162.23.39.120/dezaweb/ressources/resource_en_24708.pdf [17 September 2007].

Gobierno de Chile (2001). Instrucción Presidencial para el Desarrollo de la Política Nacional de Infocentros. Santiago de Chile, 25 Septiembre 2001. Available at www.modernizacion.cl/1350/articles-65998_instructivo_infocentros.pdf [17 September 2007].

Gobierno de Chile (2007) SII ofrece este año más de 14560000 propuestas de declaración. Servicio de Impuestos Internos, 29 March 2007. Available at www.sii.cl/pagina/actualizada/noticias/2007/290307noti01jo.htm [17 September 2007].

Grupo de Acción Digital (2004). Chile 2004–2006 Agenda Digital. Te Acerca al Futuro. February 2004. Available at www.conicyt.cl/bancomundial/documentos/ct_chile/Agenda_Digital_Chile_2004-2006.pdf [17 September 2007].

Harris R and Rajora R (2006). Empowering the poor: information and communications technology for governance and poverty reduction. A study of rural development projects in India. UNDP-APDIP ICT4D Series, Bangkok. Available at www.apdip.net/publications/ict4d/EmpoweringThePoor.pdf [17 September 2007].

Hasan M (2006). Impact of ICT carried livelihood information delivery on rural community. Project presentation. D.Net, Bangladesh. Available at www.pallitathya.org/en/publication/pdf/Impact_of_ICT_Carried_Livelihood_Information.pdf [17 September 2007].

IIM (2003). An evaluation of Gyandoot. Centre for Electronic Governance, Indian Institute of Management, Ahmedabad. Available at http://unpan1.un.org/intradoc/groups/public/documents/APCITY/UNPAN015131.pdf [17 September 2007].

ITU (2007). Definitions of world telecommunication/ICT indicators. Final version, April 2007. Available at www.itu.int/ITU-D/ict/material/IndDef_e_v2007.pdf [17 September 2007].

Kumar R and Best M (2006). Social impact and diffusion of telecentre use: a study from the sustainability access in rural India project. *Journal of Community Informatics*. Available at http://ci-journal.net/index.php/ciej [17 September 2007].

Mandana A and Amuriat G (1996). Emerging gender dimension for pro-poor community driven networks. Power Point presentation. Pro-poor Community-driven Networks Seminar, August 2006, Kampala, Uganda. Available at http://wougnet.org/ICTpolicy/ug/docs/cinug.html [17 September 2007].

Mardle E (2003). Telecentres: how did we lose the plot? *Development Gateway*, 3 February 2003. Available at http://topics.developmentgateway.org/ict/sdm/previewDocument.do~activeDocumentId=440944?activeDocumentId=440944 [17 September 2007]

OECD (2005). New perspectives on ICT skills and employment. Working Party on the Information Economy. DSTI/ICCP/IE(2004)10/FINAL, 22 April 2005, Paris. Available at www.oecd.org/dataoecd/19/7/34884388. pdf [17 September 2007].

Pallitathya (2006). Livelihood case studies: Mr. Nurul Islam Khan. Available at http://pallitathya.org/en/case_studies/index.html [17 September 2007].

Parkinson S and Ramírez R (2006). Using a sustainable livelihoods approach to assessing the impact of ICTs in development, *Journal of Community Informatics* Vol.2, no. 3. Special Issue: Telecentres. Available at http://www. ci-journal.net/index.php/ciej/article/view/310/269 [17 September 2007].

Pe-PP (2007). Partnerships for e-Prosperity for the Poor (PE-PP). Update as of January 2007.

PNUD (2006). Desarrollo Humano en Chile: Las Nuevas Tecnologías: ¿Un Salto al Futuro? Santiago de Chile, June 2006. Executive summary available at www.desarrollohumano.cl/ [10 October 2006].

Proenza F J (2003). A public sector support strategy for telecentre development: emerging lessons from Latin America and the Caribbean, October 2003, in Badshah A, Khan S and Garrido M (eds.).

Proenza F J (2006). Guatemala: Programa de Acceso Rural a Internet. Apoyo a la Inversión en el Desarrollo de Tecnologías de la Información y Comunicación para Combatir la Pobreza Rural en América Latina y el Caribe. 5° Informe de la Serie. Centro de Inversiones de FAO. Rome, 9 May 2006. Available at www.e-forall. org/pdf/GuatemalaAccesoRural_9mayo2006.pdf [17 September 2007].

Pun M, Shields R, Poudel R and Mucci P (2006). Case study and evaluation report Nepal Wireless networking project, August 2006, Nepal. Available at www.nepalwireless.net [17 September 2007].

Roman R and Colle R (2002). Creating a participatory telecentre enterprise. Cornell University, July 2002, http:// wsispapers.choike.org/creating_participatory_telecenter.pdf [17 September 2007].

SUBTEL (2005). *Estudio: Sistematización de la Información del proceso de implementación de Telecentros Comunitarios en Chile.* Gobierno de Chile. Subsecretaria de Telecomunicaciones. Available at www.infocentros.gob.cl [17 September 2007].

SUBTEL (2007). Bases específicas del concurso público para la asignación de subsidios para la instalación y operación de la red de telecentros comunitarios de información y comunicación. Fondo de desarrollo de las telecomunicaciones. Available at www.subtel.cl/prontus_subtel/site/artic/20070801/pags/20070801005746. html [17 September 2007].

UNCTAD (2006a). *Information Economy Report 2006.* Geneva, November 2006. Available at www.unctad.org/ ecommerce [17 September 2007].

UNCTAD (2006b). UNCTAD questionnaire. ICTs and poverty reduction. Case study: Chilean telecentres network. August 2006.

UNCTAD (2007). UNCTAD questionnaire: promoting men's and women's livelihoods through ICTs. The case of telecentres. April 2007.

UNDP (2006). A telecottage handbook. UNDP and Hungarian Telecottage Association, February 2006. Available at http://europeandcis.undp.org/?menu=p_cms/show&content_id=1EAEB3D3-F203-1EE9-BD0C279330992AC8 [17 September 2007].

Notes

1. As of 2006, 44 per cent of Governments had introduced a national ICT plan, and another 20 where in the process of preparing one (UNCTAD 2006a).

2. See chapter 3, Pro-poor ICT policies and practices, in UNCTAD's Information Economy Report 2006.

3. For a detailed presentation of the sustainable livelihoods framework, see DFID (1999)

4. For further information on how ICTs can support sustainable livelihoods, see Chapman, Slaymaker and Young (2001) and Ducombe (2006).

5. See chapter 3, Pro-poor ICT policies and practices, in UNCTAD>s Information Economy Report 2006.

6. For instance, for the Government of Chile community access points are a key strategy for ensuring universal access to ICT (Grupo de Acción Digital, 2004; Gobierno de Chile, 2001).

7. For further information on best practices to support telecentres> sustainability, see telecentre.org, Bell (2006) and Fillip and Foote (2007).

8. Annex 1 provides an overview of the respondents.

9. Telecentre.org is a collaborative initiative that supports and strengthens the telecentre movement. Launched in 2005 with the support of Canada's International Development Research Center (IDRC), Microsoft and the Swiss Agency for Development and Cooperation (SDC), it acts as a «connecting point» among dozens of telecentre networks.

10. A telecentre network is an interconnected group or system of telecentres.

11. Suggested by telecentre.org.

12. The percentages provided refer to the telecentre networks that responded to the individual question.

13. Figures provided are based on a simple average per network, and not on absolute figures.

14. Weighted average (per number of telecentres) among the 15 networks that provided data.

15. Dial-up access is when the total Internet capacity in both directions is below 256 Kbps. Broadband access is when the sum of the Internet connection capacity in both directions is equal to, or greater than, 256 Kbps (based on ITU definition; see ITU, 2007).

16. Training to develop basic ICT user skills to use generic tools such as e-mail, web browsing, work processors, spreadsheets and presentation tools (see OECD, 2004).

17. To exploit business opportunities provided by information and communication technologies (for example to buy and sell online).

18. See www.najuqsivik.com/gateway/arts-crafts/index.htm.

19. Others, notably Proenza (2003), have highlighted the importance of independence from political interference.

20. Mali, Mozambique, Uganda, South Africa and Senegal.

21. See, for example, www.tradepoint.org.

22. See e-Choupal (2006).

23. See chapter 2 of UNCTAD's Information Economy Report 2006, e-Choupal (2006) and Conroy (2006).

24. See the recommendation to expand e-Choupal>s activities into one-stop shop where villagers can also obtain other services, such as e-government services, e-education and e-health services (Fillip and Foote, 2007, p. 66).

25. A study (CTO, 2005) of a project funded by DFID in India, Mozambique and the United Republic of Tanzania shows how communication flows are much slower to change than communication technologies and highlights the importance of established and trusted communication patterns.

26. For more details on this programme see Proenza (2006).

27. See www.iie.cl.

28. See Bell (2006), Fillip and Foote (2007) and UNDP (2006).

29. For a better understating of the 12 Cs framework and practical examples of how it is used to assess to what extent an ICT policy or programme supports poverty alleviation, see chapter 3 of UNCTAD>s Information Economy Report 2006.

30. The date in square brackets at the end of most entries is the date on which the text was accessed.

Chapter 8

HARMONIZING CYBER LEGISLATION AT THE REGIONAL LEVEL: THE CASE OF ASEAN

A. Introduction

An increasing number of developing countries are adapting their legislation to e-commerce to remove barriers to online services and provide legal certainty to business and citizens. The adoption of cyberlaws is an essential step to widen market access to small and medium-sized enterprises (SMEs) at the regional and the international level. Issues such as security, privacy, consumer protection and intellectual property rights need to be addressed in a digital economy.

The impact of the introduction of legislation on the expansion of e-commerce activities is reported by countries to be positive, leading to increased ICT-related business opportunities and foreign direct investment, according to a survey on e-commerce legislation in developing countries carried out by UNCTAD in 2007.[1] The survey revealed that 93 per cent of the 32 responding countries have prepared or are in the process of adapting it in order to benefit from market opportunities arising from ICT and the economic growth and development opportunities that accompany them. A majority of responding countries chose to consider the United Nations Commission on International Trade Law (UNCITRAL) model laws when they revised their legislation. This shows the importance of international harmonization in view of the need for uniformity of the law applicable in the information economy and the development of harmonious international economic relations at the regional and global levels.

Developing countries within their region and subregion are also considering the development of a basic harmonized legal framework for e-commerce to make their region competitive and help boost e-business and economic growth. The harmonization of e-commerce legal frameworks is expected to lead to larger internal and external consumer and business markets by facilitating cross-border e-commerce.

Regional initiatives for the harmonization of cyberlaws are developing and are facing challenges, including the different legal, social and economic systems of countries in a particular region. Other barriers include the lack of human resources, the lack of public awareness about the scope and application of the law and its benefits, and the legal security of e-business through the legislation adopted. More broadly, countries reported consumers' lack of trust in the security of e-commerce transactions and privacy protection, the difficulty in setting up the technical infrastructure, and the need to accompany legal reform by a broader national reform of the information economy in order to seize the benefits of e-commerce.

The Association of Southeast Asian Nations (ASEAN)[2] is the first regional organization in the developing world to adopt a harmonized e-commerce legal framework consistent across jurisdictions. By the end of 2008, all ASEAN member countries will have enacted consistent national e-commerce legislation. This chapter presents a case study of the Harmonization of E-Commerce Legal Infrastructure in ASEAN Project–a major four-year project to help the ten ASEAN member countries develop and implement a harmonized e-commerce legal infrastructure.[3]

The experience of the ASEAN member countries in the E-Commerce Project may be helpful for other regional associations in the developing world that are currently considering the harmonization of e-commerce legal infrastructure. This is the case, for instance, of the East African Community and the Latin American Integration Association (ALADI) member countries, which receive technical assistance from the United Nations Conference on Trade and Development (UNCTAD) to help design e-commerce legislation at the national and regional levels,[4] as well as other regional organizations in Africa, Asia and the Pacific.

The ASEAN experience might also prove useful for developing countries that are formulating their own e-commerce legislation, and developing a comprehensive legal infrastructure, including regulations, standards, training and education.

The adaptation of legal frameworks to e-commerce is a key step among ICT-related policy measures that Governments should take in order to foster the use of ICTs and development of e-commerce. In previous editions of the *E-commerce and Development Report* and the *Information Economy Report*, UNCTAD focused on specific e-commerce legal issues, for which references are provided in this chapter, and more generally on ICT strategies to help them take full advantage of ICT for development. In this edition, the chapter aims to provide guidance for developing countries and regions when they start developing their e-commerce legal framework.

On the basis of the ASEAN case, the chapter will set out the modalities for the implementation of law reform, as well as possible options and potential challenges awaiting countries in the development of a common regional and national e-commerce legal framework. Such challenges include different e-readiness levels and the development stage of e-commerce legislation, which can vary from one country to another.

The chapter is divided into four sections. Section A provides background information on the e-ASEAN initiative and E-Commerce Project. Section B presents the challenges of regional and domestic implementation of e-commerce legal infrastructure and addresses cyberlaw coverage in ASEAN member countries. Section C presents the lessons learned vis-à-vis regional harmonization of e-commerce legal infrastructure, as well as at the national level. Section D proposes policy recommendations to help regional organizations and developing countries prepare their legal framework.

1. Background

ASEAN was created in 1967 to promote regional cooperation among its member States with the objective of (a) accelerating economic growth, social progress and cultural development, and (b) to promoting peace and stability in the region. It currently has 10 member countries: Brunei Darussalam, Cambodia, Indonesia, the Lao People's Democratic Republic, Malaysia, Myanmar, the Philippines, Singapore, Thailand and Viet Nam.

With the advent of ICTs, ASEAN member countries endorsed the e-ASEAN initiative in 1999 as a result

of the ASEAN Vision 2020, defined two years earlier and aimed, with regard to economic development, at creating a stable, prosperous and highly competitive ASEAN Economic Community in which there is a free flow of goods, services and investment, a freer flow of capital, equitable economic development and reduced poverty and socio-economic disparities in 2020 (the target date for establishing the ASEAN Economic Community has since been brought forward to 2015). The purpose of the e-ASEAN initiative is to complement national ICT strategies, and promote economic growth and competitiveness for better integration of ASEAN member countries into the global information economy. The e-ASEAN initiative sets out an action plan focusing on physical, legal, logistical, social and economic infrastructure to embrace the development and use of ICTs.

In 2000, the ASEAN member countries entered into the e-ASEAN Framework Agreement to facilitate the establishment of the ASEAN Information Infrastructure – the hardware and software systems needed to access, process and share information – and to promote the growth of electronic commerce in the region. The framework comprises five main elements: information infrastructure, e-society, e-government, a common marketplace for ASEAN ICT goods and services, and the creation of an e-commerce friendly environment. This chapter will focus on the implementation of the E-Commerce Project.

The e-ASEAN Framework Agreement is to be implemented by a series of measures set out in the Roadmap for Integration of e-ASEAN Sector (the e-ASEAN Roadmap).[5] The E-Commerce Project has been assisting ASEAN in meeting two key targets in the Roadmap:

- Measure 78: Enact domestic legislation to provide legal recognition of electronic transactions (i.e. cyberlaws) based on common reference frameworks (deadline: 31 December 2008);

- Measure 79: Facilitate cross-border electronic transactions and the use of digital signatures (deadline: 31 December 2009).

These targets and deadlines were confirmed by the ASEAN economics ministers in Cebu, Philippines, in December 2006.

The coordinating body for this work in ASEAN is the ASEAN E-Commerce and ICT Trade Facilitation

Working Group, a group that reports to ASEAN TELSOM (Telecommunications and ICT Senior Officials Meeting), which in turn reports to TELMIN (Telecommunications and ICT Ministers). The Working Group's agenda covers most aspects of e-commerce. Other relevant Working Groups include those on e-society and ICT Capacity Building, universal access, digital divide and e-government, and ASEAN information infrastructure.

2. E-commerce project goals

The overall goal of the e-commerce project is to assist ASEAN to integrate into one market for goods, services and investment. The e-commerce project has had three phases.

The focus of Phase 1 was the harmonization of e-commerce legal infrastructure. This phase was conducted between 2004 and 2005, and included the establishment of a broad, high-level harmonized legal, regulatory and institutional infrastructure for e-commerce. Phase 2, which was conducted in 2006, examined the potential establishment of a harmonized legal, regulatory and institutional infrastructure for electronic contracting and online dispute resolution (ODR). Phase 3, which was conducted in 2007, examined the possible establishment of a harmonized legal, regulatory and institutional infrastructure for the mutual recognition of digital signatures, so as to facilitate cross-border trade.

Each project phase built on the achievements of previous phases and the outputs of each phase became more detailed and technical in nature as the overall project progressed.

3. Harmonization structure

Harmonization projects are designed to align individual member country laws in order to remove unwanted gaps, overlaps and duplication.[6] E-commerce harmonization projects aim to increase legal certainty for parties engaged with more than one member country – for example, multinational businesses that are attempting to expand their business in a new region. Harmonization projects usually fall into one of two categories – soft harmonization (based on training and capacity-building) and hard harmonization (based on model or uniform laws). Most e-commerce legal harmonization projects are soft harmonization projects, in that there is no intention or requirement

for countries to adopt the same (or even model) laws and regulatory systems. All that is undertaken are training and capacity development activities, aimed at ensuring a common (or harmonized) understanding of e-commerce legal requirements.

Examples of soft harmonization projects include e-commerce law harmonization projects in:

- The United Nations Economic and Social Commission for Asia and the Pacific (UNESCAP), which has 62 member countries.[7] They are undertaking a soft harmonization project called the Technical Assistance Project on Harmonized Development of Legal and Regulatory Systems for E-commerce in Asia and the Pacific: Current Challenges and Capacity Building Needs.[8]

- The South Asian Association for Regional Cooperation (SAARC)[9] comprising eight member countries, which is undertaking a soft harmonization project called Harmonization of E-Commerce Laws and Regulatory Systems in South Asia.[10]

- The Pacific Islands Forum (PIF),[11] which has 16 member countries. They have developed a Cyberlaws Strategy (part of the Pacific Plan for strengthening regional cooperation and integration[12]). It is designed to include the harmonization of e-commerce laws as one of its goals. Overall, the strategy is based on soft harmonization, although some specific cyber laws (e.g. spam legislation) may be based on sample laws and subject to hard harmonization.

There are some benefits to the soft harmonization approach. One of them is the considerable potential for integration and coordination with other regional capacity-building activities, resulting in low costs and useful collaboration with regional neighbours. In addition, some assistance is already available in the form of training materials and kits on e-commerce laws. For example, UNCITRAL is developing training materials regarding the United Nations Convention on the Use of Electronic Communications in International Contracts 2005. Third, in theory, consistency in training should deliver reasonable consistency in outputs, including the laws, regulations and other aspects of e-commerce legal infrastructure. However, in practice, consistency in training has not always delivered consistency in outputs. The UNCITRAL model laws, for example, have been implemented very differently in numerous countries, despite the availability of uniform training and capacity-building assistance.

ASEAN was concerned that it would not necessarily have control and ownership of the development and delivery of training and materials, especially implementation guides that could help to ensure consistency. The only training materials that were available when the project started were generic materials that had not been customized for an Asian audience or for the particular political and legal systems in ASEAN member countries.

With respect to hard harmonization projects, examples include e-commerce law harmonization projects in:

- The Southern African Development Community (SADC).[13] Its 14 member countries are undertaking a hard harmonization project based on a customized model e-commerce law.

- The European Union (EU).[14] Its 27 member countries have standardized domestic e-commerce legislation based on the EU Directive on Electronic Commerce, 2000.[15]

Ultimately, ASEAN chose to pursue hard harmonization, based on implementation guidelines rather than simple capacity-building.

The hard harmonization model offers five main benefits. First, the guidelines are an inclusive instrument, which ensures the participation of less developed ASEAN member countries. They also contain implementation steps that ensure greater consistency across ASEAN member countries. In addition, the guidelines include a timetable, which helps to ensure an orderly, phased development of e-commerce legal infrastructure in ASEAN. Moreover, they leave less room for interpretation and reduce inconsistencies. Finally, they remain a flexible instrument that can be reviewed and updated (for example, every three years).[16]

Under this hard harmonization option, E-Commerce Project Guidelines have been developed that build on the common objectives and principles for e-commerce legal infrastructure. The guidelines include more prescriptive information on implementation steps, and a timeline. They were developed by an Australian private consulting firm[17] in collaboration with project participants and technical experts from ASEAN member countries. They have been endorsed by the ASEAN E-Commerce and ICT Trade Facilitation Working Group.

B. Regional and domestic implementation

The E-Commerce Project sought to identify and address both regional and domestic implementation issues. Regional issues were important since the project had a trade facilitation focus, and so it was important to address any barriers to cross-border trade that might arise from gaps or inconsistencies in e-commerce legal infrastructure. Domestic implementation issues were important since ASEAN has set certain goals and target dates for economic integration that require all 10 member countries to implement e-commerce legal infrastructure at the domestic level.

1. Regional implementation

The project included numerous activities and outputs over a four-year period, including eight full ASEAN workshops with an average of 35 participants per workshop, 14 country visits and ongoing liaison with other international organizations such as Asia Pacific Economic Cooperation (APEC), UNCITRAL, UNCTAD and UNESCAP.

A range of implementation guides and checklists have been produced during the project, including both generic (regional) guides and country-specific guides.

Several surveys were conducted during the E-Commerce Project in order to gain a more detailed understanding of developments and issues in the region, and to give member countries an opportunity to provide detailed input at key stages of the project. Questionnaires were sent to government representatives in each of the ASEAN member countries and the results were collated and published during the project. Surveys regarding the following issues were conducted:

- E-commerce legal infrastructure in ASEAN (October 2004, Phase 1).

- Implementation issues and constraints for a harmonized e-commerce legal infrastructure in ASEAN (February 2005, Phase 1). This survey is the most relevant survey for other developing nations and regional associations. It collected data on the implementation challenges faced by ASEAN member countries: some of the survey results are discussed below in subsection 3.

- Cyberlaws in ASEAN (October 2005, Phase 1). This survey revealed some significant trends and gaps in cyberlaw coverage, which are discussed below in subsection 4.

- Current legal, regulatory and institutional approaches to e-contracting and ODR infrastructures in ASEAN (February 2006, Phase 2).

2. Domestic implementation

As is the case in other regional groupings, the 10 ASEAN member countries are very diverse in areas such as culture, religion, legal systems, types of government and level of development.

When ASEAN reaches a consensus decision to take action – for example, its decision to implement harmonized e-commerce laws – the next challenge is to implement this decision at the domestic level. In the area of e-commerce legal infrastructure, the diversity amongst member countries is most noticeable with regard to the different levels of development, since ASEAN includes both highly developed countries with a mature e-commerce infrastructure (such as Singapore) and developing countries with only a rudimentary e-commerce infrastructure (such as the Lao People's Democratic Republic).

However, a positive aspect of the E-Commerce Project is that it continues to have a direct impact on the development of domestic e-commerce legal

infrastructure in ASEAN member countries, sooner than anticipated in the project design. This impact can be seen in the table 8.1.

It is encouraging that both developed and developing member countries have been able to enact domestic e-commerce laws in a relatively short period. The three countries, as shown in the table 8.1, which are yet to enact e-commerce laws all have draft laws available and are on track to meet the e-ASEAN deadline of December 2008. However, the scope of the project goes beyond enactment of e-commerce laws alone, and some significant challenges remain in implementing a comprehensive e-commerce legal infrastructure in all 10 ASEAN member countries. For example, several member countries are still developing detailed regulations for the implementation of their e-commerce laws in areas such as the accreditation of digital signature service providers.

In addition, those member countries that did have e-commerce laws in place at the beginning of the project may still need to review their laws for consistency with its guidelines; in particular, the guidelines include a recommendation that member countries consider amending their domestic legislation to ensure harmonization with the United Nations Convention on the Use of Electronic Communications in International Contracts 2005. This issue is discussed in greater detail below (section C).

Table 8.1

E-commerce legislation in ASEAN member countries

Member country	Status of e-commerce laws – project inception (January 2004)	Status of e-commerce laws – current (October 2007)
Brunei	Enacted	Enacted
Cambodia	None	Draft
Indonesia	None	Draft
Lao People's Democratic Republic	None	Draft
Malaysia	None	Enacted
Myanmar	Draft	Enacted
Philippines	Enacted	Enacted
Singapore	Enacted	Enacted
Thailand	Enacted	Enacted
Viet Nam	None	Enacted

Source: ASEAN E-Commerce Project, internal project materials, ASEAN secretariat and Galexia, October 2007.

3. Domestic implementation challenges

Priority challenges for all member countries

During the project a survey was conducted to examine some of the specific implementation challenges faced at the domestic level.[18] It revealed that there are some common ASEAN-wide challenges to the implementation of e-commerce legal infrastructure. The main challenges are limited government policy or support, limited infrastructure, limited funding; and a shortage of skills and training.

Lack of private and public sector support has also been identified by some member countries as posing a barrier to the implementation of e-commerce legal infrastructure. Ultimately, it is in the interest of the public and the private sector if they can complete more efficient transactions in the online environment. During implementation, initiatives can be employed by Governments to increase awareness and support on the part of the private and public sectors. Those initiatives have the added advantage that they also help to increase the uptake of e-commerce.

Priority challenges for CLMV member countries

The implementation challenges faced by CLMV countries (Cambodia, the Lao People's Democratic Republic, Myanmar and Viet Nam – the less developed ASEAN member countries) are slightly different from those faced on an ASEAN-wide basis. This is because of the different needs of these countries in implementing their e-commerce legal infrastructure. The results of the survey reveal that the main implementation issues faced by CLMV countries are (in order of priority): limited government policy and support, the lack of a strong market for e-commerce products and services, limited funding, limited infrastructure, and a shortage of skills and training.

The survey revealed that limited government policy support appears to pose a significant and shared barrier to implementing a harmonized e-commerce legal infrastructure. There are many reasons why there may be only limited government policy in support of a particular aspect of e-commerce law. In some jurisdictions other areas of regulation may be prioritized before a harmonized e-commerce legal infrastructure is established. Also, Governments may perceive that there is no market for e-commerce at the present time,

or that the necessary e-commerce infrastructure is not yet in place.

Without supportive government policy it is difficult for implementation to occur. Although there is clear support for a top-level e-commerce law in each member country, a comprehensive e-commerce legal infrastructure requires that training, capacity-building and awareness activities be carried out. These aspects of legal infrastructure require financial support and will often compete for funding with other government priorities. Government policy provides the primary impetus to implement e-commerce legal infrastructure; once there is government support, other resources such as funding, infrastructure, and skills and training can be allocated.

The results of the survey also revealed that the lack of funding, skills and training presents a significant implementation challenge. This indicates that capacity-building initiatives and other forms of assistance can play a vital role in enhancing the e-commerce infrastructure.

Funding is generally required from the initial scoping and policy formulation stage right through to implementation and follow-up work. Appropriate resources will need to be allocated by member country Governments to implement e-commerce legal infrastructure. Some member countries indicated that they did not have the resources or access to funds to enhance existing legal infrastructure. There may be a need for some member countries to seek assistance in this area from external sources such as ASEAN, the United Nations, and other regional and international organizations.

Capacity-building projects play an important role in educating government officials of ASEAN member countries and ensuring a common understanding among them about the requirements of a harmonized e-commerce legal infrastructure. Capacity-building will not only improve the skills and knowledge of professionals but would also help ensure a common understanding of the legal requirements of e-commerce infrastructure if projects were to be undertaken on an ASEAN-wide basis.

ICT infrastructure is generally still emerging in CLMV countries. Less than 1 per cent of the population in Cambodia, the Lao People's Democratic Republic and Myanmar, and less than 5 per cent of the population in Viet Nam have access to the Internet.[19] The market for e-commerce in these countries is therefore small.

At present there is no perceived need (or no market) in those countries to prioritize e-commerce legal infrastructure.

ASEAN is involved in several projects to narrow the digital divide in the region. The Initiative for ASEAN Integration (IAI) is specifically aimed at doing that. Essentially, the IAI provides a framework for regional cooperation through which more developed ASEAN members could help those member countries that most need it. It focuses on education, skills development and worker training.

The IAI Work Plan for the CLMV countries focuses on the priority areas of infrastructure development (transport and energy), human resource development (public sector capacity building, labour and employment, and higher education), information and communications technology, and promoting regional economic integration (trade in goods and services, customs, standards and investments).

Projects such as the IAI and other development initiatives in the region and in individual member countries are all geared towards improving the ICT and telecommunications infrastructure in ASEAN member countries. The existence of such infrastructure can be an important building block for the implementation of a more advanced e-commerce legal infrastructure. It should be pointed out that measures addressing the digital divide are not relevant only to CLMV countries, the ASEAN secretariat having noted that there are also pockets of underdevelopment in the ASEAN-6 countries (Brunei Darussalam, Indonesia, Malaysia, Philippines, Singapore and Thailand).[20]

A lack of skills and training was identified in the survey responses of the CLMV countries as a barrier to implementation of the recommendations. Such a lack was also identified as a challenge to implementation by other countries, including Indonesia and Thailand. One of the concerns is that professionals in those countries are not exposed to developed ICT regulatory and policy frameworks. This lack of exposure means that professionals in developing countries have not had the chance to develop or refine skills and knowledge in that area.

Resources required

In order of greatest need, survey respondents identified the following resources needed to assist member countries in implementing their e-commerce legal infrastructure: funding for implementation activities; an ASEAN-commissioned integration project to help channel assistance from more developed member countries to less developed member countries; legal and technical training for key implementation staff such as policymakers and legal drafters; and awareness-raising for the Government and the private sector.

4. Cyberlaw coverage

In order to identify the broader legal context for the development of e-commerce legal infrastructure in ASEAN, the ASEAN E-Commerce and ICT Trade Facilitation Working Group asked for detailed information on the overall cyberlaw coverage in ASEAN member countries.

In 2005 the cyberlaws survey was conducted as part of the ASEAN E-Commerce Project. The results were briefly updated in June 2007.

The term "cyberlaws" generally refers to laws that address the legal issues that arise from the Internet and other communications technology. The survey examined laws enacted specifically for the online environment. These include laws where only a portion, for example, a part, a chapter section or chapter subsection, contains cyberlaw provisions. Laws of general application, for example consumer or intellectual property laws, are not intended to form part of this analysis of cyberlaws. However, it is noted that those general laws may nevertheless have some application in an online setting.

The list of cyberlaws that could be covered by this survey was limited by time and resource constraints, and so the survey was restricted to a selection of laws considered by the ASEAN E-Commerce and ICT Trade Facilitation Working Group to have a high priority.

Cyberlaw coverage is varied across ASEAN.[21] Table 8.2 summarizes progress made in ASEAN member countries in enacting the nine identified cyberlaws as of August 2007.[22]

The survey revealed the following trends in cyberlaw coverage in ASEAN:

Consumer protection

There is limited coverage in ASEAN of consumer protection laws for e-commerce. Coverage does not extend beyond a section in Malaysia's Communications and Multimedia Act 1998[23] that provides that network service and application providers must deal reasonably with consumers and adequately address consumer complaints; a subsection in the Philippines Electronic Commerce Act 2000[24] that states that violations of the Consumer Act 1991[25] committed through the use of electronic transactions will be subject to the penalties contained in that Act; and a proposed provision in the Indonesian Bill on Electronic Information and Transactions,[26] which states that consumers have the right to receive complete information on contract requirements and product details.

Despite the limited coverage in ASEAN of consumer protection laws for e-commerce, most jurisdictions have general consumer protection legislation in place. These, in most circumstances, will apply to goods and services sold online.

Online consumer protection is a high priority in some key ASEAN trading partners, including Australia, the European Union and Japan, and this has resulted in growing interest in online consumer protection in ASEAN. In early 2007 the ASEAN E-Commerce and ICT Trade Facilitation Working Group added the harmonization of online consumer protection laws to its working agenda. This is likely to result in further discussion of online consumer protection laws

in ASEAN Member countries, although no specific project or timetable has been considered at this stage

Privacy and data protection [27]

At present no laws have been enacted in ASEAN member countries that deal with privacy and personal data protection. Three member countries have draft privacy legislation – Indonesia, Malaysia and Thailand. However, it is uncertain in some member countries when legislation will be enacted and the drafts were developed some time ago. Instead of implementing laws, Singapore has chosen to adopt a self-regulatory approach through the E-Commerce Code for the Protection of Personal Information and Communications of Consumers of Internet Commerce.

The majority of ASEAN member countries are also members of APEC, which has been developing and implementing the APEC Privacy Framework since 2004. This has a strong influence on privacy developments throughout Asia, and it is unlikely that ASEAN member countries will make further progress on privacy legislation until some key APEC privacy issues have been resolved, particularly APEC's work on cross-border privacy rules. APEC expects to conduct some "pathfinder" projects on cross-border privacy rules in 2008.

Table 8.2

ASEAN cyberlaw coverage

	Consumer protection	Privacy	Cybercrime	Spam	Online content regulation	Digital copyright	Domain name regulation	Electronic contracting	Online dispute resolution
Brunei	None	None	Enacted	None	Enacted	Enacted	Enacted	Enacted	None
Cambodia	None	None	None	None	None	Enacted	Enacted	Draft	None
Indonesia	Draft	Draft	Draft	None	None	Enacted	Draft	Draft	None
Lao People's Democratic Republic	None	None	None	None	None	Planned	None	Draft	None
Malaysia	Enacted (voluntary)	Draft	Enacted	Enacted	Enacted	Enacted	Enacted	Enacted	None
Myanmar	None	None	Enacted	None	Enacted	Planned	None	Enacted	None
Philippines	Enacted	None	Enacted/draft	Draft	None	Enacted/draft	None	Enacted	Enacted
Singapore	None	Enacted (voluntary)	Enacted	Enacted	Enacted	Enacted	None	Enacted	None
Thailand	None	Draft	Enacted	Planned	None	Planned	None	Enacted	Planned
Viet Nam	None	None	Enacted	None	None	None	Enacted	Enacted	None

Source: ASEAN E-commerce Project, Survey of Cyberlaws in ASEAN (internal project materials), ASEAN secretariat and Galexia, October 2005 (updated August 2007).

Meanwhile, the ASEAN E-Commerce and ICT Trade Facilitation Working Group is maintaining a watching brief on privacy laws in the region, and there is some interest in developing a project on harmonization of privacy laws in ASEAN in the future.

Cybercrime [28]

At present seven out of ASEAN's ten member countries have cybercrime laws in place, and Indonesia has a draft law. Legislation adopted in member countries varies slightly; however it contains similar offences. Offences in cybercrime laws in ASEAN centre on unauthorized access, unauthorized access with the intent of committing an offence and unauthorized modification of computer material. These mirror offences contained in the Computer Misuse Act 1990 (United Kingdom).[29] Cybercrime legislation in most ASEAN member countries also contains provisions that make it an offence to disclose a computer access code without authorization.

Cybercrime is of particular concern to ASEAN and the ASEAN E-Commerce and ICT Trade Facilitation Working Group has added cybercrime laws to its list of potential harmonization projects.

Spam

Singapore is the only ASEAN member country that has enacted an anti-spam law,[30] although there is also a provision in a Malaysian Act that may be of some limited assistance in prosecuting spammers. There are plans to enact laws in two other member countries.[31]

There may be greater cyberlaw coverage in ASEAN in the future as spam becomes more of a nuisance and lawmakers become more familiar with the legal controls needed to regulate it. The enacted legislation in Singapore may provide lawmakers with a convenient reference point for drafting their own dedicated spam laws.

Online content regulation

Four out of ten ASEAN member countries have content-regulation regimes in place. They control the publication of online material on the basis of social, political, moral and religious values of the Governments of those countries. The regimes have similar content restrictions that apply to print media.

As use of the Internet increases, it is likely that other ASEAN member countries will introduce content-regulation regimes reflecting their social, moral and political values. Online content regulation is now commonplace in most developed countries, as Governments attempt to regulate the availability of illegal content such as child pornography and racial vilification sites.

While government policy plays a significant role in establishing online content-regulation regimes, it might be some time before establishment of such regimes becomes a priority in some CLMV countries. The fact that the ICT infrastructure in those countries is still developing means that it will take a number of years before there is a significant rate of Internet use, and CLMV member countries have higher priorities at this stage.

Digital copyright

The coverage in ASEAN countries of laws on digital copyright is quite extensive. Six member countries have laws in place and a further two have plans to enact legislation.[32]

The scope of coverage of digital copyright issues varies across member countries. Some have detailed provisions addressing a number of digital copyright issues, while others do not go beyond recognizing that copyright subsists in computer programs and the granting of associated rights. This limited recognition is often to be found in less developed Member countries and it is unlikely that regulation of digital copyright issues will advance greatly beyond that level.

Domain name regulation [33]

Currently, four out of ten ASEAN member countries – Brunei, Cambodia, Malaysia and Viet Nam – have laws on domain names. The level of detail of those laws varies greatly, and only Cambodia has an entire legal instrument devoted to that cyberlaw topic.[34]

The limited coverage of laws regulating domain names does not mean that there are no administrative procedures or alternative mechanisms in member countries to regulate this particular cyberlaw. Member countries have generally an organization assigned as the chief registrar of domain names, which controls the granting of and conditions for use of, domain names. This body often has a number of policies and procedures in place regarding the granting and use of its country-level domain.

Additionally, domain name registrars in ASEAN countries often have in place online dispute resolution policies to aid in resolving disagreements about the use of domain names.

Electronic contracting [35]

Electronic contracting is the area where full coverage in ASEAN countries seems likely. All member countries have enacted legislation or have draft legislation in place.[36] There is also a great degree of similarity in the electronic transactions laws of member countries since the legislation in most countries is based either on the UNCITRAL Model Law on Electronic Commerce[37] or on the project recommendations from Phases 1 and 2 of the current ASEAN E-Commerce Project.

ASEAN has set a December 2009 deadline for member countries to facilitate cross-border electronic transactions and the use of electronic signatures through better practice guidelines for electronic contracting. Those guidelines are based on the new UN Convention on the Use of Electronic Communications in International Contracts 2005[38] and were developed during Phase 2 of the current ASEAN E-Commerce Project.

Online dispute resolution (ODR) [39]

There are very few laws on online dispute resolution (ODR) in ASEAN countries. At present only the Philippines has laws in place, and Thailand is the only member country with plans to enact laws. Singapore has established ODR facilities and Malaysia has plans to establish an International Cybercourt of Justice; however, neither country has laws in place.

ODR initiatives are a growing area of IT and e-commerce development in ASEAN. Once more ODR mechanisms have been put in place, it is likely that laws will follow. As mentioned above, ASEAN has set a December 2009 deadline for the development of guiding principles for ODR services in ASEAN.

A good example of an ASEAN-based ODR service is the Philippines MultiDoor Courthouse described in box 8.1.

5. Jurisdiction

Online jurisdiction is a significant policy issue with regard to cyberlaws and international trade that has been considered in some detail in ASEAN. Although online jurisdiction did not form part of the cyberlaw coverage survey discussed above, it did form part of Phase 2 of the ASEAN E-Commerce Project.

Online jurisdiction is likely to be of interest to many regional and multinational organizations that are attempting to harmonize e-commerce legal infrastructure. Unfortunately, the issue does not have a "neat" solution and further work is required in ASEAN and elsewhere on harmonization of approaches.

Legally effective electronic transactions require a method of resolving cross-border jurisdiction issues that is properly defined and certain. There are a number of benefits to having these complex issues resolved, including an increase in confidence and uptake of e-commerce by businesses and consumers.

Achieving harmonization of jurisdiction and of law rules on e-commerce is important in ASEAN in order to help achieve the region's economic integration goals. Jurisdiction in business-to-consumer (B2C) e-commerce transactions is a very high priority issue for ASEAN.

The ASEAN project examined numerous international developments in this field, including developments in the EU, the OECD and the Hague Conference on Private International Law. It also examined international case law and some sample approaches to online jurisdiction in Australia and Japan.[40] At this stage there is no international consensus approach to jurisdiction for e-commerce transactions, although discussions on this important issue are continuing in various forums.

Ultimately, the project guidelines provide recommendations that would assist ASEAN in seeking a consistent, harmonized, pro-consumer approach to B2C jurisdiction in order to build consumer confidence and trust in e-commerce.

For example, the project recommended that ASEAN consider developing a guideline on the harmonization of jurisdiction for B2C e-commerce transactions. The guideline should encourage member countries to adopt measures that ensure that proceedings against consumers are brought only in the State where the consumer is domiciled, and measures that ensure that consumers cannot waive their consumer protection rights in contracts.[41]

Box 8.1

Online dispute resolution in the Philippines

The Philippine MultiDoor Courthouse [42] (PMC) is a Philippines-based ODR service. It was launched in 2004 by the Cyberspace Policy Centre for the Asia-Pacific (CPCAP),[43] a non-profit organization concerned with issues pertaining to the overlap between law and technology.

The ODR service offered by the PMC includes online arbitration, mediation, neutral evaluation and blind bidding services to disputants. All services are available to the general public, as well as to merchants, retailers and commercial establishments. However, disputes that are filed must be of a commercial nature. They include claims for failure to deliver goods or delivering faulty goods, insurance claims and commercial contract disputes.

A dispute resolution process is initiated when a party files a dispute by logging in to its user account on the PMC website (new users can create an account online) and then proceeding to supply the particulars relating to the dispute. This would typically include information about the complainant and respondent and the relief sought.

The complainant is asked to upload a copy of any contractual arrangement between the parties that requires them to submit the dispute to the PMC. If such an agreement exists, an email is sent to the respondent, informing him that the dispute has been filed with the PMC. If there is no agreement, an e-mail is sent inquiring whether the respondent agrees to participate in the dispute resolution process. If the respondent consents, PMC staff then evaluate the particulars of the dispute and make a recommendation as to which of the specific services offered by the PMC should be used by the parties to resolve the dispute. The parties are, however, able to agree to use whichever mode of dispute resolution they prefer.

In cases where parties choose arbitration as their preferred method for resolving the dispute, they are able to select the arbitrator or a panel of arbitrators from a roster of accredited arbitrators provided by the PMC. Similarly, if mediation is chosen, the parties are able to nominate their preferred mediator from a list provided by the PMC.

As a general rule, the fees for the ODR service would be much less than what a party would have to pay when compelled to litigate. During the pilot phase of the PMC's existence, no fee has been charged for the use of its services. However, the PMC states on its website that as the volume of disputes filed begins to increase, it will start charging a fee to cover the costs of services provided to disputants.

Source: ASEAN E-Commerce Project, Documented Analysis of ASEAN Activity in Electronic Contracting and Online Dispute Resolution (ODR) - Discussion Paper 3 (internal project materials), ASEAN secretariat and Galexia, May 2006.

Further work is required in ASEAN and elsewhere to resolve online jurisdiction issues in order to provide greater certainty in electronic transactions across borders.

C. Lessons learned

1. Regional lessons

The ASEAN E-Commerce Project provides key lessons on regional harmonization of e-commerce legal infrastructure that might be of interest to other regional associations with regard to regional harmonization of e-commerce infrastructure and trade facilitation

Global harmonization focus

ASEAN is just one of a number of regional and multinational organizations that have expressed an interest in regional harmonization of e-commerce legal infrastructure. As discussed above, other groups

interested in harmonization in this field include UNESCAP, the South Asian Association for Regional Cooperation, the Pacific Islands Forum, the European Union and the Southern African Development Community.

However, when selecting a model for harmonization, ASEAN member countries also wanted to ensure that their legal infrastructure would be compatible with international developments.

An excellent example of ASEAN considering international models is its selection of the UN Convention on Electronic Contracting as the core model for electronic contracting law (both regionally and domestically). There are some difficulties in using model laws as a basis for harmonization, as they have been interpreted differently in different jurisdictions. There are obvious advantages in using binding treaties as a model for harmonization.

UNCITRAL finalized the United Nations Convention on the Use of Electronic Communications in International Contracts in 2005. It is known in ASEAN by its short title – the UN Convention on Electronic Contracting. It is the first binding UN convention addressing legal issues created by the digital environment.

The convention seeks to enhance the legal certainty and commercial predictability of international electronic transactions by setting out a number of interpretive rules for the use of electronic communications in negotiating and drawing up contracts.

The convention also seeks to harmonize national laws regarding how electronic contracts can be drawn up. Harmonized domestic legislation will reduce the legal uncertainty in business transactions when contracting parties are from different countries. While the previous UNCITRAL model laws are texts that the United Nations recommends national legislators incorporate into domestic law, a United Nations convention once signed and then ratified becomes legally binding on the ratifying country. Model laws provide a greater degree of flexibility than conventions in allowing enacting States to adapt legislation best suited to their existing domestic law, but at the risk of a reduction in cross-border regulatory uniformity or consistency.

Since the convention is loosely based on the model laws, its adoption and ratification by countries with existing legislation deriving from those laws should be straightforward, even though some model law principles have been modified in the convention to meet the need for the greater certainty required in a binding treaty.

Additional principles have been incorporated to address new legal concerns that have arisen in electronic contacting since the first Model Law was issued in 1996, notably invitations to make offers,[44] use of automated information systems[45] and errors in electronic communications.[46] Those provisions would be useful additions to national laws, giving more certainty to e-commerce.[47]

For ASEAN member countries currently without any e-commerce-enabling legislation, it has been suggested that it would be more convenient and progressive to consider one set of new laws to give effect to both model laws and the extra provisions of the Convention.

Whether or not a particular country signs the convention, the text is likely to reflect minimum standards for most cross-border transactions. The text of the convention will likely to become the default standard for e-commerce law, and ASEAN member countries may wish to ensure that their own laws and practices are compatible with it. Becoming a party to the convention will further trust and certainty in cross-border contract formation and electronic transactions.

Ultimately the ASEAN E-commerce Project recommended that member countries consider signing the convention. It also recommended that they consider amending domestic electronic contracting laws to ensure consistency.[48] There has been considerable interest in the convention as many countries in the region have signed it, including China and the Russian Federation. Singapore was the first ASEAN member country to sign the convention, in 2006, and the Philippines signed it in 2007. Most ASEAN member countries are currently considering either signing the convention or ensuring that their domestic laws are compatible.

Trade facilitation focus

Trade facilitation issues are important in ASEAN as the region is attempting to boost the pace of development in member countries by expanding regional and international trade opportunities. ASEAN is conducting a wide range of trade facilitation projects and activities. The ASEAN E-Commerce Project includes a strong trade facilitation focus and attempts to identify and address any barriers to cross-border trade that might arise from gaps or inconsistencies in e-commerce legal infrastructure.

In that context, there were strong practical links between the development of e-commerce legal infrastructure and trade facilitation in some of the activities of the ASEAN E-Commerce Project. Those links included:

- Project recommendations to remove e-commerce legal exemptions that might restrict trade (e.g. customs documentation exemptions);

- Project recommendations to expand domestic e-commerce legislation to allow recognition of digital signatures across borders in order to facilitate trade;

- Consideration of overlaps with e-commerce chapters in free trade agreements; and

- The use of practical trade case studies to illustrate the links between e-commerce legal infrastructure and trade facilitation in project workshops and documentation

One of the most interesting practical trade case studies in the E-Commerce Project was the development of the Pan-Asia E-Commerce Alliance. Box 8.2 highlights the link between e-commerce legal infrastructure and trade facilitation.

Importance of developing a comprehensive e-commerce legal infrastructure

At the regional level the E-Commerce Project found that it was necessary to develop a comprehensive legal infrastructure – not just legislation. This required work on:

- **A. Laws**
  The law itself, as passed by Parliament.

- **B. Regulations and codes**
  Any additional regulatory instruments that provide more specific guidance on how to apply the law, such as regulations and codes of conduct.

- **C. Regulators**
  The regulatory agencies that are responsible for administering and enforcing the law, regulations and codes.

- **D. Registration, licensing and accreditation**
  The system, if applicable, for registering, licensing and/or accrediting individuals and organizations that may provide services under the law. This may include the establishment of licensing and accreditation agencies (although sometimes the regulator may also take this role).

- **E. Standards**
  Government and/or industry may develop technical standards that play a particular role in ensuring compliance with the law. In some instances the law may require compliance with the standard, or it may be a licensing or registration requirement.

- **F. Enforcement and review**
  Some legal infrastructures will include a specific forum for review and enforcement of the particular law (such as a specialized tribunal). In many situations this role will be undertaken by the general court system.

- **G. Training, education and awareness-raising**

The provision of training and education to ensure that the law is understood and complied with, and the raising of awareness of the legal requirements amongst service providers and end-users.

A good example of the application of this comprehensive approach to legal infrastructure is set out in the table 8.3, which compares the required components of legal infrastructure across four common elements of e-commerce.[49]

2. Domestic lessons

The ASEAN E-Commerce Project also provides some interesting lessons on the domestic implementation of e-commerce legal infrastructure that might be of interest to other developing nations that are developing their own e-commerce legal infrastructure

Implementation tools

Detailed implementation guides, implementation checklists, timelines and target dates all proved useful in the development of e-commerce legal infrastructure in ASEAN member countries.

The use of implementation checklists will be of interest to developing nations and regional associations. These checklists incorporate all of the E-Commerce Project recommendations and allow member countries' progress to be tracked against each implementation task. A sample from the generic (regional) guide (with data removed) is provided in table 8.4.

Member countries complete the implementation checklist at regular intervals and a summary of progress is presented to the ASEAN E-Commerce and ICT Trade Facilitation Working Group. This has proved to be a useful mechanism for encouraging member countries to review their own progress and to share information with other member countries.

In addition to the generic guide and the implementation checklist, country-specific guides were prepared for Phase 1 of the E-Commerce Project. They provided more detailed recommendations and advice as they were customized to the particular needs and stage of development of each country. A sample from a country guide (with data removed) is provided in table 8.5.

The country guide repeats the key implementation tasks from the generic guide, but includes observations

Box 8.2

The Pan-Asian E-Commerce Alliance

The Pan-Asian E-Commerce Alliance (PAA) was established in July 2000 by three leading e-business service providers: CrimsonLogic of Singapore, Tradelink of Hong Kong (China), and Trade-Van of Taiwan Province of China. A further six companies,[50] from China, the Republic of Korea, Japan, Malaysia, Macao (China), and Thailand have joined it. These nine members now reportedly represent a total combined membership of more than 150,000 organizations, including most active trading enterprises in the Asian market.[51]

The goal of the PAA is to "promote and provide secure, trusted, reliable and value-adding IT infrastructure and facilities for efficient global trade and logistics".[52] To achieve that goal, the PAA has three primary areas of focus:

- The secure and reliable transmission of trade and logistics documents (including the mutual recognition of digital certificates in exchanged electronic documents);

- The interconnection of network services to provide e-commerce transaction application services for the business community; and

- The establishment of a Pan-Asian portal to facilitate communication between global businesses.

The PAA's major initiatives are outlined below:[53]

- **Secure cross-border transaction services**

This project facilitates the secure electronic exchange and filing of trade-related documents and logistics information across borders, such as electronic applications for trade permits and tracking of consignments. In 2004, the PAA launched a cross-border customs declaration service and extended its use to freight forwarders.[54]

- **Mutual recognition of public key infrastructure (PKI)**

The PAA has established an infrastructure to support both end-to-end digital signatures and digital signatures between service providers.

- **Cargo tracking service**

The PAA is working to incorporate a cargo-tracking service into the secure cross-border transaction services that will provide information for freight forwarders on the status of their cargo.

The PAA is seen as a successful and growing initiative to facilitate cross border trade. Its significance is recognized by both ASEAN and APEC.

The ASEAN E-Commerce Legal Infrastructure project has used the PAA as a case study on the links between e-commerce and trade facilitation. The APEC Trade Facilitation Action Plan (2002) encourages member economies to adopt common frameworks for trade-related procedures among enterprises, and specifically names the PAA as an example.[55] The Action Plan seeks to implement the "Shanghai Goal" of reducing business transaction costs by 5 per cent.

While there are no statistics available to demonstrate the exact economic impact of its initiatives, the PAA estimates that savings from electronic transactions in trade chains could reach 20 per cent or more.[56]

The PAA case study demonstrates that there is a clear business demand for electronic services to enable cross-border trading and that, in the absence of government action, businesses are ready and able to take joint action to meet that demand.

Although the PAA has been effective in enabling transactions, there are certain risks and challenges associated with this kind of private-sector-led initiative. Lack of legal certainty regarding the use of electronic documents and digital signatures may mean that Governments or courts will not recognize their validity in the event of a dispute. Even where Governments in the economies represented by PAA members accept the use of electronic documents for certain purposes, such as Customs filing declarations, this does not amount to a government-wide recognition that can be relied on in other commercial contexts.

Source: ASEAN E-Commerce Project, International Developments on Mutual Recognition of Digital Signatures – Discussion paper 6 (internal project materials), ASEAN secretariat and Galexia, July 2007.

and suggestions about appropriate coordination and drafting tasks, and commentary on the suitability of specific laws, provisions and terms.

Lastly, an implementation timeline provided an excellent source of motivation for implementation in member countries. The *ASEAN E-Commerce Project* benefited from two key deadlines. First, ASEAN

Member countries should have enacted domestic legislation to provide legal recognition of electronic transactions by 31 December 2008. Second, by 31 December 2009, ASEAN member countries should have facilitated cross-border electronic transactions and the use of digital signatures through the adoption of better-practice guidelines for electronic contracting and guiding principles for online dispute resolution

Table 8.3

Components of an e-commerce legal infrastructure

E-commerce legal infrastructure components	1. Online drawing up of contracts	2. Jurisdiction	3. Authentication	4. Electronic payment
Summary	The ability to draw up contracts via electronic means, free of legal restrictions which would require paper records or handwritten signatures.	The ability to determine the jurisdiction of any e-commerce transaction, including determining the applicable law and the appropriate forum for the resolution of disputes.	The ability to reliably authenticate the parties to an e-commerce transaction and the content of any messages exchanged in the transaction.	The ability to complete an e-commerce transaction via a secure and reliable electronic transfer of value.
A. Laws	A broad law may be required which confirms that contracts can be drawn up via electronic means. This law may contain some limited exceptions.	On this issue national laws will not play a significant role. Public international law and international and regional agreements and treaties will be more significant.	National laws, based on the UNCITRAL model laws and the UN Convention, may be required in order to ensure that digital signatures can be used with confidence in e-commerce transactions.	National laws regulating banking, payment systems and the provision of credit may have to be amended to include coverage of electronic payment systems.
B. Regulations and codes	Usually, no further regulation or codes are required for this aspect of e-commerce law.	Regulations and codes are unlikely to be required.	Additional regulations and codes may be required which provide more specific guidance on particular forms of authentication. In addition, a regulation or code may be required in order to establish the rules for recognizing digital certificates issued in other jurisdictions.	Most of the detailed regulation of electronic payment systems will be contained in specific regulations and codes of conduct. Laws on this topic tend to be very broad.
C. Regulators	Usually, no specific regulator is required for this aspect of e-commerce law. There is no specific regulator in most jurisdictions for contract law.	Usually, no specific regulator is required for this aspect of e-commerce law. In any case, there is no appropriate regional or international regulator that could fulfil this role.	Either an existing regulator or a new regulator may need to take responsibility for establishing and maintaining the integrity of the public key infrastructure. However, this role may also be undertaken by a non-governmental agency under certain conditions.	Most electronic payment providers will fall within the existing jurisdiction of national financial system regulators, depending on the nature of the service provided.
D. Registration, licensing and accreditation	Usually, no specific registration, licensing or accreditation is required for this aspect of e-commerce law. All parties are presumed to have the capacity and ability to enter into online contracts without the need to be registered or obtain particular accreditation.	Usually, no specific registration, licensing or accreditation is required for this aspect of e-commerce law.	Some service providers and parties within a public key infrastructure will need to be registered. Mandatory or voluntary licensing may be required. Some form of accreditation will be essential in order to ensure that high standards are maintained in the issuing of digital certificates.	Most electronic payment providers will require a licence at the national level.
E. Standards	Technical standards for the formation of online contracts may be useful, but are not essential. They should not be required in jurisdictions where a "technology neutral" approach has been adopted.	Some international standards have already been developed for the allocation of jurisdiction in online contracts.	Technical standards will play an important role in ensuring interoperability. These standards may be referred to in the law or regulations, or their use might be left to market forces.	There are a plethora of international and national technical standards which apply to the electronic transfer of funds. These help to ensure interoperability and to encourage best practice.

Table 8.3 *(continued)*

E-commerce legal infrastructure components	1. Online drawing up of contracts	2. Jurisdiction	3. Authentication	4. Electronic payment
F. Enforcement and review	The general court system is likely to provide adequate enforcement and review of the online formation of contracts. However, there may need to be some minor changes to court rules and procedures regarding the admission of evidence etc.	The courts will have a significant role in enforcement and review of jurisdiction. In the absence of international and regional agreements and treaties, the courts will decide the jurisdiction of e-commerce transactions.	The general courts will play a limited role in the enforcement and review of the use of digital certificates. Some enforcement and review activities may also be undertaken by the regulator.	The general courts will play a limited role in the enforcement and review of the use of electronic payment systems. Some enforcement and review activities may also be undertaken by the relevant regulator/s.
G. Training, education and awareness-raising	Education regarding the correct drawing up of online contracts may be important amongst industry and consumers. There may also need to be some general awareness-raising activity to overcome any assumption by parties that all contracts must be "in writing".	Education regarding jurisdiction may become an important requirement, especially if no international or regional agreement is reached on this matter.	Training, education and awareness raising will all play an important role in promotion of the use of digital certificates.	Training, education and awareness raising will all play an important role in promotion of the use of electronic payment systems. This is essential in developing consumer trust and confidence in e-commerce.

Source: C Connolly and P van Dijk, What is e-commerce legal infrastructure? Galexia, June 2004.

services in ASEAN, as well as a mutual recognition framework for digital signatures.

The use of deadlines worked well in ASEAN as it provided a source of motivation. Deadlines have also helped staff that are responsible for implementing e-commerce legal infrastructure at the domestic level to obtain funding and support from senior government officials

Importance of aligning domestic and international e-commerce laws

It is important for countries not to have two e-commerce laws (one for domestic transactions and one for international transactions). That situation may arise where a country signs international agreements (e.g. the United Nations Convention on Electronic Contracting or a specific free trade agreement) that apply a particular legal standard only to cross-border contracts.

This issue was considered in detail in the E-Commerce Project and the Singapore case study in box 8.3 provides useful lessons for other countries.

Information-sharing

Information-sharing was seen as one of the most important tools in ASEAN, as its member countries include both developed and developing nations. Information-sharing required the publication of legislation and draft legislation in a common language (English) and the availability of legislation and related documents in electronic form.

During the project, an electronic repository of all relevant E-commerce primary materials (Acts, legislation, conventions and guidelines) was developed and hosted by Galexia on a secure project extranet. All ASEAN member countries provided an official English translation of at least one major legal instrument for inclusion in that repository and the materials are regularly accessed and searched by hundreds of government representatives across the region.

ASEAN is considering options for the permanent hosting of those materials after completion of the project.

Implementation assistance

Many developing nations are not able to develop effective e-commerce legal infrastructure without some form of external assistance. Several ASEAN member countries have benefited from external assistance, including training programmes and advisory services on the legal aspects of e-commerce provided by UN organizations such as UNCTAD,[57] UNCITRAL,[58] and UNESCAP.[59]

Table 8.4

Sample generic implementation guide

Phase 1. Guideline 3: E-Commerce Law Availability
E-commerce legislation (including drafts where publicly available) should be made available in English and in electronic form.

Implementation Task (from the Phase 1 Generic Implementation Guide (GG1))	Brunei	Cambodia	Indonesia	Lao People's Democratic Republic	Malaysia	Myanmar	Philippines	Singapore	Thailand	Viet Nam
3A. Include e-commerce legislation in an online legislation collection	☐	☐	☐	☐	☐	☐	☐	☐	☐	☐
3B. Ensure that the e-commerce legislation is available in English	☐	☐	☐	☐	☐	☐	☐	☐	☐	☐
3C. Ensure that the e-commerce legislation is easily searchable	☐	☐	☐	☐	☐	☐	☐	☐	☐	☐
3D. Promote e-commerce legislation through prominent links from a well-used business or trading portal	☐	☐	☐	☐	☐	☐	☐	☐	☐	☐
3E. Ensure that a plain-language summary of the e-commerce legislation is available online	☐	☐	☐	☐	☐	☐	☐	☐	☐	☐

Source: ASEAN E-Commerce Project, Generic Implementation Guide 1 (internal project materials), ASEAN secretariat and Galexia, January 2006.

Table 8.5

Sample country implementation guide

Progress on Guideline 9: Recognition Criteria		
Coordinating agency		
Coordination tasks		
Drafting agency		
Drafting tasks		
Implementation task	**Notes**	**Check**
9A. Develop recognition criteria		☐
9B. Assess compatibility with APEC guidelines		☐
9C. Ensure that recognition criteria are available in English		☐
9D. Ensure that recognition criteria are available online		☐
9E. Use definitions that are consistent with international models		☐

Source: ASEAN E-Commerce Project, Country Guide 1 (internal project materials), ASEAN secretariat and Galexia, December 2005.

UNCTAD's technical assistance includes the delivery of training activities and the provision of policy advice to help in the preparation of e-commerce laws taking into account national legislation and priorities, as well as regional harmonization initiatives.

With regard to ASEAN, UNCTAD has organized a series of awareness-raising workshops in Lao PDR and Cambodia since 2003 in cooperation with the Science Technology and Environment Agency of the Lao People's Democratic Republic and the Cambodian Ministry of Commerce, in an effort to help those countries move forward in passing a law on electronic commerce before 2008, the deadline set by the e-ASEAN initiative. The objective was to create greater awareness and better understanding of the legal implications of e-commerce among policymakers and the business community.

The workshops were followed by an assessment of the need for law reform in the light of the current state of e-commerce development of the two countries. In the Lao People's Democratic Republic and Cambodia, technical assistance was provided to the drafting groups who are currently working on a draft decree to reflect the national legal system, as well as ASEAN

and international best practice. It was agreed that the most appropriate approach was one of relative simplicity compared to that of its ASEAN partners, bearing in mind the current state of development in Lao PDR and Cambodia, particularly concerning electronic commerce. The drafting process reflected five principles: (a) simplicity, reflecting the current state of development in the countries; (b) flexibility, recognizing that in the future more complex rules may be required; (c) reflecting regional best practice of ASEAN partners, based on existing laws and draft recommendations; (d) consistency with international harmonization initiatives from UNCITRAL; and (e) consistency with existing national laws.

Box 8.3

Singapore and the UN Convention

The UN Convention on the Use of Electronic Communications in International Contracts 2005 [60] seeks to enhance the legal certainty and commercial predictability of international electronic transactions by setting out a number of interpretive rules for the use of electronic communications in negotiating and drawing up contracts.

In 2004 the Infocomm Development Authority of Singapore and the Attorney-General's Chambers began a joint public review of the Electronic Transactions Act 1998 [61] (ETA) and the Electronic Transactions (Certification Authority) Regulations 1999,[62] partly in the context of substantial progress towards concluding a text for the subsequently adopted Convention.

Singapore was the first ASEAN member country to conduct such a review, and its experience should be helpful for other ASEAN members when considering whether to become a party to the convention (or at least to make their e-commerce legislation consistent with it). The written and publicly available submissions received from legal and commercial experts ensure that government attention is informed by the widest range of relevant expert stakeholders. The consultative process is also assisted by encouraging readers of the submissions to respond to a series of structured policy questions.

Stage 1 of the public consultation concerned possible amendments to the ETA relating to electronic contracting which might be desirable in the light of the convention. The consultation paper entitled *Stage I: Electronic Contracting Issues* [63] highlights the main changes and issues that might arise from adopting the convention, notably:

- The recognition of e-signatures – particularly whether the requirements in relation to function and reliability in the convention are consistent with current legal provisions;

- The rules on time and place of dispatch and receipt – particularly whether the present rules should be amended to be consistent with the convention's rules on controlling and retrieving electronic messages;

- Automated systems – particularly the status of e-contracts resulting from interaction with automated systems, as well as issues relating to errors made by a person in communication with an automated system; and

- Miscellaneous issues – including the validity of incorporating terms and conditions by reference in electronic communication, and whether legislation relating to the sale of goods in the physical world also applies to electronic goods.

The consultation paper of the joint review, *Stage III: Remaining Issues,*[64] recognizes that if the convention is widely adopted by other countries, it will set an international standard.[65] To enhance Singaporean commerce, it would be in Singapore's interest to have laws which are consistent with international standards. If Singapore accedes to the convention, it is likely that the regime applicable under the convention will be adopted for all contractual transactions, whether local or international. The Electronic Transactions Act 1998 will therefore have to be amended to be consistent with the provisions of the convention. To have two separate legal regimes for local and international contracts is considered confusing, especially since in electronic transactions it is often difficult to determine whether one is contracting with a local or foreign party.

Singapore signed the Convention on 6 July 2006.[66]

Source: ASEAN E-Commerce Project, Framework Document (internal project materials), ASEAN secretariat and Galexia, September 2006.

D. Concluding remarks and policy recommendations

Although the ASEAN E-Commerce Project has had some success in delivering rapid progress in the development of a harmonized e-commerce legal infrastructure in ASEAN, it also reinforces some of the significant challenges faced by organizations in implementing harmonization projects of this type or implementing domestic e-commerce legal infrastructure.

During the project, ASEAN member countries noted a number of implementation challenges that they faced, and that are likely to be common to many other countries, especially developing countries. Key challenges include securing government policy support, identifying sufficient funding, and obtaining relevant training and assistance.

Thus, many developing nations may need some form of external assistance to develop effective e-commerce legal infrastructure. Several ASEAN member countries have benefited from external assistance, including training programmes and advisory services on the legal aspects of e-commerce provided by various United Nations organizations, such as UNCITRAL and UNCTAD.

Some challenges remain regarding in particular online jurisdiction, which is a significant policy issue in the development of cyberlaws and international trade. Although online jurisdiction was considered in some detail in ASEAN, it remains a major challenge. Online jurisdiction is likely to be of interest to many regional and multi-national organizations that are attempting to harmonise e-commerce legal infrastructure. Unfortunately, the issue of online jurisdiction does not have a "neat" solution and further work is required both in ASEAN and elsewhere on harmonization of that issue.

Policy recommendations to Governments and regional institutions considering harmonization of e-commerce legal infrastructure

The following policy recommendations, drawn from the ASEAN experience, could help Governments and regional institutions considering harmonization of e-commerce legal infrastructure. Some of the recommendations may also be relevant to developing countries that are in the process of reviewing existing laws or drafting new ones.

1. Focus on international interoperability

Part of the success of the ASEAN E-Commerce Project is due to its focus on global harmonization and international interoperability, rather than merely on regional harmonization. This focus on international interoperability included the selection of international models and templates, particularly the UN Convention on Electronic Contracting, for the implementation of domestic e-commerce law in ASEAN member countries. This ensured that ASEAN's e-commerce legal infrastructure would also be compatible with international developments, providing greater certainty for consumers and greater consistency for businesses.

2. Focus on trade facilitation

ASEAN has a strong focus on trade facilitation and is conducting a range of trade facilitation projects and activities. Trade facilitation was also included as an objective of the ASEAN E-Commerce Project. This resulted in constant testing of the project outputs against trade facilitation objectives. Some key recommendations that resulted from this focus were the recommendation to limit e-commerce law exemptions that impact on trade and the recommendation to adapt e-commerce laws to provide for cross-border recognition of electronic transactions.

3. Implementation tools

The ASEAN project shows that there is a need for detailed implementation tools to help developing countries implement e-commerce legal infrastructure, rather than simply high-level recommendations or generic discussion papers. The implementation tools used in the ASEAN project included regional implementation guides, country-specific implementation guides, implementation progress checklists and implementation timelines. All of these implementation tools were considered vital by project participants in achieving harmonization within the ambitious timelines set by ASEAN.

4. Comprehensive legal infrastructure

The ASEAN project shows the importance of developing comprehensive legal infrastructure, not just written laws. In the project work was undertaken on laws; regulations and codes; regulators; registration, licensing and accreditation; standards; enforcement and review; and training, education and awareness-raising.

5. Aligning domestic and international e-commerce laws

The ASEAN project shows the importance of aligning domestic and international e-commerce laws to avoid overlaps and inconsistencies. Although this is perhaps a lower priority issue than some of the other policy recommendations in this section, it is still important for countries to minimize inconsistencies and duplications in order to create a smooth, consistent legal platform for businesses engaging in e-commerce in the region.

Annex 8.1

ASEAN's countries' legislation

Electronic Transactions Order 2000 (Brunei)

[Draft] Law on Electronic Commerce 2007 (Cambodia)

[Draft] Bill of Act of the Republic of Indonesia on Electronic Information and Transactions (Indonesia)

[Draft] Law on Electronic Commerce 2006 (Lao People's Democratic Republic)

Electronic Commerce Act 2006 (Malaysia)

Electronic Transactions Law 2004 (Myanmar)

Electronic Commerce Act 2000 (Philippines)

Electronic Transactions Act 1998 (Singapore)

Electronic Transactions Act 2002 (Thailand)

Electronic Transaction Law 2005 (Viet Nam)

Annex 8.2

Survey on e-commerce legislation in developing countries

To take stock of initiatives by developing countries in the area of e-commerce and law reform, the UNCTAD secretariat sent out a questionnaire focusing on the action undertaken by developing countries to adapt their legislation to e-commerce. Governments were invited to complete the questionnaire, and to provide a copy of their e-commerce legislation. UNCTAD also took advantage of a training course on the legal aspects of E-commerce organized for Latin American Integration Association (ALADI) member States in August 2007 to gather responses.[67]

The questionnaire asked whether countries had prepared national or regional e-commerce legislation. Questions were designed to identify the approach chosen and the priority given to e-commerce legal issues in the legislative process. Other questions focused on issues related to the enforcement of e-commerce legislation and the impact of such legislation on trade and development.

Responses to the questionnaire were received from Albania, Angola, Argentina, Bolivia, Brunei Darussalam, Chile, China, Costa Rica, Croatia, Cuba, Djibouti, Ecuador, El Salvador, Lao People's Democratic Republic, Lebanon, Malaysia, Mauritius, Mexico, Nicaragua, Pakistan, Paraguay, Peru, the Philippines, Qatar, the Republic of Korea, the Russian Federation, Senegal, South Africa, Sri Lanka, Turkey, Uruguay and Venezuela (Bolivarian Republic of).

It emerges from the 32 responses that 20 countries have already adapted their national legislation to ecommerce and 8 are in the process of doing so. In order of priority, Governments have focused on e-transactions, information security law (including public key infrastructure and cybercrime), consumer protection, intellectual protection rights, ISPs' liability, privacy, dispute resolution and e-contracting. To prepare their legislation, 20 countries have considered the UNCITRAL Model Law on E-commerce; 25 have considered the UNCITRAL Model Law on Electronic Signatures and 11 have considered the UNCITRAL Convention on the Use of Electronic Communications in International Contracts, while a few of them have actually adopted, signed or ratified the UNCITRAL e-commerce instruments (2 the Model Law on E-commerce; 11 the Model Law on Electronic Signatures and 5 the United Nations Convention). Eight countries have based their legislation on the European Directive on E-commerce, and 2 on the US e-commerce laws, while other countries have considered legislation adopted by France, Germany, India, Spain, Singapore and the United Arab Emirates.

The survey revealed that in most cases Governments, through the Ministry of Commerce, took early steps to adapt the legal and regulatory framework for e-commerce (between 1997 and 2002), while national ICT master plans were developed at a later stage or are being developed. When the legislation was not yet drafted, it became part of the national ICT plans. The survey reported eight countries that included it in their national ICT master plan.

As well as the adoption of e-commerce legislation, countries are concerned as regard its application and effectiveness, and one of the issues stressed in the survey was the building of capacities of legal professionals. To ensure enforcement of laws and regulations, countries such as Chile, Cuba, Ecuador, the Philippines and Sri Lanka have put in place programmes to train judges and civil servants in various e-commerce legal issues such as digital signatures, security and data protection. In Sri Lanka, capacity development through seminars and workshops was conducted by the Ceylon Chamber of Commerce and the Bankers' Association, and an ICT legal capacity-building programme through the Intellectual Property and Law Advanced Diploma Programme of Sri Lanka Law College has been developed. A large majority of reporting countries have not yet put in place any capacity-building programmes but intend to develop policies and initiatives in order to ensure the enforcement of e-commerce laws.

The main challenges reported by countries in terms of enforcement of e-commerce legislation included not only the lack of human resources in the legal community but also the lack of public awareness about the scope and application of the law and its benefits, and the legal security of e-business through the legislation adopted. More broadly, countries reported consumers' the lack of trust in the security of e-commerce transactions and privacy protection, the difficulty in setting up the technical infrastructure (public key infrastructure, for example) and the need to accompany legal reform by a broader national reform of the information economy in order to seize the benefits of e-commerce.

For example, following the adoption of the Electronic Commerce Act in 2000, the Government of the Philippines has organized seminars and briefings on various e-commerce-related issues and concerns. Moreover, different groups have also been created among the agencies concerned, for example, technical working groups to discuss and address specific issues and concerns, including legal issues reagrding e-commerce, such as but not limited to, data protection, e-payment, public key infrastructure and consumer protection. The adoption of ecommerce legislation by the Philippines was accompanied by initiatives to create awareness of e-commerce opportunities, as well as the definition of an ICT roadmap for 2006–2010 entitled "Empowering a Nation through ICT: The Philippine Strategic Roadmap for the Information and Communications Technology Sector", issued by the Commission on Information and Communications Technology in October 2006.

The survey enquired about existing law cases since the implementation of e-commerce legislation. Only the Republic of Korea reported cases related to e-commerce issues such as data protection, domain names, electronic payment, consumer protection, patents, electronic contract formation, system integration and web-hosting. Sri Lanka mentioned ongoing cases before the Commercial High Court of Appeal, and the Philippines reported a criminal case in the Metropolitan Trial Court (MTC) in Manila in 2005 in which a hacker pleaded guilty to hacking the government portal "gov.ph" and other government websites.

The survey asked about the impact of the introduction of e-commerce legislation on the development of e-commerce activities. A majority of countries responded that the establishment of a legal and regulatory framework had led to increased ICT-related business opportunities (15 countries) and attracted foreign direct investment (7 countries). However, in most countries, there were no baseline studies on e-commerce activities prior to the introduction of e-commerce legislation. According to the Government of Croatia, while recent studies show an increase in ICT-related business opportunities and increased in foreign investments, it would be hard to attribute those developments to the introduction of e-commerce legislation only. The Republic of Korea reported that e-commerce transactions in 2005 had increased by 14.1 per cent compared with 2004, and according to the government, e-commerce legislation plays a substantial role in the development of the country's e-commerce economy, providing a supportive and conducive environment for e-commerce. In less developed countries, for example the Lao People's Democratic Republic, the authorities reported that legislation creates confidence for e-commerce users and facilitates electronic-based business activities.

References and bibliography

Allende LA. and Miglino MA, Internet Business Law Services, *Internet Law – International Electronic Contracting: The UN Contribution*, 6 March 2007, http://www.ibls.com/internet_law_news_portal_view.aspx?s=latestnews&id=1610%20.

ASEAN secretariat, *Roadmap for Integration of e-ASEAN Sector*, 29 November 2004, http://www.aseansec.org/16689.htm.

ASEAN secretariat, *e-ASEAN Framework Agreement*, 29 November 2004, http://www.aseansec.org/6267.htm.

ASEAN secretariat, *e-ASEAN Reference Framework For Electronic Commerce Legal Infrastructure*, 2001, http://www.aseansec.org/EAWG_01.pdf.

Chong KW and Suling JC, Singapore Academy of Law, *United Nations Convention on the Use of Electronic Communications in International Contracts – A New Global Standard*, 2006, http://www.sal.org.sg/Pdf/2006-18-SAcLJ-116%20ChongChao.pdf.

Connolly C, and Ravindra P, *First UN Convention on E-Commerce Finalised*, Computer Law and Security Report, 2005, http://www.galexia.com/public/research/articles/research_articles-art35.html.

Connolly C and Ravindra P *Fantastic Beasts and Where to Find Them – A Guide to Exemptions in the Electronic Transactions Act (ETA) in Australia*, Internet Law Bulletin (September 2004), http://www.galexia.com/public/about/news/about_news-id33.html.

Duggal P, Advocate, Supreme Court Of India, *Harmonization of Ecommerce Laws and Regulatory Systems in South Asia*, Regional Expert Conference on Harmonized Development of Legal and Regulatory Systems for E-Commerce, 7-9 July 2004, Bangkok, Thailand, http://www.unescap.org/tid/projects/ecom04_s3dug.pdf.

Infocomm Development Authority (IDA) and Attorney-General's Chambers (AGC), *Joint ISA-AGC Review of Electronic Transactions Act Stage III: Remaining Issues,* 22 June 2005, http://www.agc.gov.sg/publications/docs/ETA_StageIII_Remaining_Issues_2005.pdf.

Luddy Jr WJ and Schroth PW, Academy of Legal Studies in Business, *The New UNCITRAL E-Commerce Convention in the Mosaic of Developing Global Legal Infrastructure*, 8-12 August 2006, http://www.alsb.org/proceedings/copyright/UNCITRAL_William_Luddy_Peter_Schroth.pdf.

Polanski PP, *Convention of E-Contracting: The Rise of International Law of Electronic Commerce Law*, 19th Bled eConference, 5-7 June 2006, http://domino.fov.uni-mb.si/proceedings.nsf/0/48ecfcae60f83bf6c12571800036bae9/$FILE/49_Polanski.pdf.

UNCTAD, *E-Commerce and Development Report 2003,* United Nations publication, New York and Geneva, http://www.unctad.org/ecommerce.

UNCTAD, *E-Commerce and Development Report 2003,* United Nations publication, New York and Geneva, http://www.unctad.org/ecommerce.

UNCTAD, *E-Commerce and Development Report 2004*, United Nations publication, New York and Geneva, http://www.unctad.org/ecommerce.

UNCTAD, *Information Economy Report 2005*, United Nations publication, New York and Geneva, http://www.unctad.org/ecommerce.

UNCTAD, *Information Economy Report 2005*, United Nations publication, New York and Geneva, http://www.unctad.org/ecommerce.

Notes

1. See annex II.

2. http://www.aseansec.org

3. The Harmonization of E-Commerce Legal Infrastructure in ASEAN Project (referred to in this chapter as the ASEAN E-Commerce Project) is funded by the ASEAN Australia Development Cooperation Program (AADCP). AADCP is funded by the Australian Government through AusAID and implemented in close collaboration with the ASEAN secretariat, and is managed by Cardno Acil. The project is being implemented by Galexia, a private consulting company, with the assistance of several regional experts.

4. More information on UNCTAD's assistance on e-commerce and law reform is provided at http://www.unctad.org/ecommerce.

5. ASEAN Secretariat, *Roadmap for Integration of e-ASEAN Sector*, 29 November 2004, <http://aseansec.org/16689.htm>.

6. All materials in this section on harmonisation structure are from ASEAN E-Commerce Project, *Documented Analysis of International Developments– Discussion Paper 2 (DP2)* (internal project materials), ASEAN secretariat and Galexia, November 2004.

7. http://www.unescap.org.

8. http://www.unescap.org/tid/projects/ecom04_conclu.pdf.

9. http://www.saarc-sec.org.

10. *Harmonisation of Ecommerce Laws and Regulatory Systems in South Asia*, Pavan Duggal, Advocate, Supreme Court of India, Regional Expert Conference on Harmonised Development of Legal and Regulatory Systems for E-Commerce, 7–9 July 2004, Bangkok, Thailand, <http://www.unescap.org/tid/projects/ecom04_s3dug.pdf>.

11. http://www.forumsec.org/.

12. http://www.pacificplan.org/.

13. http://www.sadc.int.

14. http://europa.eu/.

15. *Directive 2000/31/EC of the European Parliament and of the Council of 8 June 2000*, 8 June 2000, http://eur-lex.europa.eu/LexUriServ/site/en/oj/2000/l_178/l_17820000717en00010016.pdf.

16. ASEAN E-Commerce Project, *Options Paper 1* (Internal project materials), ASEAN Secretariat and Galexia, February 2005.

17. http://www.galexia.com.

18. All materials in this section are from ASEAN E-Commerce Project, *Survey of Implementation Issues and Constraints for a Harmonized E-Commerce Legal Infrastructure in ASEAN (S2)* (internal project materials), ASEAN secretariat and Galexia, February 2005.

19. See chapter 1, statistical annex.

20. ASEAN secretariat initiatives for ASEAN Integration Unit, *Bridging the Development Gap among Members of ASEAN*, http://www.aseansec.org/14683.htm.

21. The material in this section is from: ASEAN E-commerce Project, *Survey of Cyberlaws in ASEAN* (S3) (Internal project materials), ASEAN Secretariat and Galexia, October 2005 (updated August 2007).

22. The summary table contains some double entries where part of a law has been enacted and part of a law remains draft. References to "enacted / voluntary" refer to situations where a Member Country has established coverage of a cyberlaw via self-regulatory mechanisms (e.g. an industry Code of Conduct).

23. Communications and Multimedia Act 1998 (Malaysia), http://www.mcmc.gov.my/mcmc/the_law/ViewAct.asp?lg=e&arid=900722.

24. Electronic Commerce Act 2000 (Philippines), http://www.mcmc.gov.my/mcmc/the_law/ViewAct.asp?lg=e&arid=900722.

25. Consumer Act 1991 (Philippines), http://www.dti.gov.ph/uploads/files/Forms1_File_1104836450_RA7394.pdf.

26. Bill on Electronic Information and Transactions 2006 (Indonesia), http://www.depkominfo.go.id/.

27. For more information on privacy and data protection, see the *E-Commerce and Development Report 2004*.

28. For more information on cybercrime, see the *Information Economy Report 2005*.

29. Computer Misuse Act 1990 (UK), http://www.hmso.gov.uk/acts/acts1990/Ukpga_19900018_en_1.htm.

30. Spam Control Act 2007 (Singapore), http://statutes.agc.gov.sg/.

31. The Philippines and Thailand plan to enact spam legislation.

32. Brunei, Cambodia, Indonesia, Malaysia, the Philippines and Singapore have laws in place. The Lao People's Democratic Republic and Thailand have plans to enact laws.

33. For more information on domain name regulation, see the *E-Commerce and Development Report 2003*.

34. Regulations on Registration of Domain Names for Internet under the Top Level 'kh' 1999 (Cambodia), http://www.mptc.gov.kh/Regulation/DNS.htm.

35. For more information on electronic contracting, see the *Information Economy Report 2006*.

36. ASEAN Framework Agreement for the Integration of Priority Sectors, 29 November 2004, http://www.aseansec.org/16659.htm.

37. UNCITRAL Model Law on Electronic Commerce, 1996, http://www.uncitral.org/pdf/english/texts/electcom/05-89450_Ebook.pdf.

38. United Nations Convention on the Use of Electronic Communications in International Contracts 2005, http://www.uncitral.org/uncitral/en/uncitral_texts/electronic_commerce/2005Convention.html.

39. For more information on ODR, see the *E-Commerce and Development Report 2002*.

40. ASEAN E-Commerce Project, *Documented Analysis of International Activity in Electronic Contracting and Online Dispute Resolution (ODR)–Discussion Paper 4* (DP4) (internal project materials), ASEAN secretariat and Galexia, May 2006.

41. ASEAN E-Commerce Project, *Framework Document* (internal project materials), ASEAN secretariat and Galexia, September 2006.

42. http://www.disputeresolution.ph/.

43. http://cpcap.org/.

44. Article 11.

45. Article 12.

46. Article 14.

47. See: C Connolly and P Ravindra, *First UN Convention on E-Commerce Finalised, Computer Law and Security Report, 2005*, <http://www.galexia.com/public/research/articles/research_articles-art35.html>.

48. ASEAN E-Commerce Project, *Framework Document* (internal project materials), ASEAN secretariat and Galexia, September 2006.

49. This table was originally developed in early 2004 but was updated for an ASEAN project workshop in Manila in May 2006. , http://www.galexia.com/public/research/articles/research_articles-pa04.html.

50. China International Electronic Commerce Centre, CIECC (China), Korea Trade Network, KTNET (Republic of Korea), Trade Electronic Data Interchange, TEDI (Japan), Dagang Net (Malaysia), TEDMEV (Macao (China)) and CAT Telecom Public Company Limited (Thailand).

51. See <http://www.paa.net/paaweb/paa/About.htm>

52. See <http://www.paa.net/paaweb/paa/Charter.htm>.

53. See: < http://www.paa.net/paaweb/paa/projects.htm >.

54. See PAA Press Release, 19 August 2004, at http://www.paa.net/paaweb/paa/newsroom.jsp#15.

55. http://www.apec.org/apec/apec_groups/committees/committee_on_trade.html.

56. See PAA Press Release, 28 August 2003 at http://www.paa.net/paaweb/paa/Press11.htm.

57. More information on UNCTAD's technical assistance programmes can be obtained from http://r0.unctad.org/ecommerce/ecommerce_en/ecomlaw.htm.

58. More information on UNCITRAL's technical assistance programmes can be obtained from http://www.uncitral.org/uncitral/en/tac/technical_assistance.html.

59. More information on UNESCAP's technical assistance project can be obtained from http://www.unescap.org/pdd/projects/TC-transition/index.asp.

60. United Nations Convention on the Use of Electronic Communications in International Contracts 2005, http://www.uncitral.org/uncitral/en/uncitral_texts/electronic_commerce/2005Convention.html.

61. Electronic Transactions Act 1998 (Singapore), http://statutes.agc.gov.sg/.

62. Electronic Transactions (Certification Authority) Regulations 1999 (Singapore), http://www.ida.gov.sg/ Policies and Regulation/20061220102423.aspx.

63. Infocomm Development Authority (IDA) and Attorney-General's Chambers (AGC), Joint IDA-AGC Review of Electronic Transactions Act Stage I: Electronic Contracting Issues, 19 February 2004, http://www.agc.gov.sg/publications/docs/ETA_StageI_Electronic_Contracting_Issues_2004.pdf .

64. Infocomm Development Authority (IDA) and Attorney-General's Chambers (AGC), *Joint ISA-AGC Review of Electronic Transactions Act Stage III: Remaining Issues*, 22 June 2005, <http://www.agc.gov.sg/publications/ docs/ETA_Stageiii_Remaining_Issues_2005.pdf>.

65. Ibid. at page 73.

66. United Nations, *China, Singapore, Sri Lanka sign UN Convention on Use of Electronic Communications in International Contracts,* Press Release, 6 July 2006, <http://www.un.org/News/Press/docs/2006/It4396.doc.htm>.

67. For more information on the training for ALADI member States, see http://www.unctad.org/ ecommerce.

Questionnaire

Information Economy Report, 2007-2008

In order to improve the quality and relevance of the Information Economy Report, the UNCTAD secretariat would greatly appreciate your views on this publication. Please complete the following questionnaire and return it to:

ICT and E-Business Branch, SITE UNCTAD, Palais des Nations, Room E.7075 CH-1211 Geneva 10, Switzerland	Fax: +41 22 917 0052 E-mail: ecommerce@unctad.org

The questionnaire can also be completed on-line at: http://r0.unctad.org/ecommerce/ecommerce_en/edr05_en.htm
Thank you very much for your kind cooperation.

1. How do you judge the quality of the 2007-2008 Information Economy Report?

	Excellent	*Good*	*Adequate*	*Poor*
Overall assessment	[ ]	[ ]	[ ]	[ ]
Policy conclusions and recommendations	[ ]	[ ]	[ ]	[ ]
Quality of analysis	[ ]	[ ]	[ ]	[ ]
Originality	[ ]	[ ]	[ ]	[ ]
Timeliness of chosen themes	[ ]	[ ]	[ ]	[ ]

2. How useful was the Report for your work?

Very useful	*Useful*	*Of some use*	*Irrelevant*
[ ]	[ ]	[ ]	[ ]

3. What is your main use of the Report? (click one or more)

   [ ] Academic work or research
   [ ] Government policy work
   [ ] Legal or regulatory activities
   [ ] Journalism or media work
   [ ] NGO support activities
   [ ] Education and training
   [ ] Business / commerce
   [ ] Other

4. Please list any topics that you would like to see covered in future editions of the report. If you have any suggestions for improvement, both on the Report's substance and its form, please make these here:

5. Will you be interested in next year's report?

Yes	Perhaps	Unlikely
[ ]	[ ]	[ ]

Answering "yes" or "perhaps" and fully completing this survey will reserve a copy of the 2008-2009 edition for you.

In this case, Please provide us with your mailing details.

Tile (Mr., Mrs.)	
Names and Surname	
Function or profession:	
Company or institution:	
Street address or PO Box:	
City:	
Postal code or Zip code:	
Country:	